Russian Space Probes

Scientific Discoveries and Future Missions

Brian Harvey with Olga Zakutnyaya

Russian Space Probes

Scientific Discoveries and Future Missions

 Springer

Published in association with
Praxis Publishing
Chichester, UK

Brian Harvey, FBIS
2 Rathdown Crescent
Terenure
Dublin 6W
Ireland

Dr Olga Zakutnyaya
Space Research Institute
Russian Academy of Sciences
Profsoyuznaya 84/32
Moscow
GSP-7 117997
Russia

SPRINGER–PRAXIS BOOKS IN SPACE EXPLORATION
SUBJECT *ADVISORY EDITOR*: John Mason, M.B.E., B.Sc., M.Sc., Ph.D.

ISBN 978-1-4419-8149-3   e-ISBN 978-1-4419-8150-9
DOI 10.1007/978-1-4419-8150-9
Springer New York Dordrecht Heidelberg London

Library of Congress Control Number: 2011921111

Cover design: Jim Wilkie
Project copy editor: Christine Cressy
Typesetting: BookEns, Royston, Herts., UK

Printed on acid-free paper

Springer is part of Springer Science + Business Media (www.springer.com)

Contents

Introduction by the authors

Russia launched the first Earth satellite in 1957 and the first scientific laboratory into Earth orbit the following year – Sputnik 3. Most accounts of Russian and Soviet space achievements have, understandably, focused on manned spaceflight, the cosmonauts, the rockets, the politics, and the engineering achievements of the Russian and Soviet space programs. There has not yet been an examination of what Russian space science has actually achieved in building our knowledge of the space environment and the solar system. This is a largely untold story. During the days of the space race, the scientific outcomes of Soviet space missions were not well known and reached only eminent scientists at international gatherings of their peers. In the English-language-speaking world, media coverage of Russian scientific discoveries was limited, some was even dismissive (and, in the Cold War period, suspicious), and the Soviet Union lacked the channels like *National Geographic Magazine* to communicate its message. Their leading scientists were little known.

As we will see, these discoveries were substantial. This book attempts to build a comprehensive picture of the record and story of Russian space science, before, during, and after the Soviet period. It focuses not just on the higher-profile missions to Mars, Venus, and the Moon, but on a broad range of missions from astrophysics to the ionosphere, from solar studies to the plasmasphere. The questions we attempt to answer are: *What scientific missions were undertaken? How? Why? What instruments were used? What was learned? What discoveries were made? Where were the greatest gains in our knowledge? How important was science within the Soviet and Russian space program? Who were the key personalities? What were the principal decisions and priorities?*

Defining what is and what is not "space science" is not as straightforward as it may first appear, especially in the area of space-based applications looking back towards Earth. Here, we have generally excluded the use of space-based instruments to map the Earth and its ground features, as well as other applications of spaceflight, such as communications and navigation. This book includes space-based research to improve our knowledge of the relationship between our atmosphere, water, and land, as well as space biology and the analysis of substances in microgravity in Earth orbit. This book covers not only unmanned robotic probes, but scientific work undertaken on board orbital space stations, such as Salyut, Almaz, Mir, and the International Space Station.

Although the Soviet Union and Russia have engaged in many international collaborative missions, especially around the space station Mir, the focus here will be on Soviet/Russian space science, rather than on international equipment carried on Russian satellites and space stations. Where international equipment was used as an integral part of Russian space science projects, it is, of course, very much included.

Brian Harvey
Dublin, Ireland

Olga Zakutnyaya
Moscow, Russia
2010

Acknowledgments

Many people helped to make this book possible by providing access to information, documentation, and papers as well as permission to use photographs, diagrams, and illustrations. We especially wish to thank, in Britain, the late Rex Hall; Suszann Parry, Mary McGivern, and Ben Jones for access to the library of the British Interplanetary Society; in Swindon, Doug Stimson of the Library of the Science Museum; Andrew Ball (Open University); in Belgium, Bart Hendrickx, especially for the information he provided on the Elektron, MS, and DS missions; in the Netherlands, Bert Vis; in Denmark, Øjvind Hesselager; in Paris, Dr Aaron Janovsky, for providing access to the papers of COSPAR; COSPAR; in Moscow, the Director of the Institute for Space Research of the Russian Academy of Sciences, Dr Lev Zelenyi; and his colleagues there, Yuri Zaitsev, Dr Alexander Zakharov, Dr Oleg Vaisberg, and Dr Tatiana Mularchik; Dr Sergei Pulinets; Dr Natasha Khisina; Dr Viktor Khalipov; Dr Vladimir Temnyi; Dr Oleg Bartunov of the Sternberg Astronomical Institute of Moscow University; in the Czech Republic, Dr Jaroslav Syýkora, Dr Pavel Triska, Eva Vlčková, and Ivana Kolmašová; and in Canada, Joel Powell.

Glossary

AIS	Automatic Interplanetary Station
ARAKS	Artificial Radiation and Aurora between Kerguelen and the Soviet Union
ARCAD	ARC Aurorale et Densité
AU	Astronomical Unit (distance of the Earth from the Sun = 1 AU)
AUOS	*Avtomaticheskaya Universalnaya Orbitalnaya Stantsiya* (Automatic Universal Orbital Station)
Aureole	AURora and EOLus
CNES	Centre National des Études Spatiales (French space agency)
KORONAS	Comprehensive Orbital Near Earth Observations of the Active Sun
COSPAR	Committee on Space Research
DS	Dnepropetrovsky Sputnik
GAISh	State Astronomical Institute in memory of P.K. Sternberg of Moscow State University
GEOKHI	Vernadsky Institute for Geochemistry and Analytical Chemistry of the Russian Academy of Sciences
IGY	International Geophysical Year
IKI	Institute for Space Research of the Russian Academy of Sciences, Moscow
IZMIRAN	Pushkov Institute of Terrestrial Magnetism, Ionosphere and Radio Wave Propagation of the Russian Academy of Sciences
KNA	*Konteyner Nauchnoy Apparatury*, or Scientific Equipment Container
KOMPASS	Complex Orbital Magneto Plasma Autonomous Small Satellite
KREEP	Potassium, Rare Earth Elements and Phosphorus (type of Moon rock)
KS	Korabl Sputnik
LMC	Large Magellanic Cloud
MKA	*Maly Kosmicheski Apparat* (Small Space Apparatus)
MS	*Maly Sputnik* (Small Satellite)

NASA	National Aeronautics and Space Administration (United States)
NIIYaF	Skobeltsyn Institue for Nuclear Physics, Moscow State University
OKB	*Oputno Konstruktorskoe Byuro* (Experimental Design Bureau)
OSOAVIAKHIM	Society for the Promotion of Defence, Air Travel and Chemistry
RIFMA	Röntgen Isotopic Fluorescent Method of Analysis
PrOP	*PRibori Otsenki Prokhodimosti* (Penetrometer to test the terrain)
SIGNE	Solar International Gamma Ray and Neutron Experiment

Terminological and translation notes

The term "weightlessness" is used to describe the gravity environment in which people (or other life forms) find themselves during the course of space journeys. Although "microgravity" is more correct, the term "weightlessness" is generally well understood.

Politically, the term "Russia" is used as shorthand for "the Russian federation" in the period from January 1992. In the case of "Germany", the term "Germany" will be used, for convenience, to refer to both the Federal Republic of Germany before 1991 (often then known as "West" Germany) and to the reunited country after 1991. The state known as "East" Germany for 1949–1989 will be called by its formal title, the GDR (German Democratic Republic).

For temperatures, two units of measurement will be used: Celsius and Kelvin. Celsius, which runs from $0°$ (the freezing point of water) to $100°$ (boiling) is the most popularly understood and is cited as $°C$. Where measurements are much larger, many scientists use Kelvin (K), which begins at absolute zero, which is $-273°C$. Accordingly, both are used and indicated appropriately and readers should add or subtract 273 to make the necessary conversion.

It was a normal habit of the first of a series of Russian spacecraft to have a simple name, without a number. Yuri Gagarin was launched in Vostok, not Vostok 1. Thus, we have the first Moon rover Lunokhod, not Lunokhod 1. There were exceptions of course, when more than one were launched together (e.g. Elektron 1 and 2) or when they were clearly going to be part of a long series (e.g. Cosmos 1) or when they were retrospectively renamed (e.g. Venera 1).

There is a difference in academic degrees between Russia and Europe. In Russia, the first stage is Candidate of Science, equivalent to Ph.D. The second stage is Doctor of Science, which does not have a direct equivalent.

Russian names are transliterated into English in what is called the simplified form (British Standard).

Every effort has been made to ensure that the reproduction of photographs and illustrations is of the highest quality. Readers are asked to make allowances for the fact that in the case of some historic photographs, especially from the early Soviet period, original negatives were not always available and reproduced versions may have aged over time.

Reference notes

The general sources for the research are reported and discussed in more detail in the bibliographical note at the end. In the case of more specific chapter references, rather than disrupt the flow of the narrative by numerous references after each individual point, the scientific results of each mission are given a composite set of references. Where there are multiple authors (three or more), the first named is normally given.

Tables

Illustrations

Figures

1

Early space science

The story of Russian space science may be traced to 28th January 1724, when the Academy of Sciences was founded in St Petersburg. Eighteenth-century Russia had few indigenous scientists at the time, so the Tsar Peter the Great staffed it with scientists from Germany and Switzerland – so the early Academy had a very Germanic feel. The Academy was to become, in the course of our story, one of the gathering places and mobilizing forces of Russian space science. A physics section, with its own instruments, was established in the Academy that year. The 18th century thus became a time of considerable scientific development in Russia. Tsarina Elizabeth II sent imperial sledges to observe the transit of Venus in 1761 to as far away as Tobolsk. The most famous astronomer of the age was Mikhail Lomonosov (1711–1765), also a chemist, cartographer and poet, the champion of Newton and Copernicus in Russia. He was the first student of aurorae and, participating in the studies of that transit of Venus, was the first to determine that Venus had an atmosphere at least equal to and possibly greater than Earth's atmosphere. He was co-founder of Moscow State University (1755), where, in its physics department, Peter Lebedev later measured the effect of solar light pressure on a comet's tail.

Returning to St Petersburg, a key date was 1839, the completion nearby of Pulkovo observatory on the initiative of the Tsar Nicholas I. The first director was Friedrich Struve (1793–1864), whose crowning achievement was a map of 3,000 pairs of double stars. He was succeeded by his son Otto Struve (1819–1905) and, in turn, by his son Karl Struve (1854–1920), who specialized in planetary satellites and whose brother Gustav Struve (1858–1920) went to direct solar studies from Kharkov observatory [1].

Astronomy was only one discipline. Russia produced many scientists who, due to cultural, language and political barriers, are poorly known elsewhere, such as Dmitri Mendeleev (1834–1907), who made the periodic table; embriologist and pathologist Ilya Mechnikov (1845–1916); the world's leading seed and plant geneticist Nikolai Vavilov (1887–1943); and physiologist Ivan Pavlov (1849–1936). The list of claimed "first inventions" and discoveries made in Russia – although many are disputed – include powered flight (Mozhaisky, 1882), radio (Alexander Popov, 1895), and television (Boris Rosing, 1911). Despite the difficult political circumstances in which

Pulkovo observatory

they worked, 20th-century Russia produced some of the world's greatest physicists, such as Lev Landau (1908–1968), Peter Kapitsa (1894–1984), Andrei Sakharov (1921–1989), Yakov Zeldovich (1914–1987), and Ukrainian-born American George Gamow (1904–1968). Antarctica was discovered by Faddei Bellingshausen and Mikhail Lazarev during their expedition there in 1819–1822 [2].

From the point of view of our story, one of the most important was Vladimir Vernadsky (1863–1945), a St Petersburg geographer, geologist, crystallographer and chemist who traveled the length and breadth of Europe and Asia exploring volcanoes and searching for minerals and meteorites. He was the first to describe radioactivity (1908), the Tunguska impact in Siberia (1908), and set up the State Radium Institute (1922), but his greatest achievement was the idea of the Earth having its own, interdependent ecosystem in which the relationship between oxygen, nitrogen, and carbon dioxide was potentially fragile (*The Biosphere*, 1926). As far back as 1932, he explained the great Siberian impact of 1908 as a comet that had grazed the Earth, split, caused a huge explosion, and that the remnant had then headed back into solar orbit – a theory confirmed 75 years later [3]. Allied to the scientists was a group of writer philosophers, like Nikolai Fedorov (1828–1903), who described how humans would launch expeditions across the cosmos and then settle and evolve on other worlds. By the early 20th century, Russia had a wealth of scientific talent, although some of this was squandered first, when Lenin exiled many of the intelligentsia in 1922 (including astronomer Vsevolod Stratonov) and, second, when Stalin carried out his purges in the following decade [4].

Cosmonautics emerged from neither St Petersburg nor Moscow but, perhaps improbably, from the sleepy provincial town of Kaluga, where a shy, deaf schoolteacher, Konstantin Tsiolkovsky, lived for 43 years and there outlined in the late 19th century how to build multi-stage rockets, construct space stations, and travel through the solar system. He wrote of the joy of exploring new worlds and on his interplanetary spaceships, cosmonauts learned to grow their own food in a greenhouse, an *oranzheriya* (his preference was for carrots, cabbage, sugar beet, rice, and bananas). His work was inspirational, theoretical, and practical – indeed, his formula for rocket propulsion is the basis for all rockets and is still called

Vladimir Vernadsky

"Tsiolkovsky's formula". His most famous quotation was that human-kind would not remain forever on the Earth, but in pursuit of light and space would emerge from the bounds of the atmosphere and then conquer the whole of circumsolar space.

The revolution in 1917 was followed by an explosion of interest in space exploration.

The driving force was a mixture of rocket enthusiasts and designers, science fiction writers, thereoticians and popularizers. Whilst the improvement of scientific knowledge was a feature of their grand design, most had much greater ambitions, nothing less than the conquest of the cosmos itself. Among the bodies established in the 1920s were the Society for the Study of Interplanetary Communications (Moscow, 1924), the Society for the Study and Conquest of Space (Kiev, 1925), the Gas Dynamics Laboratory (Leningrad, 1927), which developed rocket engines and the Society for Interplanetary Communications (Leningrad, 1928). Although to the outside, they might appear to be amateur, they included many of the Soviet Union's leading scientists, like Professor Nikolai Rynin, who compiled and published the first multi-volume encyclopedia of cosmonautics (1928). The first World Exhibition of Interplanetary Machines and Mechanisms ran in Moscow (April–June 1927), attracting huge crowds [5]. What would now be considered a rocket club, led by a young aviation engineer, Sergei Korolev, fired rockets from the forests outside Moscow, the first liquid-fuelled rocket taking off in August 1933. Guest speaker at the 1935 Red Square parade in Moscow was none other than the elderly Konstantin Tsiolkovsky, who told the thousands marching past that space travel would begin in their lifetime. When he died later that year, some say it was the biggest funeral of Stalin's Russia.

SPACE SCIENCE BY BALLOON

Pending the development of the rocket, the only other avenue open to space science was the balloon. Here enters Russia's greatest pre-space age astronomer, Gavril Tikhov (1875–1960). During his apprenticeship in France, to escape the thicker, cloudy layers of the atmosphere, on 15th November 1899, he ascended in a balloon over Paris with fellow Russian astronomer Alexei Gansky and French colleagues to observe meteors.

МЕЖПЛАНЕТНЫЕ СООБЩЕНИЯ

Н.А. РЫНИН

ЛУЧИСТАЯ ЭНЕРГИЯ

· 1 · 9 · 3 · 1 ·

Book by Nikolai Rynin

In Russia after the revolution, ballooning was developed by the popular society, OSOAVIAKHIM (Society for the Promotion of Air Travel), in what we may consider the beginning of space science. OSOAVIA-KHIM built the first modern scientific balloon in Russia, equipping the aeronauts on board with instruments to study cosmic rays, atmospheric temperature, winds, the ozone layer, the Sun, and the stars. They rode in a sealed, pressurized, heated gondola with oxygen bottles and carbon dioxide removers, the precursor to a modern spaceship, with eight 80-mm windows through which they carried out their scientific observations. There was even a landing pad underneath to cushion the touchdown on return to Earth. Long-range radios enabled them to maintain contact with ground control. Building the 2.3-m-diameter, 280-kg gondola and its 25,000-m^3 canopy was a major undertaking, involving new technologies in balloon envelope manufacturing, riveting, and welding [6]. Previewing the subsequent rivalry between space design bureaus, there were two rival balloon enterprises, the first led by Georgi Prokoviev (1902–1939) of the First Airship Division (the USSR had an extensive airship fleet) and the second by the OSOAVIAKHIM society.

The balloons were important for marking the appearance of one of the giants of Russian space science, Sergei Vernov (1910–1982), who developed the cosmic ray detector for the balloons. Sergei Vernov was born in Sestroretsk near St Petersburg on 11th July 1910 [7]. His father was a post office worker, his mother a mathematics teacher, so it was no surprise that in 1926, he finished as "best graduate in maths" at the United Work School in Leningrad, going on to the Physical Mechanical Department of Leningrad Polytechnic. This department had been created in 1919 by Abram Ioffe and so quickly established a reputation that it was known in the education world as "fizmekh" for short. Here, the young Vernov became entranced by the study of cosmic rays, which he saw as opening the door to the mysteries of the universe and he went on to study them in the Leningrad Polytechnical Institute in 1931. Although named "rays", they are not electromagnetic rays, but rather rarefied gas of rarely interacting high-energy particles, charged and neutral. More than 90% are ionized hydrogen atoms (protons). Cosmic rays are effectively absorbed by the Earth's atmosphere, so that only secondary particles can be registered at the surface, originating from the primary particle hitting the atmosphere from space. He fitted his cosmic ray detectors to the first balloon, called the *USSR 1*. Balloon flights were important for this first stage of cosmic rays studies, as they allowed measurements at altitudes otherwise inaccessible.

The first ascent, in the *USSR 1*, was made by Georgi Prokofiev, Ernest Birnbaum (pilots), and rubber technologist Konstantin Godunov on 30th September 1933 from Moscow's Frunze Airport, the balloonists being equipped with protective suits, a powerful radio transmitter, and communications softhats, an earlier version of the suits cosmonauts later wore. They brought instruments in their gondola for the taking of samples of the atmosphere and for trapping cosmic rays as well as barographs, thermometers, altimeters, and variometers. At 3,000 m, they closed the hatches and sealed the pressure. As they rose, they watched the sky turn to deep violet and they could see 80 km into the distance. The aeronauts studied the Sun using special light filters. The cosmic rays' flux intensity was measured, confirming the idea of their extraterrestrial origin. Air samples were collected, which showed that air content at the 18,000-m altitude was similar to that near the Earth's surface.

Sergei Vernov

Although these results may seem modest now, they were the first attempts to reach space previously unknown. With their balloon expanding to 100% of its volume, they broke the previous world altitude level, 17.2 km, set by the great Swiss explorer Picard and, with the envelope completely filled and spherical at 19 km, began a slow descent to land 71 km east of Moscow. The ripping line failed to drop and at 6,000 m, Godunov had to climb out of the gondola to retrieve it. After 8 hr 20 min aloft, the three aeronauts landed in Kolomna, 100 km southeast of Moscow, pursued by a rescue crew in a bus. The connection between what they did and the first flights into space may not have been apparent to everyone, but it was to Konstantin Tsiolkovsky, who was overjoyed and sent them an immediate telegram. Nor was it lost on the *New York Herald Tribune*, which hailed the flight of the *USSR 1* as "a historic achievement of Soviet science". The pilots received the Order of Lenin.

The flight of *USSR 1* not only created records, but established a crucial link between the practitioners of space travel and the Academy of Sciences. Following the revolution, there had been some changes. Exactly 201 years after its formation, it had been renamed the Soviet Academy of Sciences (or, to be more precise, the Academy of Sciences of the USSR). Whatever the politics, following the flight of *USSR 1*, the Academy of Sciences convened a conference on stratospheric studies. Here, Academician Abram Ioffe, who had been involved in the balloon flights, presented a paper on cosmic ray studies.

USSR 1 balloon completed

USSR 1 balloon crew

USSR 1 balloon take-off

USSR 1 balloon landed

Sergei Korolev persuaded the Academy to let him present a paper in which he argued that the best way forward in stratospheric studies was to fly a high-altitude rocket there. The Academy obviously warmed to the idea, because the following year, it organized a follow-up conference on the theme of "The Uses of Rocket-Propelled Craft for the Exploration of the Stratosphere". Here, Korolev outlined a rocket ascent into the atmosphere, with not just instruments on board, but pilots as well. The Academy of Sciences established an Astronomy Council in 1936, given the shorthand name of Astrosoviet (1937–1939), and also had a commission on the stratosphere.

As the academicians deliberated, further balloon ascents were made. The second ascent was by pilot Pavel Fedoseenko, engineer Andrei Vasenko, and young Leningrad physicist Ilya Usyskin four months later. Fedoseenko had undergone ground tests in Leningrad not only to test the gondola, but to set baselines for measurements in aviation medicine. They brought up a new set of scientific instruments, flying the balloon called *OSOAVIAKHIM 1*, and reached a new record of 22 km. A scientific program was devised by Academician Abram Ioffe, including cosmic ray measurements. On the way down, though, the gondola tore free of the balloon and plunged to the ground at a great speed, so quickly that the crew was unable to parachute free (Figure 1.1). The three died on impact and were buried in the Kremlin Wall two days later amidst scenes of national grief.

Despite this, the first balloon in the series, the *USSR 1*, flew a second time as *USSR 1*bis, incorporating design improvements learned from the crash. The gondola had a special parachute whereby the crew could descend separately (an idea resurrected many years later for the first cosmonaut flights). The key personality was Leningrad physicist Alexander Verigo. Son of a professor of medicine, Alexander Verigo was born in St Petersburg in 1893, went to Kiev University to study physics and mathematics, and, in 1923, entered Vernadsky's Radium Institute, where he made the study of cosmic rays his passion. Like other cosmic ray scientists, such as Indian Vikram Sarabhai in the Himalayas, he knew that cosmic rays were best trapped at altitude, so he made his first expedition to Mount Elbruz in the Caucasus in 1928, making observations from as high as 5,400 m. Verigo's companions were Kristian Zille, commander, and Yuri Prilutsky, with Georgi Prokoviev as mission supervisor. The gondola included a short-wave radio station; optical instruments, including a spectrograph for studies of the sky spectrum and instruments to study the sky's luminosity; instruments for sampling the air; cameras; two Hess electrometers; two cloud chambers (also called Wilson chambers); a barograph; a thermometer to measure the outside temperature; a mercury barometer; and two altimeters. On the outside were 15 vessels for air samples, a variometer, alcohol barometer, and two antennae.

The flight took place on 26th June 1935, the balloon reaching its ceiling height of 16 km 90 min later. Now, Alexander Verigo was able to make his cosmic ray measurements at an altitude at which cosmic rays were 240 times more intense than at sea level. No sooner had he done so than the balloon began to leak hydrogen and began a violent descent. Verigo found himself having to jump for his life – he had never parachuted before – followed by Prilutsky, but Zille rode the balloon down to a rough but survivable landing at Trufanovo so as to save the scientific instruments. Verigo remained at his post in the State Radium Institute in Leningrad during the great siege, developing radon baths to treat wounded soldiers and making luminous paint in the lightless city, living to 1953 [8].

A final ascent was made by a third balloon, the *Komsomol*. This was a project of the USSR Academy of Sciences and the aviation company Aeroflot, with a focus on cosmic rays. On board were Aeroflot commander A. A. Fomin, pilot A. F. Krikun, and scientific engineer M. I. Volkov and they reached an altitude of 16.8 km during their 6-hr ascent on 12th October 1939. Science engineer Krikun measured cosmic rays at 47 points, photographed the ground below, and operated an electric soldering apparatus at ever greater altitudes. Returning to the Earth was a problem again, the gondola catching fire and the crew having to parachute out. The unmanned,

OSOAVIAKHIM preparations

OSOAVIAKHIM take-off

Pavel Fedoseenko

Andrei Vasenko

Illya Usyskin

smoking gondola landed in a snowy peat bog spitting flames. The missions are summarized in Table 1.1.

Sergei Vernov, meantime, continued with unmanned balloons. In April 1935, he launched an automated balloon with a 30-kg payload from Leningrad with equipment to measure cosmic rays. Its radio relayed back information on cosmic rays to a height of 13.6 km, the first time this had ever been done from a distance. A balloon was launched to 23 km with a biological payload the following year, to test hereditary changes to fruit flies caused by radiation (none was detected). More projects were planned. The *USSR 2* would have lifted two pilots in pressure suits to 30 km, at which height they would exit the cabin for what we would now call a spacewalk. Launch problems foiled attempts to launch both it and the *USSR 3*, the last attempt being made in 1939.

Balloon memorial stamps

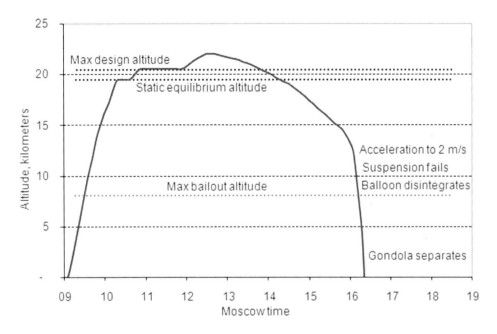

Figure 1.1. Chart of *OSOVIAKHIM* flight.

Table 1.1. The balloon flights.

30 Sep 1933	*USSR 1*	Georgi Prokofiev, E.K. Birnbaum, K.D. Godunov
30 Jan 1934	*OSOAVIAKHIM*	Pavel Fedoseenko, Andrei Vasenko, Ilya Usyskin
26 Jun 1935	*USSR 1*bis	Kristian Zille, Yuri Prilutski, Alexander Verigo
12 Oct 1938	*Komsomol*	A.A. Fomin, A.F. Krikun, M.I. Volkov

As for Sergei Vernov, he successfully defended his candidate dissertation in 1936, "Studying Cosmic Rays in the Atmosphere with Radio Probes". Later on, he continued his studies of cosmic rays with the help of balloons launched from the ship *Sergo Ordzhonikidze*. They confirmed the idea that cosmic rays were charged particles rather than photons. He made measurements at the equator, Yerevan, and Leningrad, which together showed that the flux of cosmic rays at the equator was four times lower than at higher latitudes (the so-called latitude effect). This could happen only if cosmic rays were charged and the trajectories deflected by Earth's magnetic field. In 1939, Sergei Vernov defended his doctoral dissertation, "Cosmic Rays' Latitude Effect in the Stratosphere and a Test of Cascade Theory in the Case of Electrons Penetrating Through the Media with a Small Atomic Number".

In the late 1930s, Stalin unleashed the great purge, which brought the work of popular societies and balloons to an effective end. Few of the leading scientists or rocketeers of the day escaped: some were shot, many (like Sergei Korolev) were sent to the gulag, others continued to work but under house arrest (*sharashka*). Talking about space science risked the accusation of treason for diverting the attention of citizens away from meeting the threat of counter-revolution. In Leningrad, 36 astronomers were arrested, including the director, and most were shot.

SCIENTIFIC FLIGHTS INTO THE ATMOSPHERE: THE *AKADEMIK* SERIES

With the war over, space science could progress once more. The Soviet Union was quick to appreciate the rapid advances in rocketry made by Germany. There, the A-4 rocket, developed by Wernher von Braun, was able to reach 150 km in altitude, well above our atmosphere, and travel 200 km downrange. In October 1946, all the German hardware, technology, and personnel were transferred to the Soviet Union and Stalin decided to build up a rocket industry in the USSR itself. New facilities were built in Moscow, a launching range opened near Stalingrad (now Vologograd), called Kapustin Yar, and a council of designers was formed, led by Sergei Korolev, head of the country's leading experimental design bureau, OKB-1, including also A.F. Bogomolov, Mikhail Ryazanski, Nikolai Pilyugin, Valentin Glushko, Vladimir Barmin, and V.I. Kuznetsov. Korolev was appointed chief designer. On 18th October 1947, the Soviet Union launched a captured German A-4 rocket and a year later, on 10th October 1948, a domestically built version, the R-1 ("R" for "Raketa", or rocket).

A version of the R-1 was adapted for scientific purposes and this was the first example of rockets, albeit ballistic ones, used for scientific purposes. Although Stalin was primarily interested in these rockets as an instrument of war, Korolev was able to fly them for scientific purposes on the basis that something useful should be put into the nose cone. Being a rocket designer rather than a scientist, Sergei Korolev sought help as to what scientific instruments should be carried. Sergei Korolev approached the Academy of Sciences for assistance on the instrumentation to be used, where he met the president of the Academy, Sergei Vavilov (president from

Anatoli Blagonravov

July 1945 to January 1951). The senior academician who took the most interest in space research was a rapidly rising star of the Academy, Mstislav Keldysh, who suggested turning the existing Commission on the Stratosphere into a new Commission for the Investigation of the Upper Atmosphere, headed by former artillery general, subsequently academician Anatoli Blagonravov.

These scientific launches were called the *Akademik* tests and there were five subsets, named A, B, V, D, and E after the first five letters of the alphabet. The term *Akademik* was applied as far back as 2nd November 1947, when scientific instruments were installed on a German V-2. As for the main series, they were also called V tests, V for "Vertikal". For the R-1D and R-1E versions, side containers weighing 65 kg were fitted called *Geofan*, which measured the chemical composition of the atmosphere. Once the rocket reached altitude, the containers would separate and parachute down, the two falling from the sky parallel to one another. The nose cone would also detach from the rocket. The first of these *Akademik* launches, on the R-1A, took place on 21st April and 24th May 1949 and were followed by the R-1B (1950), R-1D, and R-1E (1955). The *Akademik* series carried experiments for solar, ultraviolet and X-ray measurements, chemical, and mass spectrometer analysis of the atmosphere and tested for micrometeorites. Scientists were interested in such experiments because they made it possible to measure cosmic rays, the atmosphere, and its ionospheric parameters (content and temperature). Such information was also important for rocket engineers, as winds, temperature, gaseous content, and radio wave propagation were crucial for rocket and missile development.

The rockets in the *Akademik* series provided about 10-min flight time above the atmosphere. Here, Sergei Vernov resumed the work he began on the balloons in the 1930s. He had now moved to the Institute of Nuclear Physics in Moscow. During the war, he had been evacuated to Kazan and as soon as it was over, he sailed around the world in the ship *Vityaz* with physicists N.A. Dobrotin and Naum Grigorov to release more cosmic ray balloons and test for rays across different parts of the world. His genius was quickly recognized abroad and photos from the period show him in the company of the Curies. For the *Akademik* series, Vernov was joined by the man who was to be his main collaborator over the years, Alexander Chudakov (1921–2001), and together they made the first measurements of cosmic rays outside the

R-2A rocket

atmosphere. The *Akademik* series was also important for the appearance of Konstantin Gringauz (1918–1993), who developed ion traps, and he was, with Vernov, to become one of the great scientists of the space age.

An upgraded rocket, the R-2, a lengthened version of the R-1, made its first flight on 26th October 1950. A geophysical version was developed, the R-2A, also with side containers, able to reach 200 km, first flying on 16th May 1957. A third ballistic rocket followed. This was called the R-5 and was first flown on 15th March 1953. Although intended to carry the Soviet nuclear warhead, Korolev was quickly able to adapt it as a sounding rocket, the R-5VA, and a shorter version, the V-11. The R-5 was 21 m long, had a payload of 600 kg, and was able to reach 480 km. The payload normally separated at 71 km on the way down.

The *Akademik* flights reached ever greater altitudes: 102 km (1949); 192 km (1952); 200 km (1957); and 480 km (1958). A total of 160 ascents were made from 1949 to 1960. The mission of 16th May 1957 was the first to carry a camera, taking images of the limb of the Earth from an altitude of 200 km and measuring the composition of the atmosphere. *Pravda* ran a full-page article on the series, with photographs, and the Academy of Sciences held an all-Union conference on rocket research into the atmosphere in April 1956, "On Rocket Research into the Upper Layers of the Atmosphere". The outcome of the *Akademik* flights was reported in a dedicated two-volume issue of *Achievements of Physical Science*, September 1957, and dedicated to Konstantin Tsiolkovsky, marking his 100th anniversary [9]. The main result was a new model of the upper atmosphere, up to 450 km. Reaching above 400 km was important, because the structure of the atmosphere had a number of key altitude markers, as Table 1.2 illustrates, and this point was the start of the upper ionosphere.

Solar observations were one of the functions of the *Akademik* missions. Instruments confirmed that the standard temperature of the Sun was in the order of 1,000,000 K, in line with predictions by Iosif Shklovsky, then the USSR's emerging leader of solar studies, who obtained his doctorate on the Sun and its corona in 1949.

The mission of 21st February 1958 was especially significant. It reached 470 km, carrying instruments for measuring air pressure and spotting meteorites. Moreover,

Table 1.2. The ionosphere.

400–900 km	*Upper ionosphere*
70–400 km	*Ionosphere*
70–90 km	D layer
90–150 km	E layer
150–250 km	F1 layer
250–400 km	F2 layer

(Source: Glushko, 1970)

it marked the first attempt to make *in situ* measurements of the point at which the ionosphere extended into the geomagnetic field to form the protonsphere. At that time, the word "magnetosphere" had not yet been introduced and it was assumed that the upper atmosphere interfaced directly with the interplanetary medium. Konstantin Gringauz's ion traps measured electron densities there of $2 \times 10^6/cm^3$. Electrons were measured by altitude in the course of three sounding rockets in 1958 (Figure 1.2). By way of a side discovery, it was found that the rocket generated its own mini electrical field.

The *Akademik* series continued in the form of a program of sounding rockets with the R-5 until 1983. This trusty rocket was joined by the Vertikal built by another design bureau, OKB-586, directed by Mikhail Yangel. Although "Vertikal" was the name of the rocket, it was also the name of the series and the last two R-5 launches listed were also part of the Vertikal series. Table 1.3 gives details of the *Akademik* program and subsequent sounding rocket launches. Disentangling different dates can be difficult, both a function of incomplete records being kept at the time and because the missions were officially classified. Results of some individual missions are available. The first mission in June 1963, for example, made a profile of atmospheric density compared against Sputnik 3, showing a substantial decrease over five years. Temperatures were also plotted (Figure 1.3) [10].

The *Akademik* series was paralleled by the beginning of the meteorological rocket program. This was developed by design bureau 2 (KB 2), headed by Russia's leading solid rocket designer, Alexander Nadiradze (1914–1987). His first rocket, the MR-1 (*Meteorologicheskaya Raketa 1*), was 480 mm in

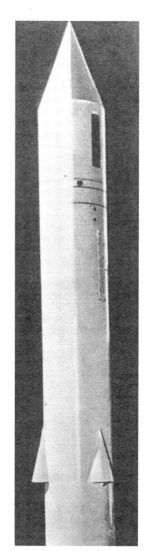

Cosmic ray slits on
Akademik rockets

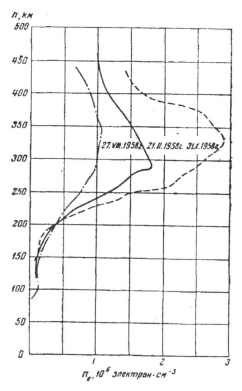

Figure 1.2. Measurement of electrons by three 1958 sounding rockets.

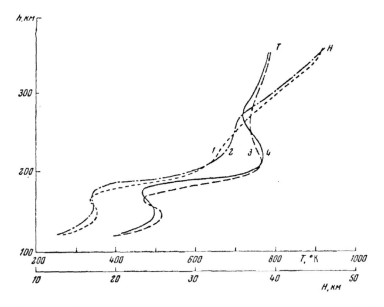

Figure 1.3. Measurement of temperature by sounding rocket, June 1963.

Table 1.3. *Akademik* and subsequent scientific launchings.

1949–1952	12 launches (all dates not available)
1953–1956	18 launches (all dates not available)
16 May 1957	
24 May 1957	211 km, biology; meteorite detectors
24 Aug 1957	Biology
21 Aug 1957	
25 Aug 1957	Meteorite detectors
31 Aug 1958	
9 Sep 1957	Biology
21 Feb 1958	473 km, geophysics, payload 1,515 kg
2 Aug 1958	Geophysics, meteorological
13 Aug 1958	Geophysics, meteorological
27 Aug 1958	Biology
13 Sep 1958	
19 Sep 1958	Meteorological
4 Oct 1958	110 km, V-11
10 Oct 1958	470 km, V-11
21 Oct 1958	
31 Oct 1958	
23 Dec 1958	110 km, V-11
25 Dec 1958	110 km, V-11
2 Jul 1959	200 km, biology
10 Jul 1959	
19 Jul 1959	200 km
21 Jul 1959	100 km, V-11, X-ray detectors
21 Jul 1959	100 km
28 Jul 1959	
15 Apr 1960	200 km
24 Jun 1960	200 km
6 Sep 1960	100 km, V-11
19 Sep 1960	100 km, V-11
21 Sep 1960	100 km, V-11
23 Sep 1960	200 km
15 Feb 1961	100 km, V-11, solar eclipse
15 Feb 1961	100 km
23 Sep 1961	430 km, V-11
15 Nov 1961	430 km
18 Oct 1962	500 km
6 Jun 1963	563 km, temperature and density
18 Jun 1963	400 km, temperature and density
20 Sep 1965	500 km, geophysics and solar
1 Oct 1965	500 km
31 Oct 1965	500 km
3 Oct 1970	500 km, solar, ultraviolet and X-ray studies
24 Sep 1971	500 km, solar, ultraviolet and X-ray studies
9 Oct 1971	500 km
28 Nov 1970	487 km, solar and ionospheric probe
21 Aug 1971	463 km

All from Kapustin Yar

diameter, 7.8 m long, weighing 680 kg, with a payload of temperature and pressure gauges, cameras, air samplers, ozone detectors, and ultraviolet spectrometers, and able to reach 100 km [11]. The first launch was in October 1951 from Kapustin Yar and the series concluded in 1958 after 50 launches. In 1956, the MR-1 was joined by a second meteorological rocket, the smaller MMR-05 built by OKB-3 directed by Dominik Sevruk (1908–1994), which made 260 launches by 1959 from Heiss Island, Novaya Zemlya. Their smaller size meant that they could be launched from ships, and several were fired from the *Ob* in Antarctica.

SCIENTIFIC TEST FLIGHTS WITH ANIMALS

The idea of putting animals into rockets came from chief designer Sergei Korolev. In 1949, he approached the Moscow Institute of Aviation Medicine, originally founded in 1934, closed during the war, but reopened in 1947 [12].

There, army doctor Vladimir Yazdovsky was put in charge of the space biomedical program, to be joined in 1955 by the man who emerged as the leading Russian biomedical scientist, Oleg Gazenko (1908–2007). The Russians spent some time considering which animals would be the most useful in paving the way for human spaceflight and many types of animals were considered: rabbits, rats, mice, and reptiles. The institute was well aware of experimental American research with monkeys, but the institute considered them nervy, capricious, and difficult to train and handle. By contrast, Russian animal physiology and psychology were based around the dog, thanks to the work of pioneer Ivan Pavlov. Dogs had the advantage of being small in size, light in weight, easy to handle, and the best predictor as to how humans might respond in such key areas as breathing, blood circulation, and reaction to stress. They were small enough (6–8 kg) to fit inside rockets. The only disadvantage was their individualism: dogs had unique personalities. So, in the Pavlovian tradition, it was decided to fly two dogs at a time, thereby evening out the differences between them.

Accordingly, the Moscow Institute of Aviation Medicine began to recruit dogs from December 1950. The dogs were obtained from owners wanting to give away unwanted pets and some strays were also recruited from the city dog pound, the argument being that dogs with such a tough life experience would more likely be "the right stuff" for space travel, rather than over-pampered pets. Twenty-four dogs were selected in the first round, of which 17 were trained for missions, some making more than one. The specialists had a preference for mongrels rather than more nervous thoroughbreds. Not all were as well behaved as they should have been: Smelaya must have had second thoughts about her mission, for she ran away the day before her launch, but was recovered and made a successful flight. Another, Rozhok, did run away and was never found, but was replaced at short notice by another runaway from the barracks canteen, called Zib (in Russian, "substitute"). For their flights into the atmosphere, cameras were fitted to observe their responses and see how they would react to acceleration, deceleration, noise, weightlessness, and vibration. A typical mission gave up to 180 sec of weightlessness in a 370-sec flight.

Space biology began shortly before dawn on 22nd July 1951 with the launch of an R-1V from Kapustin Yar, the cosmonaut dogs being Tsygan and Dezik, watched by Korolev and Blagonravov. They were both the first dogs to fly into space and to be recovered. The pre-sunrise launch improved tracking, for the timing enabled the early-morning light to illuminate the rocket as it ascended to 100 km. Soon after the main rocket impacted near the horizon in a bright bang, the parachute with the nose cone and dog container could be seen floating gently down. Although the scientists may have been unaware of this, they beat the United States to the first biological launch by two weeks. A second launch a week later went wrong, the parachute failing to open, killing the two dogs on board, Lisa and Dezik, though the film survived. Blagonravov seized his favorite dog, Tsygan, and brought her home, lest she suffer a similar fate. Later that summer, two more dogs died due to a valve failure and loss of pressure [13].

Four more successful missions were carried out in summer 1951. The animals were filmed: their agitation during ascent was obvious as they endured five times the force of gravity (5 G), as was their disorientation during 3-min weightlessness, but they seemed to endure no lasting ill effects and were always in good condition when recovered, enjoying the attention lavished on them and a reward of sausage meat. Typically, heart levels spiked to 260 beats per minute, three times normal during ascent, breathing up to four times the normal rate. Previously flown dogs could be seen readying themselves for the jolt of the parachute coming out.

The second series of dog missions began three years later, running over 1954–1956, using the R-1D and R-1E, a total of nine missions. Several innovations were introduced. First, instead of parachuting back in a nose cone, the animals were ejected, one at high altitude (80 km), the other at low altitude (40 km, with parachute opening at 3 km). Second, the dogs were put into spacesuits, with their own helmets, and one dog was routinely placed in an unpressurized part of the cabin, so it had to function properly. Thereby, early work was done on the technology of spacesuits that cosmonauts would eventually wear. Third, radio telemetry was used to transmit biological information live from the animals, such information covering breathing rate, blood pressure, pulse, and temperature – a system called *Tral*.

The results of these animal missions were published at an international conference in Paris in December 1956 as "Vital Activity of Animals during Rocket Flights into the Upper Atmosphere", the presentation being given by Alexei Pokrovsky, the director of the Institute of Aviation Medicine. Pictures showed the Earth from over 100 km, in anticipation of what humans would soon see. Pokrovsky told the delegates how the dogs' pulses and respiration had risen during vibration, ascent, and free fall, but stabilized during coast, with temperatures constant. The missions provoked protests from dog protection societies in Britain. To respond, the Institute of Aviation Medicine held a press conference in June 1957 through the aegis of the State Committee for Cultural Relations with Foreign Countries. The Institute showed films of Soviet space dogs, presenting an alert Kozyavka and Linda and a somewhat bored Malyshka.

From the two series of flights on the R-1 variants, it was clear that dogs could survive sub-orbital flights to an altitude of 100 km with no ill effects. Now, Soviet scientists decided to push back the medical frontier further, using a new and more

Space dog Kozyavka

powerful rocket for higher and longer missions. The R-2 rocket, whose scientific version was called the R-2A, was able to reach 180 km, orbital height. The nose cone was sufficiently large for two dogs to be installed there in a hermetically sealed cabin, observed by a film camera overhead. Blood pressure, pulse, respiration, and electrocardiograms were transmitted by *Tral*, the animals receiving a full range of before-and-after medical tests (e.g. blood and urine analysis). The first flight, by dogs Ryzhaya and Damka on 16th May 1957, was successful, doubling the period of weightlessness to 6 min. The second was disastrous, with the loss of Ryzhaya and Dzhoyna. The descents from the R-2A caused much higher G forces, up to 8G, which caused the animals some bleeding – a warning of possible effects on cosmonauts. Overall, the third series of flights was successful. The fourth and final set of ballistic flights took place in 1958, using the new R-5A rocket. This was able to bring animals to the altitude of 400 km, far out into space, and extend the period of weightlessness to 9 min. Details are given in Table 1.4.

The dog tests indicated that it would be possible to contemplate putting dogs into orbit, which would give even better information as to how humans might respond to space travel. Now, more than 60 years after the first human voyage into space, it is easy to criticize the approach of scientists in the 1950s as over-cautious in their approach to human spaceflight, but there were very real fears that humans would not make good space travelers. Some scientists feared that the first cosmonauts would suffer badly from acceleration and deceleration, would panic during weightlessness, and, disorientated by the three-dimensional environment, go mad. Tales of insane cosmonauts lost in space were the fare of several popular press stories at the time.

Table 1.4. Sounding rocket flights with animals.

First series, 1951

22 Jul 1951	R-1V	Tsygan, Dezik
29 Jul 1951	R-1B	Dezik, Lisa (fail)
15 Aug 1951	R-1B	Mishka, Chizik
19 Aug 1951	R-1V	Smelaya, Ryzhik
28 Aug 1951	R-1B	Mishka, Chizik (fail)
3 Sep 1951	R-1B	Neputevy, Zib

Second series, 1954–1956

26 Jun 1954	R-1D	Lisa 2, Ryzhik
2 Jul 1954	R-1D	Damka, Mishka
7 Jul 1954	R-1D	Damka, Ryzhik
25 Jan 1955	R-1E	Rita, Lisa 2
5 Feb 1955	R-1E	Lisa 2, Bulba (fail)
4 Nov 1955	R-1E	Malyshka, Knopka
31 May 1956	R-1E	Malyshka, Linda
7 Jun 1956	R-1E	Albina, Kozyavka
14 Jun 1956	R-1E	Albina, Kozyavka

Third series, 1955–1960 (R-2A)

16 May 1957	R-2A	Ryzhaya, Damka
24 May 1957	R-2A	Ryzhaya, Dzhoyna (fail)
25 Aug 1957	R-2A	Belka, Modnitsa
31 Aug 1957	R-2A	Belka, Damka
6 Sep 1957	R-2A	Belka, Modnitsa
2 Aug 1958	R-2A	Kusachka, Palma
13 Aug 1958	R-2A	Kusachka, Palma
2 Jul 1959	R-2A	Otvazhnaya, Snezhinka, rabbit Marfusha, 241 km
10 Jul 1959	R-2A	Otvazhnaya, Zhemchuzhnaya, 211km
15 Jun 1960	R-2A	Otvazhnaya, Malek, 221km
24 Jun 1960	R-2A	Otvazhnaya, Zhemchuzhnaya
16 Sep 1960	R-2A	Palma, Malek

Fourth series, 1958 (R-5A)

21 Feb 1958	R-5A	Palma, Pushok, 480 km
27 Aug 1958	R-5A	Belyanka, Pestraya, 452 km
31 Oct 1958	R-5A	Zhulba, Knopka (fail)

THE IDEA OF A SCIENTIFIC EARTH SATELLITE

The idea of a Soviet Earth satellite took root during the 1950s, accelerating in the period after Stalin's death. The political and engineering side of this story is now well known and has been told many times, so is not repeated here [14]. Sputnik was not the original satellite the USSR planned to put into space. Sputnik was, we now know, hurriedly put together to beat the Americans by getting a satellite in orbit first

– an objective in which it spectacularly succeeded. The original satellite, what should have been the first Sputnik, was a large scientific laboratory weighing over a tonne, the design of which began in 1955, with a full-scale model made by year's end. At the time, America's first satellite, the Vanguard, was weighing in at 1 kg, so the contrast in ambitions between the two space powers could not have been greater.

The original satellite blueprint was done by Korolev's leading designer, Mikhail Tikhonravov. On 16th September 1953, approval was given for Tikhonravov to form a satellite study group.[1] In May 1954, Tikhonravov and his colleagues presented to the government the "Report on an Artificial Satellite of the Earth" in which he made proposals for a 3-tonne satellite called "object D". Object D1 would be an orientated satellite with a television system but able to send film down to Earth (though a simpler, unorientated version could be launched first), while object D2 would carry an animal, the next stage being to proceed to human orbital flight [15]. The report included studies of insertion trajectories, the influence of main disturbances on the satellite's orbit (e.g. Sun and Moon, atmospheric drag), stabilization, and orientation. Lidia Soldatova was responsible for the satellite's energy supply, which led to the idea of photoelements, the common source of energy for many spacecraft today.

At this stage, the team's proposals converged with plans to mark the International Geophysical Year (IGY), to run from 1st July 1957 to 31st December 1958, and focus on the ionosphere, cosmic rays, the Earth's magnetic field, the atmosphere, and the impact of the Sun. The IGY was designed as a worthy successor to the International Polar Years, 1882–1883 and 1932–1933. Indeed, the IGY was originally called the Third International Polar Year, but its brief became much broader [16]. It was a concerted effort to re-build science in the post-war world and turned out to be the most successful international collaborative scientific venture of the middle of the 20th century. Countries the world over warmed to the idea and by the time it was under way, the IGY had 66 nations involved and 60,000 scientists participating, their international links enduring long after the IGY was over. Both rocket engineers and geophysical scientists considered this a golden opportunity to use new techniques and approaches to study the Earth and its environment, an artificial satellite of the Earth holding out considerable promise. It may be difficult to appreciate now the sudden explosion of interest in space science in the mid 1950s. Popular writings on spaceflight by authors such as Ari Sternfeld were repeatedly republished. In 1956, the Soviet Academy of Sciences convened a conference on life on other planets in Leningrad, which led to a five-year plan for the development of astrobiological research, which included not only the study of other planets for life, but also the growth of plants under the simulated planetary conditions [17].

---

[1]   The members were Mikhail Tikhonravov, Vladimir Galkovsky, Gleb Maksimov, Lidia Soldatova, Grigori Moskalenko, Oleg Gurko, and Igor Bazhinov.

Table 1.5. Scientific objectives, object D, as set in 1956.

Measure the density, pressure, ion composition of atmosphere from 200 to 500 km
Research solar corpuscular radiation
Measure ion concentrations
Measure electrical charges
Study cosmic rays
Measure Earth's magnetic field, 200–500 km
Investigate ultraviolet, X-ray parts of solar spectrum

SCIENTIFIC OBJECTIVES OF THE FIRST EARTH SATELLITES

Here, the link between rocketeers and the Academy of Sciences, first built in the 1930s, restored in the *Akademik* missions, was used to construct the science around the first Earth satellites. Korolev invited the Academy of Sciences to elaborate both the program and the instruments, notably Academician Anatoli Blagonravov (who had been in charge of the *Akademik* atmospheric program) and Academician Leonid Sedov (who had a responsibility for bringing together scientists concerned with spaceflight issues). Korolev suggested scientists contribute instruments for a scientific program to incorporate the ionosphere, cosmic rays, the Earth's magnetic field, the upper atmosphere, the Sun, and other natural phenomena. Over the winter of 1955–1956, the vice-president of the Academy, Mstislav Keldysh, consulted widely with the scientific community to refine the scientific package [18]. This was an important moment, for it is where and when the early Soviet program of Earth orbit-based space science was defined.

The ambition of a 3-tonne spacecraft was probably running ahead of its time, so in July 1955, Tikhonravov presented a more modest proposal for a satellite in the 1,000–1,300-kg range – one more likely to win governmental approval. In its 1955 iteration, object D weighed in the order of 1,000–1,400 kg, of which 200–300 kg would be for scientific instruments. Soviet leader Nikita Khrushchev visited Korolev's OKB-1 in January 1956 and, learning from Korolev that object D was almost 1,000 times heavier than America's planned first satellite, persuaded the government to approve the satellite in Resolution #149-88 of 30th January 1956. A launch date was set for July 1957, to mark the start of the IGY, or the end of 1957 at the latest, but certainly before the Americans. Work began on object D on 25th February 1956 and at a meeting in the Academy of Sciences, leading experts in physics of the Earth upper atmosphere, magnetic fields, ionosphere, and cosmic rays were given the task of preparing experiments. The design was signed off on 24th July that summer, with scientific objectives as outlined in Table 1.5.

KEY PEOPLE

Here, it is useful to pause for a moment to consider the key people who determined the nature of Soviet space science at this crucial turning point: Mstislav Keldysh, Leonid Sedov, and Anatoli Blagonravov.

Mstislav Keldysh (1911–1978), later called the "chief thereotician of cosmonautics", was the Soviet Union's leading mathematician and by this time had become vice-president of the Academy of Sciences, to the outside world its guiding force. Mstislav Keldysh had a distinguished background, for his father was Vsevolod Keldysh (1878–1965), inventor of the formula for reinforced concrete, designer of the Moscow canal, the Moscow metro, and the Dnepr aluminum plant. Young Mstislav was educated by his older sister Lyudmilla, who became a noted mathematician. Mstislav joined the Zhukovsky Central Institute for Aerohydrodynamics, TsAGI, as a young mathematician in the 1920s and quickly made an impression by combining mathematics with engineering to solve practical problems in aircraft design such as propeller flutter and landing gear failure, winning Stalin prizes in 1942 and 1946. His family was persecuted during the purges and his historian brother Mikhail was shot as a German spy.

Very much an all-rounder, he branched into physics, being awarded his doctorate in 1938. He ran a lecture course in Moscow University from 1938 to 1958, focusing on complex variables, partial differential equations, and functional analysis, being considered the father of the theory of function approximation in the complex domain. Keldysh became a corresponding member of the Academy of Sciences in 1943 (aged 32) and a full member in 1946. His favorable disposition to spaceflight and its scientific potential closed the circle of engineers like Korolev, scientists like Vernov and Gringauz, with the political and academic community that approved and endorsed the idea of an Earth satellite.

Looking at the other two members of the trio, Anatoli Blagonravov (1894–1974) was a military officer who rose to general, taught artillery in military academies, became academician in 1943, and, later, in 1953, head of the Institute of Engineering

Mstislav Keldysh (right) with Sergei Korolev (left) and Igor Kurchatov

Science. Leonid Sedov (1907–1999) was a gas and hydrodynamicist who taught at Moscow State University and TsAGI, becoming a corresponding member of the Academy of Sciences in 1946 and full member in 1953. He became the official Soviet representative in the International Geophysical Year and to the international media was "Mr Space Science" for the USSR, even though his professional background was not in the area at all.

Soviet space science was effectively led by Keldysh, although only the vice-president at the time. The actual president during this period (January 1951 to May 1961), Alexander Nesmeyanov, admitted that he knew little about space research, although he was well disposed to it, both privately and publicly. On Keldysh's initiative, the Academy in 1954 established a body with the long-winded title of the "Interdepartmental Commission for the Coordination and Control of Scientific–Theoretical Work in the Field of Organization and Accomplishment of Inter-planetary Communications" of 27 academicians under Leonid Sedov, a second commission to join Blagonravov's commission for investigation into the upper layers of the atmosphere (1949). Sedov's commission reviewed the proposals for the instrumentation on object D and the early satellite program. All was now set for the launch of this large satellite in time for the opening ceremony of the International Geophysical Year, on 1st July 1957 [19].

INSTEAD, *PROSTEISHY SPUTNIK*

When it became apparent that object D would, because of its complexity and the slowness of sub-contractors, not be ready in time and that there was a real danger than the United States would put a satellite in orbit first, Korolev decided on a change of course. In December 1956, he decided that there be a new, simplified design called the PS, *Prosteishy Sputnik*, or Preliminary Satellite, in Russian *Sputnik*, which would be built entirely within his own design bureau, OKB-1 ("prosteishy" means basic, a diminutive derived from "prostoi", "simple"). The two PS satellites were approved by government on 25th January 1957, for now called PS 1 and PS 2. Object D was set aside for the moment.

PS was little more than a radio transmitter encased in a box within a steel ball, weighing only 83.4 kg and filled with dry nitrogen. There was no time to consider scientific instrumentation, though, by default, the radio transmitter filled a scientific purpose. Korolev was determined that as many people as possible should hear the signal from Sputnik, so he asked his engineers to attach four whiplash aerials on the back (two pairs of 2.4 and 2.9 m), transmitting on the long-wave bands of 7.5 and 15 m, which he hoped could be heard over wide distances, one transmitting a steady tone (40 MHz), the other a beeping tone (20 MHz).

The question of the transmission frequency was resolved as follows. Until then, the only bands used to link with rockets were ultra-short waves, which permitted the transmission of large amounts of information, but could only be received over a limited area. Now, the designers needed to put a relatively large transmitter on board Sputnik, so that it could be received by a simple radio set on Earth. This task was

The first Sputnik showing how the instrument box was fitted into a polished steel ball

assigned to the Laboratory of Radio Wave Propagation in Radio Technical Institute, led by Konstantin Gringauz, who had earlier reported on radio transmissions from sounding rockets in a series of classified reports. Gringauz now proposed the use of the decameter band to transmit the signal. Some engineers predicted that the decameter signal from the satellite would be reflected from the F layer into space and would never reach the Earth, but Gringauz took the view that, based on the measurements from sounding rockets, inhomogeneities in the ionosphere would ensure the signal was received. A complete description of these transmitters was published in the June issue of the *Soviet Journal for Radio Amateurs*.

In autumn 1957, the stream of announcements about a forthcoming satellite intensified. The centenary of the birth of Konstantin Tsiolkovsky was marked on 16th September 1957, a day duly commemorated in Kaluga with predictions that his dreams were about to turn to reality. The PS 1 was duly launched on 4th October 1957 into an orbit out to 947 km, causing joy all over the Soviet Union but consternation verging on panic in the United States. Korolev worried needlessly about his signals not being audible, for the *beep! beep! beep!* of Sputnik was heard up to 12,000 km away on every pass and became the iconic sound of the new age of space. They were even picked up in the Soviet Antarctic base from a distance of 15,000 km. They were the sound that, in the immortal words of American television anchor Walter Cronkite, "forever separated the old world from the new".

Sputnik is remembered for the hundreds of thousands of amateurs who tuned in to its transmissions or went out into their streets and backyards to watch the new creation pass overhead (although they probably saw the rocket carrier, rather than the much smaller satellite). The scientific value lay in following the effect of the thin atmosphere on its orbit and measuring the gradual slowing down in its orbital period. For this, precise measurement was required and Sputnik was followed by observatories all over the Soviet Union. By the time of the launch, the Astronomical Council of the Academy of Sciences, the Astrosoviet, had built a network of 70 observation posts, from Archangel in the north to Yerevan in the south, from Kaliningrad in the west to Vladivostok in the east. Observers were trained and equipped with Latvian AFU-75 cameras and bulletins were published to synthesize the results [20].

One of the first countries to realize the scientific value of Sputnik was China, where the Chinese Academy of Sciences coordinated observation made by observatories in Beijing, Guangzhou, Wuhan, Changchun, Yunan, and Shaanxi, the principal station being the Zijin Shan Purple Mountain observatory of Nanjing. Other countries, too, set up their observation points and between them, they enabled the building up of a global picture. The transmitter battery gave out after 21 days, the spherical Sputnik itself burning up on 4th January after 92 days and 1,440 circuits of the Earth (the cylindrical rocket carrier lasted 60 days, until 2nd December). The frequency of Sputnik's transmissions was also affected by the temperature inside the pressurized container, so ground controllers were able to calculate how much temperature rose or fell on the basis of these changes. The pattern of the signals through space enabled scientists to measure the F layer of the ionosphere from 200 to 320 km. Vladimir Kurt of the Sternberg Astronomical Institute recalled later how surprised they were at the high density of the air at altitude (10^8 atoms/cm^3) and its composition (atomic oxygen, rather than hydrogen, as expected) [21]. The gradual decay of its orbit enabled air density to be measured. Although the scientific outcomes might seem, at this distance, modest, they paved the way for more ambitious missions.

PS 2

It was normal practice, even this early in the space age, to build satellites two at a time, the second being available should the first fail, or as a body from which parts could be scavenged in the event of doubts about parts on the original model. PS 2 had already been built, but such was the concentration on launching PS 1 that no attention had yet been given to when it would fly. No sooner had Korolev and his colleagues arrived on the Black Sea for their first holiday for years than Soviet leader Nikita Khrushchev, delighted at the enthusiastic popular response to Sputnik in his own country and the discomfort of his foes, ordered a second launching, to take place before the anniversary of the revolution, only three weeks away, and exhorted Korolev to do something even more spectacular. Korolev's satellite designer Mikhail Tikhonravov had already thought of a purpose-built biological satellite to fly an animal in orbit for a day and recover the cabin, but it had not yet been built (indeed, some had campaigned for the first Sputnik to carry a dog). The master of improvisation, Korolev requisitioned a container used from the R-5A atmospheric dog missions. So, beneath the PS 2 was installed a dog container, which meant that the payload would be the entire upper stage and the nose cone would not drop off.

PS 2 carried the first scientific instruments to make direct measurements in orbit. Professor Sergei Mandelstam of the Lebedev Institute of Physics prepared a solar ultraviolet and X-ray detector, which were installed in the nose cone above the spherical PS. Meanwhile, Sergei Vernov had already been preparing cosmic ray detectors for object D – they were declared ready in September 1957 – and he was furious to learn that a satellite had now been launched without any instruments at all. At once, he contacted Sergei Korolev to ask for the KS-5 cosmic ray detector, originally planned for object D, to be put on the next satellite, persuading him that cosmic rays must be measured to test whether cosmonauts would be adversely affected. He agreed, but the only space available for the counter was on the final rocket stage, where it was given its own battery and *Tral* channel. So as to save battery power, the counter would not be activated until shortly before take-off – a task accomplished manually on the fueled-up rocket stage [22]. The detector was put together by a team of four people who would thereafter guide Russian cosmic ray research: aside Vernov, there were, from Moscow University, Naum Grigoriev, Alexander Chudakov, and Yuri Logachev.

There was no time to train a dog especially for the mission, so a dog who had already flown on an atmospheric mission was selected – Laika. As was the case with the flights into the atmosphere, the *Tral* telemetry system was installed, redesigned to transmit physiological data on the health of the animal in a 15-min burst of data while over Russian ground stations. PS 2 used the same open frequencies as its predecessor (20 and 40 MHz), to which amateurs were encouraged to listen, but scientific and biomedical data were relayed back on classified frequencies (66 and 70 MHz).

Sputnik 2 was duly launched on 3rd November 1957 into an even more elliptical orbit, out to 1,671 km. The launching of an animal into space duly astounded the world, though it led to vocal protests by animal lovers. Sputnik 2 opened the first opportunity to obtain information on the condition of an animal in prolonged weightlessness, *Tral* relaying real-time data on the first orbit. The first animal to experience weightlessness for more than a few minutes, Laika's condition was gradually returning to normal, though three times slower than following centrifuge tests on Earth. Relays from electrodes showed her respiration rise during launch but then settle down in orbit.

Sadly, the insulation covering Sputnik 2 was torn off at the point of orbital insertion, causing the temperature of the cabin to begin to rise and Laika must have died painfully. *Tral* indicated that she was moving, barking and agitated until data ceased between the third and fourth orbits. The *Tral* battery lasted until 10th November, long after the unfortunate animal was dead. Her death was announced only obliquely in the Soviet media, which gave the impression that the air had run

Sputnik 2, showing the dog container (bottom), reiteration of PS-1 (middle), and instrumentation (top)

out after a week. The full story did not emerge until the 1990s. Still, the flight of Laika showed that, in principle, animals could live and survive in orbit. There seemed to be no reason to believe that had not the thermal system failed, then she would have survived as long as the air on board lasted. Years later, with the perspective of time and in a world more conscious of the rights of animals and ethical issues, there was regret about putting a dog in space at a time when there was no prospect of recovery.

As was the case with the PS Sputnik, ground observers were able to follow the Sputnik in orbit and follow its transmissions. Sputnik 2 intentionally did not separate from its carrier rocket, making the combined structure easier to follow. In its final orbits, the rocket body tumbled end over end, flashing brightly, and was eventually incinerated over the north Atlantic on 14th April 1958 after 2,370 revolutions over 162 days.

Its mission was the opening scene of the first drama of science in the early space age. Sputnik 2's cosmic ray detector transmitted as long as its battery lasted, which was only a week (9th November). This was a simple set of glass Geiger tubes, identical exemplars of which an American scientist was able to buy in a scientific shop store in Leningrad on the day following the launch [23]. The cosmic ray detector noted an increase in high-energy charged particles at the highest latitudes of its orbit, but scientists were unsure what to make of this. The drama began coming up to 5 am Moscow time the day after launch. At 0430, at latitude 40°N, the radiation count was a normal 18 pulses/sec, but it was beginning to rise. By 0440, it had reached 36 at 55°N, falling back briefly to 32 at 60°N, before making a steep climb to 72 at 0445 at 65°N, then falling back, with a dip to 18 at 0450. They had seen nothing like it and did not know what to make of it. Reporting the results in two articles (10th December 1957 and 27th April 1958), *Pravda* noted that the radiation level increased whenever the satellite approached apogee by 40%, sometimes as much as 50%, more so over polar latitudes. This was probably the first intimation that a radiation belt surrounded the Earth. Russian ground controllers could receive Sputnik 2's data only when it was overflying the Russian landmass, so they had data for only those parts of its orbit. The picture was incomplete and indeed Sputnik 2 was actually underneath most of the radiation belt during its northern hemisphere perigee, where its altitude was between 250 and 700 km over Russia. Australian observers asked for the code to interpret the data over Australia, but the request was turned down. As for the other experiment, the outcomes from the solar, ultraviolet, and X-ray experiment were impossible to interpret: only years later was it realized that it had been calibrated at such a level that its measurements were swamped by the Earth's radiation belt [24].

Sputniks 1 and 2 between them enabled Soviet scientists to calculate the density of the atmosphere through the altitudes through which they passed and measure heating due to radiation. The principal report was given by one of the scientists involved, Lidia Kurnosova, to the IGY:

- The outer ionosphere over 300km to 400km varied according to day/night, north to south and east to west;
- Electrons were concentrated in the middle part, at up to $3m/cm^{-3}$;
- Although the effective upper limit of the atmosphere was 200km, density levels varied according to solar conditions;
- Charts and diagrams were published showing the curve of atmospheric density and height, with the strange, sharp rise in radiation in northern latitudes. [25]

OBJECT D: THE FIRST LARGE SCIENTIFIC SATELLITE IN ORBIT

By now, OKB-1's "object D" was at last almost ready, the outstanding problem being a troublesome tape recorder. One of the most important features of object D was this tape recorder. An irony of the new space age was that although engineers were, in the course of space exploration, to develop new, cutting-edge technologies, an existing device was to give space programs the world over so much grief. Because satellites flew over their countries of origin for only a small proportion of their mission, normally only several minutes, receiving data from satellites was a real problem. If satellites transmitted data continuously, there was no one to pick up these signals, especially when passing over deserts and oceans, unless one built a worldwide network of ground stations and tracking ships – something that did not yet exist. The best alternative, in the meantime, was to have satellites collect data on a tape recorder, store it, and then transmit the data, at high speed, over a tracking station – what is called "dumping" data or downlink. In those days, most tape recorders used spools that, even on Earth, are still known to jam. In space, tape recorders must work for months on end, without jamming, collect a volume of data, and dump the information every 95 min or so when over a ground station. If the tape recorder jammed, there was no one to free the spool with a handy screwdriver. For the first decades of the space age, more scientific missions were lost through faulty tape recorders than any other cause. Even when digital recorders came in, these problems persisted.

Object D was impressive – a shiny cone with aerials and instruments to measure micrometeorites, radiation, cosmic rays, solar radiation, the density of the atmosphere, and high-energy particles. Object D was 3.57 m tall, 1.74 m in diameter, weighing 1,327 kg but of which 968 kg were payload. It had high-density silver zinc chemical batteries to prolong its working life. There was a small solar battery made by N.S. Lidorenko and 16 thermal control louvers to stabilize temperature. The solar battery was connected to the cosmic ray counter and the radio transmitter. One of the most important instruments was the magnetometer, although its inclusion was almost by chance, arising from a 1956 meeting between chief designer Sergei Korolev and the first head of the Space Magnetic Research Laboratory, Shmaia Dolginov (1917–2001), head of the laboratory in the Institute of Terrestrial Magnetism (IZMIRAN). He had mapped the Earth's magnetic field by sailing around the world in a wooden ship, the *Ob*, using no metallic, magnetic parts. Now, he had the opportunity to put his equipment into orbit.

Object D was duly launched on 27th April 1958, but the R-7 rocket exploded 88 sec into the mission and the laboratory was destroyed. Thankfully, the backup version was complete and ready, though trouble with the tape recorder had still not been resolved. The meteorite detector was on the bottom, the cosmic ray detectors on the side, and the ion trap and magnetometer on the top. Konstantin Gringuaz made the ion traps, Vadim Istomin the mass spectrometer, Vera Mikhnevich the ionization gauges, Tatiana Nazarova the micrometeorite detector, Lidia Kurnosova the Cherenkov particle detector, Valerian Krassovsky the scintillation counters, and Sergei Vernov and Alexander Chudakov the cosmic rate counters [26]. These are detailed in Table 1.6.

Table 1.6. Sputnik 3 instruments.

Magnetometer
Photomultiplier (Sun's corpuscular radiation)
Cosmic ray detector
Ion trap
Mass spectrometer
Micrometeorite counter
Heavy cosmic ray counter
Electrometer

Nikita Khrushchev again rushed the launching and ordered Korolev to get the backup version aloft within two weeks, timed so as to coincide with parliamentary elections in Italy, where, Khrushchev reasoned, another Soviet space triumph would impress floating voters so much that they would flock to the Communist Party. Hoping that the tape recorder problems had at last been sorted, the engineers put Sputnik 3 into orbit on 15th May 1958. The troublesome tape recorder broke down almost at once and this was to have a profound effect.

Sputnik 3 transmitted using its batteries until 17th June 1958. Some of the instruments worked longer than others, the meteorite detector, for example, only between 15th and 25th May. When the chemical battery ran out, the small and experimental solar battery enabled the radiation counter and transmitter to continue to work and send data until Sputnik 3 crashed out of its orbit on 6th April 1960 after 10,037 revolutions (691 days). The cosmic ray counter broadcast on a *Mayak* transmitter, which lasted until 15th August, its beacon being received far afield.

The failure of the tape recorder meant that there was no way of dumping the data during the passes over the Soviet Union. Sputnik 3 transmitted throughout its orbit, including long periods when it was not over the Soviet land mass, but no one could interpret its signals. As it passed over the Soviet Union, Sputnik 3 transmitted details of a radiation belt around the Earth – but without tracking stations abroad, they had no idea as to the extent of the belt nor could they determine whether this was a local detection of radiation or a worldwide phenomenon. Sputnik 3 noted a sharp increase in the level of charged particles when it transited through latitudes 55–65°N over the Soviet Union. Sputnik 3 detected solar electrons of 10 keV when they reached the Earth's environment and noted how they accelerated once they reached Earth's northern latitudes, more so by night. In effect, Sputnik had identified the way in which charged particles streamed down into the polar regions, reflecting the shape of Earth's magnetic field. Sputnik 3 reached an altitude of 1,864 km – enough to reveal the outer radiation belt.

When Sputnik 3's tape recorder failed, the Russians appreciated the importance of trying to obtain southern hemisphere data, especially as Sputnik 3's apogee was in the southern hemisphere. The Soviet Antarctic base began to collect in the data. The Russians cabled the codes to the School of Physics in the University of Sydney, Australia, in June. The scientific ship *Ob* was sent south to the Southern Ocean, though it did not arrive on station until September. Sharp increases in radiation, up to 40 times, were quickly noted [27].

The Russians released the preliminary results of Sputnik 3's mission in July 1958 at the Moscow meeting of the committee for the International Geophysical Year, but at that time, they had northern latitude data only. It was too late, for the Americans had already announced the discovery of the radiation belts. Explorer 1, the first United States satellite early in 1958, had also noted that as soon as it climbed to the high point of its orbit, the number of radiation hits went up and up. Explorer 1 did not have a tape recorder, so this increase in radiation could only be identified whenever Explorer's signals could be picked up in real time, during its passes over the United States. Explorer 3, though, on 26th March 1958, had a tape recorder, which dumped the previous 2 hr of data in just 5 sec as it passed over its main ground station. The Americans were in no hurry to analyze the data. Head of the physics department in the University of Iowa, James van Allen was a long-standing expert in cosmic rays, having flown cosmic ray detectors on American high-altitude rockets since the 1940s and his cosmic ray detectors on the two Explorers had been designed to pick up cosmic rays from deep space. He and his colleagues waited until the summer, when the two Explorers had ceased transmissions, before analyzing what they had expected to be routine data of cosmic rays from deep space. They nearly fell over when they did. Van Allen gasped when he went through the tapes: "Space is radio-active!", exclaimed E. Ray, his colleague.

Scientists from both countries had difficulty interpreting their data – the localized data from Sputnik 2 and 3 and Explorer 1, and the global picture from Explorer 3. The Americans first thought the radiation might have been caused by high-altitude nuclear tests, both superpowers then being in the habit of exploding nuclear test bombs in the atmosphere. The Russians thought that electrons might be bouncing off the satellite's shell. James van Allen announced the preliminary results of his Explorer 3 analysis in summer 1958 at an atomic conference in Europe, making the important interpretative breakthrough that the radiation was not just local, but surrounded the whole planet. He himself made no proprietary claim to the belt – indeed, he acknowledged that the Russians had obtained the raw data first – but the press promptly named it the "Van Allen radiation belt". Both van Allen and his opposite number Sergei Vernov soon met in Moscow (the IGY assembly in Moscow in September 1958) and each acknowledged the other's contribution to finding what had turned out to be a complete surprise – a strong radiation belt surrounding Earth. In the Western press, though, the "discovery" was attributed exclusively to the scientifically more competent Americans – a memory that has endured. This was not the doing of the modest van Allen, who was far from combative and who kept up his contacts with Soviet colleagues during the Cold War when it was far from fashionable to do so.

Although Sputnik 3 was most remembered for its role in determining the radiation belt, that was only part of its scientific haul. Although the radiation belt or belts were the primary focus of scientific interest, the mission gave tantalizing glimpses of more radiation mysteries [28]. Valerian Krassovsky's instrument suggested that particles were not static, but could rain down. Particles with energies of about 10 keV were of sufficient intensity and pitch angle distribution to be captured into the geomagnetic trap as ring current particles, which was a sensation at that time, contradicting existing views that they came from high-energy cosmic ray particles. Sputnik 3 noted the way in which molecules turned into atomic hydrogen at 500 km. Konstantin Gringauz's ion traps found ion densities of $1,000/cm^3$ at 1,000 km. Indeed, there seemed to be not one radiation belt, but an outer electron belt (100 keV +) and an inner proton belt (100 MeV +). Indeed, profiles of ion density showed that the ionosphere was inhomogeneous, confirming Gringauz's earlier view on decameter wave propagation. Sputnik 3 suggested that parts of the Earth's surface had areas of much stronger magnetism, later called geomagnetic anomalies. Polish scientists under Ludoslav Cichowicz in the Warsaw Institute of Technology used the signals from Sputnik 3 to model the Earth's gravitational field, the beginnings of satellite geodesy. Because of the signaling issue, this could only be done over Soviet territory, of course. Sputnik 3 detected the first solar flare from orbit and the first cosmic ray burst from orbit on 7th July 1958. The Cherenkov integral detector installed by Vitaly Ginzberg and Lidia Kurnosova made the first measurements of the chemical composition of cosmic rays and the first ever detection of cosmic rays emanating from the Sun.

Instruments designed and built by Vera Mikhnevich (b. 1919) recorded gas outside the Earth's atmosphere. The density of the Earth's atmosphere was measured at varying altitudes (e.g. 1/10,000 of an atmosphere at 266 km), far greater than anticipated, and a profile was compiled (Figure 1.4). The first three Sputniks flew at a time of a solar maximum, when the Sun heats up the electrons in the high atmosphere, increasing drag. Research over 1957–1959 began to suggest that drag, far from being constant, went up and down considerably to reflect solar maxima and minima. Sputnik 3 found high-altitude aerosol particles in a dust cloud between 100 and 300 km on one orbit. The micrometeorite detector measured velocities of passing micrometeorites of 11–70 km/sec every 100 sec on average.

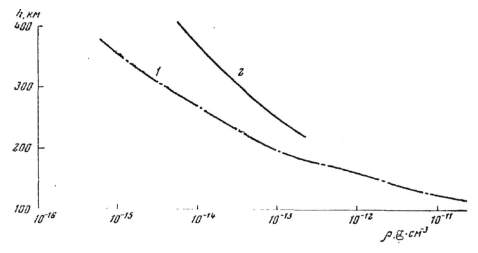

Figure 1.4. Air density measurements taken by Sputnik 3.

Table 1.7. The first Sputniks: missions.

Sputnik (PS 1)	4 Oct 1957	83.6 kg	228–947 km, 65°, 95 min
Sputnik 2 (PS 2)	3 Nov 1957	508 kg	225–1,671 km, 65°, 103 min
Sputnik 3	15 May 1958	1,327 kg	226–1,880 km, 65°, 106 min

R-7 launcher from Baikonour

Details of the first missions are given in Table 1.7. The outcomes from the three missions were submitted to the committee for the International Geophysical Year as "Preliminary Report on Launching in the USSR of the First and Second Artificial Earth Satellites" (February 1958), with further reports in August 1958 and March 1959, as well as at the Moscow IGY assembly in August 1958, while a final full scientific report on Sputnik 3 was issued in May 1959. A final international IGY meeting on the three Sputnik missions and the American Explorers and Vanguards concluded in July 1960.

During the period before, during and immediately after the geophysical year, the level of contact between Soviet and Western space scientists was relatively open and cordial, the principal barrier being that of language rather than politics. The IGY had intended to bring together scientists in a coordinated scientific effort, the first since the war, and the numbers indicated that it more than succeeded. James van Allen was honored by the Academy of Sciences, gave talks to its members and accepted an invitation to contribute an article to the Soviet journal *Progress in Physical Science*. Soviet scientists received the relevant English-language journals of the day. Leonid Sedov gave a lecture tour on the Sputnik results in the United States in November 1959 and Sergei Vernov visited the following year. In order to read its science reports, Van Allen's University of Iowa even subscribed to *Pravda*, where the articles were translated by a post-doctoral student. Russian scientists were offered visiting professorships in American universities [29].

EARLY SPACE SCIENCE: WHAT WAS LEARNED?

Early Russian space science built on a tradition of scientific and astronomical research dating to the early 18th century. The theoretical work of Konstantin Tsiolkovsky, combined with the activities of the rocketeers in the 1930s, brought closer the possibility that scientific observations could be made from the atmosphere and, later, from Earth orbit. The 1930s saw a combination of balloon exploration of the stratosphere and scientific research – one that brought together engineers and the scientific community. This relationship was renewed after the war with, first, the *Akademik* flights (1948) and then the biological missions (1951). The alliance of chief designer Korolev with rising academician Mstislav Keldysh in the Academy of Sciences meant that as the idea of an Earth satellite developed in the 1950s, scientific objectives could be developed and appropriate instrumentation carried. The strength of commitment to scientific objectives is evident when we look at the payload and instrumentation of object D. Although the scientific outcomes of the first Sputnik

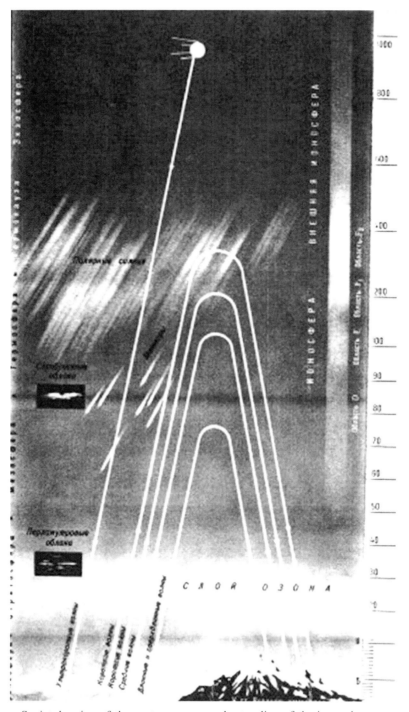

Soviet drawing of the contemporary understanding of the ionosphere

missions might seem obvious and modest by today's standards, they were important breakthroughs for their time and set the background for an expanded program of scientific space research. If we look at what was learned, the following were the main outcomes:

- Radiation belt detected as satellites approach northerly latitudes;
- The measurement of magnetism and magnetic anomalies;
- Detection of cosmic rays;
- The atmosphere varied according to solar seasons and could slow satellites down;
- Animals can survive in orbit.

REFERENCES

[1] Moore, Patrick: *The Guinness Book of Astronomy*. Guinness, Enfield, 1995.

[2] Parry, Albert: *The Russian Scientists*. MacMillan, London, 1973.

[3] Drobyshevski, E.M.; Galushina, T.Y.; Drobyshevski, M.E.: *A Search for a Present Day Candidate for the Comet P/Tunguska, 1908*. Ioffe Physical Technical Institute, St Petersburg and State University, Tomsk, 2009. For an account of Vladimir Vernadsky, see The 125th anniversary of the birth of Vladimir Vernadsky. *Earth & The Universe*, No. 2, March–April 1988.

[4] Chamberlain, Lesley: *The Philosophy Steamer – Lenin and the exile of the intelligentsia*. Atlantic Books, London, 2006.

[5] Winick, Lester E.: Birth of the Russian rocket program. *Spaceflight*, Vol. 20, No. 5, May 1978.

[6] Karmanov, B.I.: Taking the first steps toward the stars. *Earth & the Universe*, No. 1, 1986. For an account of the early balloon missions, see Shayler, David: Where blue skies turn black – Soviet stratospheric balloon program in the 1930s. *Journal of the British Interplanetary Society*, Vol. 50, 1997; Sergevev, A.A.: *Essays on the History of Aviation Medicine*. NASA, TTF 176, 1965.

[7] Assovskaya, A.S.: Sergei Vernov, Soviet cosmo-physicist. *Earth & the Universe*, No. 1, 1991.

[8] Assovskaya, A.S.: AB Verigo, tireless researcher of cosmic rays. *Earth & the Universe*, undated.

[9] Bulkeley, Rip: The Sputniks and the IGY, in Launius, Roger D.; Logsdon, John; Smith, Robert, eds: *Reconsidering Sputnik – forty years since the Soviet satellite*. Harwoood, Amsterdam, 2000.

[10] Central Intelligence Agency (CIA): Scientific Intelligence Report, Soviet Space Research Program, Monograph IV, *Space Vehicles*. Washington, DC, 1960; Gdalevich, Gennadiy: *Measurement of Electrostatic Field Strength at the Surface of a Rocket Flying in the Ionosphere*. NASA, TTF 8,324, 1962; Mikhnevich, Vera: Density and temperature of the atmosphere based on measurement results obtained on high altitude geophysical stations in 1963, in Skuridin, G.A., *et al.*, eds: *Space Physics*, papers from conference held in Moscow, 10–16 June 1965. NASA, TTF 389.

[11] Lardier, Christian: Soviet meteorological rockets, a history 1946–1991. Presentation to the International Astronautical Congress, Glasgow, 2 October 2008.

[12] Burgess, Colin; Dubbs, Chris: *Animals in Space – from research rockets to the space shuttle*. Praxis and Springer, Chichester, 2007; Gazenko, Oleg: Achievements of Soviet space medicine, in Sagdeev, Roald Z.: The principal phases of space research in the USSR, in USSR Academy of Sciences, History of the USSR, New Research, 5, *Yuri Gagarin – to mark the 25th anniversary of the first manned spaceflight*. Social Sciences Editorial Board, Moscow, 1986.

[13] Gazenko, Roman: *Space Dogs*. BBC Four television, 6th July 2009; Pokrovsky, Alexei: Vital activity of animals during rocket flights into the upper atmosphere, in Kreiger, F.J., ed.: *Behind the Sputniks – survey of Soviet space science*. Washington, DC, Rand Corporation, 1960.

[14] Siddiqi, Asif A.: Korolev, Sputnik and the IGY, in Launius, Roger D.; Logsdon, John; Smith, Robert, eds: *Reconsidering Sputnik – forty years since the Soviet satellite*. Harwoood, Amsterdam, 2000.

[15] Siddiqi, Asif A.: Before Sputnik – early satellite studies in the Soviet Union. Part I: *Spaceflight*, Vol. 39, No. 10, October 1997. Part II: *Spaceflight*, Vol. 39, No. 11, November 1997.

[16] Walsh, Tom C.: Communicating science in the Sputnik era. Unpublished Master's degree, Dublin City University, 2002.

[17] Central Intelligence Agency (CIA): Scientific Intelligence Report, the Soviet Space Research Program, Monograph X, *Space Biology and Astrobiology*. Washington, DC, Author, 1959.

[18] Siddiqi, Asif A.: Korolev, Sputnik and the IGY, in Launius, Roger D.; Logsdon, John; Smith, Robert, eds: *Reconsidering Sputnik – forty years since the Soviet satellite*. Harwoood, Amsterdam, 2000.

[19] Barry, Willam: The missile design bureaux and Soviet manned space policy, 1953–1970. PhD thesis, University of Oxford, 1996.

[20] Masevich, Alla: First Sputnik, early years of observing artificial Earth satellites, early results, in Zakutnyaya, Olga, ed.: *Space, the First Step*. IKI, Moscow, 2007.

[21] Kurt, Vladimir: The first steps in our space astronomy, in Zakutnyaya, Olga, ed.: *Space, the First Step*. IKI, Moscow, 2007.

[22] Logachev, Yuri: The beginning of the space era at the Skolbeltsyn Institute of Nuclear Physics, in Zakutnyaya, Olga, ed.: *Space, the First Step*. IKI, Moscow, 2007.

[23] Walsh, Tom C.: Communicating science in the Sputnik era. Unpublished Master's degree, Dublin City University, 2002.

[24] Panasyuk, Mikhail: Radiation reflections, in Zakutnyaya, Olga, ed.: *Space, the First Step*. IKI, Moscow, 2007; Hess, Wilmot N.: *The Radiation Belt and Magnetosphere*. Blaisdell, 1968; Zhdanov, G.; Tindo, I.: *Space Laboratories*. Foreign Languages Publishing, Moscow, 1960.

[25] Kurnosova, Lidia, ed.: *Artificial Earth Satellites – results of the investigations carried out according to the International Geophysical Year program with the help*

of the first and second artificial Earth satellites. Plenum Press, New York with Chapman & Hall, London, 1960, Vol. 1, with *Results of Sputnik 2*, Vol. 2.

[26] Mitchell, Don: Group for the study of jet propulsion. Don P. Mitchell, *www.mentallandscape.com* (accessed 19 February 2008).

[27] Grahn, Sven: Sputnik 3 – its flight and radio systems. *www.svengrahn.pp.se* (accessed 25 March 2007); Vakulov, P.V., *et al.: Earth's Radiation Belts.* NASA Goddard Space Flight Centre, SEV PF 10,335; Vernov, Sergei: *Nuclear Physics and the Cosmos.* NASA, Goddard Space Flight Centre, ST PR 10,210.

[28] Krassovsky, Valerian, *et al.*: Discovery of approx 10keV electrons in the upper atmosphere, in Kurnosova, Lidia, ed.: *Artificial Earth Satellites – results of the investigations carried out according to the International Geophysical Year program with the help of the first and second artificial Earth satellites.* Plenum Press, New York with Chapman & Hall, London, 1960; Lewis, Richard S.: *Illustrated Encyclopedia of Space Exploration – a comprehensive history of space discovery.* Salamander, London, 1983; Lemaire, J.F.; Gringauz, Konstantin: *The Earth's Plasmasphere.* Cambridge University Press, Cambridge, 1998; Panasyuk, M.I.: Cosmic rays are wanderers of the universe, in Zakutnyaya, Olga; Odinstova, D., eds: *Fifty Years of Space Research.* Institute for Space Research, Moscow, 2009; Zielinski, Janusz Bronislav: My personal consequences of Sputnik 1, in Zakutnyaya, Olga, ed.: *Space, the First Step.* IKI, Moscow, 2007; Vakulov P.N., *et al.*: Investigation of cosmic rays, in Muller, P., ed.: *Space Research*, Vol. IV. COSPAR, Paris, 1963.

[29] Walsh, Tom C.: Communicating science in the Sputnik era. Unpublished Master's degree, Dublin City University, 2002.

2

Deepening our understanding

Although Sputniks 2 and 3 had identified the presence of the radiation belt, the Americans won both the scientific interpretation and the media battle over the discovery of the radiation belts. This was painfully obvious even in the Soviet Union itself, where a cruel joke was circulated about the country's chief radiation scientist: *Q: What did Sergei Vernov discover? A: Van Allen's radiation belt!* The Soviet Union was stung by the crediting of Earth's radiation belts to the American Van Allen, with claim and counter-claim following, a Cold War battle that the scientists of neither country encouraged. The Americans observed that of all the scientific disciplines followed in the USSR, the geophysical sciences were the most important, with more scientists engaged there than any other country in the world and characterized its record as outstanding [1]. When the International Geophysical Year (IGY) was over, the Academy of Sciences turned the IGY organizing committee into a new Interdepartmental Geophysical Committee. Repeated statements in the Soviet press emphasized the importance of geophysical space research. A small example of the relative priority of geophysics between the two countries was that the Russians had two all-wood ships to study geomagnetism: the *Ob* and the *Zarya*. By contrast, when the similar American ship, the *Carnegie*, was destroyed by fire in 1926, it was not even replaced.

FOLLOWING SPUTNIK: THE MS SERIES

Accordingly, Vernov made the proposal to follow Sputnik 3 with a set of high-apogee satellites that would explore the radiation belt and the Sun–Earth relationship. An important influence here was Alexander Chizhevsky (1897–1964) and an associate of Tsiolkovsky. Although 19th-century scientists had made connections between sunspots, polar lights, and magnetic storms, interest in the area had waned in the first half of the 20th century, but it was given a new lease of life by the International Geophysical Year. Chizhevsky's book, *The Terrestrial Echo of Solar Storms* (1936), made the first comprehensive set of connections between changes in the Sun, its 11-year cycle, and its effects on the Earth. He came close to suggesting the idea of the solar wind and indeed he was the first person to coin the phrase "space weather". His book had certain mystical elements – he was regarded with great skepticism and might be considered "new age" in today's terminology – but he was fundamentally right about the Sun–Earth relationship [2]. There are even some objective studies now linking stress and heart attacks to geomagnetic disturbances (in medical terms, "adaptive stress reaction").

Although the set of high-apogee satellites made progress in OKB-1, later that year, Korolev suspended the project because of the high demands on him to develop the first manned spaceship [3]. The leading scientists at the time were concerned that science would now be downgraded to a secondary role in space exploration. The person who came to the rescue was Mstislav Keldysh, who firmly believed that space science must be a prominent objective of the space program. He invited the leading scientists of the day to make proposals for scientific space missions and these were forwarded on to the design bureaus for concept development and design. Keldysh also convened an Interdepartmental Scientific and Technical Council on Space Research (in Russian, the MNTS po KI). This managed to restore at least some meaningful role for science in the rapidly developing space program and in particular approved:

- four small satellites, later called the MS series; and
- four high-apogee satellites, later called the Elektron series.

The project was agreed by the government on 9th May 1960 and all were developed by Korolev's OKB-1 in Moscow. A knowledge of the belts was necessary to judge the likely effects on unmanned spacecraft passing through and to anticipate effects on cosmonauts. The Americans also made the belt a priority, several of the small Explorer satellites being devoted to the purpose.

The MS series used the new Cosmos rocket developed by Mikhail Yangel's design bureau, OKB-586, in Dnepropetrovsk in the Ukraine. Four were launched, but one was lost at launch. They were called MS, for *Maly Sputnik* ("Small Sputnik"). The Dnepropetrovsk bureau either did not know or did not realize what MS stood for and thought it meant *Moskovsky Sputnik* (Moscow Sputnik, where OKB-1 was located), so they named their own satellites DS, for *Dnepropetrovsky Sputnik*, which are described shortly. In the meantime, 1MS was allocated to Konstantin Gringauz in the Radiotechnical Institute while 2MS went to Valerian Krassovsky and Yuri Galperin in the Laboratory for Auroral Physics. Table 2.1 provides details of the series.

Table 2.1. MS series.

1MS series		
Cosmos 2	6 Apr 1962	212–1,560 km, 49°, 102.3 min
Launch failure	25 Oct 1962	
2MS series		
Cosmos 3	26 Apr 1962	228–719 km, 49°, 93.8 min
Cosmos 5	28 May 1962	190–1,587 km, 49°, 102.7 min

Cosmos 2 rocket from Kapustin Yar

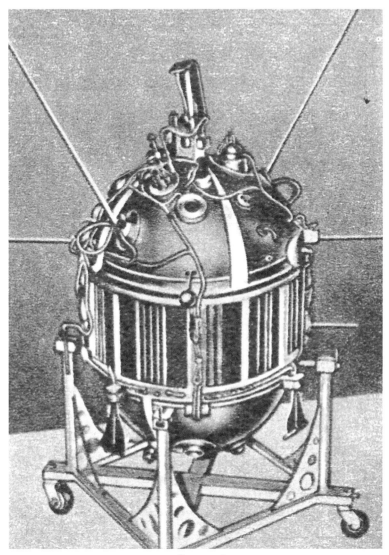

Cosmos 2

The series was designed to take forward the exploration of the radiation belts begun by Sputniks 2 and 3. Although they had design similarities, their purposes were slightly different. The 1MS series was intended for "studying primary cosmic rays and the effects of radiation on spacecraft", specifically investigating the absorption of short-wave radiation, while the 2MS mission was listed as to "study auroræ, the ionosphere, photo-electrons, super-hot particles with energies of tens and hundreds of eV" [4]. The objective of Cosmos 2 was to measure electrons up to 600 km and ions up to the high point of the orbit, 1,560 km (perigee was 212 km). Cosmos 2 carried three photo-electron analyzers to study electrons and ions in the

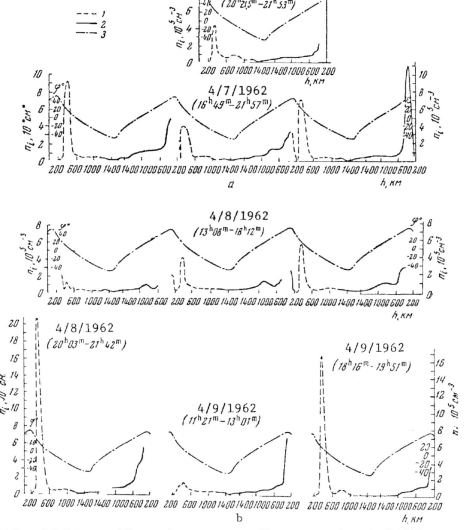

Figure 2.1. Printout of Cosmos 2 measurements of ion concentrations, April 1962.

ionospheric plasma, honeycomb ion traps, and a cylindrical Langmuir probe, rotating every 2 min. Following the debacle with Sputnik 3's tape recorder, a new memory system was installed. The other objectives were to measure ion and electron temperatures and concentrations at a time of declining solar activity.

Cosmos 2 provided significant scientific results. A chart of short-wave radiation was published. Cosmos 2 was credited with establishing the link between solar activity and the expansion and contraction of the Earth's ionized, gaseous envelope, its geosphere, which it suggested extended out to 20,000 km. It duly measured ions at different altitudes between 1,000 and 2,000 km using ion traps (Figure 2.1), finding

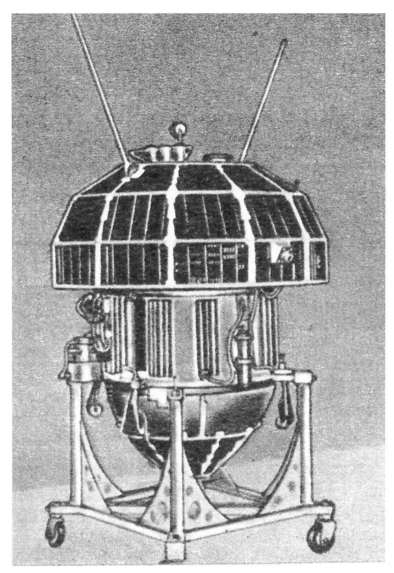

Cosmos 3

irregularities in the ionosphere, some 600 m long, some 150 km, others 250–350 km long. Helium ions were found as high as 550 km. Cosmos 2 found that the F2 layer of the ionosphere was characterized by instability between its ions and electrons. A chart was published showing the fall of ion concentrations plotted against latitude 33–46°N and altitude. Between 200 and 300 km up, there were ionospheric winds faster than the Earth's rotation (super-rotation) [5].

Cosmos 3 transmitted scientific results for a scheduled two weeks and was turned off on 10th May. Cosmos 5 was designed for longer data transmission and operated until 17th October. Cosmos 3 and 5 found low-energy electrons, diurnal variations in electrical fields of 40 keV, and hydrogen emissions from protons and that high winds in the atmosphere traveling at over 100 km induced electrical fields, proof that photo-electrons were capable of penetrating from one hemisphere to another by magnetic field lines.

Cosmos 3 and 5 provided data about the density of the upper atmosphere matched against latitude and time of day. Although their primary aim was auroral studies, they could not perform direct measurements in the auroral zones due to their inclination (only 49° to the equator). Their measurements of particle fluxes were therefore taken in the equatorial and middle latitudes, hitherto not known. They confirmed the existence of electric fields in the magnetosphere, finding how the electromotive forces of ionospheric winds transported photo-electrons from the sunlit to the dark side of the ionosphere along geomagnetic field lines. Cosmos 3 and 5 found three unrelated groups of intense fluxes of corpuscles at 700 km at $\pm 49°$, one with protons of energies of 50 MeV, a second of 100 keV, and a third of 10–20 keV. They found electrons of less than 5 keV in the sunlit atmosphere and 7–50 keV in high latitudes and altitudes. Cosmos 3 and 5 found electric fields in the magnetosphere that were perpendicular to the magnetic field. Streams of electrons found by Cosmos 5 were plotted against those found by Cosmos 3. They tested the stability of the F2 layer as a reflecting layer for short-wave spacecraft communications. Although Cosmos 3 and 5 were not specifically identified, a 1965 account of the discoveries of Soviet spacecraft about the radiation belts gives a considerable detail of information likely to have come from Cosmos 3 and 5.

A particular point of interest for Cosmos 3 was the South Atlantic magnetic anomaly, which had first been mapped by Korabl Sputnik 2 in August 1960 (see Chapter 6). This was a zone of intense radiation around lower Argentina, stretching across the South Atlantic. Cosmos 3 mapped the contours of the South Atlantic magnetic anomaly from 650 km from 24th April to 10th May. A new map was published, showing its magnetic contour lines in cross-sections at 200, 300, 400, and 500 km. Cosmos 3 and 5 marked out areas of low-energy particles (e.g. over Indonesia) and hard particles (the South Atlantic) including "forbidden" regions where certain types of particles perished. Finally, Cosmos 3 and 5 measured atmospheric density in the 180–320-km range. The search for magnetic anomalies continued the following year with Cosmos 12 and 15, both of which were fitted with a spherical analyzer to record low-energy particles and which found an anomaly with fluxes over 1 keV near New Zealand, especially pronounced in the late evening local time (Figures 2.2–2.5) [6].

COSMOS 5 AND *STARFISH*: INTRODUCING YURI GALPERIN

Perhaps the most important outcome of the 2MS series was an unintentional one. Cosmos 3 and more so 5 were associated with *Starfish*. This was the high-altitude nuclear explosion let off by the United States on 9th July 1962 at a time when the superpowers saw nothing wrong in testing nuclear devices in deserts (e.g. Nevada,

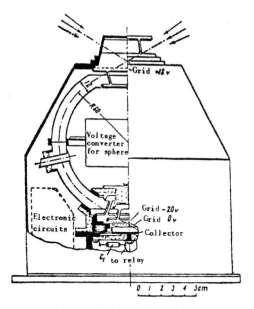

Figure 2.2. Cosmos 12 analyzer.

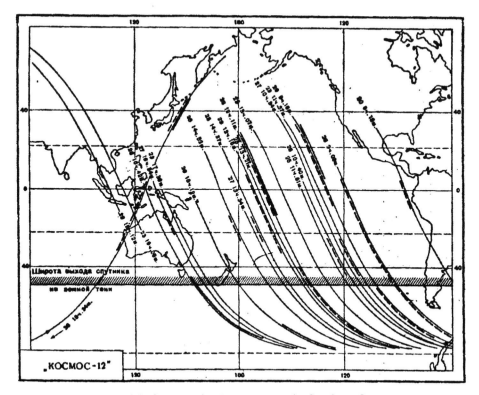

Figure 2.3. Cosmos 12, 15 passes over the Southern Ocean.

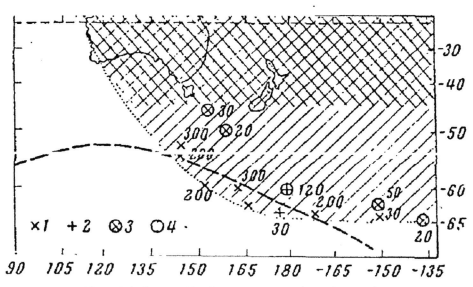

Figure 2.4. Cosmos 12, 15 measurement points of anomaly.

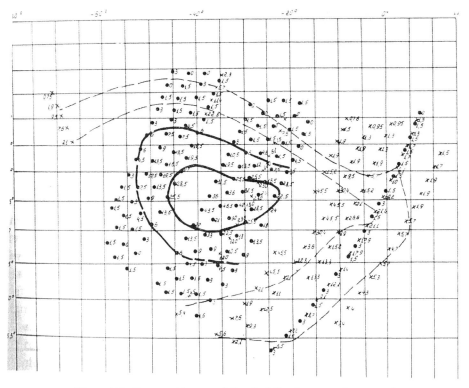

Figure 2.5. Cosmos 15 map of South Atlantic magnetic anomaly.

Australia), islands (Novaya Zemlya), or the atmosphere. *Starfish* was not the first high-altitude nuclear explosion, but the fourth, following three originally secret *Argus* blasts high over the Atlantic in 1958. *Argus* was the idea of elevator engineer turned Livermore laboratory military scientist Nicholas Christofilos (1916–1972) (called "the crazy Greek" in the press) who persuaded the Atomic Energy Commission that clouds of space radiation would disrupt the arming and fusing systems of incoming Soviet warheads and cause them to go off, long before they reached the ground. Explorer 4 was launched by the Americans just before the first *Argus* went off on 27th August 1958 over the South Atlantic to measure the effects of what was called "the Christofilos effect". *Starfish* was the fourth and most powerful explosion – one that lit up the night sky over the Pacific, causing considerable worldwide apprehension.

The *Argus* and *Starfish* explosions had the effect of creating artificial radiation belts that were of especial interest to scientists, who could measure and calibrate them against the natural radiation belts. As a military experiment, they disappointed because it was later ascertained that they were too weak to detonate missiles passing through, but they were sufficiently strong to disable peaceful satellites like the world's first active communications satellite then orbiting, Telstar.

Cosmos 5 happened to be one of five spacecraft in orbit at the time of *Starfish*, the others being Britain's Ariel, Telstar (launched the day before the explosion), and the American Injun 1 and TRAAC. Cosmos 5, though, was in the perfect position, being over China at the time of the explosion, 7,500 km distant high over Johnson Island in the Pacific. Although it was below the horizon at the time of the blast, Cosmos 5 noticed a 100-fold spike in radiation as it flew through the blast zone. There, it registered 2 min of hard radiation, followed by gamma and neutron radiation, as well as electrons in the 50–300-keV range causing polar lights. As it continued to circle the Earth, it observed a rise in orbital radiation levels, up 5.6 times more radiation at 300–400 km, up four times at 200 km, using the baseline radiation levels measured two years earlier by Korabl Sputnik 2 in August 1960. Later, Cosmos 5 scientists made a map of the *Starfish* blast, showing how it created a vast cloud and then two jets extending upward into space from 700 km (see Figure 2.6). The printout of the radiation meter shows a flat, normal Geiger count of 7/sec until the blast, when the needle soars in an instant to 20,000/sec, falling away to 500/sec 3 hr later – a dramatic incident. Cosmos 3, which had been turned off, was put back on air the day after the explosion. Although it had been far away from the blast, at 44°S, 3.1°W, it quickly recorded an increase in radiation around the Earth of 260%. It was finally turned off on 22nd July.

This is a good moment to introduce the scientist who made his name from the Cosmos 5 mission in general and *Starfish* in particular. Yuri Galperin was born on 14th September 1932, both his parents being famous linguists. He himself learnt how to speak English and French at a young age and as a schoolboy read Shakespeare in its original English. As a child, he was evacuated with his family beyond the Urals during the war. He entered the Department of Mathematics and Mechanics at Moscow State University in 1950 to study astronomy, where his mentor was the great astronomer Iosif Shklovsky. Shklovsky (1916–1985) was a somewhat eccentric

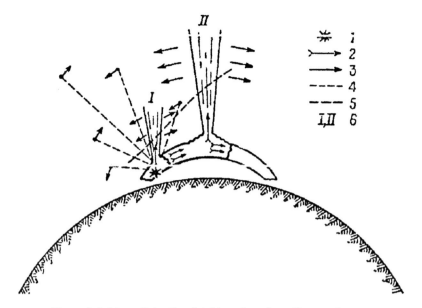

Figure 2.6. Map of the *Starfish* blast, based on Cosmos 5.

scientist: he was a serious astronomer, but somewhat jeopardized his reputation when he speculated in 1959 that Mars's moons Phobos and Deimos were hollow, left there by a traveling civilization. But some of his questioning mind, lateral thinking, and irreverent manner rubbed off on Galperin, who likewise loved the idea of science being popular and thrived on discoveries and sensations that interested ordinary people, although this never compromised his academic integrity. Shklovsky was later involved in an epic 1960 right-to-publish dispute in the Academy of Sciences, defending Konstantin Gringauz's discovery of the plasmasphere, leading eventually to its publication. Shklovsky was a good teacher, great at involving new young scientists, and is fondly remembered many years later. He predicted the existence of natural astrophysical lasers and masers long before they were actually discovered.

On graduation, Yuri Galperin was appointed to the Institute of Atmospheric Physics, where he worked under the head of the upper atmosphere department Valerian Krassovsky, the man who had developed infrared sensors for the Red Army during the war and later head of the upper atmospheric physics department of the Geophysics Institute (GEOFIAN). Galperin had hoped to get a post in Moscow, but he was turned down, possibly out of his association with Shklovsky. In disappointment, he took a posting to the institute's base in Loparskaya, near Murmansk in the Arctic Circle, where he observed the northern lights – a subject then very much a scientific backwater, but, with the International Geophysical Year, beginning to attract fresh scientific interest. Studying the northern lights in person inspired him and his fresh reinterpretation of the phenomenon became the basis of his Ph.D. These studies of auroral phenomena were the beginning of magnetospheric research before the space age. Indeed, the Loparksaya experience seems to have been

Yuri Galperin   Iosif Shklovsky

Yuri Galperin with Iosif Shklovsky

2

Making observations in the far north

a formative one for many others, too. Meantime, Valerian Krassovsky was responsible for the 10-keV particle detector installed on Sputnik 3 and even though Galperin was only in his twenties, he made him his deputy for the experiment. As construction of the DS and MS series got under way in 1961, Valerian Krassovsky's department, and with it Yuri Galperin, was given responsibility for the instruments on the 2MS series. According to one of his colleagues, Oleg Vaisberg, "we were all carried away with Krassovsky's enthusiasm ... he was the only scientist in the USSR who had a clear understanding of the connection between the polar auroræ and the precipitation of energetic charged particles into the Earth's upper atmosphere".

Some histories say that Cosmos 5 was put into orbit with the specific purpose to monitor American nuclear atmospheric tests, but this is not correct. In fact, the scientists had to persuade the military to use the satellite to study *Starfish*, not the other way around. Unlike *Argus*, which was not announced at the time, the upcoming *Starfish* test was known in advance: it was Galperin who persuaded the skeptical military that it would be useful for Cosmos 5 to monitor it. The problem was that Cosmos 5's tape recorder could take data for only 205 min, after which it would stop accepting fresh information and Cosmos 5 would not be over a Soviet ground station within 205 min of the expected explosion time. Although only 30 at the time, Galperin found himself before a panel of senior military officers trying to find a way around the problem. The answer was: "Tell us what you need and we will do what is required." The solution was for Cosmos 5 to make a data dump to a ship off the British coast, presumably a fishing trawler equipped with signals gear.

Later spacecraft returned to study the artificial radiation belt, finding that it polluted near-Earth space for a number of years and that its shape had some similarities with the ring current. Cosmos 6 tracked the persistence of fission fragments at high altitudes weeks after the explosion and measured how the satellite itself absorbed the *Starfish* radiation. Measurements of the newly formed artificial radiation belt went on into spring 1963 and were followed by Electron the following year (below). Cosmos 137 found what were clearly *Starfish* particles in the inner radiation belt more than four years later. Galperin's doctoral dissertation was devoted to interpreting these experiments. He modeled the physical pattern of formation of the artificial radiation belt that appeared as the result of the explosion, its characteristics, and decay time. Galperin discovered a number of new effects, such as "gamma dawn" (the scattering of gamma radiation beyond the horizon), the long-term conservation of heavy ions due to vertical drift in the equatorial ionosphere anomaly, and hot plasma breaking through upward in the equatorial zone. Yuri Shafer described how the blast "cracked" the magnetosphere, creating protons of up to 190 MeV, disturbing the radiation belts and creating an 80% spike in radiation levels [7]. For comparison, Cosmos 259 and 262 later passed through the Chinese nuclear blast of 27th December 1968 10 and 23 hr later, respectively, recording its dispersal in the upper atmosphere over Mongolia, where it had drifted from the western desert of Chinese territory.

FOLLOWING SPUTNIK: ELEKTRON

The Elektron missions came at a time when both the Moon race and the related man-in-space race were hotting up and, as a result, they received little press attention. Scientifically, though, they were important. In terms of data returned, they were amongst the most productive missions of the early Soviet space program. Using instruments first tested on the MS series, these satellites had a unique design and the most appropriate orbit for them was carefully considered [8].

At this stage, a much improved knowledge of the Earth's environment had emerged. As radiation from the Sun reached the Earth, it streamed into the Earth's doughnut-shaped magnetic field. Arriving over the polar regions, the reaction of radiation with the Earth's magnetosphere triggered off auroral displays and electrical storms. As the Earth traveled in its orbit, its magnetic field created a tail trailing behind. The Elektrons would take a set of irregular orbits to map the radiation belts surrounding Earth and related physical phenomena. The two Elektrons had different shapes: the first of each series was a cylinder with paddle wings, the second a cylinder with solar cells on its structure. Elektron 1 was the smaller, being 355 kg, 75 cm in diameter, 2 m long, and the panels having an area of 4.3 m^3. Elektron 2 was heavier, at 465 kg, 1.8 m in diameter and 2.4 m long, its solar cells having an area of 4.8 m^3. Shutters ventilated heat from the satellites 8 min every hour. Tape recorders were carried, so as to later relay data collected when the spacecraft were outside Soviet territory.

The Elektron pairs were scheduled for 61° orbits, with perigees set in the northern hemisphere over the USSR, so as to obtain maximum data of the thickest portion of the radiation belts. Elektron 1 (and later 3) was set to orbit from 400 to 7,000 km

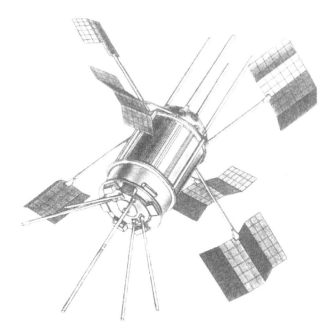

Elektron 1

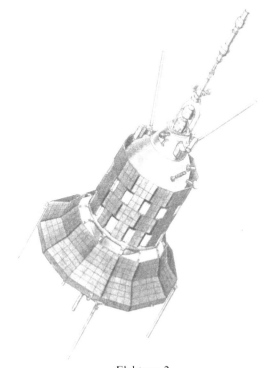

Elektron 2

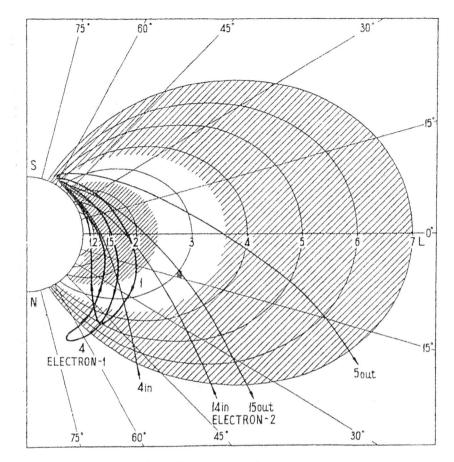

Figure 2.7. Elektron trajectories. (Credit: COSPAR)

(1 Earth radius), into the area where auroræ were formed and just into the bottom part of the outer belt. Elektron 2 (and later 4), in an orbit of 450 by 66,000 km (10 Earth radii), would pass from the upper regions of the inner belt to just beyond the outer region of the outer belt and cross the region beyond the radiation belt with unstable low-energy electron fluxes (Figure 2.7). Instrumentation was as shown in Table 2.2.

After three years of design and construction, the first launching took place on 30th January 1964. Deploying the satellites was tricky, the first time that the Soviet Union had put two satellites into orbit – and quite different orbits – from the same rocket. Elektron 1 was jettisoned while the third stage was still operating, while Elektron 2 dropped into its higher orbit some 20 sec later. Although the maneuver was unprecedented and the satellites of an entirely new design, almost perfect orbital insertions were achieved. The only disappointment was that the radiation belts degraded the solar cells more quickly than expected, with Elektron 1 ceasing transmission after only two months (27th March 1964) as it was frequently exposed

Table 2.2. Instrumentation on Elektron satellites, 1964.

Elektron 1, 3	Elektron 2, 4
Radiation detector	Radiation detector
Instrument to detect corpuscular radiation	Magnetometer
Mass spectrometer	Low-energy particle detector
Proton detector	Cosmic ray chemical composition detector
Micrometeorite detector	Electron detector
	X-ray detector
	Mass spectrometer

to proton fluxes in the inner radiation belt, which is more energetic than electron particles in the outer belt, where Elektron 2 spent most of its time. Elektron 2 lasted its intended lifetime, broadcasting until 30th July.

Sometime during the early summer, it was decided to launch the backup spacecraft on an identical mission, both to extend the lower orbit data and to obtain higher orbit data from two spacecraft simultaneously. With Elektron 2 still on air, the second pair was launched on 10th July. The solar cells were hardened and the electrical system recalibrated to use less power. As a result, Elektron 3 lasted six months, until 13th January 1965, and Elektron 4 even longer, until 23rd May 1965. The missions coincided with 1964: International Quiet Sun Year, with a solar minimum that July, which meant that solar minimum conditions could now be baselined. The principal scientists on the Elektrons were Sergei Vernov, Alexander Chudakov, Konstantin Gringauz, P.V. Vakulov, Yuri Galperin, and Olga Khorosheva (Table 2.3).

Table 2.3. Elektron series.

Elektron 1	30 Jan 1964	406–7,100 km	2 hr 49 min	60.8°	329 kg
Elektron 2		460–68,200 km	22 hr 40 min	59.7°	440 kg
Elektron 3	10 Jul 1964	405–7,040 km	2 hr 48 min	60.9°	350 kg
Elektron 4		459–66,235 km	21 hr 54 min	60.9°	444 kg

R-7 Vostok launcher from Baikonour

ELEKTRON FEAST

The results of the two missions were presented by mission scientists, principally Sergei Vernov, at a Soviet conference on the physics of outer space for the purpose, held in Moscow in July 1965. Elektrons 1–4 returned a true deluge of data on the radiation belts and the space environment. Papers from the mission were published into the 1980s [9]. First, the Elektron missions between them enabled nothing less than a new picture to be painted of the Earth's magnetic environment and its constituent parts (Figures 2.8–2.12, showing the new maps of Earth's magnetic field compiled by the Elektrons). The Elektrons were able to make a cross-section of the entire magnetosphere and a new map of Earth's magnetosphere was published. Charts showed how the Elektrons moved in and out of the magnetosphere boundary. A book based on the results, *Models of Space*, became the reference for space engineers. B.A. Tverskoy subsequently published a full model and theory of the radiation belts based on the Elektron studies:

- The outer radiation belt ranged from 3 to 9 radii out from Earth, comprising mainly ions in the order of 1 MeV, the amount fluctuating, being replenished by solar activity from time to time;
- The inner radiation belt extended from 1.1 to 1.3 radii and comprised protons in the 10–100-MeV range, reinforced by cosmic rays bouncing off the upper atmosphere and was much more stable in nature. The main principle, as developed by Yuri Logachev, was that the spatial distribution of trapped particles depended strongly on their type and energy. The more energetic particles were closer to the Earth, comprising the greater bulk and the higher intensity.

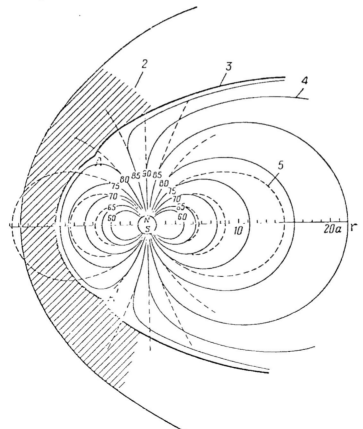

Figure 2.8. New map of the magnetosphere.

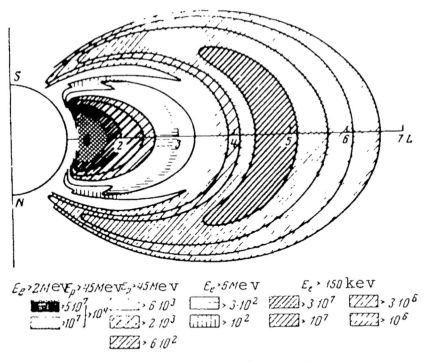

$E_e > 2Mev$ $E_p > 45Mev$ $E_p > 45Mev$ $E_e > 5Mev$ $E_e > 150$ kev

Figure 2.9. New map of the distribution of electrons.

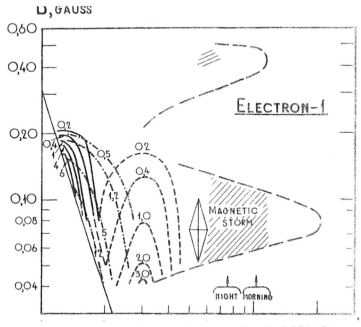

Figure 2.10. Map of a magnetic storm. (Credit: COSPAR)

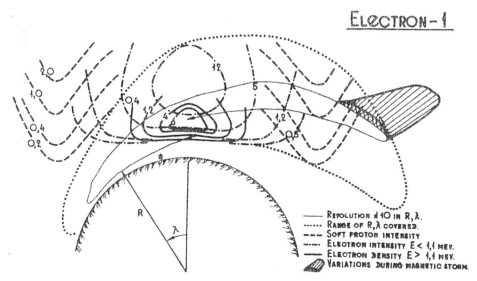

Figure 2.11. Elektron map of soft protons. (Credit: COSPAR)

Figure 2.12. Elektron map of the sporadic zone.

Within this, two distinct magnetically active regions were identified: an outer one at 10–15 radii (where the Earth's magnetic field met the Sun's radiation) and an inner one at 3–5 radii, where magnetic storms took place. The Elektron data of 1964 were later compared to information collected by Cosmos 3 and 5 in 1962, finding that the inner radiation zone was relatively constant and appeared to have some form of stabilizing mechanism.

Maps of intensity contours were published. The ionosphere was irregular, with concentrations of electrons at 300-km altitude and in the F2 layer, as dense as $10^9/$ cm^2/sec. The maps were tested against a series of morning and evening, ascending and descending orbits and a typical storm was modeled. Each time the satellite approached Earth on an inbound orbit, there were sharp increases in ion currents between 20,000 and 15,000 km, ion fluxes outside the belts themselves, and a "soft electron component" in the outer radiation belt. They found a zone of high-intensity electrons of 0.1–10 keV outside the belt of captured electrons. In effect, the Elektron missions led to the conclusion that the height of the ionosphere was not 1,000 km as originally thought, but extended to 25,000 km. This was called the plasmasphere, a light ion extension of the ionosphere. Elektrons 2 and 3 were the first to detect the magnetic field of a quiescent current ring.

Second, they showed how the radiation belts reacted to the Sun. Instruments from Elektron charted the ups and downs of magnetic quietness that spring and recorded magnetic storms on 31st January, 12–13th February, 20th–21st February, and 8th April 1964, following the intrusion of particles into the upper atmosphere. The belt was relatively undisturbed during the solar minimum, with little movement for periods of four to six days at a time. Earth's radiation belt was found to move and adapt according to the level of magnetic perturbation and could also fluctuate even in quiet conditions. The radiation belts as a whole were found to be rather stable, the changes between solar maximum and minimum not being enormous. The Elektrons found that the Earth's magnetic field formed and reshaped itself considerably from night to day, Elektron 2 finding what was called a "morning electron region". They found sporadic fluxes lasting several hours at high magnetic latitudes at midnight. The whole radiation belt moved to slightly higher latitudes over 1964. During periods of quiet Sun, it appeared that Earth's magnetic field had insufficient stocks of electrons to set off polar lights, but that Earth gradually rebuilt its stock once the quietest period was over. From 800 to 1,000 km, the composition of the ionosphere changed, with the oxygen–nitrogen envelope becoming a hydrogen one, with about 10% of helium. They detected and measured hydrogen gas high in the atmosphere. Helium concentrations were found at 580-km altitude by night and 780 km by day. They found that there were considerable variations according to the time of day: atmospheric density rose daytime at 200 km, 40% more during peak solar activity. Electrons of 40–50 keV were found – but only at night. They increased towards an altitude of 1,500 km, where they had an intensity of 10^8 cm^2/sec. Electron temperatures were measured at between 1,800 and 3,000 K daytime between 250 and 550 km.

Third, the Elektrons found many anomalies and variations with the radiation belts. Elektrons 1 and 3 found big electron fluxes 6,000 km high over the South

Atlantic as big as polar magnetic storms, where there were fluxes of hundreds of keV and hitherto unknown fluxes around 49°N and 49°S between 200 and 1,600-km altitude. Electrojets were found striking out from the poles [10]. On Elektron 2, on 16th February 1964, Olga Khorasheva discovered an additional radiation sub-zone, connected by magnetic coils from the North Pole to the South Pole (see figures above). Returning to the mission of Cosmos 5, the Elektrons re-examined the way in which radiation fragments from *Starfish* had spread to a great height, at 1,200 km.

Fourth, Elektron 2, confirmed by Elektron 4, found that Earth itself was an emitter of radiation, in the order of 1 MeV. Earth emitted its own radio emissions, rather like Jupiter, coming from soft electrons streams in the ionosphere at 30–50° latitude. The American probes Interplanetary Monitoring Platforms 6 and 8 later discovered what was called "auroral kilometric radiation". A retrospective look later found it in old Elektron data – sporadic radio sources 2–8 Earth radii out at 1,110 kHz between 60 and 76°S [11].

Fifth, the Elektrons found how Earth reacted to radiation from further afield. The Elektrons measured the rate of cosmic ray radiation, which rose 2% a month from February to May, only 1% during the summer and stabilized during the autumn, coinciding with the decline in solar activity to the minimum of July 1964. Cosmic rays were made of lithium, beryllium, and borum nuclei, suggesting that their sources were "very far away in the depths of the universe". Just as picking up the Earth's signals was a surprise, so, too, was the detection of signals from deep space. Radio-receivers on the Elektrons picked up radio waves of 200–400 m in length coming from deep space. These may have contributed to a sensational story run in the Soviet press on cosmonautics day, 12th April 1964, that a civilization transmitting from deep space had been discovered and later that year, an all-USSR conference on extraterrestrial civilizations was held in Byurakan, Armenia (October 1964). Ultimately, the alien civilization turned out to be a quasar, which received the more prosaic name of CTA-102. The story, though, did have benign consequences, for it sparked off the first meeting of American and Soviet scientists to discuss extraterrestrial civilization, held in September 1971 in Byurakan observatory, where the participants included Carl Sagan and Frank Drake. Prominent in the conference was Iosif Shklovsky, who had earlier written a book on the possibilty of life in the universe, *The Universe. Life. Mind* (1962), and published in the West with Carl Sagan as *Intelligent Life in the Universe*.

Finally, although the Elektron missions were primarily about radiation, useful data were also returned in other areas. Meteor detectors encountered three meteor showers – on 31st January, 11–13th February, and 23rd–25th February – but little in between (Figure 2.13). They were tiny particles, leading scientists to conclude that Earth passed through periodic concentrations of dust as it orbited the Sun, these clouds being 3,000,000–5,000,000 km in diameter, Elektron 2 passing through three such clouds, sustaining 185 hits during one. Elektrons 2 and 4 detected eight episodes of heavy cosmic ray nuclei coming from the Sun, but doing so independently from other known solar activity at the time.

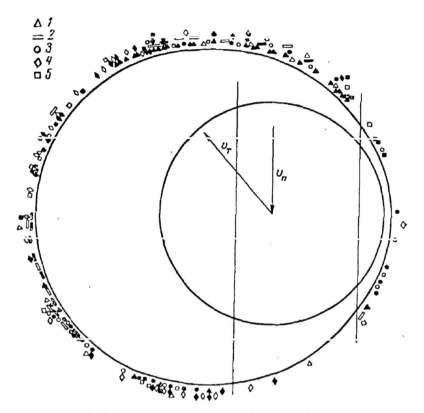

Figure 2.13. Elektron detection of meteor showers in January 1964.

INTRODUCING THE COSMOS PROGRAM

The MS and Elektron programs in effect represented unfinished business from the first three Sputniks from 1957 to 1958. In the early 1960s, it was reasonable to expect a more comprehensive, systematic program of space science to emerge. This began in 1962 as the Cosmos program, extended internationally seven years later as the Intercosmos program.

The Cosmos program, although originally intended to be purely for this new series of scientific satellites in Earth orbit, actually came to be used as a catch-all designator for a huge number of quite different types of satellites. Although the first three satellites were indeed scientific satellites in the original Cosmos program, from Cosmos 4 onwards, they began to be used as a cover name for primarily military missions. "Cosmos" became, in the course of time, a flag of convenience for failed lunar, Venus, and Martian missions as well as tests of manned spaceships, new propulsion systems, civilian Earth resources satellites, applications missions, and large space station modules. Much later, in the period of *glasnost*, these specialized missions were taken out of the Cosmos program and civilianized, so that from the

1990s, "Cosmos" became used only for military and semi-military navigation satellite missions, the exact opposite of where the program had started. By then, the number of Cosmos missions was well past the 2,000 mark. Although the Western popular press accused the USSR of designing the Cosmos program as a flag of convenience, it was in reality intended purely as a science program.

The Cosmos program was assigned not to Sergei Korolev's design bureau, OKB-1, which had built the first three Sputniks, MS, and Elektron, but to another leading bureau, that of Mikhail Yangel in the Ukraine. Here, a word of explanation about the role of the design bureau is helpful. The organizational core of the Russian space program is the design bureau. This could be classified by design bureau (*Konstruktorskoye Buro*, KB) or experimental design bureau (*Opytnoye Konstruktorskoye Buro*, OKB). The design bureau was the middle element in the three-part chain – a system developed in Stalin's time [12]. First, concepts were tested in a scientific research institute (NII) (*Nauchno Issledovatelsky Institut*). Once deemed possible or desirable, hardware was designed, built, and tested by an OKB or KB. Once perfected, it was put into production in the third part: the factory. The operation of the system was actually more complex than this, because some design institutes grew up with factories alongside and were closely associated with one another. Furthermore, a design product of one OKB could be sent for production in a factory affiliated to a rival design bureau.

Western impressions of "the Soviet system" were that it was a command-and-control system in which the directors of the state – the government and the Communist Party – issued orders to design bureaus to carry out centrally decided policies. In fact, there was a high level of rivalry between the design bureaus, each importuning central government and the political leadership to adopt pet projects. The government, for its part, saw competition between the bureaus as a means of driving up standards and ideas. This form of socialist market competition was fine in theory, but in practice, it operated in an undisciplined way, decisions often being remade, with contracts going to favorites, with a premium on design bureau chiefs who had an inside track to the central committee.

Despite the command-and-control appearance, the Soviet system was actually quite decentralized. Research institutes were the heart of all Soviet science. There were then up to 600 major institutes in the USSR, connected either loosely or more formally to the Academy of Sciences, which had 10 divisions, guided by 500 full and corresponding members, their research programs being a matter of negotiation with either republican or all-Union government. This European-based system was quite different from the United States, where research was centered on the universities, guided through competitive grants and supplied directly with undergraduates or graduates from these or other universities. Soviet institutes had no direct relationships to the universities, which were primarily teaching institutions for undergraduates. Little space-related research was carried out in the universities (the great exception being Moscow Lomonosov State University).

Mikhail Yangel is one of the least well known of the Russian space designers and as the architect of the Soviet scientific satellite program, we should say a little more about him. Born on 25th October 1911 near Irkutsk in Siberia, he became an aircraft designer at the Polikarpov aircraft design bureau. An active Communist Party member, he was trusted to go the United States from 1938 to 1941 to study mass-production techniques in the American aviation industry [13]. The government charged him in 1951 with responsibility for developing a volume-production rocket and missile plant in Dnepropetrovsk, giving him an old car factory as his base and assigning him 25 top engineers, drawn reluctantly from both Korolev's bureau and that of engine designer Valentin Glushko. Although the emphasis of his work was on the mass production of rockets and missiles, he was also permitted to develop a design bureau alongside the enterprise, OKB-586. In the event, his design-to-development periods were the shortest and most successful of all the Soviet Union's rockets. Although he built his reputation around mass production, he also designed and built the Soviet lunar module, the LK, a small but sophisticated design that made three entirely successful test flights. His mass production ideas were applied, with success, to the Cosmos program. He died in 1971, aged only 60.

Yangel was charged with the development of a rocket using storable fuels, which became the single-stage R-12, Cosmos 1 rocket and this made its first flight on 22nd June 1957. Khrushchev and the military soon realized that although Korolev's larger R-7 rocket had been developed as a missile, its real value was in space exploration. It was quite unsuitable as a missile, being large, visible, and taking hours to fuel up. The R-12, by contrast, used storable fuels at room temperature (nitrogen tetroxide), could be fueled up within a couple of hours, and kept ready to go for long periods. It was small enough to be based in an underground silo. The R-12 development history went like a dream and it was accepted into the armaments within nine months, several being installed in military silos around Kapustin Yar as a strike force. Khrushchev and the much impressed military later asked him to develop a more powerful version, the R-14, and work on this began in July 1958.

Indeed, Yangel had so impressed some key figures in the Soviet establishment that when OKB-1's object D fell behind schedule, consideration was briefly given to asking Yangel to build a first satellite and launch it on an R-12 [14]. In the event, the idea went no further, but it obviously sowed the seed of the concept of a satellite and launcher program led by OKB-586 in Dnepropetrovsk. The technical requirements for such a program were drawn up by an interdepartmental scientific technical board that first met under the chairmanship of Mstislav Keldysh in December 1959 and that included Mikhail Yangel.

DNEPROPETROVSKY SPUTNIK (DS)

The idea of a satellite program developed by the Yangel bureau was formalized in two government decrees, on 23rd June and 8th August 1960, a little later than the MS satellites. This specified the need for a series of "small satellites" ("*malye Sputniki*"), cheap and suitable for a large number of launches, with the aim of conducting scientific studies of characteristics of the upper atmosphere, ionospheric radiation, the magnetic fields of the Earth, and so on. The government approved the conversion of the R-12, Cosmos 1 rocket into a satellite launcher. Yangel was given facilities at the Kapustin Yar cosmodrome, the site of the post-war A-4, *Akademik*, sounding rocket tests and dog flights. Kapustin Yar was much closer to Dnepropetrovsk than the Baikonour cosmodrome and Kapustin Yar became, in effect, the Yangel cosmodrome. With a second stage and satellite on top, the R-12 was actually too tall for the closed silos from which it was normally launched. Accordingly, the new version was still placed in the silo, with the top end sticking out above the ground and launched in the normal way but called the Cosmos 2 launcher.

The decrees ordered an initial batch of 10 satellites. The Cosmos 2 made possible the launch of a considerable number of small, light satellites, in the 200–400-kg class, much smaller than object D, but more than sufficient for basic space science research. They were, broadly speaking, the equivalent of the Explorer series in the United States. As we saw earlier, four of the OKB-1 satellites acquired the title MS while the OKB-586 Dnepropetrovsk bureau called its satellites DS, *Dnepropetrovsky Sputnik* (also shortened as *Dneprovsky Sputnik*), in English "Dnepropetrovsk satellite". It is not known whether the two bureaus worked off the same template, but it is possible that some preliminary studies had been done by OKB-586 in the mid 1950s. The actual designer was Vyacheslav Kovtunenko, later head of the bureau after Yangel.

Two technology demonstrators were built first, DS-1. Indicating its importance in Soviet space planning, the first launch was on 21st October 1961, set to mark the XXII Congress of the Communist Party of the Soviet Union. The launching failed, as did a second on 21st December 1961. Despite the two failures, Yangel proceeded to a new version, DS-2, this time fitted with scientific equipment, two being built. Happily, this third launch, on 16th March 1962, was successful and marked the introduction of the program, the Cosmos 2 launcher, Kapustin Yar as a satellite-launching cosmodrome and the DS satellite, so it was an important date in Soviet space science. Being a first test, Cosmos 1 was actually only a small, 47-kg test satellite with a radio beacon called *Mayak* (broadcasting on 20.003, 20.005, 90.018, and 90.0255 MHz) and one of Sergei Vernov's Geiger counters and transmitted until

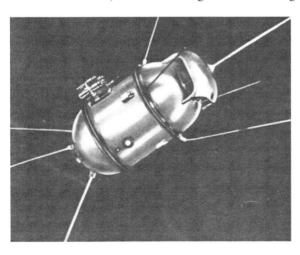

The DS design (MT version)

25th May 1962. Despite being a test mission, useful scientific data were returned, especially on the structure and characteristics of the ionosphere, electron concentrations (measured every 1–2 km), and their temporal, latitudinal, and longitudinal variability. Ionospheric inhomogeneities were found for the first time. Moreover, this mission and its successors began at a time of decreasing solar activity, which contrasted with the first three Sputniks, which

Table 2.4. Aims of the Cosmos program, announced 16th March 1962.

Exploration of the upper and lower atmosphere (including cloud formation)
Measurement of charged particles and their concentration
Detection and measurement of cosmic rays and corpuscular fluxes
Study of the Earth's magnetic field
Measurement of short-wave emissions from celestial bodies
Propagation of radio waves in the atmosphere
Measurement of Earth's radiation belts
Solar radiation
Meteorites
Study of the ionosphere

---

flew when solar activity was increasing. Radio Moscow heralded the mission as the beginning of a vast program of scientific research in Earth orbit. Some Western commentators dismissed the statement as hyperbole, but the scientific program embarked on that day did turn out to be extensive. The first results of the series were given at the 7th COSPAR symposium in Florence, Italy, in 1964 and later in international journals. The value of the announcement has sometimes been overlooked, so it is worth paying it some attention (Table 2.4).

The DS satellites used a common design, or "bus", which was a pressurized cylindrical hull with a domed top and bottom, with different types of instruments protruding from each end. This presented economies of scale, which meant that it was possible to build a DS spacecraft in less than three months. The Cosmos 2 rocket was replaced by a more reliable successor, the Cosmos 2M, introduced on 16th March 1967. Between them, the Cosmos 2 and 2M made 144 successful launches between March 1962 and June 1977, flying originally from Kapustin Yar and then from 1967 from the northern cosmodrome of Plesetsk.

DS spacecraft weighed 47–321 kg, with the weight of scientific equipment ranging from 4.5 to 44 kg. Batteries were sufficient for 10–15 days. The first missions were undertaken in phases, the DS-2, A1, and K8 missions approved first, followed by MT, MG, and concluding with MO in March 1967. Normally, two satellites were built for each mission, the second to be used if the first failed. It also happened that if the first mission was entirely successful, then the second version might be adapted for a repeat mission for a second dataset or a different mission (the backup for Cosmos 1 was modified to fly as Cosmos 51). Although the initial order was for a batch of 10 scientific satellites, the DS model was quickly adapted for military missions, such as radar calibration missions for the Soviet air defense system (DS-P1-Yu series), which will not be considered here. Starting in 1968, the Soviet Union also began the first of two series of geodetic missions to map the Earth's gravity field (*Sfera*, followed by *Musson*), the purpose being primarily military.

The DS program was sub-divided into categories: DS-MG to study the magnetic field, DS-MT for meteorites, A1 for radiation, and so on. The DS series ran from March 1962 to July 1965, with two later additions, the MO sub-series, as shown in Table 2.5. Table 2.6 shows the DS scientific missions.

Table 2.5. DS series categories.

Test	Cosmos 1
A1	Cosmos 11, 17, 53 *Spin*, 70
K8	Cosmos 8
MG	Cosmos 26, 49
MT	Cosmos 31, 51
MO	Cosmos 149, 320 (*Opticheski/Strela*)

Table 2.6. DS series, first round, science missions.

Launch test, DS-2		
Cosmos 1	16 Mar 1962	204–967 km, 49°, 96.4 min
DS-K8: meteorites		
Cosmos 8	18 Aug 1962	244–598 km, 49°, 93 min
Cosmos DS A1 Science, magnetosphere, artificial radiation		
Cosmos 11	20 Oct 1962	234–901 km, 49°, 96 min
Cosmos 17	22 May 1963	260–788 km, 49°, 95 min
Cosmos 53 *Spin*	30 Jan 1965	218–1,180 km, 49°, 99 min
Cosmos 70	2 Jul 1965	215-1,147km, 49°, 98min
DS-MT series: meteorites, ultraviolet and gamma radiation, the stellar background		
Cosmos 31	6 Jun 1964	222–492 km, 49°, 91.7 min
Cosmos 51	9 Dec 1964	262–533 km, 49°, 92.5 min
DS-MG series: the magnetic field		
Cosmos 26	18 Mar 1964	266–387 km, 49°, 91 min
Cosmos 49	24 Oct 1964	264–466 km, 49°, 91.8 min
Cosmos 2 from Kapustin Yar		
*DS-MO series (*Opticheski*)*		
Cosmos 149	21 Mar 1967	245–285 km, 48°, 89 min
Cosmos 320	16 Jan 1970	247–326 km, 48°, 90 min
Omega series		
Cosmos 14	13 Apr 1963	252–499 km, 49°, 92 min
Cosmos 23	13 Dec 1963	240–613 km, 49°, 93 min)

We have results and outcomes from a number of these missions. Cosmos 11 made a profile of electron concentrations by altitude (Figure 2.14), finding them to be in an S-shape. P.V. Vakulov's Cosmos 17 was one of the most important. Its mission, which lasted from 22nd to 30th May 1963, had the task of measuring the distribution of charged particles in the 250–780-km range, 49°N to 49°S. It had an extensive range of instrumentation to measure cosmic rays, looking at their variations in intensity plotted against solar radiation cycles. High-density memories and data transmission systems were used for the first time, with particle hits recorded on photographic film and transmitted to the ground [15]. It had two scintillation

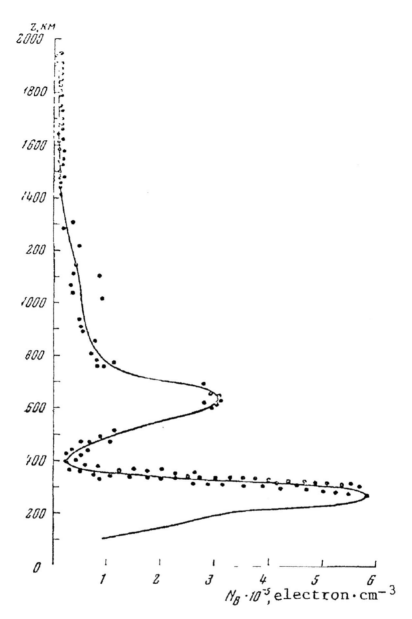

Figure 2.14. Cosmos 11 data on ion concentrations by altitude.

counters to measure electrons in the 50–180-keV range and one gas discharge counter to measure protons in the 600-keV and 5.4–8.5-MeV range. Cosmos 17 mapped the movement of high and low-altitude particles in the geomagnetic field, especially over the South Atlantic. Measuring particle distribution at low altitudes is of especial interest to magnetospheric scientists because at these altitudes, the Earth's magnetic field is rather different from the dipole. Particle behavior is more intricate: as predicted, particles died out as they met the atmosphere at 150 km, but this was confirmed by Cosmos 17. Higher up, analysis suggested the existence of a constant belt of high electrons varying little according to altitude. Cosmos 17 found electron streams left by the *Starfish* explosion and by Soviet explosions before nuclear testing in the atmosphere was banned. It explored magnetic anomalies. It found streams of soft electrons in the 50–100-keV range and corpuscular streams under the lower boundary of the inner radiation zone in a geomagnetic trap. Cosmos 17 made it possible to compile a map of soft electrons around the Earth, the main concentrations being in a belt over southern latitudes from the South Atlantic across Africa to the Southern Ocean (Figure 2.15). P.V. Vakulov wrote his candidate thesis on the results, which took a year to analyze, called "Research on Earth's Radiation Belts and Cosmic Rays on Cosmos 17".

Cosmos 26 and 49 were important missions whose objective was to compile a magnetic map of the Earth, following a large program of magnetic mapping by both air and sea (by special non-magnetic ship), initiated by Nikolai Pushkov, the first director of IZMIRAN. They were part of a little-advertised program of Soviet–American cooperation. Talks on areas of collaboration were held between Anatoli Blagonravov of the Soviet Academy of Sciences and NASA deputy Hugh Dryden in Geneva, Switzerland, from 29th May to 7th June 1962, leading to a formal agreement signed on 8th June 1962, where it was agreed that there would be cooperation in the area of weather satellites, passive communication satellites (the

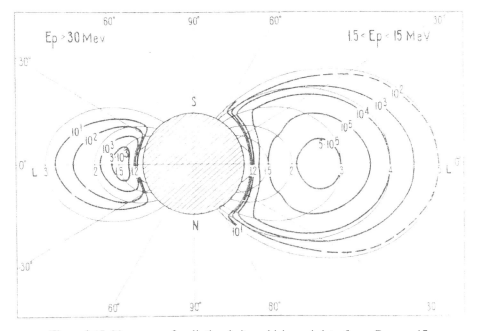

Figure 2.15. New map of radiation belts, which used data from Cosmos 17.

American Echo series), and mapping the Earth's magnetic field [16]. Biology was added later. Cosmos 26 and 49 followed in the footsteps (or wake) of the famous wooden schooner *Zarya*, which, in 1956, began the World Magnetic Survey project led by the USSR, United States, and Canada by sailing 500,000 km across the oceans of our planet making a magnetic map.

The MG spacecraft were ellipsoids 1.8 m long and 1.2 m in diameter, with a boom 3.3 m long for the two proton precision magnetometers, 90° apart, which collected data every 33 sec, designed by Marat Chinchevoy of Kiev Radio Factory. The readings were stored as magnetograms (Figure 2.16) in a tape recorder with up to 18,000 distinct data points and 800 min of tape and then relayed to the ground. Principal Investigator was Shmaia Dolginov of the Institute for Terrestrial Magnetism, IZMIRAN, and they were the first satellites associated with the institute. They operated from 18th to 30th March 1964 and 24th October to 6th November 1964, respectively, the outcome being a magnetic chart plotted against latitude from 49°S to 49°N (Figure 2.17), with detailed maps of individual areas and continents (Figure 2.18). Cosmos 49 found intermediate-sized magnetic anomalies and a number of irregularities at 120°E, possibly due to the nature of the Earth's upper mantle. Natalya Benkova presented Earth's new geomagnetic reference catalog in 1965, 75% of Earth [17].

The pioneering MO series, developed by the Institute of Atmospheric Physics (IFA) by G.V. Rosenberg, used a spacecraft aerogyroscopic attitude control system for the first time in the world. Cosmos 149 was an unusual mission developed by Vyacheslav Kovtunenko called the "space arrow" and carried what was variously

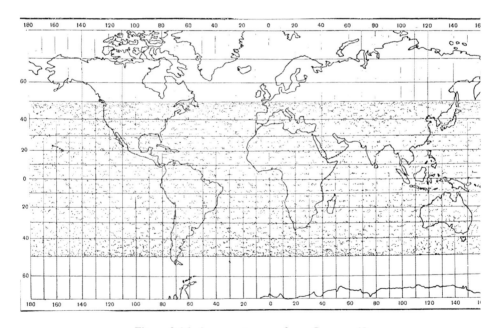

Figure 2.16. A magnetogram from Cosmos 49.

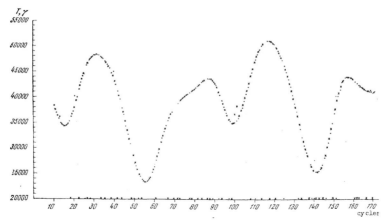

Figure 2.17. The part of Earth covered by Cosmos 26 and 49 mapping, each dot representing a measuring point.

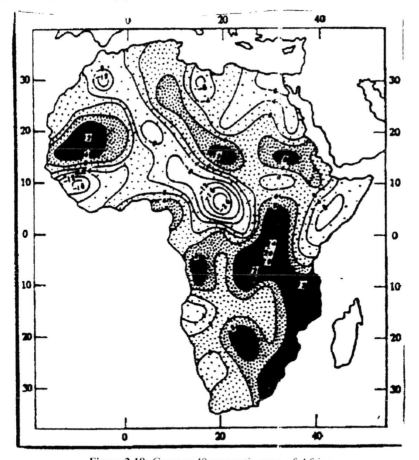

Figure 2.18. Cosmos 49 magnetic map of Africa.

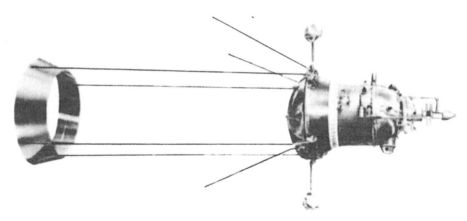

Cosmos 149 *Kosmicheski Strela*

described as an "extended skirt" or aerodynamic stabilizer and aerogyroscopic stabilization system, to ensure the spacecraft was constantly pointed towards Earth for weather observations. Cosmos 149 verified the concept of the aerodynamic system: it did indeed save fuel, but only really worked at low altitudes where the orbit would last as little as two to three months. There were some problems with the instruments and satellite stabilization, so the experiment was re-flown on Cosmos 320, which decayed on 10th February 1970 after a successful mission. The venture was so successful that a film was made, *Space Arrow* (TsentrNauchFilm, 1967, produced by K. Dombrovsky).

Leaving aside the engineering aspects, the "space arrow" had a television set called Topaz 25M and a set of instruments called *Atkin 1* comprising scanning telephotometers and radiometers. The satellite was designed for optical sensing of the atmosphere and determination of its structure. It was able to determine the temperatures of sea surfaces, clouds, the height of cloud tops, their moisture and temperature, and the interaction of solar and Earthly radiation. The details were as follows:

- The narrow angle infrared radiometer (principal investigator: A. Gorodetsky) determined the thermal structure of clouds, detected aerosols and heat loss from the atmosphere;
- The telephotometer (principal investigator: A. Malkevich) measured the structure of clouds, aerosols, assessed cloudtop heights, and determined the mass of atmospheric vapor by scanning in 20 × 30-km bands with a resolution of 10 km;
- Radiation balance was measured by wide-angle radiometers developed by G.P. Forapanova of the Institute of Atmospheric Physics;
- The upper atmosphere ion analyzer used an electron beam to separate particles in the spacecraft's flight path (principal investigator: N. Dzhordzhio).

The spacecraft's low altitude, skimming the upper layers of the atmosphere, enabled the building up of an early model of Earth's atmosphere [18].

Cosmos 51 was the first to attempt to measure the brightness of the night sky. Using a wide-angle photometer, it measured its brightness at 2,300 Å (visible waveband) and 3,500–6,500 Å (ultraviolet) [19]. Cosmos 53 was part of the *Spin* program, which was a system of photometric observations led by V.M. Grigorevsky of Chisnau observatory. He used the Astrosoviet network to track its gently spinning booster to measure rotation and air density against the effects of the Sun and Earth's magnetosphere (hence *Spin*), while the main part of the mission, led by Yevgeni Gorchakov, focused on cosmic rays. The satellite had a small U_{235} fission chamber to measure the flux of neutrons inside its cabin [20].

This brings us to one of the main objectives of the DS missions: to make systematic measurements of air density in a program that ran well into the 1970s. The first long-term measurement of air density had been made by Korabl Sputnik over 1960–1963 (see Chapter 6, Figure 6.1). The measuring of air density was important in understanding how long satellites could stay in orbit before being slowed down; equally, it was important to measure variations and what caused such changes so as to predict them in the future. Air density instruments were later added to DS and unrelated satellites in order to build a comprehensive picture (see Figure 2.19).[1] Mikhail Marov, the lead scientist on this project, found that air density varied not only according to the solar cycle, but also according to time of day and latitude, being highest between 0200 and 0500 above 70° and between 1300 and 1600 lower than 70°. He and his colleagues concluded that:

- both the shape and orientation of a satellite do matter in determining the extent to which it is dragged down by air;
- air density varies according to the solar cycle;
- air density effects can be measured up to 350 km;
- densities are influenced by westerly rotational winds at an altitude of 180–200 km; and
- one could model air density with five key variables of solar activity, season, latitude, time of day and altitude. Densities are more stable in April and October, least so January and July. [21]

Marov found that the atmosphere cooled during the early 1960s, confirmed by the cold northern latitude winters of that period. Most of the DS missions were designed in OKB-586, but some were developed in collaboration with scientific institutes (e.g. Cosmos 31 and 51 with the AF Ioffe Crimean Astrophysical Observatory).

Another pair, Cosmos 14 and 23, belonged to the related *Omega* series developed by VNIEM (the All-Union Scientific Research Institute of Electromechanics) and had large solar panels: their purpose was to test equipment for atmospheric research and to pave the way for operational weather satellite systems. They later led to Cosmos 144, precursor to the Meteor series of weather satellites. It was also an

---

[1]   For the record, the missions were Cosmos 17, 19, 25, 26, 31, 36, 38, 39, 42, 43, 49, 51, 53, 76, 101, 106, 116, 135, 137, 142, 144, 145, 148, 151, 152, 156, 163, 173, 176, 184, 191, 196, 204, 215, 219, 221, 222, 230, 233, 242, 283, 307, 311, 321, 324, 327, 334, 347, 357, 362, 369, 378, 388, 393, 408, 421, 423, and 435.

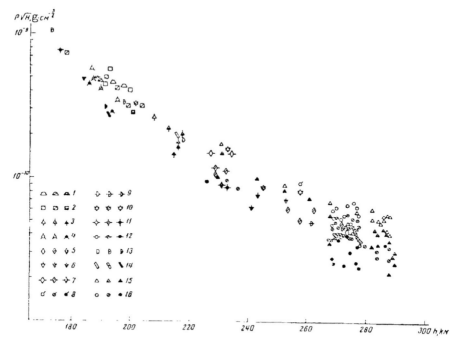

Figure 2.19. Air density measurements of the early Cosmos satellites, with density on the vertical axis and altitude on the horizontal axis.

important atmospheric satellite in its own right. Cosmos 144 (principal investigator was V.G. Boldyrev) carried four radiometers for the hydrometeorological service, two narrow-angle and two wide-angle, working in 2,500-km-wide strips, with the narrow angle scanning at an angle of 66°. As a result, the meteorological service was able to produce radiation temperature maps of the atmosphere, the charts also being sent on the "cold line" to the United States from March 1967 to March 1968 for analysis and archiving.

DS-U SERIES

The DS series was sufficiently successful to lead to a larger, successor series, called the DS-U, "U" standing for "universal". Approval was given for 18 such satellites on 22nd June 1965. The basic U bus had three versions: U1 (chemical batteries, unorientated), U2 (solar batteries, unorientated), and U3 (solar batteries, orientated) (Table 2.7). This standardized system greatly reduced the cost of the program, which had multiple sub-divisions according to the type of mission (Table 2.8). Although the initial batch was for only 18 satellites, in the end, 30 eventually flew in the DS-U program and 19 in its extensions up to 1971.

DS-U design

Table 2.7. DS-U satellites in the Cosmos program – outline.

DS-U1-G	*Geofizchevski*	Cosmos 108, 196
DS-U2-I	*Ionosferni*	Cosmos 119, 142, 259
DS-U2-MP	*Meteoritni*	Cosmos 135, 163
DS-U2-D	*Dosimetri*	Cosmos 137, 219
DS-U1-Ya	*Yaderni*	Cosmos 225
DS-U3-S	*Solnetsi*	Cosmos 166, 230
DS-U1-A	*Astronomichevski*	Cosmos 215 *Zybalik* (first astronomical observatory)
DS-U2-GF	*Heliofizichevski*	Cosmos 262 (first stellar observatory)
DS-U1-R		Cosmos 335
DS-U2-M	*Molekulyarni*	Cosmos 97, 145
DS-U2-GK	*Geofizichevsky Kompleks*	Cosmos 261, 348
DS-U2-MG	*Magnetni*	Cosmos 321, 356
DS-U2-MT	*Meteoritni*	Cosmos 461
DS-U2-V	*Vibratsioni*	Cosmos 93, 95, 197, 202
DS-U2-K	*Geofizichevsky*	Cosmos 426
DS-U2-IP	*Ionosferni*	Cosmos 378

The DG-U1 model measured 2.4 m tall by 2.3 m in diameter. Typical lifetime was one (DS-U1) to three (U2 and U3) months. These satellites ranged in weight from 500 kg (orbits up to 320 km) to 230 kg (higher orbits out to 2,000 km), but around 302 kg was typical. Once again, they were complimented by an even larger batch of military satellites for radar calibration, 79 in the DS-P1-Yu series and 20 in the DS-P1-I series (99 in all). Two other designs were completed: the DS-U4, a recoverable satellite for applications, biological, and technical experiments; and DS-U5, a maneuverable satellite able to change orbits between 4,000 and 10,000 km. The first one was to study the ionosphere (DS-U5-1), but even though the design was signed off, the project was abandoned in 1967 due to overwork in the company.

As was the case with DS, some were designed in collaboration with other institutes, such as the PK Sternberg institute for Cosmos 215 and 348; the LN Lebedev institute for Cosmos 225, the AF Ioffe for Cosmos 135 and 163; and the Sergei Vavilov institute for Cosmos 261 and 348. Table 2.8 gives a more detailed breakdown of the missions and their parameters.

Table 2.8. DS-U series – detailed breakdown.

DS-U1-G (Geofizichevski) series: density of upper atmosphere and ultraviolet solar activity
Cosmos 108	11 Feb 1966	219–855 km, 49°, 95.4 min
Cosmos 196	19 Dec 1967	223–860 km, 49°, 95.5 min

DS-U2I (Ionosferni) series: astronomy, ionosphere, and ultra-long radiation
Cosmos 119	24 May 1966	208–1,292 km, 48°, 99.8 min
Cosmos 142	14 Feb 1967	207–1,336 km, 48°, 100.3 min
Cosmos 259	14 Dec 1968	215–1,331 km, 48°, 100.3 min

DS-U2-MP (Meteoritni) series: micrometeorite dust
Cosmos 135	12 Dec 1966	253–649 km, 48°, 93.6 min
Cosmos 163	5 Jun 1967	244–611 km, 48°, 93.1 min

DS-U2D (Dosimetrichevski): streams of charged particles, radiation belts, inner belt protons
Cosmos 137	21 Dec 1966	219–1,719 km, 49°, 104.5 min
Cosmos 219	26 Apr 1968	215–1,745 km, 48°, 104.7 min

DS-U1-Ya (Yaderni): streams of charged particles and cosmic rays
Cosmos 225	11 Jun 1968	255–512 km, 48°, 92.2 min

DS-U3S: astronomy and solar radiation
Cosmos 166	16 Jun 1967	281–553 km, 48°, 92.9 min
Cosmos 230	5 Jul 1968	285–543 km, 48°, 92.9 min

DS-U1-A (Astronomichevski) Zyablik: astronomy
Cosmos 215	18 Apr 1968	255–403 km, 48°, 91.1 min

DS-U2-GF (Heliofizichevski): geophysical studies, ultraviolet, and X-ray radiation
Cosmos 262	26 Dec 1968	259–798 km, 48°, 95.2 min

DS-U1-R: ultraviolet studies and atmospheric research
Cosmos 335	24 Apr 1970	250–401 km, 48°, 91 min

*DS-U2-M (*Molekulyarni*): theory of relativity and electromagnetic oscillations*
Cosmos 97   26 Nov 1965   200–2,098 km, 49°, 109 min
Cosmos 145   3 Mar 1967   215–2,116 km, 48°, 108 min
All Kapustin Yar

*DS-U2-GK (*Geofizichevski Klampleksni*)*
Cosmos 261   20 Dec 1968   207–642 km, 71°, 93 min
Cosmos 348   13 Jun 1970   201–651 km, 71°, 93 min

*DS-U2-IP (*Ionosferni*)*
Cosmos 378   17 Nov 1970   234–1,742 km, 74°, 105 min

*DS-U2-MG (*Magnitni*): magnetic field, cosmic rays, and northern lights*
Cosmos 321   20 Jan 1970   272–479 km, 71°, 92 min
Cosmos 356   10 Aug 1970   231–573 km, 82°, 92 min
Both from Plesetsk

*DS-U2-MT (*Meteoritni*) astronomy, X-, and gamma-rays*
Cosmos 461   2 Dec 1971   488–511 km, 69°, 94.6 min
Cosmos 3M from Plesetsk

DS-U2-D scientific
Cosmos 137   21 Dec 1966   219–1,718 km, 48.8°, 104 min
Cosmos 219   26 Apr 1968   215–1,745 km, 48.4°, 105 min
Cosmos 3 from Kapustin Yar

*DS-U2-V (*Vibratsionni*)*
Cosmos 93   19 Oct 1965   216–513 km, 48°, 92 min
Cosmos 95   4 Nov 1965   211–521 km, 48°, 92 min
Cosmos 197   26 Feb 1967   217–486 km, 48°, 91.5 min
Cosmos 202   20 Feb 1965   213–483 km, 48°, 91.4 min

*DS-U2-K (*Geofizichevski*) scientific*
Cosmos 426   4 Jun 1971   389–1,997 km, 74°, 109 min
Cosmos 3M from Plesetsk

MOLECULAR OSCILLATOR IN ORBIT

We have further details on some of these missions. Some were quite exotic: the *Molekulyarni* and the *Meteoritni*. One of the most unusual was the theory of relativity mission, Cosmos 97 *Molekulyarni*, which carried a molecular oscillator (the term "quantum ammonia generator" was also used), the first one having been built by Academician Nikolai Basov and Alexander Prokhorov in 1954. Nikolai Basov (1922–2001) came from Voronezh, where his father was a professor of forestry. After serving on the Ukrainian front in the war, Nikolai Basov joined the Lebedev institute, obtaining his doctorate in 1956 entitled "A Molecular Oscillator", based on both theoretical and practical work with ammonia beams. Five years later, he built the first powerful lasers and he was inventor of the laboratory maser. He is considered one of the founders of quantum radio physics,

Nikolai Basov

was one of the first to realize the potential of the superconductor, and was a Nobel physics prizewinner.

He was supported by some leading scientific personalities and the same year, Vitaly Ginzburg proposed that a satellite be used to test whether Einstein was right and whether gravity truly displaced light waves. Nikolai Basov first proposed the mission at the annual meeting of the Academy of Sciences in 1960. Being the guest speaker of honor, he had a strong platform. The original idea was to build two atomic clocks, which would take the form of a quantum mechanical oscillator or stabilized quartz oscillator or hydrogen mass driver. His proposal was to put a molecular generator in orbit, recover it, and compare its time and speed with one on Earth, which, if Einstein was right, would be different. He worked out a formula to test the time difference (for the record, $\Delta T \approx 1/2\beta^2 T$, where $\beta = v/c$).

It would seem, though, that they had to test the theory without a recoverable spacecraft. Nikolai Basov and his colleague Professor M.I. Borisenko installed a maser on the top of Cosmos 97 with a liquid-ammonia pressure cylinder and a four-month supply on the inside. The way it worked was that the maser generated beams of $N^{14}H_3$ molecules at 23.870 MHz while the quartz oscillator generated beams at 2,525 MHz. Three control masers were installed on the ground. Once in orbit, the equipment was turned on and stabilized. Transmissions were made in the course of 13 passes. Relativity was tested by measuring the degree to which the satellite-borne clocks were affected by the speed at which the spacecraft was traveling, but it proved difficult to separate out these differences from their Doppler shift. The mission report was issued on 29th December 1966 and confirmed that quantum frequency standards could be used in space. Nikolai Basov thought that such clocks could be very helpful for a system of navigation satellites. The mission, though, was probably ahead of its time and happened only because it was successfully pressed by Basov in the academy. As for Basov, he moved on to other projects, developing lasers to measure the distance from the Earth to the Moon [22].

THE GREAT HUNT FOR ANTI-MATTER

Cosmos 135 and 163 were some of the most unusual missions of the early space age. Anti-matter was a new subject of 20th-century physics. It was discovered by the physicist Dmitri Skobeltsyn in Petrograd in 1923, when, in the course of trapping

Cosmos 97 *Molekularni*

gamma rays in a cloud chamber, he found anti-electrons, now called positrons, or the anti-matter opposite of eletrons, though the full exposition of the concept of anti-matter belongs to Cambridge mathematician Paul Dirac, who published *The Principles of Quantum Mechanics* in 1930.

Iosif Shklovsky later recalled being summoned by Mstislav Keldysh to a meeting in the Academy of Sciences. There, he found the meeting addressed by Boris Konstantinov (1910–1969) of the Leningrad Institute for Physics and Technology (called *FizTech* by those in the field). This had begun life as a laboratory in Gatchina in 1954 and he had built it up as a cutting-edge physics institute, installing there a nuclear research reactor in 1959. Konstantinov had in 1960 persuaded Khrushchev to allocate him millions of rubles to research anti-matter, on the basis that it would counter American developments of the neutron bomb. Konstantinov came from an agricultural background, helping his mother to run the farm when his father died when he was very young. He entered the institute when only 17 and stayed there until his death in 1969. He was a learned man, a versatile scientist who had made his name in musical acoustics and wind instruments, branching out into isotopes, electrophoresis, thermonuclear fuel processing, holography, astrophysics, and gamma-ray astronomy. In all this, he struggled with endocarditis throughout his adult life. Konstantinov told the meeting that comets had left behind meteors and smaller trails of dust and anti-matter as they flew around the solar system. A spaceship would find what they had left in Earth's vicinity.

Boris Konstantinov

Shklovsky told the meeting that the concept was crazy, but Keldysh knew better than to argue with Khrushchev, so the missions went ahead. They had the double, related objective of detecting anti-matter and to see whether comets and primitive matter had left a cloud of dust in orbit around the Earth.

Konstantinov defended himself. He accepted that the concept of there being anti-matter around the Earth might be considered "somewhat extravagant", but he cited a number of prominent astronomers in support, such as Sir Bernard Lovell, director of the Jodrell Bank radio telescope near Manchester, and solar physicist Hannes Alfven, not to mention experts on the 1908 Tunguska impact who believed that the Earth was hit by anti-matter. At a practical level, he quoted his own studies of meteor streams from balloons 18 km high in August 1961, which showed an increase in gamma radiation, well known to be associated with anti-matter. Cosmic ray expert Naum Grigorov had searched for anti-nuclei impacts on Korabl Sputnik 2 (see Chapter 6).

Cosmos 135 and 163, the *Meteoritni* missions, used microphones to test for cosmic dust (impacts were expected at 100 a day). Cosmos 135 operated from 12th December 1966 until 12th April 1967 and Cosmos 163 from 5th June 1967 until 11th October. Far from being bombarded with cometary dust, no impact was recorded until orbit 87 after 130 hr! Yet, tests showed that the detector was working properly. By orbit 200, there had been 30 impacts and 141 by 630 hr. Impacts remained low, even at a time when Earth passed through the trail of three meteor showers: the

Geminids (13–14th December), *Ursids* (22nd December), and *Quadrantids* (2nd–3rd January). In the experimental results, many readings had to be eliminated on account of electronic noise or, mundanely, crackling of the plates from temperature changes, so that the definitive meteor hit count was only four.

Boris Konstantinov immediately and humbly admitted that his hypothesis of a dust cloud was "erroneous" but felt that useful observations could be made by launching the backup, Cosmos 163. Additional efforts were made to reduce noise and erroneous readings. In his final report on 16th April 1969, Konstantinov related how it met an even greater desert, with only three impacts in 1,370 hr and only 11 in its mission altogether. They disproved the then current theory that there was a cosmic dust cloud around the Earth: routine meteors, yes, but no dust showers. The level of the meteorite flow was measured at 5.5×10^{-6} particles/m^2/sec.

As for anti-matter, two detectors were fitted. To find anti-matter, Konstantinov fitted a scintillation γ spectrometer and a 64-channel amplifier for the range 0.3–2.5 MeV. The spectrometer attempted to measure the level of gamma rays, for the theory of anti-matter held that cosmic radiation, on reaching Earth's environment, would split into gamma rays and anti-matter positrons, the presence of the latter being informed by that of the former. Specifically, gamma rays with an energy of 0.511 MeV would result from the annihilation of micrometeorite particles reaching the Earth's atmosphere, so preliminary measurements were made of the intensity of the gamma-ray flux. The instrument duly measured the flux of gamma rays at 0.35 cm/sec at 250–650 km. The 64-channel amplitude analyzer was used to measure annihilation radiation during the meteor showers (Figure 2.20) and did find radiation being generated at 0.511 MeV, which was considered to be the annihilation frequency at the time. Konstantinov estimated that about 20 mg of anti-matter was generated each day, more during meteor showers, less otherwise.

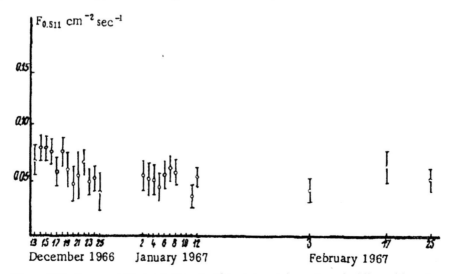

Figure 2.20. Cosmos 135 measurements of gamma-ray impacts, coinciding with meteor showers.

Konstantinov is revered to this day, a bust being erected in his memory at the entrance to the institute he dominated for half a century, the institute later being renamed in his memory. His final contribution to science was a book inspired by Gavril Tikhov, *Inhabited Space*, which was published posthumously, containing essays by Tikhov ("The Possibility of Life on Mars") and Iosif Shklovsky ("The Multiplicity of Inhabited Words") [23].

Another, later *Meteoriti* mission, Cosmos 461, also carried an omnidirectional gamma-ray scintillator to make spectra every 6 min over 28 keV to 4.1 MeV and measured both gamma-ray bursts and the gamma-ray background. It was an unusually long mission, transmitting from 2nd December 1971 until 21st February 1979, over eight years. The most important event during its mission took place on 17th January 1972, when it recorded a 37-sec, three-pulse gamma flash, sufficiently long for it to be identified as coming from far away, deep in either the Milky Way or another galaxy.

This was the last anti-matter mission for a long time, the search not being resumed until 15th June 2006, when an instrument called PAMELA was carried aloft on the Resurs DK Earth resources mission. In a joint project with Italy, PAMELA stood for Payload for Anti Matter Exploration and Light nuclei Astrophysics, a precision calorimeter designed to detect high-energy particles that may be the result of dark-matter interaction with Earth's atmosphere. The Americans developed an interest in anti-matter and the last cargo manifested to be ferried up to the International Space Station by the space shuttle was an Alpha Mass Spectrometer for this very purpose.

FIRST ASTRONOMY AND HELIOPHYSICAL MISSIONS

Cosmos 215 was the first dedicated astronomy mission called the *Zybalik* (the Russian word for the bird chaffinch) after its scientific payload, which was developed by N. Dimov and V. Prokoviev in the Crimean observatory and their colleagues Vladimir Kurt at the Sternberg institute and V. Tiyt in Estonia, its aim being to examine ultra-violet and X-ray radiation. Operating for 40 days, Cosmos 215 had eight parallel 70-mm telescopes and swept the sky in the spectral region of 1,250–2,700 Å, focusing on the Milky Way, studying about 10–25 hot stars in the course of each revolution of the Earth. It had an X-ray telescope (0.5–5 Å) and photometers. Cosmos 215 did not have a pointing device, but had a damping system to keep it stable as it swiveled across the sky with a 1° field of view. Cosmos 215 measured the luminescence of the hydrogen shell of our planet and confirmed that it stretched over 20,000 km out from Earth and found, simultaneously with the Americans, tropical glow, which possibly resulted from a recombination of electrons and atomic hydrogen. It was followed by Cosmos 262, which had three 16-channel photometers to study the stars until it fell out of orbit on 18th July 1969. Cosmos 262 had instruments to study ultraviolet and soft X-ray radiation of the Sun, stars, interstellar medium, and Earth's upper atmosphere, providing data on radiation related to solar activity, such as solar flares and the galactic radiation background.

The first two solar observatories were Cosmos 166 in 1967 and 230 in 1968, the first Soviet probes to image the Sun directly. Each carried eight mirror telescopes, an X-ray telescope, and two photometers to study the radiation in hot stars in various wavebands, while IZMIRAN equipped them with instruments to detect ultraviolet and gamma rays. The spectrometer measured ionized helium in the 304-Å range, the X-ray solar image tracer covered the 44–60-Å range, and the X-ray counter the 2–8, 8–18, and 44–60-Å ranges. These were aligned to scan the Sun each orbit and between them, they obtained 1,000 X-rays, finding hot and cold zones on the Sun. Looking at the X-rays coming from the Sun, they found two cycles: 3–4 and 24–48 hr, with remarkable ranges within them. One of the main targets of the solar experiments was solar flares, which were accompanied by an increase in X-ray radiation and accelerated particle fluxes. The missions lasted 130 and 72 days, respectively. The X-ray telescope focused on the hot regions of the Sun (5,000,000–10,000,000 K), which generated dense plasma and intense magnetic fields for several hours at a time. The precise nature of the *Yaderni* mission (Cosmos 225) is unclear, but it carried a telescope and mission scientist Lidia Kurnosova put forward the subsequent hypothesis that electrons tended to bunch at the 300-MeV range and become the basis of subsequent radiation belts [24].

NEW GEOMAGNETIC MAP

Cosmos 321 *Magnitni*, led by Shmaia Dolginov, continued the mapping of the Earth's magnetic field begun by Cosmos 26 and 49 (and before them, by Sputnik 3). Cosmos 321 and 356 were the second set of IZMIRAN satellites, this time operating in near-polar orbit and had the secondary objectives of exploring polar electrojets and magnetospheric ionospheric currents. Cosmos 321 had a quantum cesium vapor scalar magnetometer to provide magnetic field data from a 3.6-m-long boom. It operated for two months, its tracking moving 2° each day, and re-measured 94% of the Earth's magnetic field at a frequency of 0.5 Hz from an altitude of 237–507 km. It made 12,000 individual measurements with an accuracy of 2 γ. This was combined with the American OGO 6 to present a new geomagnetic chart for 1970, showing variations and changes since the 1965 one (Figure 2.21). Individual areas were re-mapped, such as the auroral oval over the poles (Figure 2.22). This research was a major contribution to geology, enabling a remodeling of the Earth's crust. Cosmos 321 also recorded the geomagnetic storm of 8–10th March 1970 and observed the magnetic field of equatorial electrojets for the first time. In the 1980s, Soviet scientists hoped to make a fresh, ever more accurate magnetic map of the Earth using an Intercosmos satellite in a project called DIDEX with Poland and the GDR, using two satellites orbiting 150 m apart. Lack of finance prevented it from happening and the project was eventually realized by the Americans in the GRACE program [25].

Results from the *Ionosferni* missions Cosmos 142 and 259 were published in *Pravda* on 29th October 1969. Each carried four measuring poles 15 m long. Project scientists L. Zhekulin, V. Mikhailov, V. Aksenov, and I. Lishin reported that there were high day-to-day variations in daytime electron density between 50 and 100 km.

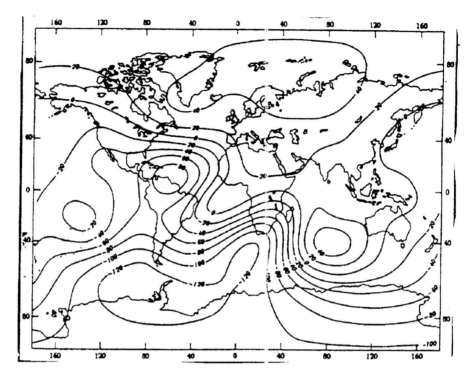

Figure 2.21. Cosmos 321 new world geomagnetic map.

They found geomagnetic ducts in the ionosphere above 200 km and that particles moved up and down in fibrous streams. They had findings that interested people who listened to long-wave radio. They measured its transmitivity at 70% by night, but two to three times less by day. Cosmos 259 and 262 operated simultaneously with Cosmos 261, which focused on auroral studies (see below) and contributed measurements of hard radiation [26].

SUPPLEMENTARY SCIENCE

The Cosmos program was, as noted earlier, intended purely as a scientific program. This noble objective was compromised in two ways. First, as already noted, the extensive DS-P series was used as a radar calibration system for the military. Second, the Soviet Union began its military photo-reconnaissance program on 26th April 1962 with the flight of its first Zenit spacecraft. Because the USSR had made such a fuss about the American militarization of space, it was faced with a problem of how to identify military missions of its own. The Soviet Union's only previous experience of a difficult-to-explain Earth orbital mission was the failed Venus launch of 4th February 1961, which was unconvincingly called *Tyzhuly Sputnik*, or "heavy satellite". Faced with the prospect of a series of implausibly heavy satellites, the

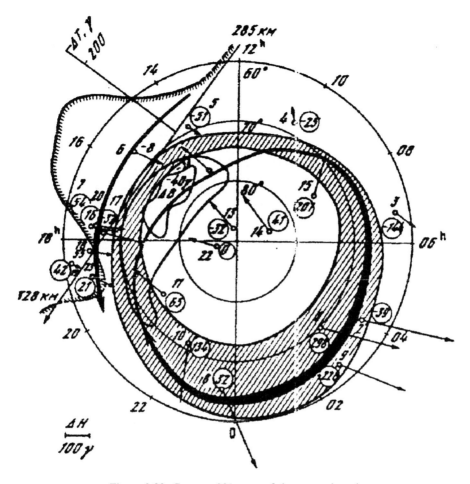

Figure 2.22. Cosmos 321 map of the auroral oval.

temptation to use the Cosmos label for the Zenit and other series proved too strong, so it acquired the name Cosmos 4.

In the event, the distinction between the science and military Cosmos program was far from forensic, for in his relentless quest to map radiation levels, Sergei Vernov persuaded the authorities to fly his radiation meters on many of the early military satellites, such as Cosmos 4. Its images were also made available to civilian meteorologists to make a first map of Earth's cloud deck. Vernov's Geiger counters subsequently flew on Cosmos 4, 6, 7, 9, 10, 12, 13, 15, 16, and 18.

Cosmos 4 began the tradition of fitting scientific payloads onto military, applications, or other missions. We can now catalog these missions as "supplementary science missions" (Table 2.9).

Radiation was the primary field of interest in the supplementary science program. Radiation detectors were fitted onto Cosmos 4, 7, 9, and 15 to map, in more detail,

Table 2.9. Scientific instruments flown on military and other Cosmos missions.

Radiation	Cosmos 4, 7, 9, 15, 19, 25, 41, 45, 140, 213
Inner radiation belt	Cosmos 1549, 1623, 1648, 1730, 1790, 1810, 1843, 1983, 2019, 2028
Biology (plants)	Cosmos 92, 94, 109, 368
Earth's radio signal	Cosmos 1110, 2293
Earth's atmosphere	Cosmos 45, 65, 115, 1076, 1939
Earth's radiation	Cosmos 156, 243
Astronomy	Cosmos 251, 731
Ionosphere	Cosmos 184, 274, 438, 481, 721
Airglow	Cosmos 45, 92, 115
Night sky	Cosmos 92, 213
Solar	Cosmos 650, 1066
Charged particles	Cosmos 906, 1520, 1524
Meteorites	Cosmos 213, 470, 502, 541
Gamma rays	Cosmos 2326
Protons	Cosmos 60, 159
Electrons	Cosmos 127, 137, 143, 256
Cosmic rays	Cosmos 19, 140, 213, 259
Solar cosmic rays	Cosmos 480
Charged particles	Cosmos 906, 2344

the South Atlantic (40°W, 35°S) and Bering Sea magnetic anomalies first identified by Korabl Sputniks 2 and 3. Radiation levels in 1963 were four times higher than in 1960 – a feature of either natural change or of *Starfish*. Cosmos 4, 7, 9, and 15 were equipped with tube and scintillation counters that allowed the flow of charged particles over the surface of the Earth from 65°N to 65°S to be recorded, with daily transmissions of data. They measured a trough in the level of cosmic rays, bottoming out at 10°S, 60°W. Cosmos 4, 9, 12, and 15 all had cosmic ray counters and they found a relatively stable level of intensity of cosmic rays over 1962–1963. Reassuring for forthcoming manned flights, they found that average radiation levels for the early manned missions were acceptable for several weeks at least, with no dangerous spikes. Radiation levels measured by the early Cosmos missions ranged from 14 millirads on Cosmos 18 to 55 millirads on Cosmos 7 [27]. Cosmos 19 operated from 6th August to 31st December 1963 and measured cosmic rays from an altitude of 350 km and ascertained the velocity of the corpuscular stream (between 320 and 420 km/sec). Within its 11-year cycle, the Sun had a 27-day sub-cycle. Solar cosmic rays were the focus of an instrument devised by Elmar Sosnovets for Cosmos 480.

Cosmos 41, which had been intended as a communications satellite (its relay transmitters froze and failed to deploy), focused on the outer radiation belt. It carried solid-state detectors, Geiger counters, and scintillators developed by Sergei Vernov and measured geomagnetically trapped low-energy protons from August to December 1964, finding that protons and electrons were trapped in a similar way and that the outer proton belt was comparatively stable. The first successful communications satellite, Molniya 1, carried cameras to photograph Earth's cloud deck from the 23,000-km-high point of its orbit over the northern hemisphere (see below). Sergei Vernov was not the only scientist who prevailed on the authorities to carry his instruments. Yevgeni Gorchakov was successful in installing radiation detection instruments on Cosmos 127 and 137 to detect high-energy electrons. Identification of the Earth's own radiation signal had been a breakthrough for Elektrons 2 and 4 and this work was continued by Cosmos 1110 and 2293.

Cosmos 45 (principal investigator Pavel Shargin) carried a calorimeter to measure variations in radiation on the night side of Earth. It also carried a scanning diffraction spectrophotometer that made a chart of the Earth's thermal radiation field from 65°N to 65°S. The spectrometer was turned full on for one orbit on 13th September 1964, in the course of which it took 10,000 scans. Infrared radiation was concentrated in three layers, each 10 km thick, around 280, 420, and 500 km. Cosmos 45 was part of a set with 65 and 92, making thermal profiles of Earth's atmosphere. The task of instruments on Cosmos 51 was to measure the brightness of the night sky (an experiment continued on Cosmos 213), but here the breakthrough was on Cosmos 45, which identified the phenomenon of night airglow over the western Pacific and Australia, stretching from 300 to 3,000 km, bringing back film of the event. Cosmos 92 repeated the experiment on 12 passes, finding night airglow over the western Pacific and Australia stretching from 300 to 3,000 km high, with a radiation intensity of between 1,800 and 2,500 Å (Figure 2.23). It used instruments developed by Vladimir Krasnopolsky (b. 1938) and Alexander Lebedisnky (1913–1967), two of Russia's great experts on planetary atmospheres. Cosmos 45 measured the level of water vapour in the atmosphere, finding a layer of concentration at

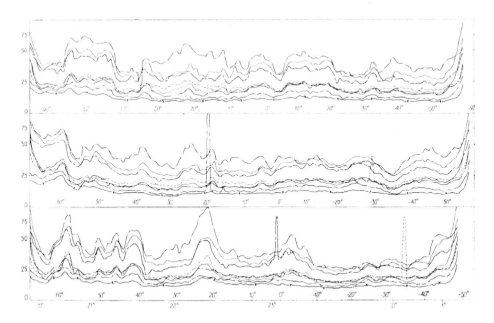

Figure 2.23. Cosmos 92 airglow measurements by intensity (left axis) and latitude (horizontal). (Credit: COSPAR)

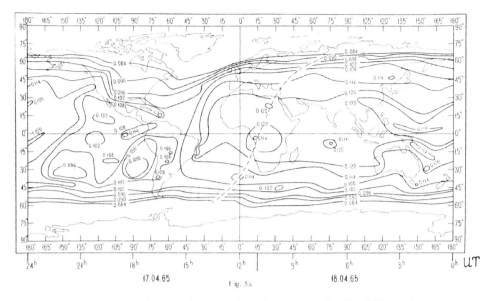

Figure 2.24. Cosmos 65 map of Earth's ozone. (Credit: COSPAR)

80-km altitude, where it condensed in the evening into noctilucent clouds. Cosmos 224 found a rise in nitric oxide in the atmosphere, but this was attributed to the solar cycle rather than any human causes.

Cosmos 65 measured ozone levels in the course of 22 passes, enabling a global ozone map to be published, finding lower levels around the equator and a decline in northern polar latitudes (Figure 2.24). Cosmos 65 carried an ultraviolet spectrometer that obtained 2,500 spectra that measured radiation coming from the atmosphere in the 2,250–3,070-Å range, measuring Earth's "radiation budget". Using a scanning spectrophotometer, Cosmos 45, 65, and 92 between them took 10,000 spectra in the 7–20 and 14–38-μm range. They established the outer boundary of the upper atmosphere at 40 km. They calculated the actual level of radiation in the atmosphere and the elements that determined it, such as temperature, humidity, and ozone [28].

These missions supplemented the already extensive work of ionospheric missions. Cosmos 274 carried a mass spectrometer that during the strong magnetic storm of 23rd March 1969 measured a 50% increase in ions at auroral latitudes that led to an increased density, wind, and temperature and a changed composition. At the other extreme, atomic ions were at their quietest in the sub-auroral trough at dawn and in middle and lower latitudes after sunset.

Precursor missions for manned flights often had spare space and weight for scientific instruments. Examples were unmanned tests of the Soyuz spacecraft, where Cosmos 140 tested a new form of a cosmic ray detector in the form of a cryogenic superconducting magnet, while three were flown on Cosmos 213, which also carried a micrometeorite detector, ultraviolet photometer, and the *Ray* apparatus to measure cosmic ray positrons and electrons. *Ray* was a 17-kg helium cryostat with an electromagnet that was spun so as to create a magnetic field that would trap and then measure charged particles of 0.1 to several GeV. Such a field was created after 13 hr but things then went wrong: the spinning caused the helium to heat up and it was depleted after 18 hr, by which time the field had fallen to 0.

Cosmos 60 and 159 were failed lunar probes. Cosmos 60 failed to leave Earth orbit, but its 16-channel gamma-ray detector was still put to good use to detect gamma rays from space in the first mission of Yuri Surkov, one of the great writers of lunar and planetary space science (*Exploration of Terrestrial Planets from Spacecraft – Instrumentation, Investigation, Interpretation*) and director of the planetary geochemistry laboratory at the Vernadsky institute. Cosmos 60 measured the level of gamma-ray background (for the record, 1.74 quanta/m^2/s, falling off sharply above 1.5 MeV). In the case of Cosmos 159, the rocket left the spacecraft stranded in a high Earth orbit (60,000 km) but Yuri Logachev's instruments measured the degree to which interplanetary protons penetrated the Earth's magnetosphere. Their data were matched against a similar proton detector on Venera 4 (see Chapter 4), which was registering the intensity of energetic solar particles (solar cosmic rays) outside Earth's magnetosphere. They were in close agreement, indicating that interplanetary protons could penetrate the Earth's magnetosphere and leave it – but, should any perturbation occur to affect the particle motion, it might become trapped to replenish Earth's radiation belt. The idea of trapped solar particles as a source of the Earth's radiation belt was later confirmed by Koronas F (see Chapter 7). In the astronomy missions, Cosmos 251 and 264 carried gamma telescopes and found a new source of gamma radiation of 100 MeV in the region α3.6° to 5°, Δ4° to 9°, probably source 3C 120.

Finally, when the Soviet lunar module, the LK, was tested in Earth orbit, it was fitted with ion density meters. The flight profile of these missions (Cosmos 379, 398, and 434) was for the lunar module to orbit the Earth at 190–280 km, after which the lunar module engine would simulate the firing required for lunar landing and take-off, bringing the spacecraft out to an elliptical orbit of 200–6,000 km. These maneuvers provided the opportunity to reach altitudes not studied since the Elektron missions several years earlier. The findings were that ion densities were at their densest at 300 km (5.6 × 10^5 ions/cm^3), after which they fell away by 50–60%, but varying according to whether they were in the nighttime or daytime parts of the orbit when they fell and rose [29].

NAUKA MODULES

The placing of scientific equipment as secondary payloads on other missions was taken a stage further by what were called, informally, *nauka* modules ("nauk" is the Russian word for science). These were small scientific packages that could be placed on the primary payload, generally a military photo-reconnaissance satellite adapted for the mission concerned. The system only came to light as the result of some detective work by Western analysts of the Soviet space program, notably Joel Powell [30]. Later, a *nauka* module appeared at the Exhibition of Economic and Scientific Achievements, the VDNK in Moscow.

Nauka actually comprised three different systems: a separated capsule from 20 to 250 kg, MKA (*Maly Kosmicheski Apparat*, or Small Space Apparatus), an attached hatbox-shaped cabin (the most common), and a container on the main recoverable spacecraft whose lid could be lifted to expose an experiment to the space environment called KNA (*Konteyner Nauchnoy Apparat*, or Scientific Container Apparatus). *Nauka* modules followed a standard design, being 1.8 m in diameter, 95 cm tall, 630 kg in weight, but able to hold 200 kg of scientific equipment. The MKAs became, many years later, the Pion series of satellites dropped off in orbit by remote sensing spacecraft. *Nauka* modules transmitted independently to the ground and they could either remain attached to the mother satellite or detached into their own orbit. In some cases, the *nauka* package separated before the main spacecraft re-entered. On the first *nauka* mission, Cosmos 208, the main spacecraft returned on 1st April 1968, but the *nauka* module separated beforehand and flew independently to study high-energy gamma rays for E.A. Pryakhin. It was a successful start, for Cosmos 208 measured high-energy cosmic rays in ranges previously investigated by Protons 1–3 and found rays of 30, 50, 90, 150, and even 500 MeV. Cosmos 208 made the first Soviet sky study of X-ray intensities, finding surges of emissions from a number of constellations [31] (Figure 2.25). The experiment was repeated with Cosmos 228, which also tried to measure the chemical composition of primary cosmic radiation.

The third *nauka* was also successful, Cosmos 243, associated with Anatoli Basharinov. The scientific part of the mission, though, was, first, an attempt to measure the heat emitted from the Earth's atmosphere, "Earth's radiation budget", determine its moisture content, see how Earth absorbed solar radiation, and look at the interaction between moisture and wind. Second, an Antarctic ice map was subsequently compiled, as were cross-sections of sea across the Pacific. Cosmos 243 covered the oceans in the 0.8, 1.35, 3.4, and 8.5-cm bands, mapping the water vapor

Nauka module   KNA system (Credit: ESA)

Loading KNAs (Credit: ESA)

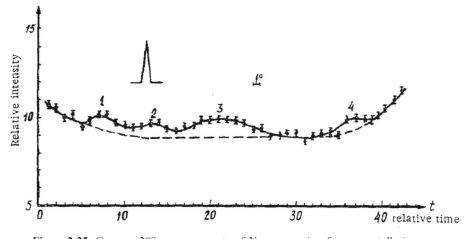

Figure 2.25. Cosmos 208 measurements of X-rays coming from constellations.

content and gradients in the Pacific, Atlantic, and Indian Oceans by latitude, finding more in the northern than in the southern hemisphere and much higher levels of moisture at the equator. A temperature profile of the Pacific was compiled, showing the rise from 0°C at 60° latitude to 25°C at the equator, with measurements for individual bodies of water (e.g. 30°C off South Africa, 16°C in the Mediterranean). Seasonal variations in the icefields of Antarctica were mapped. The volume of water moisture in the Indian, Atlantic, and Pacific Oceans was quantified (Q) and compared (see Figures 2.26 and 2.27). Cloud tops and their temperatures were measured across the Pacific (15 km high, temperature –70°C). Landmass temperatures were measured (e.g. Australia (Figure 2.28), North Africa) and peat fires identified. Cosmos 243 and 384 continued the experiments begun on the *Space Arrow*. Cosmos 384, associated with A.K. Gorodetsky and A.S. Malkevich of the Institute of Atmospheric Physics but funded by the hydrometeorological service,

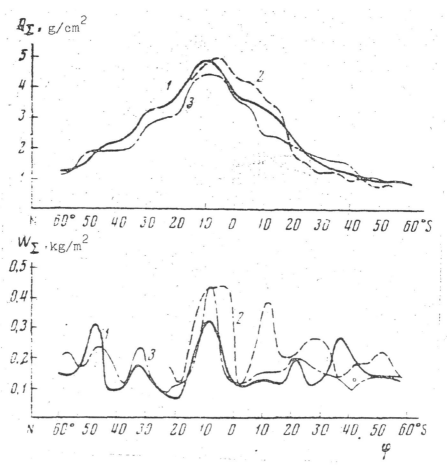

Figure 2.26. Cosmos 243 measurement of water vapor levels over the Atlantic, Pacific, and Indian Oceans.

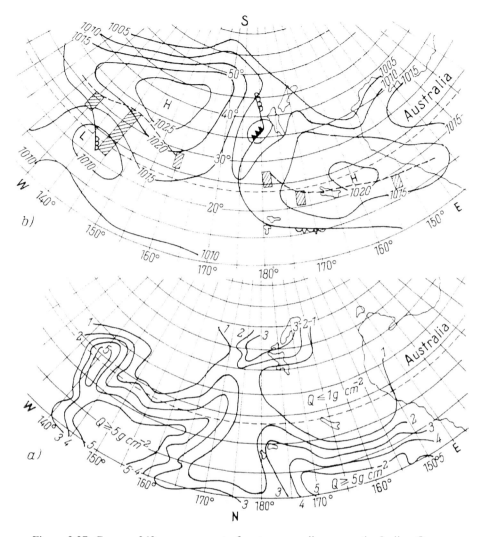

Figure 2.27. Cosmos 243 measurement of water vapor lines over the Indian Ocean.

used a narrow-angle infrared radiometer to determine the thermal profiles of clouds. One outcome was a new map of the Antarctic icefields (Figure 2.29). Cosmos 384 also carried microwave radiometers and here the aim of the experimenters, Anatoli Basharinov and Alexander Gurevich of the same institute, was to measure the geophysical parameters of the atmosphere and its clouds. They were successful in determining cloud water content at different latitudes and estimating total water content in the Earth's atmosphere. The missions provided an opening understanding of the nature of the global climate and the interaction of air, moisture, ice, landmasses, and ocean (see figures) [32].

Forty-four *nauka* modules were flown between then and 1984. Most *Nauka*

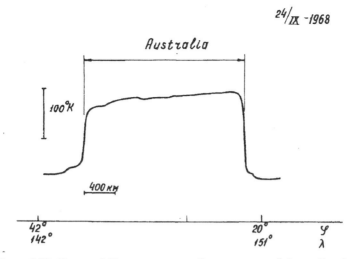

Figure 2.28. Cosmos 243 measurement of temperature of Australian landmass.

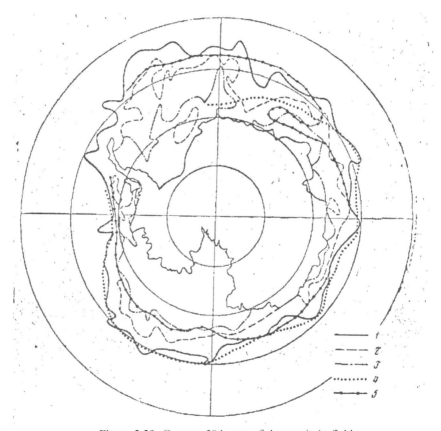

Figure 2.29. Cosmos 384 map of Antarctic icefields.

Table **2.10.** *Nauka* modules with scientific payloads, 1968–1979.

Cosmos 208	21 Mar 1968	Gamma rays (E.A. Pryakhin)
Cosmos 228	21 Jun 1968	Cosmic rays (E.A. Pryakhin)
Cosmos 243	23 Sep 1968	Microwave radiometer
Cosmos 264	23 Jan 1969	Extragalactic X-rays
Cosmos 280	23 Apr 1969	Charged particles
Cosmos 293	28 Aug 1969	
Cosmos 309	12 Nov 1969	
Cosmos 368	8 Oct 1970	Radiation and biology (watched by Pompidou)
Cosmos 384	10 Dec 1970	Microwave radio telescope
Cosmos 410	6 May 1971	Radiation measurements
Cosmos 428	24 Jun 1971	X-rays, gamma rays, bursts 40–200 keV, cosmic rays (G.I. Pugacheva and M.I. Kudryavtsev)
Cosmos 443	7 Oct 1971	Radiation measurements (like 410)
Cosmos 470	27 Dec 1971	Micro-meteorites
Cosmos 477	4 Mar 1972	Radiation and particle fluxes (like 410, 443)
Cosmos 484	6 Apr 1972	The Sun, cosmic rays, electrical fields, electrons < 30 keV (O.R. Gregoryan)
Cosmos 490	17 May 1972	Electrons, high energy cosmic rays, electrons < 80 MeV (G.I. Pugacheva)
Cosmos 502	12 Jun 1972	Micrometeorites
Cosmos 518	15 Sep 1972	
Cosmos 525	18 Oct 1972	
Cosmos 541	27 Dec 1972	Micrometeorites
Cosmos 552	22 Mar 1973	
Cosmos 555	25 Apr 1973	Electrons, solar and cosmic rays, protons (*Spektrum*)
Cosmos 561	25 May 1973	Galactic gamma rays
Cosmos 596	3 Oct 1973	
Cosmos 629	24 Jan 1974	
Cosmos 635	14 Mar 1974	
Cosmos 669	13 Jun 1974	*Obzor* microwave radiometer
Cosmos 692	1 Nov 1974	
Cosmos 721	26 Mar 1975	Solar cosmic rays, particles (S.N. Kuznetsov)
Cosmos 728	18 Apr 1975	Electron beams
Cosmos 731	21 May 1975	Gamma rays
Cosmos 747	27 Jun 1975	
Cosmos 769	23 Sep 1975	
Cosmos 776	17 Oct 1975	
Cosmos 780	21 Nov 1975	Electron beams
Cosmos 784	3 Dec 1975	
Cosmos 856	22 Sep 1976	Spectra of high-energy cosmic rays, 100 MeV
Cosmos 898	17 Mar 1977	
Cosmos 914	31 May 1977	Spectra of high-energy cosmic rays, 100 mMV (as 856)
Cosmos 966	12 Dec 1977	
Cosmos 973	27 Dec 1977	
Cosmos 1102	25 May 1979	
Cosmos 1106	12 Jun 1979	Gamma rays, X-ray sources

Principal investigators indicated in parentheses

modules were carried within the Zenit 2M *Gektor* photo-reconnaissance program, the exceptions being Cosmos 309 (Zenit 2) and Cosmos 1102 and 1106 (*Zenit 2M/ NKh*, *NKh* standing for the "national economy"). The *nauka* series ended with the *Zenit 2M* program but continued to fly on the Resurs Earth resources and Foton materials processing missions. Details are given in Table 2.10.

One was a biology mission: Cosmos 368 tested the *Oasis* system of space gardens, subsequently flown on the Salyut orbital station (Chapter 6). Several modules were concerned with the Earth's atmosphere. Cosmos 669 carried a 200-mm liquid helium sub-millimeter radiometer built by the Lebedev Physical Institute of Moscow to determine the water vapor count of the upper atmosphere. Cosmos 669 measured temperatures in the northern latitudes, finding them around 220–260 K in cloud-free zones. Water vapor sank in the atmosphere in high pressure, but wet air rose in low pressure [33].

The most information is available on deep-space observation missions. Cosmos 428 had an X-ray telescope that detected a number of sources in its two-week mission, correlating with the West's *Uhuru* satellite. The gamma-ray spectrometers of Cosmos 856 and 914 were designed to examine the spectra of gamma rays of more than 100 MeV in the Milky Way's equator. Cosmos 856 and 914 were companion missions of 265 and 290 hr, respectively, carrying gamma-ray spectrometers called GG2M for the range 100–4,000 MeV and made a survey of the gamma-ray sky. On Cosmos 1106, the crystal scintillation spectrometer was intended to record hard X-rays and gamma rays from deep space, with a sensitivity of 0.6 MeV. Closer to home, Cosmos 484 focused on electrical fields in the 30–300-keV range, finding intense fields in the boundaries of the magnetosphere. This was the first mission of electric fields expert Stanislav Klimov (b. 1937). From 600 crossings of the auroral zone, Cosmos 484 found that electrons and protons up to 10^5 eV were leaving the plasmasphere, especially in the build-up to magnetic activity, on the nightside and the night cusp. Cosmos 721 found, in the course of 78 orbits, 180 pulsations of quasi-captured electrons lasting 20–100 sec. Electron fluxes up to 1.4 MeV appeared at 66–68° for 30–40 sec. "Relativistic electrons" of up to 2.3 MeV were detected in the auroral zone [34].

APPLICATION SATELLITES WITH SCIENTIFIC INSTRUMENTS

In addition, other space missions have carried additional scientific instruments on an ad hoc basis (Table 2.11). Some were fitted to recoverable spacecraft, while others were fitted to satellites for communications or navigation in high orbit (Molniya, GLONASS), weather forecasting from polar orbit (Meteor), or communications from geostationary orbit (Raduga, Gorizont). Molniyas, for example, went through the outer-belt electrons and were in a good position to make measurements. These combinations enabled a composite three-dimensional pattern of measurements to build [35]. This is a list of satellites that have been identified with scientific instruments, but in the literature, many were not specifically named (simply "a Meteor satellite" was referred to).

Table 2.11. Applications satellites with scientific instruments.

Molniya 1-1	23 Apr 1965	Earth's cloud deck
Molniya 1-5	24 May 1967	Radiation, ionization counters
Molniya 1-14	26 Jun 1970	Ionization counter
Molniya 1-15	29 Sep 1970	Ionization counter
Molniya 1-16	27 Nov 1970	Neutron detector
Molniya 1-17	25 Dec 1970	Ionization detector
Molniya 1-19	19 Dec 1971	Cerenkov detector for cosmic rays
Molniya 1-21	14 Oct 1972	Radiation belt electron spectrometer
Molniya 1-23	3 Feb 1973	Proton spectrometer
Molniya 1-24	30 Aug 1973	Proton spectrometer
Molniya 1-27	20 Apr 1974	Instruments to analyze magnetosphere
Molniya 2-14	9 Sep 1975	Proton spectrometer
Meteor 1-22	18 Sep 1975	Electrons, cosmic rays in radiation belt
Meteor 1-25	15 May 1976	Ion mass spectrometer
		Fourier infrared spectrometer (GDR)
Meteor 1-28	29 Jun 1977	Fourier infrared spectrometer (GDR)
Raduga 3	23 Jul 1977	Proton detector, plasma layer electrons
Cosmos 1520	29 Dec 1983	Electron spectrometers
Cosmos 1554	19 May 1984	Electron spectrometers
Elektro	31 Oct 1994	Thermal mapping of the Moon

Meteor spacecraft, generally not individually identified, were used to supplement ionospheric research and the use of these spacecraft for radiation studies appears to have been extensive. For example, the Meteor launched in May 1976 carried an ion mass spectrometer to profile electrons up to 35 eV in 4-sec scans from an altitude of 900 km. In February–March 1973, Molniya 1-23 found a region of much increased proton density between 76 and 78°N. Experiments on board Molniya spacecraft led to a new picture of the ring current, which evolves during geomagnetic storms at a distance of 3–5 Earth radii and is considered responsible for the depression of the geomagnetic field. It is formed by particles penetrating the magnetosphere or the acceleration of magnetospheric plasma. Although the theory of the ring current was developed earlier, the first direct observations of the magnetic storm ring current were made in the beginning of the 1970s. In particular, Molniya experiments confirmed the theoretical prediction that the injection of particles would occur mainly in the evening sector. Later, in the second half of the 1980s, new American and Soviet (Gorizont) experiments unveiled the ion composition of the ring current, showing that it was formed by solar wind particles and ionospheric particles (oxygen ions), confirming the idea of the Earth's ionosphere as a source of magnetospheric particles.

Several experiments were carried on satellites in 24-hr geostationary orbit, especially so as to ensure their protection against radiation hazards, radiation fluxes, and electrization. Satellites on the night side of the magnetosphere cross the magnetotail, which contains considerable plasma fluxes with particle energies up to tens of keV, risking damage to instruments. This problem was tackled by the

Skobeltsyn Institute for Nuclear Physics in Moscow State University working with the manufacturers of high altitude satellites such as Gorizont, Raduga, and GLONASS.

During 1986, the Soviet Union participated in the *Promis* program of international magnetospheric studies. Scientists in the Skobeltsyn Institute persuaded the GLONASS program managers to carry electron detectors on two of their GLONASS navigation satellites, Cosmos 1521 and 1554, thereby taking advantage of their unusual 21,000-km-high orbits. Their objective was to detect the rare diffusion waves of energetic electrons in the radiation belts, both spacecraft passing through the same regions of the radiation belts almost simultaneously in June 1986 [36]. The last entry, Elektro, may seem bizarre, but here Russia's only 24-hr weather satellite was used to compile a thermal map of the Moon [37]. The observations were carried out between March and July 1995 using the satellite's visible and infrared filters. The lunar temperature was measured as 261.1 K (nighttime) and 394.9 K (daytime), the only differences being along the terminator between day and night. Daytime temperatures on the lunar surface were remarkably even between highlands and maria, varying little.

Finally, unidentified Resurs F-1 satellites were used to carry instruments to detect binary radiation, instruments later transferred onto unidentified Cosmos missions, enabling a graph to be published of proton fluxes up to 10 MeV from 1995 to 2001, showing a peak in 1999 [38].

THE THIRD RADIATION BELT

Just as Sputnik 2 discovered but the Americans interpreted the two Van Allan radiation belts in the 1950s (the inner and the outer), the Russians discovered a third radiation belt around the Earth in the 1980s – only to have the credit, once again, go to the Americans. Van Allen himself had always suspected that there might be a third, inner belt of trapped particles and their existence was formally postulated by the American scientist Bernard Blake in California in 1977. These were energetic ions of oxygen, nitrogen, and neon whose origins came from outside the solar system, being probably remnants of the big bang and supernovæ. Neutral elements were apparently able to reach the Earth, where solar radiation might strip away their electrons. They then traveled back to the heliosphere boundary and after being accelerated, returned to the inner regions of the solar system and were trapped by the Earth's magnetic field. Due to their relatively low velocity, anomalous cosmic rays did not reach Earth during periods of solar maximum. In the event, the formal discovery of the belt was announced at the American Geophysical Union meeting in Baltimore, Maryland, in 1993, using data from the NASA satellite SAMPEX (Solar, Anomalous and Magnetospheric Particle Explorer).

In reality, the story went a long way back, to Russian analysis of published data from the American Pioneer 10 mission to Jupiter in 1972. This found inexplicable variations in the flows of interstellar helium (an excess of the usually more abundant hydrogen ions), which, scientists speculated, were likely to reach the Earth during

Naum Grigorov

years of the quiet Sun (but not at solar maximum), hence the term "Anomalous" (Cosmic Rays) (ACRs). As was found later, ACRs included not only helium, but also oxygen, neon, and some others. But where were they? American scientist L. Fisk suggested that ACRs were of interstellar origin. Naum Grigorov, from the Proton missions (see below), speculated that there was a third, small radiation belt around the Earth that trapped them, his views being published as "Possibility of the Existence of a Radiation Belt around the Earth Consisting of Electrons with Energies of 100Me and Above" in *Soviet Physics* in 1977. A first attempt to actually find them was made using a mica detector instrument brought up to the Salyut 6 orbital station by Mongolian cosmonaut Jugderimdin Gurragcha on Soyuz 39. This involved a high-powered team of Sergei Vernov, Naum Grigorov, and Rikho Nymmik. This produced some results, but the instrument was redesigned using solid-state nitrocellulose detectors (the Astro instrument) and flown up on the next mission, Romania's Dmitru Prunariu on Soyuz 40. This revealed carbon, nitrogen, and oxygen traces of 10-MeV nucleon energies of these anomalous cosmic rays.

Following this, similar detectors were fitted from 1984 on Cosmos missions, using *Nauka*-style plates that were exposed during the mission, closed for re-entry, and recovered. They were solid-state track detector stacks of cellulose nitrate developed by the Kodak company and flown on military Cosmos satellites on 14-day recoverable missions in 250–400-km orbits at inclinations of 62–82°. The detectors were small, each 8.5 cm² across and piggybacked on these missions. Although the missions were short, a year's data accumulated, showing the existence of a radiation belt of anomalous cosmic rays.

There were 18 such detector flights, of which ten are identified here and four may have been the *Efir* and *Energiya* program (see below). These detectors were also used to study the effects of solar flares on the magnetosphere and the group followed storms on 25th April 1984, 21st January 1985, 6th February and 24th April 1985. The discovery of the third radiation belt was published in 1985 as *High Energy Electrons in Earth Orbit* by Naum Grigorov and colleagues, which acknowledged

Table 2.12. Cosmos missions to detect third radiation belt.

Cosmos 1549	April 1984
Cosmos 1623	January 1985
Cosmos 1648	April 1985
Cosmos 1730	February 1986
Cosmos 1790	November 1986
Cosmos 1810	February 1987
Cosmos 1843	May 1987
Cosmos 1983	December 1988
Cosmos 2019	May 1989
Cosmos 2028	June 1989

Blake's earlier work in the process. Despite the *Who-found-it?* competition, the scientists from the two sides cooperated to the point that, in true American style, they had their own mission badge and sticker called the "Anomalous Component Team". Eventually, the team published a graph of the intensity of anomalous cosmic rays over the period 1970–2002, showing their levels rise and fall [39]. In the 1990s, American–Russian experiments measured the ACR flux outside and inside the magnetosphere, showing that the initial charge of ACR particles is close to +1 and confirmed the idea of their interstellar origin. The missions are listed in Table 2.12.

INTERCOSMOS

Both space superpowers broadened out their space programs from an early stage, the United States, for example, providing opportunities for Britain and Canada. In the case of the USSR's DS program, India was invited to participate as early as 1961, before the program even got under way, to pick up ionospheric data from the DS beacons [40].

Following the arrival in October 1964 of the reforming general secretary Leonid Brezhnev, there was a substantial reorganization of the Soviet space program. Brezhnev sought to extend cooperation to the socialist countries of eastern and central Europe, with the double benefit of sharing their scientific expertise and buying them into the scientific benefits of the socialist economic group (the Council for Mutual Economic Assistance (CMEA), more popularly known as Comecon). The decision to begin such internationally collaborative work was made by the government in April 1965. A first meeting of interested countries took place in November 1965 and they agreed four areas of cooperation: physics, meteorology, biology/medicine, and communications, with a fifth, Earth resources, added in 1975. A council was formed on 31st May 1966, with an annual council meeting to review past and future programs and a Soviet-appointed chairperson, academician, and control systems expert Boris Petrov. The original countries were Bulgaria, Hungary, the GDR, Poland, Romania, Czechoslovakia, and the USSR itself. Cuba and Mongolia also participated during the early discussions and Vietnam later joined as a full member (in the event, scientific contributions by these three countries were limited).

At this stage, it is worth noting that the capacities of the eastern European countries were quite different and very uneven between the different scientific disciplines. An American assessment ranked the GDR and Czechoslovakia first, the GDR especially in optics, a legacy of German wartime skills [41]. Poland and Hungary were in the middle range, but the Polish scientific community had been exterminated during the world war, while Hungary had suffered badly from an exodus of scientists during the war and after the 1956 uprising. Bottom of the league were Romania and Bulgaria, whose scientific capabilities were very low.

A formal program of work was agreed on 13th April 1967. At their Wroclaw, Poland meeting in 1970, they formally adopted the name Intercosmos. As a program, it had no defined budget of its own, the operating principle being that each country built and supplied its own equipment, and the USSR was responsible for satellite integration and launching, with the scientific results the common property of all the countries. A global network of crypton laser tracking stations was established to follow the Intercosmos missions. These were in Zvezgorod, Russia; Ondřejov, Czechoslovakia; Poznan, Poland; Riga, Latvia; Helivan, Egypt; Patakomaja, Bolivia; Simeiz, Crimea; and Potsdam, GDR. The first nine Intercosmos used the small Cosmos 2M rocket from Kapustin Yar, while Intercosmos 10–19 used the larger Cosmos 3M rocket [42]. In effect, Intercosmos took over where the DS-U program tailed off, although there was overlap between the two.

A number of Western countries had access to this collaborative framework, especially France. As far back as 1960, scientists in the new French space agency CNES had volunteered French participation in the Soviet space science program. Nothing happened for years and there was no formal response until 1965, which led to a meeting in Moscow in April 1966 hosted by Mstislav Keldysh in his perfect French. Soon, President De Gaulle became the first Westerner to visit Baikonour, followed by his successor Georges Pompidou, who watched the launch of Cosmos 368, and French equipment soon began to fly on space science missions. France did not join Intercosmos but had a privileged relationship, with its own set of missions, Aureole (a cooperation agreement was signed in June 1966). Sweden also participated, sharing a similar interest in the geophysical problems of the northern latitudes [43].

The first fruit of the new Intercosmos program was Cosmos 261, a 400-kg DS satellite launched on 20th December 1968, involving scientists from Bulgaria, Czechoslovakia, GDR, Hungary, Poland, and Romania. Although the scientific equipment was Soviet-only, the participant countries led ground-based measurements of ionospheric and thermospheric parameters, synchronized with measurements on the spacecraft, with data being analyzed together (similarly to Cosmos 348). Cosmos 261 carried out a 53-day mission at the height of the solar maximum, detected several magnetic storms, and burned up in the atmosphere on 12th February 1969. Its aim was to study the corpuscle electrons and protons that were the cause of the northern lights, electrons of superthermal energy from 214 to 650 km, and changes of density in the atmosphere because of the aurorae. It should

Table 2.13. Early Intercosmos DS-U series.

Designator	Intercosmos		Missions
DS-U3-IK	Solar	1	Solar
		4	Solar radiation (ultraviolet and X-ray)
		7	Solar and X-ray radiation
		11	Solar
		16	Solar ultraviolet radiation
DS-U1-IK	Ionosphere	2	Ionosphere
		8	Ionosphere
DS-U2-IK	Charged particles, ionosphere and magnetosphere	3	Charged particles and ionosphere
		5	Charged particles and ionosphere, solar activity changes
		9	Charged particles and ionosphere (*Kopernik*)
		10	Magnetosphere, electromagnetism
		12	Ionosphere, solar radiation, micro-meteorites
		13	Magnetosphere, ionosphere
		14	Magnetosphere, ionosphere, low frequency oscillations

In addition, there was the category for cooperation with France, DS-U2-GKA (Geofizichefsky) (Aureole 1, 2)

have been called Intercosmos 1, but because it was launched from the then officially secret cosmodrome of Plesetsk, it was given the Cosmos 261 designation so as not to draw unwarranted attention to the launch site [44]. Later, consideration was given to renaming it Intercosmos 1, but this would have attracted even more curiosity. Its main outcomes were to find streams of upward and downward electrons, old ones mixing with new and a new auroral zone extending from the equator to the border of the auroral oval (the diffuse auroral zone, a large-scale region of the polar ionosphere where low-energy electrons precipitated equatorward from the auroral oval). It found an ionospheric trough in northern latitudes. Auroral electrons in the 100-eV range had twice the intensity in summer than winter, probably because of solar illumination. Electrons in the 15-keV range disappeared inside the auroral oval [45].

Here, the first stage of the program is reviewed: those using the DS series of satellites (Table 2.13) [46]. A more sophisticated model was introduced in the 1970s, the AUOS, and this is reviewed in Chapter 7. Note that Intercosmos 6 is reviewed separately under the *Energiya* and *Efir* programs (below). The Intercosmos series operated in families, IK standing for Intercosmos.

Different countries, teams, and organizations were associated with different sets of missions. For example, Intercosmos 3 and 5 and later 17, 18, and 19 were associated with the Skobeltsyn Institute for Nuclear Physics. Different groups had their own principal investigators: for example, P.I. Vakulov and S.N. Kuznetsov led on Intercosmos 3 and 5. Each set is now reviewed in turn.

Intercosmos 1 launch

INTERCOSMOS SOLAR MISSIONS

The Intercosmos solar missions built on the experience of Cosmos 166 and 230 and comprised a three-sided scientific team drawn from the Soviet Union (IZMIRAN, FIAN, and IKI), Poland (the Academy of Sciences in Wrocaw), and Czechoslovakia (Ondřejov observatory). All used a sun-orientated design with a stabilization system.

Intercosmos 1 studied ultra-violet and X-ray emissions from the Sun and the distribution of fine particles in the upper atmosphere. It made the first polarization measurements of three solar flares, carrying an alpha photometer (GDR), X-ray spectroheliograph, and X-ray polarimeter (both USSR). It measured the way in which solar flares upset radio communications on Earth. Intercosmos 1 identified the way in which short-wave radiation was absorbed by oxygen in the Earth's atmosphere. Intercosmos 1 found a reserve of molecular oxygen and aerosols in the upper atmosphere, but they leaked away above 100 km. The nearly identical Intercosmos 4 had a Soviet solar X-ray polarimeter to study flares in the 0.6–1.2, 1.7–1.93, and 8-Å ranges with an X-ray spectrograph, GDR ultraviolet photometer, and Czechoslovakian photometer, with an eight-channel telemetry system to transmit both real-time and stored data. Measurements were made of the characteristics of X-rays in two powerful solar emissions of several minutes' duration. Intercosmos 4 detected a Chinese nuclear test as it flew over China on 14th October 1970.

Intercosmos 7 worked in conjunction with the solar probe Prognoz 2 (see below) and had Czechoslovakian and GDR equipment. It was designed as an early summer spacecraft to match autumn data from previous ones in the series. Its purpose was to measure short-wave solar radiation in the 1,200–1,300-Å range, which did not reach the surface of Earth, being absorbed by oxygen molecules. The fourth in the series, Intercosmos 11, had instruments to study solar ultraviolet and X-ray radiation and was

Intercosmos 1 instrumentation

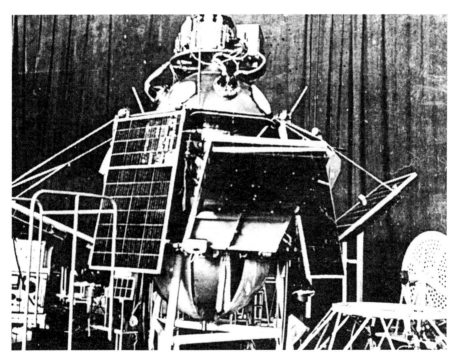

Intercosmos 7

the first to launch on the new Cosmos 3M rocket from Kapustin Yar. It also obtained data on the ozone distribution in Earth's upper atmosphere.

Intercosmos 16 involved the only failure in the program, for the original mission went off course on 3rd June 1975 and crashed 50 km downrange from Kapustin Yar. The replacement mission involved, for the first time in the program, Sweden, which supplied a solar ultraviolet spectrometer. Czechoslovakia built the multispectral photometer, which studied solar flares in the 0.3–0.6-keV range. The GDR supplied a photometer to measure molecular oxygen in the upper atmosphere. Other instruments examined ultraviolet solar radiation in the range 1,200–1,500 Å. The Polish Intercosmos 16 instruments led to a much improved knowledge of active plasma regions of the Sun, profiling their temperatures and duration as well as the presence of heavy elements (e.g. magnesium) and a small population of non-thermal electrons. The mission took place during solar minimum.

Outcomes of this sub-series of Intercosmos missions were published in 1983. The results combined data in the areas of solar X-ray radiation, dynamics, and polarization. Some had implications for atmospheric physics, especially the measurement of upper atmosphere density and the composition and the abundance of aerosols through the absorption of solar short-wave radiation. In the area of the solar–Earth environment, it was ascertained that solar flares have an important effect on the Earth's tropopause and it was even suggested that there was a correlation between solar flares and their resultant distortion of the electromagnetic field with physiological behavior of humans, ranging from cardiovascular disorder (50–100 times) to road accidents [47].

INTERCOSMOS IONOSPHERIC MISSIONS

Granted the high priority given to ionospheric research in the USSR, it may be surprising to find only two Intercosmos missions in this field. This may be explained by the development of a separate line of ionospheric missions (*Ionosfernaya Stantsiya*, see below) and by the allocation of the Aureole missions to this theme. These satellites had eight solar panels, four on the top and four on the bottom. Instruments on the spacecraft's domed cylinder aimed at studying ion densities and electrons more than 40 keV and protons more than 1 MeV.

Intercosmos 2, which marked Cuba's accession, worked until 12th February 1970 (50 days) and measured particles with concentrations from 20,000 to $1,000,000/cm^3$, depending on altitude, from 200 to 1,200 km (Figure 2.30). Temperatures ranged from 800 to 3,000 K. For the first time, it was tracked by a station outside the USSR, this time in Poland. Its orbit was timed to pass through the equatorial magnetic anomaly from 200 to 580 km by day, but at 700 km by night, which seemed to mark its outer limits.

Its successor, Intercosmos 8, continued its mission but in a much higher inclination orbit, 71°, so as to study the temperatures and concentrations of plasma during polar nights when there was no direct sunlight (it was launched in December). It was the first to fly Bulgarian instruments, the first launch from Plesetsk, and the

Intercosmos 2

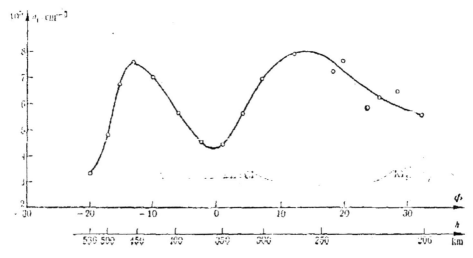

Figure 2.30. Ion concentration measured by Intercosmos 2 (left) by latitude and altitude (right).

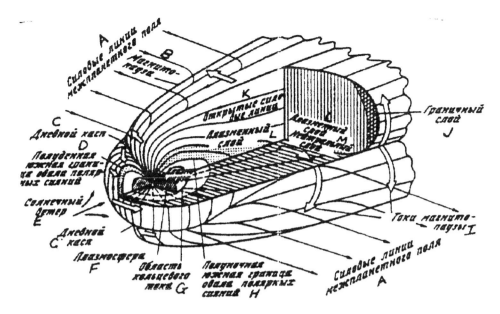

Figure 2.31. Soviet model of the magnetosphere, early 1970s.

first to use a stand, not a silo. Its observations were matched with the two previous Plesetsk launches, Cosmos 261 and 348. Also for the first time, representatives from the participating countries, Bulgaria, GDR, and Czechoslovakia, were permitted to enter Plesetsk (until then, they posted off their instruments and waited months to be told they were in orbit). Intercosmos 8 studied molecular ions in the mid-latitude trough at altitudes of 210–250 km. Intercosmos 8 made a map of the later-afternoon equatorial magnetic anomaly at 300–500 km and charted how NO ions rose in the ionosphere at 40°N and 40°S in the evening [48]. These missions, with others in the Cosmos series, enabled a new map of the magnetosphere to be compiled (Figure 2.31).

INTERCOSMOS CHARGED PARTICLE MISSIONS

The charged particle missions were the largest set: seven launches. Intercosmos 3 and 5, part-led by P.V. Vakulov and S.N. Kuznetsov, were Soviet/Czechoslovak missions dedicated to examining the composition and variations in the charged particle flux and to detecting low-frequency radio waves in the 70-Hz to 29-KHz band. Intercosmos 3 had instruments to study charged particles (Czechoslovakia), electromagnetic waves (Czechoslovakia), magnetic fields (USSR), and very low-frequency radiation. Intercosmos 3 functioned for four months. It carried an analyzer to detect natural low-frequency radiation that picked up lightning and radio waves in ionospheric plasma. Intercosmos 3 detected thunder discharges from orbit, noted the quietening of solar activity from an earlier solar maximum, and followed the solar flares emitted on 14–19th August 1971. After a solar flare, high-energy electrons reached the lower radiation belt and precipitated into the atmosphere. It followed a violent magnetic storm on 16th October 1970 and saw the way in which particles moved into the atmosphere over the Brazil anomaly and were lost. Intercosmos 3 found that there was a constant drain of electrons from the radiation belts through a "loss cone" down into the atmosphere. Finally, Intercosmos 3 and 5 carried instruments to detect low-energy cosmic rays, in contrast to the Proton missions, which focused on high-energy rays (below).

Intercosmos 5 was tracked by ground stations in the USSR (IZMIRAN, at 55.5°N, 37.3°E), Poland, Czechoslovakia (Panska Ves), and the GDR. It operated in conjunction with Cosmos 426, studying charged particles up to 2,000 km high. It carried a very long-frequency (70 Hz to 20 kHz) wave detector, an 8-m^2 circular loop antenna, which worked from 2nd December 1971 to the satellite's decay on 7th April 1972. This duly detected low-frequency emissions from the Earth's radio field and how they excited the ring current. On 7th January and 2nd February 1972, the loop discovered ion, proton, and helium cyclotron whistlers around the equator between 15°N and 15°S at altitudes of 280–330 km from 100 to 120°E (Figure 2.32). Whistlers had been known about for some time. Indeed, anyone who has tuned a radio in the short wave has probably heard a whistler – the trill, musical, up-and-down sound going up and down an octave over about 5 sec and then fizzling out at the lower end of the scale. In 1953, Cambridge graduate Owen Story figured out that they were a feature of plasma-generated electrical storms at 12,000-km altitude – a

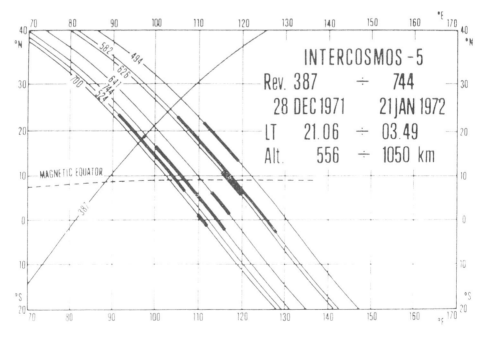

Figure 2.32. The cyclotron whistlers detected by Intercosmos 5. (Credit: COSPAR)

theory at once dismissed as impossible. But no one knew much about these whistlers or what really caused them or how they fitted into the pattern of electrical activity in the atmosphere. Intercosmos 3 and 5 discovered that low-frequency waves affected particle motion, leading to its precipitation into the atmosphere. This process was especially effective for electrons, while protons were mostly affected by very-low-frequency waves.

Intercosmos 9 was named *Kopernik 500* (Copernicus 500), in honor of the 500th anniversary of Poland's great scientist Nicholas Copernicus. The mission was first proposed at a scientific conference held in Torun, Copernicus's home town, in 1970. Mission director was Viktor Aksenov and its Polish-built receiver was designed to detect low-frequency solar and galactic frequency emissions between 6 and 0.6 MHz. This was a big electric dipole that extended 6 m from the top of the spacecraft, 15 m across, transmitting data down to IZMIRAN and Ondřejov observatory in Czechoslovakia. The payload also included a radiospectrograph. It operated for six months and recorded solar flares, radio waves, radiation, and the ionosphere feedback to solar flares. It examined solar radiation and followed the way in which it faded out in the ionosphere. *Kopernik* repeatedly detected fast bursts of about 1 min in duration that sped from the solar corona along open field lines. It also observed natural radio noise in the Earth's atmosphere emerging in high-temperature plasma.

Intercosmos 10 entered an eccentric orbit of 265–1,477 km, 74° from Plesetsk, to investigate the streams of electric currents and jets in the ionosphere in a project led by Leonid Vanyan (USSR). It had four instruments: a plasma variation low-

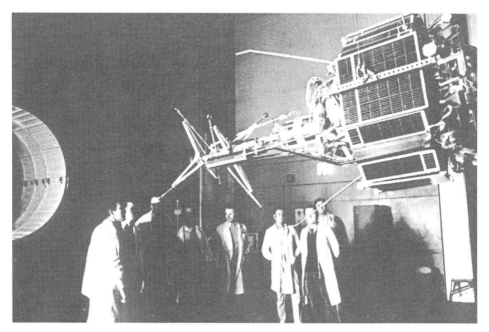

Intercosmos 10

frequency detector in the 20–22-kHz range (Czechslovakia), a plasma flow sensor for the 0.025–20-keV range (USSR), a magnetosphere variation detector in the 0.7–70-kHz range (USSR), and an ionosphere analyzer (USSR/GDR) to measure the temperature and concentration of electrons, the data being correlated with ground and sounding rocket data for the 265–1,500-km range. Intercosmos 10 had the same period as the circular orbiting Cosmos 381 *Ionosfernaya Stantsiya* (see below) and their findings may have been compared. Intercosmos 10 was the first to use the larger Cosmos 3M vehicle, which permitted a heavier payload and orbital inclinations as high as 74° and 83° and had a Czech telemetry system. Its main scientific legacy was to provide details of the lower hybrid resonance noise first found by the Canadian satellite Alouette at the transition of the plasmasphere and electron density troughs at high latitudes and at altitudes of 600–1,400 km.

Intercosmos 12 carried a mass spectrometer (USSR, Bulgaria), micrometeorite analyzer (Hungary, Czechoslovakia, USSR), and two ionosphere analyzers (Institute of Electronics, GDR; Institute of Stable Isotopes, Cluj, Romania). Micrometeorites were the main focus of interest and instruments were designed to classify them by physical character, energy, and power. Intercosmos 13, part-led by V.G. Stolpovsky, was dedicated to studying the interaction of charged particle and low-frequency electromagnetic radiation in polar regions, where data had hitherto been limited (its orbit was 83°). It found and marked the outer zone, high-latitude, low-altitude electron boundary (40 keV to 1 MeV) by local magnetic time.

Intercosmos 14 studied the ionosphere, micrometeorites, and low-frequency electromagnetic oscillations in the magnetosphere and data were sent down to a ground station in Norilsk, beyond the Arctic Circle and coordinated with ground observations. Intercosmos 14 had a four-frequency beacon to measure the composition of electrons, principal investigator being J.I. Schmilauer of Czechoslovakia, and an instrument to measure their temperatures, which he developed with Konstantin Gringauz (USSR). The main finding was that there were abrupt temperature changes in the ionosphere around the southern hemispheric oval. There was a micrometeorite detector to measure meteor showers developed by Tatiana Nazarova (USSR), I. Zakharov (Czechoslovakia), and Istvan Apathy (Hungary). The outcome was a finding that micrometeorite particles around the Earth at 1,000 km were five times denser than interplanetary space, but granted the low level there, the figure was still a low one. It also carried an extra-long-frequency receiver built by Pavel Triska in Czechoslovakia and J.I. Likhter in IZMIRAN in the USSR to measure electric and magnetic fields perpendicular to the geomagnetic field and spherical ion traps on booms built by Kirill Borissov Serafimov in Bulgaria and Gennadiy Gdalevich in IKI, USSR [49]. The series is summarized in Table 2.14.

Table 2.14. Intercosmos series.

Cosmos 2M rocket				
Cosmos 261	20 Dec 1968	Plesetsk	127–669 km, 71°, 93 min	400 kg
Intercosmos 1	14 Oct 1969	Kapustin Yar	254–626 km, 48°, 93.4 min	315 kg
Intercosmos 2	25 Dec 1969	Kapustin Yar	200–1,178 km, 48°, 98.5 min	320 kg
Intercosmos 3	7 Aug 1970	Kapustin Yar	200–1,295 km, 48°, 99.8 min	340 kg
Intercosmos 4	14 Oct 1970	Kapustin Yar	255–649 km, 48°, 93.6 min	320 kg
Intercosmos 5	2 Dec 1971	Kapustin Yar	205–1,200 km, 48.4°, 98.5 min	340 kg
Intercosmos 7	30 Jun 1972	Kapustin Yar	260–551 km, 48°, 92.7 min	375 kg
Intercosmos 8	1 Dec 1972	Plesetsk	204–649 km, 71°, 93.1 min	375 kg
Intercosmos 9	19 Apr 1973	Kapustin Yar	199–1,526 km, 48°, 102.2 min	340 kg
Cosmos 3M rocket				
Intercosmos 10	30 Oct 1973	Plesetsk	265–1,477 km, 74°, 102 min	340 kg
Intercosmos 11	17 May 1974	Kapustin Yar	484–526 km, 50.7°, 94.5 min	350 kg
Intercosmos 12	31 Oct 1974	Plesetsk	264–708 km, 74.1°, 94.1 min	350 kg
Intercosmos 13	27 Mar 1975	Plesetsk	284–1,689 km, 82.9°, 104.9 min	350 kg
Intercosmos 14	11 Dec 1975	Plesetsk	335–1,684 km, 73.9°, 105.3 min	372 kg
Intercosmos 16	27 Jul 1976	Kapustin Yar	464–517 km, 50.6°, 94.4 min	370 kg

Intercosmos 9 = Kopernik 500

IONOSFERNAYA STANTSIYA AND THE GALPERIN MISSIONS

So far, we have reviewed the main thread of scientific missions: Sputnik, the MS, and Elektron series, and then the Cosmos program and its three main sub-variants: DS, DS-U, and Intercosmos. Now, we look at a series of individual missions that developed in the next period: Cosmos 348, 381, and then the Proton, *Efir, Energiya*, and Prognoz series.

Following approval of the Sputnik, MS, Elektron, and Cosmos series, a series of specialized scientific missions was approved in a government decision of August 1964 (Decree #655-268). This is well known as the decision that committed the Soviet Union to race the United States to a manned landing on the Moon, but, as investigation of the archive has told us, this included important decisions for Russian space science as well [50]. These missions were also important in spreading the science program beyond the existing two bureaus (OKB-1 and OKB-586) to newcomers. We do not know the processes whereby these missions came to be tabled and approved, but the key decisions were:

- to fly a special ionospheric mission (*Ionosfernaya Stantsiya*);
- to develop an on-going program of sounding rockets, *Vertikal*; and
- to build a series of cosmic ray observatories, Proton 1 and 2.

In addition, it gave approval to a number of programs that either did not materialize or emerged in quite different form. These were:

- A heliophysical station, awarded to Yangel's OKB-586: this was a mission to use multiple instruments to study the Sun with extreme precision, using the R-14 launcher. This completed design in 1966, but delays by the sub-contractor led to the eventual cancelling of the mission. It was effectively resurrected, many years later, as the Koronas project.

- Astrophysical satellite *Protsion*, awarded to OKB-1 branch §3, Dmitri Kozlov's design bureau in Kyubyshev, now known as Samara, intended to study the stars, high-energy nuclei, and matter in the universe. Although a design was also completed in 1966, it was never launched. The bureau came to focus on Earth resources and photo-reconnaissance work, with which it became extremely busy, and the project fizzled out in 1970.

- Earth satellite *Plazma*, assigned to Vladimir Chelomei's OKB-52 design bureau. The purpose was to build a 4-tonne satellite to study plasma in near-Earth space in the context of solar and geomagnetic activity. The project was canceled soon after Chelomei's strongest supporter, Nikita Khrushchev, was overthrown in October 1964. The blueprints were transferred to the Lavochkin bureau in 1966, which revived it later as the Prognoz program.

- A maneuverable geophysical satellite, also awarded to OKB-52, intended to change its orbit while making a global study of the upper atmosphere. Chelomei had already designed the satellite in 1963 and hoped to build on his reputation as a builder of maneuverable satellites, of which he had already flown two, the Polyot program. It made no progress.

Cosmos 381 *Ionosfernaya*

Returning to the projects that did actually fly, the flagship was a small dedicated ionospheric satellite called Cosmos 381 *Ionosfernaya Stantsiya* (in English, "ionospheric station"). Earlier ionospheric studies had been done by the DS-Us Cosmos 321 and 356, but the intention here was to take them a stage further with a different, larger spacecraft. Remarkably, *Ionosfernaya Stantsiya* was allocated to a bureau that had never

Table 2.15. Cosmos 381 *Ionosfernaya Stantsiya* instruments.

Cosmic ray detector
VLF receiver
Solar ultraviolet detector, 3–1,500 Å
Space radiation detector
High-frequency impedance probe
Topside sounder

---

built a scientific satellite before and never did again: OKB-10, originally a spin-off branch of Korolev's OKB-1. OKB-10 was a design bureau that specialized in applications and communications satellites and was located in Krasnoyarsk, western Siberia, one of 70 closed cities in the USSR whose purpose dating to 1949 was to be a center for the nuclear industry. *Ionosfernaya Stantsiya* used what was called the KAUR 1 bus, 2.035 m in diameter, a passive single-axis magneto-gravity-stabilized, hermetically sealed cylinder with a weight of 800 kg and 3 m tall. The bus subsequently became the basis of a series of navigation and related satellites built by OKB-10, later renamed NPO-PM and now ISS Reshetnev [51]. It carried four instruments (Table 2.15).

The first *Ionosfernaya Stantsiya* was the victim of a launch failure, but all went well with the second, backup spacecraft, which flew into a circular 990-km, 74°, 104.9-min orbit from Plesetsk on 2nd December 1970. The Russians did not identify it as a scientific satellite until mid January 1971, by which time Swedish tracker Sven Grahn had long since identified it as an unusual satellite transmitting on 20.005 and 30.0075 MHz. It did not appear to last long, going off air on 31st January 1971, after only two months. Western observers were often struck by the short operational lifetimes of Soviet scientific satellites and many missions were wrongly considered failures as a result. Western satellites were generally smaller and with longer lifetimes. By contrast, the Soviet design philosophy was to return a large volume of data during a short lifetime, flying repeat and comparator missions as necessary. Suggestions that it disappointed are belied by the fact that Cosmos 381 *Ionosfernaya Stantsiya* was the centerpiece in the Soviet space pavilion at the Paris air show the following summer. Based on its outcomes, project scientists proposed there be a permanent solar patrol to warn the Earth of the effects of the Sun on the atmosphere and ionosphere.

The aim was to test ionospheric change at 1,000 km by time, latitude, and season, especially at times of magnetic storms, with signals sent down to four stations: Kharkov (50°N), Belomorsk (64.5°N), Archangelsk (64.6°N), and Murmansk (69°N). Cosmos 381's instruments included the first topside sounder for vertical pulse probing of the atmosphere and an impedance probe to measure local plasma concentrations and irregularities along the path of a satellite. Hitherto, scientists had probed the ionosphere upward, but this was the first Soviet opportunity to do so from above ("topside"). This was the first of three topside sounders (the others being Intercosmos 19 and Cosmos 1809, see Chapter 7) and followed the world's first topside sounder carried on the Canadian satellite *Alouette*. The concept was to emit a pulse of 80–100 μs to excite ions that would be back-detected, like a radar. The sounding took place over 20 frequencies in the range 2–13.4 MHz, one second-long pulse every minute, the aim being to measure electron concentrations at different altitudes. Cosmos 381 measured the density and heat of electron densities under both calm and disturbed conditions, matching them against diurnal variations, storms, latitude, and the effect of solar ultraviolet radiation, so it also contributed to the study of Sun–Earth relationships. A graph of electrons showed them at a low level during the nighttime, but rising from dawn to midday before trailing away. During a sampling period of 29 passes, daytime electrons were measured in a range 8×10^{-3} to 4×10^{-2}, but nighttime electrons were much lower, at 3.5×10^{-3} to 2.2×10^{-2}. The level fell from 45 to 60°N, where it dipped for the trough and then rose again at 70°N. Irregularities were found in the ionosphere between 250–450 km by night and 150–750 km by day, most pronounced between 2200–0100 and 0400–0500 local time. Mean electron concentration was measured at 4×10^{3} to $4 \times 10^{4}/cm^{3}$, the highest intensities being between 60 and 74°N, weakest 50–60°N, stronger again at 40–50°N, with 200-km blobs of local variations. Double charged or Auger electrons were found in the course of the solar flare of 24th January 1971. Cosmos 381 made the first systematic study of ionospheric inhomogeneities, originally discovered by *Alouette*. Cosmos 381 in effect made possible a new model of the density and temperatures of the ionosphere and how it was affected by the key variables of Sun, season, latitude, altitude, and time of day [52].

Five other ionospheric missions of this period are taken together: Cosmos 184, 261, 348, 378, and 426. The first three were the key projects, during this period, of Yuri Galperin of Cosmos 3 and 5 fame. There had been a huge output of scientific papers from the 2MS missions, for interpreting, writing, and publishing the outcomes took several years. Yuri Galperin's interpretation of the results from Cosmos 5 formed the basis of his higher Ph.D. and he was made head of department. In 1967, a group from the department was transferred to the new Institute of Space Research, IKI, where Yuri Galperin formed the Laboratory of Physics of Auroral Phenomena (also called the "Auroral Physics Laboratory"), which he headed until his death. Galperin proposed a successor to the Cosmos 3 and 5 mission. Their low 49° angle was little use for studying polar or auroral phenomena, for they spent only a few seconds in the auroral zone, so he proposed that it fly from Plesetsk, the new northern latitude cosmodrome that had opened in 1966. The first of these was the first Intercosmos mission, Cosmos 261, mentioned already. Yuri Galperin developed the backup spacecraft as a Soviet-only mission, the idea being to operate 261 and 348 as a pair, 261 being a winter project and 348 a summer one. Cosmos 348 found streams of upward and downward photo-electrons passing one another at 650 km, the upward one rising from the sunlit polar cap with energies of up to 1,000 eV and escaping into space. As a result, a fresh understanding of the magnetosphere around the polar cap began to emerge (Figure 2.33).

As he prepared Cosmos 348, Galperin did not yet have his own building and he and his colleagues worked in other institutes, such as the Institute for Applied Mathematics. His own laboratory, with the spacecraft instrumentation, was in an apartment basement that flooded during a water burst during the October revolution holiday. Everything had to be rebuilt from scratch, but in the end worked well.

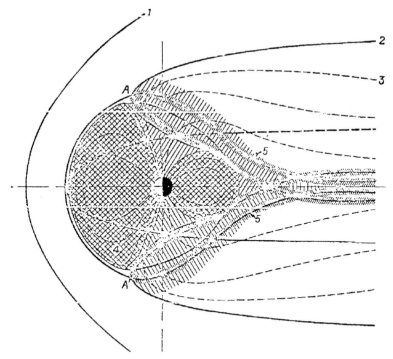

Figure 2.33. Soviet map of the magnetosphere showing the polar currents explored by Cosmos 348.

As the pair was prepared, he was meantime involved in the Cosmos 184 mission. Cosmos 184 carried a high-resolution scanning infrared radiometer to measure radiation from clouds and oceans, surface and cloud temperatures, and the exchange of energy in the Earth-atmospheric system and in a collaborative venture, data were sent to the United States until May 1968. More importantly, Cosmos 184 had an ion detector developed by V.N. Ponomarev to measure ion concentrations at 600 km by time of day and latitude (they ranged from 35,000 to 400,000 ions/cm^3) (the mission led to a global isometric chart of ions). He noted how ionospheric plasma changed abruptly during magnetically disturbed periods. This led to the discovery of what is now called the "polarization jet" in the ionosphere (a narrow stream of supersonic ions). This was announced in 1973, but so poor was the interaction between Russian and Western scientists at the time that it went unnoticed in the West, where American scientists eventually and independently discovered the same phenomenon again in 1977 but called it the Sub Auroral Ion Drift (SAID) [53].

Cosmos 378 (DS-U2-IP) was not a Galperin mission, but covered similar fields, the chief experimenters being Konstantin Gringauz, S.N. Kuznetsov, and Gennadiy Gdalevich. Its mission was to make a prolonged and continuous study of the ionosphere in general and its polar regions in particular, with an interest in low-energy electrons > 40 keV, especially in its daytime apogee in the southern latitudes. At the time of the launch, the northern latitudinal part of the orbit crossed the ionosphere during the polar night when solar wave radiation had no effect on the ionosphere and its characteristics were determined by energetic charged particle fluxes. Data were downloaded when it subsequently passed over Russian ground stations. Although its memory ceased functioning after 200 revolutions, the satellite continued to operate until 9th August 1971 and a good volume of information continued to be received during overpasses. Cosmos 378 observed how low-energy electrons streamed into the southern hemisphere high ionosphere at 1,700–1,750 km over the South Pole. Cosmos 378 measured the eastern Siberian geomagnetic anomaly at 100°E. It found sharp increases in ions both with latitude and in the early morning. Cosmos 378 made the first study of the dayside cusp of the outer ionosphere, where it found the temperature of the plasma to be quite uneven. Captured electrons > 40 keV moved outward by day. Protons > 1 MeV were found over the polar caps during storms. The gains in knowledge of the ionosphere from the missions of Cosmos 261, 348, 378, and 381 were considerable [54].

Galperin's success of Cosmos 184 and 348 led to a spin-off series of scientific satellites developed with France, Aureole, where he was the prime mover. Aureole arose from a scientific conference in Paris in 1968, being proposed by him to the government on his return. His fluency in French obviously made this an appealing prospect to the French. Originally, the series was to be called ARCADE (ARC Aurorale et Densité), but an hour before lift-off was renamed Aureole (for AURora and EOLus). The Aureole 1 and 2 were DS-U2-GKA models and launched on the Cosmos 3M from Plesetsk. Aureole 1 and 2 carried an identical mixture of French instruments to study low-energy electrons and protons, where they were expert, with Soviet instruments for the high-energy range, where they were knowledgeable. Aureole 2 focused on the way in which the atmosphere heated the regions of the polar lights in the form of a slow-motion nuclear reaction. A model was constructed of the equatorial auroral zone, called the "convection boundary model", with French plasma physicist François Cambou. Scientists established the spot-like shape of the cusp and determined the global pattern of the evolution of the energetic spectra of protons, from their injection from the tail into the inner magnetosphere, their drift in the magnetosphere, and their precipitation in the diffuse zone. Aureole 3 belonged to the next generation of OKB-586 satellites, the AUOS Z M-A-1K series (see Chapter 7).

Finally, Cosmos 426 *Geofizicheski* carried experiments UER-1 to study the angular distribution of electrons over 20 keV and TEP-3 to detect photons and electrons in the radiation belts. It was a companion to Cosmos 378, both having the same principal investigator, S.N. Kuznetsov. During a storm on 21st–26th November 1971, Cosmos 426 was able to follow what happened to the trapped electrons in the 20-keV to 1.5-MeV range, subsequently published in Leningrad by Nauka as *Substorms and Disturbances in the Magnetosphere*. It remained in orbit until May 2002, 11,299 days. The Galperin and other ionospheric missions are summarized in Table 2.16.

Table 2.16. Ionospheric satellites.

Cosmos 184	24 Oct 1967	600–638 km, 81°, 97 min	
Failure	27 Dec 1969	Launch failure	*Ionosfernaya*
Cosmos 348	13 Jun 1970	201–651km, 71°, 93 min	
Cosmos 378	17 Nov 1970	240–1,770 km, 74°, 105 min	
Cosmos 381	2 Dec 1970	961–1,013 km, 74°, 104.9 min	*Ionosfernaya*
Aureole series (DS)			
Aureole 1	27 Dec 1971	410–2,500 km, 74°, 114.6 min	
Aureole 2	26 Dec 1973	407–1,955 km, 74°, 109.2 min	

Cosmos 3M from Plesetsk

PROTON: ELUSIVE COSMIC RAYS

Proton was also approved in the government decree of August 1964, but it is possible that by then, the satellite was already well beyond the conceptual stage. Its roots lay less in science than in the need to test out a new rocket booster, the UR-500, designed by Vladimir Chelomei. This was a powerful rocket, originally intended by Soviet leader Nikita Khrushchev at the height of the Cold War as a "city-buster" rocket to destroy American cities. Just so that no one misunderstood, the first flights would carry simulated nuclear warheads. When Leonid Brezhnev came to power two months later, the rocket was scrapped as a missile, but Vladimir Chelomei persuaded him that the UR-500 (UR cunningly meant "universal rocket") could be adapted to civilian purposes, such as sending a small manned spaceship around the Moon ahead of the Americans. The UR-500 was a radically new design, using fuel tanks at the side of the rocket and powerful new engines (RD-253), so an extensive test program was warranted. Sergei Vernov, Naum Grigorov and his colleagues at the Skobeltsyn Institute persuaded both Chelomei and the government that this was an opportunity to launch a new type of scientific satellite and at the same time conveniently obscure the lunar purpose of the new rocket, which was to be called Herakles (written "Gerkules" in Russian). Herakles' lifting power enabled a large, gangly, kettle-shaped satellite with four paddle solar wings to be carried, dedicated to the study of cosmic rays and gamma radiation, hence the title "Proton".

The 1964 government decision actually approved two types of Proton satellite for the new rocket, Proton 1 and Proton 2, although the distinction between the two was not made clear until many years later. The Proton series was the first set of satellites devoted entirely to cosmic rays. The term "cosmic ray" came from the English scientist Robert Millican and the first investigation of cosmic rays was made by Victor Hess in a balloon in 1912. The principal Russian text was written by Dmitri Skolbeltsyn (1892–1990) in 1936 (*Kosmicheskie Luchi*) and he developed the cascade theory which argued that primary cosmic rays broke up in the atmosphere into secondary particles, so few reached the Earth's surface. Cosmic ray counters were flown on post-war high-altitude rockets. They found plenty of cosmic rays breaking

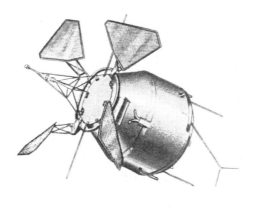

Proton

up at around 29–30 km, with a steady rate of primary cosmic rays over 50-km altitude.

Proton marked the first use of large calorimeters to measure cosmic rays in space. The idea of flying calorimeters into space had been proposed as far back as 1954 by Naum Grigorov (1915–2005), the man who became the driving force behind the program, with his colleagues Vladimir Mursin (1927–2007) and Ilya Rappoport. Calorimeters are heavyweight traps made of lead and graphite, which have a carbon surface that absorbs cosmic rays and, within a shielded chamber, breaks them up into secondaries that are absorbed by the calorimeter and can be analyzed for their composition and origin by detectors in the spaces between the target material. Every time a ray hit, there would be a flash as it entered the chamber, the bigger according to its energy. The chamber had first been tested in the Argatz Mountains in Armenia. It made possible the first ever recognition, interpretation, and analysis of cosmic rays in the range 10^{11}–10^{15} eV, which was hitherto impossible from ground-based facilities. A small nuclear photoemulsion pile had been installed on Korabl Sputnik 2 in 1960 but it was not until three years later that it was realized that the instrument had received several megaelectron impacts.

Proton 1 was intended to focus on energy particles in the 10^{10} and 10^{13}-eV ranges and the Proton 2 series objectives were for high-energy cosmic rays in the 10^{13}–10^{15}-eV range (a quadrillion volts), cosmic ray collisions in the 10^{11}–10^{12}-eV range, quarks and high-energy electrons. The intention was to study what happened when cosmic rays collided with hydrogen, carbon, and iron. This was a completely unknown zone in particle physics. The Proton 2 series, besides being bigger and aimed at higher energies, had two charge detectors, each with 16 Cerenkov detectors, meaning that each hit could be measured with great precision, details being transmitted to Earth as electric signals. The data relay rates were high and required a new computer processing system called *Artur* (particle detection rates were as high as 144,000 an hour).

These were extraordinarily heavy satellites, taking advantage of the UR-500's formidable lifting power. The Proton 1 series weighed 12.2 tonnes, Proton 2 no less than 17 tonnes. The weight of the experimental equipment was somewhat less: 3.5 tonnes for the Proton 1 series (to test the rocket, it carried up to 4.4 tonnes of ballast), but 12.5 tonnes for Proton 2. Three satellites were built for the Proton 1 series (Proton 1, 2, 3) and two for the Proton 2 series (Proton 4, 5) [55]. There were some differences within the first series. Proton 3 was charged with trying to find rays of galactic origin and was the first space mission to attempt to identify the smallest

Table 2.17. Proton series aims.

Proton reactions with carbon, steel, and lead 10^{11}–10^{13} eV with a 5% precision
Proton reactions with helium 10^{11}–10^{13} eV
Heavy protons with hydrogen, carbon, and steel between 10^{11} and 10^{13} eV
Chemical composition of protons up to 10^{15} eV
Energy intensity of primary cosmic rays 10^{8}–10^{10} eV

Table 2.18. Proton series instruments.

SEZ 14 ionization calorimeter (Proton 1–3)
IK15 ionization calorimeter
SEZ1 (cosmic rays of medium energies, Proton 1, 2)
SEZ12 to detect electron fluxes of medium energies (Proton 1, 2)
SEZ13 for particles with frictional electrical charges to detect quarks (Proton 3)
GG1 gamma-ray telescope to detect gamma fluxes with energies of 20 MeV to 2 GeV
RV (radiation environment during flight) (Proton 1–3)

particles then known to exist: quarks. The second series was not only much larger – the instrumentation was the size of a small bus – but had a slightly different design. Illustrations show that it was a tapering cylinder, with solar panels, a magnetic tracking sensor at the front, and large dipoles at the side. The series aims and instruments are reviewed in Tables 2.17 and 2.18.

SEZ 14 was the main instrument on Proton 1–3, meaning "Spectrum, Energy and Charge" in Russian (*Zaryad*), able to measure energies of 10^{14} eV. For Proton 4, the main instrument was the IK-15, meaning "Ionization Calorimeter up to 10^{15}", able to reach even higher energies.

The initial launch of the UR-500 went like a dream, although the attempt to mislead the Americans as to the true significance of the new rocket failed from the start, for they at once linked it to a round-the-Moon mission. In order to maintain the facade that the new rocket was purely intended to launch large scientific satellites, the UR-500 was also formally called "Proton", the name it retains to the present day (the title "Herakles" was abandoned even though it had already been painted on). Later, the Proton became the Soviet Union's most successful heavy rocket launcher and was still flying 45 years later, uprated as the Proton M to lift payloads of over 20 tonnes. Proton marked the first scientific satellite launching of Vladimir Chelomei, just as Cosmos 1 had for Mikhail Yangel and Spuntik for Korolev.

Proton 1 was both the heaviest and the largest scientific satellite launched at the time, 4.5 m in diameter, 9 m tall, and the panels had a span of 10 m. Proton was a cylinder inside a cylinder, the inner 3×2-m container maintained stably at room temperature and at pressures of 1.15–1.22 atmospheres. The highly original instruments comprised the heavily shielded ionization calorimeter (Figure 2.34) to measure high and super-high cosmic particles, electron spectrometers (Figure 2.35), and counters able to separate high-energy, carbon, and hydrogen ions. The

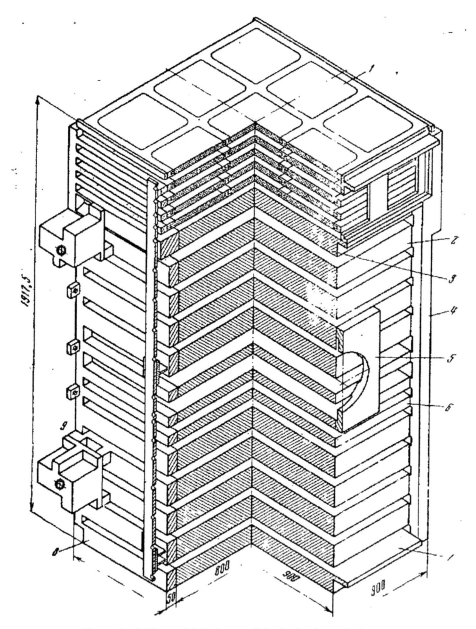

Figure 2.34. The multiple layers of the ionization calorimeter.

ionization calorimeter comprised a number of steel plates designed to separate out carbon atoms into one chamber and hydrogen atoms in the other and then isolate the protons and cosmic rays. The rays were separated by blocks of metal, paraffin, and plastic.

Low orbits were deliberately chosen so as to prevent contamination of the equipment by the radiation belts, but this meant that missions would be short. Proton 1 fell back to Earth and burned up on 11th October 1965, while Proton 2 burned up on 6th February 1965. Proton 4 lasted longest, at eight months. Proton 5 was canceled, although the reason why is not known. The SEZ instruments, though, were later flown on Cosmos 428 and 490. The missions are summarized in Table 2.19.

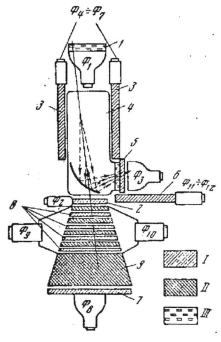

Figure 2.35. The electron spectrometer.

As for the results of the missions, the Protons succeeded in their prime task of capturing cosmic rays in the 10^{10}–10^{15} range. They were able to break down their chemical components, which ranged from hydrogen to tin to helium and iron. Most remarkably, cosmic rays accelerated notably past the 10^{12} point. Cross-sections were made of hydrogen and carbon protons in the 10^{10}–10^{12} range, finding their energies 210% higher at this point. Proton 3 made a detailed study of the solar eruption of 7th July 1966 and recorded how its fast particles streamed to Earth in the form of protons. Proton 3, with the SEZ 13 (Figure 2.36), did not find quarks, though. In the course of its 3,500 orbits, Proton 4 returned 449 sets of data, normally storing 12 hr at a time, with cross-sections of high-energy electrons and gamma rays. Proton 4 found some cosmic rays estimated to be more than 100 million years old. The further up the energy spectrum (10^{11}–10^{16} eV), the counting rate of cosmic rays declined very steadily (see figure). Their chemical composition changed at the 10^{12} mark. The outcomes of the Proton missions are illustrated in Figure 2.37 (the average intensity rates over 600 orbits), Figure 2.38 (bursts of particles), and Figure 2.39 (the number of rays by intensity, the number falling off the higher up the intensity range).

Ten scientists published the results from Proton 1 and 2 in "The Study of Cosmic Rays on the Proton Artificial Earth Satellite" in *Kosmicheskie Luchi* (Nauka Press, 1972). The level of carbon was found to increase with energy level. Forty-eight super-heavy nuclei were trapped. Six scientists published combined results from Proton 2 and Cosmos 208 in *High Energy Gamma Quanta in Primary Cosmic Rays in Cosmic Ray Studies* (Nauka Press, 1975) characterizing gamma-ray energies over 30 MeV. The Protons found a large flux of electrons more than 300 MeV near the equator and a large flux of stray particles at high latitudes. They found that between 200 and 500 km, there was a steady flux of electrons of up to 100 MeV of non-galactic and

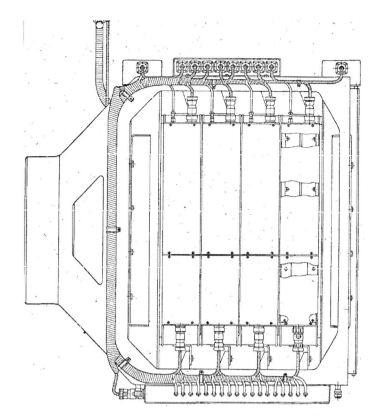

Figure 2.36. The SEZ 13, with its walls to trap cosmic rays.

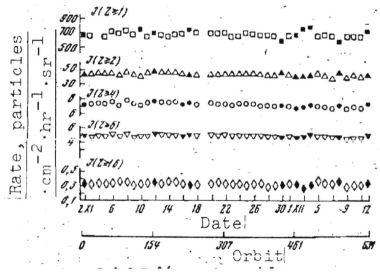

Figure 2.37. Proton 2 average rates of particles.

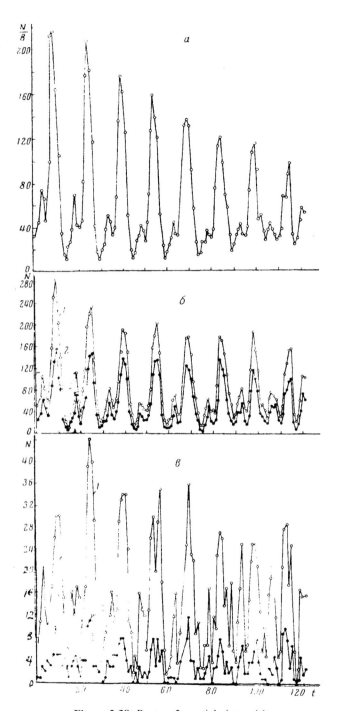

Figure 2.38. Proton 2 particle intensities.

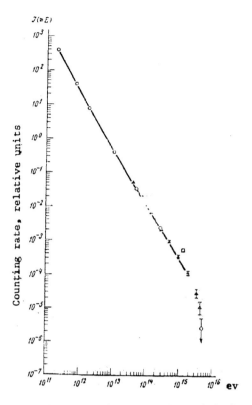

Figure 2.39. Proton 4 primary cosmic rays number (left) by intensity (right).

solar origin, whose flux was connected, albeit weakly, to the solar cycle. There was a discontinuity in cosmic rays in the 10^{12} range due to changes in their chemical composition. A chart was printed of the ionization bursts counted, with a stream of events in the first 100 orbits, steady set from orbits 300–800, but few thereafter. Naum Grigorov speculated, as a result of Proton measurements, that there was an undiscovered radiation belt around Earth of electrons over 100 MeV – a theory he elaborated over subsequent years (published as *High Energy Electrons in Earth Orbit*, Nauka, Moscow, 1985) [56].

Table 2.19. Proton series.

Proton 1 series			
Proton 1	16 Jul 1965	189–589 km, 63°, 92 min	87 days
Proton 2	2 Nov 1965	189–608 km, 63°, 92 min	92 days
Proton 3	6 Jul 1966	185–585 km, 63°, 92 min	72 days
Proton 2 series			
Proton 4	16 Nov 1968	248–477 km, 51.5°, 92 min	250 days

UR-500 rocket from Baikonour

The experiments on Proton 1–3 found a strange phenomenon in the proton's spectra, which was the bending of energies at 2×10^{12} eV, while the spectra of particles such as protons, helium, and heavy nuclei were stable. If true, this would mean that the composition of cosmic rays changed substantially at high energies. Follow-on missions were partly aimed at confirming these results (see below), such as Cosmos 208, mentioned in Table 2.10, and Cosmos 264. Whereas the Protons provided an upper limit on the gamma-ray flux, Cosmos 208 measured gamma rays with intensities of more than 50 MeV. Cosmos 264 had a spark chamber to study gamma quants with energies above 100 McV. Between them, two sky maps were made: one a gamma map, the other a background one.

SPECIALIZED SCIENTIFIC MISSIONS: *ENERGIYA* AND *EFIR*

Although Proton 5 was cancelled, some of its intended work was carried out in a series of follow-up missions. These were carried out by specialized spacecraft called *Energiya* and *Efir* developed by the Progress design bureau of Dmitri Kozlov in Kuibyshev (now Samara). These were four-day and 25-day missions, the first set recoverable, the second set non-recoverable (Table 2.20). Their true purpose, as successor to Proton, was not advertised at the time: the principal analyst of the Cosmos program, Phillip Clark, regarded them as unusual, but believed them to belong to the Zenit 4MK series of photo-reconnaissance satellites [57]. He was right, for they were derived from Zenit MK cabins and the *Efir* design was also used for the Bion series of biology missions (see Chapter 7). Although much smaller than Proton, the scientific instrumentation was still quite heavy, weighing in at 2,450 kg, most of which was the shielded shell of the calorimeter. The missions had the same science teams from the Skobeltsyn Institute.

The first was Intercosmos 6, the only recoverable Intercosmos mission, flying 7–11th April 1972. Once in orbit, it was orientated towards the oncoming stream of particles. Like the Proton 1 series, the aim was to study high-energy cosmic rays in the ranges 10^{12} and 10^{13} eV. Poland was the dominant country in the mission, assisted by Czechoslovak, Hungarian, Mongolian, and Soviet scientists in the Dubna joint institute of nuclear research. This was the first mission with Romanian and Mongolian participation, Mongolia supplying a detector of energetic nuclei. The overall project director was Professor Naum Grigorov from Proton with lead scientists Antal Somody (Central Institute of Physics Studies, Hungary), E. Fridlander (Institute of Atomic Physics, Romania), Yuri Dubinsky (Institute of Experimental Physics, Czechoslovakia), and Drs Tuvdendorzh and Chadraa (Institute of Physics and Mathematics, Mongolia). The principal experiment consisted of a sliver bromide photo-emulsion block and ionization calorimeter called BFB-C, tested out on a high-altitude stratospheric balloon in 1969. It was developed in Poland, the limiting factor being the capacity of the photo-emulsion plates to receive low-energy particle strikes and beyond a certain point, there were too many to separate out. The volume of the instrument was 48 liters, with 805 layers 450 µm thick in 10 separate sections to separate the particles, with stereo

Intercosmos 6

photographs to capture incoming rays. A scintillation counter was able to separate particles between protons, alpha particles, and heavy nuclei. The mission was set for four days: more than that and the chamber would have become too dense with particle traces.

The second experiment on Intercosmos 6 was an investigation of the density, composition, and distribution of meteorites, organized by Szabo Elik and Andras Laszlo (Central Institute of Physical Research, Hungary) working with the Astronomical Institute, Czechoslovakia, and the Institute of Geochemistry, USSR.

Intercosmos 6 was also the first mission of one of Estonia's most noted modern scientists, Rikho Nymmik. Born in Tallinn in 1936, he graduated from and spent most of his professional life at Moscow State University. He obtained his thesis on cosmic rays based on mountain observations in Armenia, worked on economic geography in Estonia, returned as an astronomer to Tartu observatory there, and then went back to the Soviet Union to design the 180-kg emulsion chamber on Intercosmos 6. This project took from 1969 to 1978.

Once the mission was over and the spacecraft recovered, it was shipped to Dubna, where the layers were separated and mounted on Czech glass plates before being transferred for further analysis to Krakow Nuclear Research Institute and Bucharest

Institute for Nuclear Physics, the purpose being to profile each trapped cosmic ray by direction, trajectory, speed, and energy level. The main outcome was that Intercosmos 6 picked up 500 hits, including three strikes of 1 billion eV (or 1 TeV) and recorded the highest ever energy electron, 5 TeV, a 1,000 billion-volt particle. Their energy level gives an idea of their source, in the case of the 5 TeV particle probably less than 1 kiloparsec away. Such events are considered rare, so they may have been lucky. Even more intriguingly, there was indirect evidence (an interaction) of a ray of 5,000 TeV, but it did not strike the stack, instead leaving 400 secondary particles behind as evidence. Of the 500 hits, it was possible to analyze 39 α particles and 70 protons. Nymmik's papers from the mission described the frequency and behavior of cosmic rays of 10^{12} eV. The recovered cabin was later put on public display [58].

The aim of the two missions of the *Efir* spacecraft (Cosmos 1513, 1713) was to determine the chemical composition, charge, and energy of cosmic rays over 1 MeV. Details of the *Efir* mission did not emerge until the design bureau that made the spacecraft, the TSSKB in Samara, published its in-house history in 1994 and described how the entire cabin was filled with a large ionization calorimeter. The experimental package was called SOKOL ("composition of cosmic rays" in Russian, also the name for a falcon). The missions lasted 27 and 25 days, respectively. In a follow-up to both missions, Cosmos 1882 (September 1987) and Cosmos 1887 (a biology mission in October 1987), at 82° and 62.3° inclinations, respectively, carried solid-state nuclear track detector stacks to detect anomalous cosmic rays in the 10–20-MeV range (C, N, O, and heavier ions) developed by Rikho Nymmik. Diagrams of the calorimeter were published in 2007 and the principal experimenters were Naum Grigorov, Igor Ivanenko (1929–1993), and Vladimir Shestoperov (1933–2002). The problem of the bending of energies, though, remained unresolved, as these missions were too short [59].

Table 2.20. *Energiya* and *Efir* series.

Energiya 13KS		
Intercosmos 6	7 Apr 1972	203–248 km, 51.8°, 89.01 min
Cosmos 1026	2 Jul 1978	207–247 km, 51.8°, 89.03 min
Recovered after 4 days		
Efir 36KS		
Cosmos 1543	10 Mar 1984	216–379 km, 62.8°, 90.6 min
Cosmos 1713	27 Dec 1985	217–398 km, 62.8°, 91 min

Decayed after 25, 27 days. All on R-7 rocket from Baikonour.

SOLAR OBSERVATORIES: PROGNOZ

The idea of a specialized solar observatory had been conceptualized in 1964 as the *Plazma* satellite and stalled but was revived by the Lavochkin design bureau as the

Oleg Vaisberg

Prognoz program under the leadership of Stanislav Karmanov. A key scientist in the project was Oleg Vaisberg, who was especially interested to develop a model of the solar wind. Oleg Vaisberg was to become one of the great plasma scientists of the Russian space program. Born in Sverdlovsk, where his father was a factory director, he had been given books on astronomy when he was a child and had developed an interest in the subject. Going on to Moscow State University, he studied astronomy under Shklovsky, moving on to Krassovsky's Institute of Atmospheric Physics, where he obtained his candidate doctorate in auroræ [60].

The original Soviet orbiting solar observatories were DS-U3S spacecraft within the Cosmos series, Cosmos 166 and 230 (see above). The new, dedicated series of orbiting solar observatories was called Prognoz, meaning "forecast". The satellite was a 860-kg box shape, 1.8 m across, with a domed top and bottom, carrying four solar panels with a span of 6 m. Its four-day curving orbit, taking it out almost to the distance of the Moon, gave it a good vantage point for studying conditions on the Sun. At apogee, it could intercept solar radiation before Earth's radiation belts began to affect it. Prognoz was assigned to the OKB-301 Lavochkin design bureau, which, from 1965, had built lunar and interplanetary spacecraft. Lead organization was the Institute for Space Research.

The mission concept was developed in 1965 by G.A. Skuridin, who argued that the magnetosphere could best be understood by a series of simultaneous observations from three different points [61]. The idea behind the treble mission, unofficially called the *Russkaya troika* (Russian for "threesome", or three horses pulling a cart or sleigh), was to put three Prognoz satellites into high-apogee orbit for simultaneous observations. Various scenarios were argued for, with one positioned on the night side, the second on the day-side cusp, and the third in the tail. In the end, resources did not permit the launch of three spacecraft in close succession, but Prognoz turned into a program that lasted over a decade and a half. After near-annual launches in the 1970s, the launching rate fell to a much slower pace in the 1980s, presumably because the spacecraft operated for longer periods. Mission planners made up for the lack of the *Russkaya troika* by operating Prognoz 1 and 2 in tandem, one being at perigee while the other was at apogee and vice versa and similar patterns were followed in subsequent missions.

Prognoz's main task was to monitor solar radiation, in particular solar flares, and forecast dangerous emissions that might disrupt Earth's weather and cause radio interference. Such warnings were also important to warn cosmonauts of potentially dangerous solar flares – indeed, one 1972 flare would have been potentially dangerous to the American Apollo landings on the Moon that year had the astronauts been walking on the lunar surface then. Leaving these applications aside, the program offered the opportunity to study a typical (albeit very close) star. The key aspect of the observatory was watching what happened when the solar wind met the Earth's magnetosphere. In addition, Prognoz carried instruments focused on the Sun, as well as instruments to study gamma and X-ray radiation from further afield.

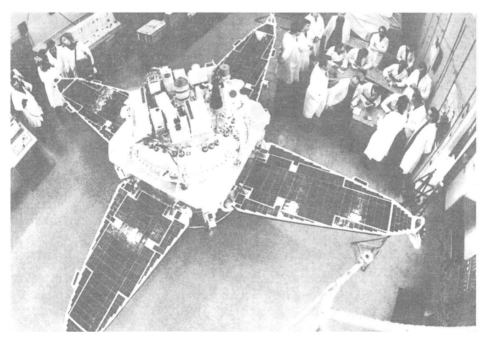

Preparing Prognoz for launch

Table 2.21. Prognoz missions and instruments.

Prognoz 1	Solar wind, neutrons, X-rays, magnetosphere (USSR)
Prognoz 2	Neutrons, gamma radiation, ions, low-energy electrons (SIGNE, France) Gamma rays, X-rays, solar plasma fluxes (USSR)
Prognoz 3	Solar flares, gamma, X-ray emissions (USSR); neutrons (France)
Prognoz 4	Solar radiation, flares, geomagnetic field (USSR)
Prognoz 5	Solar radiation, magnetic field, X-rays, gamma rays (USSR, France, Czechoslovakia)
Prognoz 6	Corpuscular and electromagnetic radiation, solar radiation, solar plasma fluxes, X-rays, gamma rays (France, Czechoslovakia) galactic ultraviolet radiation (*Galaktika*, France)
Prognoz 7	Mass spectrometer (Sweden); gamma burst detector (France); electromagnetic analyzer (Sweden); magnetometer (USSR)
Prognoz 8	Mass spectrometer (Sweden)
Prognoz 9	Gamma flash detector (France), X-rays, gamma rays, plasma fields, radio astronomy – residual radiation from "big bang" (*Relikt*), magnetic fields, ions (USSR, Czechoslovakia)
Prognoz 10	Interaction of solar wind and Earth's magnetosphere (*Intershock*)
Intercosmos 23	Cosmic rays, solar radio emissions, X-rays (Czechoslovakia)

Prognoz would enable a better reading of Earth's weather, providing clues to weather history, such as past ice ages. Sometimes, Prognoz measurements were coordinated with not only other Soviet spacecraft, but also American ones, like the International Sun Earth Explorer (ISEE).

An extensive range of scientific data were collected from Prognoz, with subsequent research papers covering the development of complex X-ray events on the Sun, the acceleration of alpha particles near the magnetosphere, the escape of high-energy solar protons into interplanetary space, the ebb and flow of energetic electrons in the outer radiation belts, gamma bursts, radiation doses, and fluxes of cosmic rays. The instrument suite varied from one mission to another, as may be seen in Table 2.21. Some missions had distinct purposes, had different teams of scientists, and involved a high level of international collaboration, mainly with France and Czechoslovakia.

An early function of the Prognoz series was to resume the work begun by the First and Second Cosmic Ships in 1959 in identifying the plasmapause or the sharp change in electron density (see Chapter 3), except that now both instruments and data transmission systems were much more sophisticated and the Prognoz could move in and out of the plasmasphere continuously, unlike the single, fast trajectories of the first Moon probes. Orbits of 940–2,000,000 km were suitable for this purpose and as a result, principal investigator Konstantin Gringauz was able to map the plasmapause in detail. One of the early conclusions from the first Prognoz was the concept of noon–midnight asymmetry, which was that the plasmapause was more extended on one side of the Earth (the day side), and correspondingly diminished on the other (the night), during quiet magnetic periods. The second was that there was an identifiable cold inner zone where ions were less than 8,000 K and a warm outer zone of more than 8,000 K.

Some of the individual missions are now described. Prognoz 1 registered its first solar flare on 20th April 1972, only a week after commencing its mission, and measured the speed of the solar wind at 300 km/sec, during a period of solar minimum. It was joined in three months by Prognoz 2 in time for what turned out to be an epic summer. The Sun produced big flares on 22nd July, 2nd (two), 4th, and 7th August. The 4th August was a super-fast surge: the normal speed of the solar plasma was 420 km/sec, but following these events, it was 1,700 km/sec for about 14 hr, at one stage exceeding an unprecedented 2,000 km/sec. Temperatures were up to 10^3 eV, never seen before. The Sun accelerated heavy particles to millions of eV in compact groups towards the Earth. The August 1972 solar flares were among the strongest ever, a real threat to cosmonauts spacewalking [62]. Prognoz 2 carried the French SIGNE 1 experiment for solar radiation studies, but could also detect cosmic gamma-ray bursts. SIGNE stood for Solar International Gamma Ray and Neutron Experiment and it detected the first gamma rays from the Sun during this violent period of 4th and 7th August 1972. This Prognoz collected almost six years of data until it lost stability in March 1978.

One of the tasks of Prognoz 3 was to measure particle fluxes with energy close to the solar wind, but none was found. Prognoz 4 began investigation of collisionless shock waves and discovered a zone of hot plasma on the periphery of the plasmasphere. Using French instruments, Prognoz 5 and 6 measured the arrival, in the Earth's environment, of interstellar wind and determined that it comprised mainly hydrogen and helium, watching the way that it met and interacted with our own solar wind. Prognoz 5 used instruments developed in the Intercosmos solar program to detect superhot (60,000,000 K) plasma within solar flares and it measured the temperature of the interstellar wind [63].

Prognoz 6 carried SIGNE II with detectors to again study solar X-ray, gamma bursts, and cosmic X-rays. It found that solar flares typically released energies of between 10^{29} and 10^{30} ergs. It was able to characterize the different phases of the solar flare from pre-flares to pulses. On 24th September, Prognoz 6 recorded flares, shocks, and then a big flare, with a series of flares and shocks to 12th October 1977, following which the Sun was quiet until the next flare on 22nd November, when it erupted again. The gamma burst detector operated in waiting mode and would only make recordings when activated, which happened on 20th October, 29th October, and 20th November 1977. Prognoz 6's *galaktika* spectrometer operated in the 1100–1900-Å range, the far ultraviolet spectra of the sky background, and studied eight dark and two bright regions, finding that some regions of the Milky Way were 10 times brighter than the darkest ones [64].

Prognoz 7 and 8 marked cooperation with Sweden, where scientists shared their interest in terrestrial magnetism, so their aim was to study the high-latitude magnetopause. Prognoz 7 was the first to make a reliable measurement of the ion mass of the solar wind, identifying the proportions of helium, oxygen, silicon, and iron. Prognoz 7 provided extensive observations of the hitherto rarely explored cusp and mantle of the Earth's magnetic field. The high point of the mission was in March 1979. First, Prognoz 7 observed an extraordinary pulsar burst on 5th March 1979. That day, a pulsar in the Large Magellanic Cloud (LMC) emitted the largest known gamma-ray burst ever, more than all its known predecessors put together, fading after 100 milliseconds, then making 12 pulses at regular 8.1-sec intervals and then stopping. Three more bursts from the same source were later picked up, one 14 hr later, the other two in the two subsequent months – a new class of flaring X-ray pulsar. Prognoz 7 observed the magnetic storm of 22nd March 1979 when proton fluxes trebled to 100 keV in the high latitudes as energy accumulated there. Prognoz 7 found and measured Dispersed Plasma Bursts (DPBs) lasting 20 min each in the tail of the magnetosphere. Prognoz 7 detected, coming from the Sun, heavy oxygen, silicon, and iron ions and noted that they had low speed, high density, and low temperatures.

Prognoz 7 made no fewer than 65,000 measurements of the solar wind between November 1978 and June 1979. Its orbit was such that 70% was in the pure solar wind before it hit Earth's magnetosphere. The main discovery was that the solar wind actually comprised five different streams of the corona, depending on whether they came from coronal holes, streamers, or ejections, all traveling at different speeds. The faster the wind, the more helium was abundant therein. The corona was identified as the specific source of the solar wind.

Prognoz 7 and 8 found that there were heavy ions in the solar wind 90% of the time. The velocity of the solar wind ranged between 298 and 374 km/sec, averaging 348 km/sec. When the Sun was disturbed, the energy of electrons from the Sun could rise from 1.1^{31} to 4.10^{32} ergs, with pressure rising from 10^{-8} dyne/cm^2. The main elements of the solar wind were helium and oxygen, 39%; silicon, 12%; iron, 8%; helium and hydrogen, 3%; oxygen and hydrogen, 6.8%; silicon and hydrogen, 8.6%; and iron/hydrogen, 5.5%. Data from Prognoz 7 and 8 were combined to make a complementary analysis. It was found that when the Sun was least steady, it emitted plasma clouds of heavy ions. Low-velocity solar wind was traced to plasma emerging from equatorial streamers and the loops of the inner corona [65].

Prognoz 8 (January–July 1981) was led by Albert Galeev. He was head of the Space Plasma Physics Department of the Institute for Space Research, IKI (1973–1988), the third director of IKI after the departure of Roald Sagdeev. Albert Galeev is an important figure in Russian space science. He was born in Ufa in 1940, entered the Moscow Energetic Institute in 1957, moving to Novosibirsk to graduate in 1963. His first work was published while still a student in 1962 and was co-authored with Viktor Oraevsky (1935–2006), the later director of IZMIRAN (1989–2003). He taught in Novosibirsk, lecturing in space plasma physics, later moving to IKI in 1973, where he built up the Space Plasma Physics Department so as to develop the theories of fundamental physics in space plasma. Later, he was to direct the Prognoz 10 *Intershock* mission and make a major contribution to Interball (see Chapter 7).

Prognoz 8 recorded 15 solar shocks from January to July 1981, the mildest arriving at 498 km/sec and the strongest, on 17th May, at 846 km/sec. Generally, a shock took 3 hr to pass and then 18 hr to die down. Prognoz 8 crossed the bow shock 10 times and made eight journeys in and out of the magnetic tail, finding that the electric fields fluctuated and had hot plasma clouds in the tail lobes. Prognoz 8 made a bow shock crossing on 2nd November 1981 into solar wind traveling at 360 km/sec and noticed, as it did so, the disturbance in particles and ion sound bursts. The characteristics of the shock point were not well understood at that point, in particular the way in which there was a "shock jump" of 150–200 m/sec, sometimes as much as 700 km/sec, and a concerted attempt was made to study the phenomenon by Prognoz 7 and 8, ISEE 1, Pioneer 10, and IMP 8. Prognoz 8 measured low-energy electrons in the magnetotail and how, during its tail crossings, disturbances caused fluctuations and ring-like ion fields. These outcomes prompted a following mission to be devoted to the bow shock issue [66].

RELIKT AND INTERSHOCK

Prognoz 9 and 10 marked a significant development of the series, the two final missions having distinct objectives: big bang radiation and the bow shock. Prognoz 9 followed an extreme, unusual orbit of 380–720,000 km, twice the distance to the Moon, so far out that it could almost be perturbed into solar orbit (it also had to be constructed carefully so that it did not fall into the Moon's gravity well). Its orbit looped 100 radii out on the tail side of the magnetopause, swinging out to the sunny side. Its mission started on 1st July 1983 and lasted until 26th February 1984.

The headline objective of Prognoz 9 was the attempt to identify fingerprint radiation left over from the "big bang" (hence the title *Relikt*), the existence of which had been theorized 30 years earlier by cosmologists A.G. Doroshkevich and I.D. Novikov (the term "relikt" was invented by Shklovsky). Actual discovery of what they called "cosmic microwave background radiation" on the 7.35-cm waveband went to American astronomers A. Penzias and M. Wilson in 1965, from which they derived the theory that the early universe was hot and dense, but when its temperature fell below 4,000 K, radiation and matter separated, the radiation being the relict of the early formation. Physicist Yakov Zeldovich calculated that its temperature had now cooled to 2.9 K. Its distribution would give us important clues to the most intriguing questions of cosmology: how the universe expanded, its direction and velocity, and how galaxies formed. It was difficult to measure such cosmic microwave background radiation from Earth, for our atmosphere affected radio wave propagation and such radiation could only be measured through a narrow window of radio waves of 3–10 mm. Only limited data could be taken from balloons and aircraft.

The idea of an instrument to measure relict radiation was first proposed by N.S. Kardashev in the early 1970s and it took many years to develop the appropriate instrument with sufficient sensitivity, as the expected anistropy of the microwave background and its non-uniformity would be very small: a hundredth or thousandth of a percent. For this, Prognoz 9 carried two antennæ and an amplifier, which scanned the sky every 2 min in the 8-mm wave band, 36 GHz. The energy required for its operation was 50 W. It was actually the smallest radio telescope ever put into orbit, weighing only 30 kg, and comprised an antenna, detector, amplifier, and switch. Data were stored for downlinks every four days (it required a lot of computer processing to eliminate noise). Every week, the attitude of the Prognoz was adjusted by 7° so as to scan the whole sky within half a year with a resolution of 5.5° (some regions were scanned twice). The *Relikt* experiment lasted from 3rd July 1983 until

Relikt experiment

Prognoz 9

February 1984. Its 27-day orbit (see Table 2.22) was adjusted to minimize the thermal influence of the Earth and Moon.

Some Western reports indicated that the experiment did not work and that the thermal environment of the Earth–Moon system crowded out the background radiation to the point that it could not be detected. Whilst there were some calibration problems, they did not stop publication, with different color levels for hotter and cooler spots and regions, of a map of the celestial sphere of the different intensity levels of background radiation at a conference in Italy when the mission was over. The 60° equatorial belt of the Milky Way could be seen to shine brightly, while in other, dimmer parts, no background radiation could be detected (suggesting the need for a follow-up mission with more sensitive instruments). Moreover, Prognoz 9 determined a final figure for the energy of the early universe, which, for the record, was

2.10^{16} GeV {m scale {4.10^{16} GeV}}.

Relikt was able to make a calculation of the velocity of our galaxy in space – 5,125 km/sec – and figure that we are traveling in the direction of the galactic cluster Virgo at an angle of 50°. It was not possible to come to a conclusion that the universe was endless. The level of cosmic microwave background anistropy was much less than expected, bringing some cosmological models into question, and it did not answer the question of the small non-uniformities of the radiation.

NASA was to take up the baton for this kind of research in 1989 with the famous Cosmic Background Explorer (COBE) (1989) and Europe many years later with the Planck and Herschel missions (2009). Disappointingly for the scientists involved in the mission, Nobel prizes were rightly awarded many years later to John Mather and George Smoot for their work with COBE – but those involved in the pioneering work of *Relikt* went unrecognized [67].

The first mission was supposed to test instruments and methods, with a view to a second *Relikt* mission to be headed by I.A. Strukov of the Institute for Space Research. It was approved in 1986 for launch in 1993, using a new instrument 20 times more sensitive and stationing the probe at the L2 point (1.5 million km on the opposite side of the Sun), sufficiently afar Earth to avoid its thermal radiation. Although *Relikt 2* received formal approval, it did not go ahead for lack of funding – a source of much regret to the scientists. By way of a footnote, the small non-uniformities of the radiation were explained by new computer algorithms. They were presented in 1991 and published the following year in *Monthly Notices of the Royal Astronomical Society* and *Pisma Astronomichesky Zhurnal* ("Astronomy Letters") just a few months after the COBE presentation in the *Astrophysical Journal*.

Prognoz 9 also carried the French SIGNE II M-9 to study cosmic and solar gamma-ray bursts in the 40–8,000-keV energy range (it had been intended to triangulate SIGNE with the earlier Venus missions Venera 13 and 14 but Prognoz was delayed). Prognoz 9 found 75 cosmic gamma-ray bursts, despite an erroneous trigger that set off many false readings. SIGNE found an extremely energetic burst on 1st August 1983 and on 11th October 1983, a burster that repeated on 13th October and thereafter (a repeating burster). Twelve repeater short bursts were analyzed in Sagittarius and they were determined to come from the same source, $10°$ from the galactic center. The precision with which these bursts were marked and located was not bettered for many years. Prognoz 9 measured the sizes and lengths of eight solar flares between July and October 1983 [68].

Prognoz 10 *Intershock* was, unlike most of its predecessors, aimed specifically at the boundary and the bow shock and how particles leaked from the bow shock into the magnetosphere. It returned to the lower apogee of the earlier satellites in the series and indeed the lowest perigee of any mission, but with its apogee well above northern latitudes so as to measure shock waves on the sunlit side of the Earth. Because of the high level of participation from the socialist countries on the mission, especially Czechoslovakia, it was made part of the Intercosmos program and also designated Intercosmos 23. The purpose of the mission was to find out what happened when solar events reached the Earth's magnetic field, what were called bow shock crossings at the point of the shock wave. What happened to the solar particles? Did some break through? Theory predicted that solar wind plasma should be decelerated, heated, and compressed in the relatively thin region before the bow shock, but the mechanisms of these processes was unclear. The aim of the project was therefore to study the internal structure of the bow shock. These tasks required much improved sensitivity and resolution in the instruments and techniques to reduce electromagnetic noise in the spacecraft. Booms with a 22-m span were developed in the Lavochkin bureau to measure electrical and magnetic fields.

The first international meeting on the project was held in 1975 in Czechoslovakia. Devising the instrumentation was slow and difficult. To test, instrumentation was first flown on Prognoz 8, the Soviet–Czechoslovakian MONITOR (for fast measurements of the energy spectrum and solar wind ion flux), and BUD 3K, to measure extremely low-frequency oscillations in the electric field and plasma flux.

Intershock was launched in April 1985 and worked until November 1995. The achievement of *Intershock* was to measure for the first time, in the course of 51 orbits, the fine structure of the collisionless shock. A special conference on the topic was held in Budapest in 1987. *Intershock* found how the ion flux decelerated before the front, split into separate beams, and then mixed again back in the magnetosheath. The solar wind ions comprised elements of oxygen, sulfur, silicon, and iron.

Prognoz 10 measured 65 bow shock crossings. There was a solar flare two days before it was launched and as soon as it got into orbit (26th April), the flare reached the Earth's environment but declined from the 27th. Its initial speed was 600 km/sec, falling to 450 km/sec in a high-speed low-density solar wind stream that caused an expansion of the bow shock 32 radii out, creating a magnetic cloud. In the next crossing, on 7th May 1985, Prognoz 10's instruments found that the bow shock front generated two sets of electrons (5–10 keV, 30–200 keV), which then accelerated away. The crossing on 11th May was transmitted in real time, instruments recording how the solar particles reached the Earth and identified spikes of energetic particles in the bow shock. As a result of Prognoz 10, it was possible to make a map of the shape of the bow shock, the areas of plasma heating behind it, changes in electric fields, and a new area called "the ramp". Finally, Prognoz 10 monitored the way in

Table 2.22. Orbiting solar observatories (Prognoz program).

Prognoz 1	14 Apr 1972	845 kg	965–200,000 km, 65°, 5,782 min
Prognoz 2	29 Jun 1972	845 kg	550–200,000 km, 65°, 5,849 min
Prognoz 3	15 Feb 1973	845 kg	590–200,000 km, 65°, 5,782 min
Prognoz 4	22 Dec 1975	905 kg	634–199,000 km, 65°, 5,740 km
Prognoz 5	22 Nov 1976	930 kg	510–199,000 km, 65°, 5,728 min
Prognoz 6	22 Sep 1977	910 kg	498–197,900 km, 65°, 5,688 min
Prognoz 7	30 Oct 1978	950 kg	464–202,970 km, 65°, 5,881 min
Prognoz 8	25 Dec 1980	950 kg	979–197,369 km, 65°, 5,689 min
Prognoz 9 *Relict*	1 Jul 1983	1,000 kg	380–720,000 km, 65°, 38,448 min
Prognoz 10 *Intershock*	26 Apr 1985	1,000 kg	349–200,000 km, 65°, 5,785 min

All on R-7 rocket from Baikonour. Prognoz 10 was also known as Intercosmos 23.
For Prognoz M, also called Prognoz 11 and 12, see Interball, Chapter 7.

which diurnal radiation levels rose from an average of 19 rads a day at the start of the mission to 28 by November 1985 – an issue that would potentially affect manned spaceflight [69].

Prognoz 10 concluded a decade and a half of observations in which spacecraft charted the structure and dynamics of the magnetic fields; the way that particles precipitated from radiation belts into the atmosphere; the proportions of hydrogen, helium, and oxygen ions; the volume of helium reaching us from the Sun; and the welling of oxygen out of the Earth's atmosphere. Prognoz mapped the convection of particles, tail dynamics, the reconnection and stabilization of the magnetic field. Later, Prognoz was succeeded by the Prognoz M program (see Chapter 7). The missions are summarized in Table 2.22.

EARLY SOVIET SPACE SCIENCE: WHAT WAS LEARNED?

This chapter reviewed the program of unmanned Soviet space science that began in the 1950s. We have seen how the initial explorations of Sputniks 1–3, especially 3, gave way to two dedicated programs to explore the radiation belts (MS and Elektron) and then the beginning of the dedicated Cosmos program of space science in 1962. Although this was not apparent at the time, the Cosmos program had numerous subsets of missions (DS, DS-U) and, in addition, scientific payloads were flown on military missions, the *nauka* modules, and other satellites. Later, dedicated science missions emerged as the Proton series, *Energiya*, *Efir*, *Ionosfernaya*, and Prognoz as well as the collaborative missions with the socialist countries, Intercosmos, which had its own subsets.

But what was actually learned? In terms of the activities actually undertaken, an emphasis on the magnetosphere and the atmosphere has been obvious, not surprising granted the importance of both to the economic development of northern Russia. According to *Soviet Space Achievements* [70], the early Cosmos program enabled a remodeling of the atmosphere, the ionosphere, and the entire radiation belts. The following were the main gains from the deepening of the science programs in the 1960s:

- a full mapping of the Earth's radiation belts, their electrical composition, with their latitudinal variations, anomalies, and peculiar features (e.g. jets); later, discovery of an inner radiation belt;
- a first understanding of the dynamics of the relationship between the Sun and the Earth's radiation belts, especially the points at which that interaction begins, with a specific knowledge of the bow shock;
- the Earth's magnetosphere as complex, dynamic, living, breathing environment;
- composition, density, temperature profile of the space environment above the Earth;
- the Earth as an emitter of radiation;
- detection of a range of energy sources from deep space, such as gamma bursts, pulsars, X-rays, and ultraviolet rays;
- the first detection of high-energy cosmic rays, through the Proton series;
- a growing understanding of the Sun as a complex star, with cycles and patterns of behavior, detailed information on the solar cycle, and the emergence of the concept of "space weather";
- a picture of the atmosphere's dynamics, composition, radiation, and heat budget with estimates of its levels of water vapor, aerosols, ozone, and other features such as nightglow;
- a measurement and map of relict radiation and the big bang remnant;
- identifications of streams of meteors, meteorites, and micrometeorites and their patterns;
- experimental missions, such as the search for quarks, the testing of relativity, and cosmic dust;
- the development of the first model of a space environment reference book covering all known space phenomena (1st edition, 1965; 8th edition by 2007).

REFERENCES

[1] Central Intelligence Agency (CIA): Scientific intelligence report, long range capabilities of the Soviet Union in major scientific fields 1957–67, Monograph I, *Summary Estimate*. Washington, DC, 1959; Central Intelligence Agency (CIA): Scientific intelligence report, long range Soviet scientific capabilities 1962–70, Monograph I, *Geophysical Sciences*. Washington, DC, 1961; Central Intelligence Agency (CIA): Scientific intelligence report, Soviet space research program, Monograph II, *Objectives*. Washington, DC, 1959.

[2] Hugged by the Sun, in *Science & Life, http://nauka.relis.ru* (accessed 26 December 2007).

[3] Temny, Vladimir: *Report of the Annual Conference of the Sergei Vavilov Institute for the History of Natural Sciences and Technology*. Russian Academy of

Sciences, Moscow, 2008; Grigorov, Naum, *et al.*: *Search for Anti-Matter in Cosmic Rays and in Cosmic Space*. NASA, TTF 8,164, 1962.

[4] Roscosmos, *www.roscosmos.ru*, translation by Bart Hendrickx; Gorzhankin, B.N.; Gringauz, Konstantin; Shutte, N.M.: Absorption of ultraviolet solar radiation in the upper atmosphere, in Smith Rose, R.L., ed.: *Space Research*. COSPAR, Paris, 1966.

[5] *Soviet Space Achievements*. Novosti, Moscow, 1965; Gringauz, Konstantin, *et al.*: Some results of measurements carried out by means of charged particle traps, in King-Hele, D.G.; Muller P.; Righini, G., eds: *Space Research*. COSPAR, Paris, 1965; Lewis, Richard S.: *Illustrated Encyclopedia of Space Exploration – a comprehensive history of space discovery*. Salamander, London, 1983; Glushko, Valentin P.: *Kosmonautika, the Great Encyclopedia*. Moscow, 1970, Soviet Encyclopedia collection, 2nd edn; Galeev, Albert; Tamkovich, G.M., eds: *35th Anniversary of the Institute of Space Research of the Russian Academy of Sciences*. Author, Moscow, 1999; Gringauz, Konstantin, *et al.*: Changes of distribution in charged particle intensity in the outer ionosphere since the solar activity maximum according to Cosmos 2, in Muller, P., ed.: *Space Research*, Vol. IV. COSPAR, Paris, 1963; Afonin, V.V., *et al.*: Brief Survey of the Results of Physical Experiments on Cosmos 2 in the Ionosphere, in Skuridin, G.A., *et al.*, eds: *Space Physics*, papers from conference held in Moscow, 10–16 June 1965. NASA, TTF 389; Krassovsky, Valerian: Ionospheric winds and anomalies in the distribution of charged particles in the geomagnetic field. *Cosmic Research*, Vol. 3, 1965; Krassovsky, Valerian: Certain problems of upper atmosphere physics and space near the Earth, in Skuridin, G.A., *et al.*, eds: *Space Physics*, papers from conference held in Moscow, 10–16 June 1965.

[6] Galperin, Yuri; Krassovsky, V.I.: Investigation of the atmosphere using Cosmos 3 and 5, in Muller, P., ed.: *Space Research*, Vol. IV. COSPAR, Paris, 1963; Krassovsky, Valerian: Certain problems of upper atmosphere physics and near Earth space, in Skuridin, G.A., *et al.*, eds: *Space Physics*, papers from conference held in Moscow, 10–16 June 1965; *Soviet Space Achievements*. Novosti, Moscow, 1965; Vernov, Sergei, *et al.*: Measurements of low-energy particle fluxes from the Cosmos and Elektron satellites, in King-Hele, D.G.; Muller, P.; Righini, G., eds: *Space Research*. COSPAR, Paris, 1965; Glushko, Valentin: *Development of Rocketry and Space Technology in the USSR*. Academy of Sciences, Moscow, 1973; Bolyunova, A.D.: Radioactivity from satellite Cosmos 3 after the explosion of 9th July 1962. *Cosmic Research*, Vol. 4, 1966; Vernov, Sergei, *et al.*: Recording charged particles with an energy of 0.1–10keV with a spherical electrostatic analyzer, in Skuridin, G.A., *et al.*, eds: *Space Physics*, papers from conference held in Moscow, 10–16 June 1965; Stern, D.: Yuri Galperin, Russian space research pioneer in the auroral phenomena and solar–terrestrial relations. Proceedings of the Conference in Memory of Yuri Galperin, Boulder, 2004.

[7] Vaisberg, Oleg: Sputnik 1 and something else, in Zakutnyaya, Olga, ed., *Space, the First Step*. IKI, Moscow, 2007; Shafer, Yuri: Radiation effects of the American high altitude nuclear explosion *Starfish* as measured by Cosmos 6.

Cosmic Research, Vol. 5, 1967; Vernov, Sergei, *et al.*: Results of measurement of fast charged particles from Cosmos 137. *Cosmic Research*, Vol. 7, 1969; Shafer, Yuri, *et al.*: Measurement of radiation effects of the thermonuclear explosion in the People's Republic of China on 27th December 1968 measured by Cosmos 259 and 262. *Cosmic Research*, Vol. 9 (4–6), 1971; Savun, O.I.; Shavrin, P.I.: Investigation of electron spectra in the inner radiation belt by Cosmos 219. *Cosmic Research*, Vol. 9 (4–6), 1971.

[8]   Hendrickx, Bart: Elektron – the Soviet response to Explorer. *Quest*, Vol. 8, No. 1, 2000.

[9]   *Soviet Space Achievements*. Novosti, Moscow, 1965; Dolginov, Shmaia, *et al.*: A survey of the Earth's magnetosphere in the region of the radiation belt (3-6 Re) from February to April 1964, in King-Hele, D.G.; Muller, P.; Righini, G., eds: *Space Research*. COSPAR, Paris, 1965; Yeroshenko, Yevgeni: A survey of the Earth's magnetosphere at distances of 7 to 11.7 Earth radii by the Elektron satellites, in King-Hele, D.G.; Muller, P.; Righini, G., eds: *Space Research*. COSPAR, Paris, 1965; Bolyunova, A.D., *et al.*: Corpuscles on Elektron Satellites, in King-Hele, D.G., *et al.*, eds: *Space Research*, Vol. V. COSPAR, Paris, 1965; Vernov, Sergei, *et al.*: Measurements of low-energy particle fluxes from the Cosmos and Elektron satellites, in King-Hele, D.G.; Muller, P.; Righini, G., eds: *Space Research*. COSPAR, Paris, 1965; Galperin, Yuri; Temny, Vladimir: Model intensity distribution of electrons trapped in the inner zone, in Mitra, A.P.; Jacchia, L.G.; Newman, W.S., eds: *Space Research*, Vol. VIII. COSPAR, Paris, 1967; *Spoutnik d'observation du soleil*. Novosti, Moscow, 1968; Sagdeev, Roald Z.: The principal phases of space research in the USSR, in USSR Academy of Sciences, History of the USSR, New Research, 5, *Yuri Gagarin – to mark the 25th anniversary of the first manned spaceflight*. Social Sciences Editorial Board, Moscow, 1986; Panasyuk, Mikhail: Radiation reflections, in Zakutnyaya, Olga, ed., *Space, the First Step*. IKI, Moscow, 2007; Panasyuk, Mikhail: *Cosmic Journeys over 50 Years*. Moscow State University, Moscow, 2007; Moore, T.E.: 50 years observing plasmas in space, in Zakutnyaya, Olga; Odintsova, D., eds: *Fifty Years of Space Research*. Institute for Space Research, Moscow, 2009; Konyakhina, S.S.: *Generation of Nuclei of Cosmic Rays of Solar Origin*. NASA, TTF 14,773, 1973; Dolginov, Schmaia: *Research on the Geomagnetic Field*. NASA, TTF 20,344, 1988; Nazarova, Tatiana: Study of meteoric matter, in Skuridin, G.A., *et al.*, eds: *Space Physics*, papers from conference held in Moscow, 10–16 June 1965.

[10]   Vernov, Sergei, *et al.*: Measurements of low-energy particle fluxes from the Cosmos and Elektron satellites, in King-Hele, D.G.; Muller, P.; Righini, G., eds: *Space Research*. COSPAR, Paris, 1965.

[11]   Benediktov, Yevgeni: Results of radio emission strengths in 725 and 1,525 mc/s by equipment on Elektron 2, in Skuridin, G.A., *et al.*, eds: *Space Physics*, papers from conference held in Moscow, 10–16 June 1965; Artemeva, G.M., *et al.*: Positrons of sources of sporadic radio emission in 0.7–2.3MHZ observed by Elektron 1 and 2. *Cosmic Research*, Vol. 20, 1982.

[12]   Barry, Willam: The missile design bureaux and Soviet manned space policy,

1953–1970. Ph.D. thesis, University of Oxford, 1996.

[13] Zheleznyakov, Alexander: Russian rocket man Mikhail Yangel – an important personality who deserves wider recognition. *Spaceflight*, Vol. 47, May 2005.

[14] Siddiqi, Asif A.: Korolev, Sputnik and the IGY, in Launius, Roger D.; Logsdon, John; Smith, Robert, eds: *Reconsidering Sputnik – forty years since the Soviet satellite*. Harwoood, Amsterdam, 2000.

[15] Logachev, Yuri I.: *40th Anniversary of the Space Age in the Research Institute of Nuclear Physics of Moscow University*. Moscow State University, Moscow, 2009; *Spoutnik d'observation du soleil*. Novosti, Moscow, 1968; Vernov, Sergei, *et al.*: Measurements of low-energy particle fluxes from the Cosmos and Elektron satellites, in King-Hele, D.G.; Muller, P.; Righini, G., eds: *Space Research*. COSPAR, Paris, 1965; Skuridin, G.A., *et al.*: Investigation of radiation intensities from artificial Earth satellite Cosmos 17. *Cosmic Research*, Vol. 4, 1966.

[16] Portree, David: *30 Years Together – US/Soviet cooperation in space*. NASA, 1993.

[17] Regan, R.D.; Davis, W.M.; Cain, J.C.: Detection of intermediate size magnetic anomalies in Cosmos 49, OGO 2, 4 and 6 data, in Rycroft, M.J.; Runcorn, S.K., eds: *Space Research*. COSPAR, Paris, 1972; Dolginov, Shmaia: The first magnetometer in space, in Haerendel, G., *et al.*, eds: *40 Years of COSPAR*. COSPAR.ESA, Paris, 1998; Dolginov, Shmaia, *et al.*: *Experiments on a Program of World Magnetic Survey*. NASA, TTF 14,407, 1972; Dolginov, Schmaia: *Research on the Geomagnetic Field*. NASA, TTF 20,344, 1988.

[18] Skuridin, G.A., ed.: *Mastery of Outer Space in the USSR*. NASA, technical translations, TTF 773; Galeev, Albert; Tamkovich, G.M., eds: *35th Anniversary of the Institute of Space Research of the Russian Academy of Sciences*. Author, Moscow, 1999; Malkevich, M.S.: *Optical Investigations of the Atmosphere using Artificial Satellites*. NASA, TTF 15,186, 1974.

[19] Dimov, N.A.; Severney, A.B.: On determination of night sky brightness from a space vehicle, in Champion, K.S.W.; Smith, P.A.; Smith-Rose, R.L., eds: *Space Research*, Vol. X. COSPAR, Paris, 1968.

[20] Masevich, Alla: First Sputnik, early years of observing artificial Earth satellites, early results, in Zakutnyaya, Olga, ed., *Space, the First Step*. IKI, Moscow, 2007; Hess, Wilmot N.: *The Radiation Belt and Magnetosphere*. Blaisdell, 1968; Sidneva, S.N.; Strelkov, A.S.: Measurement of neutron flux on Cosmos 53. *Cosmic Research*, Vol. 5, 1967.

[21] Marov, Mikhail, *et al.*: Semi-annual density variations in the atmosphere at 200km–300km and structure and motion of the thermosphere deduced from satellite drag, in Bowhill, S.A.; Jaffe, L.D.; Rycroft, M.J.: *Space Research*. COSPAR, Paris, 1971; Marov, Mikhail; Alpherov, A.M.: Diurnal variations in the thermosphere, in Rycroft, M.J.; Runcorn, S.K., eds: *Space Research*. COSPAR, Paris, 1972; Marov, Mikhail: Dynamic nature of atmospheric density at altitudes of 200–300km, in Skuridin, G.A., *et al.*, eds: *Space Physics*, papers from conference held in Moscow, 10–16 June 1965.

[22] Ginzburg, Vitaly: *Experimental Verification of the General Theory of Relativity*

using an Artificial Earth Satellite. Priroda, Moscow, September 1956, in Krieger, F.J., ed.: *Behind the Sputniks – a survey of Soviet space science*. Rand Corporation, Washington, DC, 1960; Rukman, G.I.; Yukhvidin, Yuri A.: Possible experimental verification of relativistic slowing down on a traveling clock by using quantum frequency and time standards. *Cosmic Research*, Vol. 4, 1966; Basov, Nikolai, *et al.*: Operating equipment with a satellite-borne maser oscillator. *Cosmic Research*, Vol. 5, 1967.

[23] Mazets, E.P.: Cosmic dust meteor showers, in Kondratyev, Kirill; Mycroft, M.J.; Sagan, Carl: *Space Research*. COSPAR, Paris, 1970; Leontyev, L.V., *et al.*: Some peculiarities of cosmic dust distribution, in Bowhill, S.A.; Jaffe, L.D.; Rycroft, M.J.: *Space Research*. COSPAR, Paris, 1971; Konstantinov, Boris: The possible anti-matter nature of meteorites. *Cosmic Research*, Vol. 4, 1966; Micrometeroid investigations on the satellite Cosmos 135. *Cosmic Research*, Vol. 5, 1967; Micrometeors in circumterrestrial space observed by Cosmos 163. *Cosmic Research*, Vol. 7, 1969; Investigation of γ radiation by Cosmos 135 and investigation of variations of annihilation by Cosmos 135 in connection with possible anti-matter. *Cosmic Research*, Vol. 8 (4–6), 1970. For an account of the origins of the mission, see Shklovsky, Iosif: *Five Million Vodka Bottles to the Moon*. Norton, New York and London, 1991. For an account of astrophysical lasers and masers, see Letokhov, Vladilen; Johansson, Sveneric: *Astrophysical Lasers*. Oxford University Press, Oxford, 2009. Boris Konstantinov's posthumous book is Konstantinov, Boris; Pekelis, V.D., eds: *Inhabited Space*. NASA, TTF 819, originally published as *Naselennyy Kosmos*, Nauka, 1972.

[24] Mazets, Y.P., *et al.*: *Flash Up of Cosmic Gamma Radiation from Observations On Board Cosmos 461*. NASA, TTF 15,790; Ivanov, V.D., *et al.*: Non-flare solar X-ray emissions shorter than 4Å, in Kondratyev, Kirill; Mycroft, M.J.; Sagan, Carl: *Space Research*. COSPAR, Paris, 1970; Kurnosova, Lidia: The nature of radiation background at 250–500km. *Cosmic Research*, Vol. 9 (1), 1970.

[25] Zelinsky, Y.B.; Tatevian, S.K.: Contribution of space techniques to scientific progress in geodesy and geodynamics, in Zakutnyaya, Olga; Odintsova, D., eds: *Fifty Years of Space Research*. Institute for Space Research, Moscow, 2009.

[26] Aksenov, V.I., *et al.*: Investigation of transmission of ultralong radio waves through the Earth's ionosphere – preliminary results from Cosmos 142. *Cosmic Research*, Vol. 8 (4–6), 1970.

[27] Vernov, Sergei, *et al.*: Investigation of Earth's radiation belts in the region of the Brazil magnetic anomaly at 200–400km, in King-Hele, D.G.; Muller, P.; Rinhini, G., eds: *Space Research*, Vol. V. COSPAR, Paris, 1965; and Some results of radiation measurements carried out over 1960–3 at 200–400km. *Cosmic Research*, Vol. 2, 1964; Kuzhevsky, B.M., *et al.*: Cosmic research on Cosmos 19 and 25. *Cosmic Research*, Vol. 5, 1967; Vasilova, R.N.: Study of cosmic rays during flights of spacecraft satellites and the Cosmos artificial Earth satellites, in Skuridin, G.A., *et al.*, eds: *Space Physics*, papers from conference held in Moscow, 10–16 June 1965; Aksenov, V.I., *et al.*: Investigation of VLF radio propagation in the ionosphere by Cosmos 142 and 259, in Kondratyev,

Kirill; Mycroft, M.J.; Sagan, Carl: *Space Research*. COSPAR, Paris, 1970.

[28] Romanovsky, Y.A.; Katyushina, V.V.: Thermospheric composition and temperature variations during magnetic disturbances, in Rycroft, M.J.; Reasenberg, R.D., eds: *Space Research*. COSPAR, Paris, 1973; Romanovsky, Yuri A., *et al.*: Mass spectrometer measurements of F2 region ion composition from Cosmos 274, in Rycroft, M.J., ed.: *Space Research*. COSPAR, Paris, 1974; Romanovsky, Yuri, *et al.*: One some features of the bottomside F2 region ion composition in the equatorial ionosphere, in Mycroft, M.J., ed.: *Space Research*, Vol. XVI. COSPAR, Paris, 1975; Anashkin, O.P.; Kurnosova, Lidia, *et al.*: Trial of superconducting magnetic system on board artificial Earth satellites Cosmos 140 and 213. *Cosmic Research*, Vol. 7, 1969; Babkov O.I., *et al.*: Measurement of ion density in the Earth's atmosphere from 200km to 6,000km, and Borisov, S.I.; Nikolayev, V.D.: Ion density fluctuations at heights of 200–1,300km in the ionosphere. *Cosmic Research*, Vol. 10 (4–6), 1972; Krasnopolsky, Vladimir: Concentrations of nitric oxide and atmospheric nitrogen in Earth's atmosphere as a function of solar activity. *Cosmic Research*, Vol. 17, 1979; Iozenas V.A.; Krasnopolsky, Vladimir: Some ozonosphere characteristics determined from satellite observation data, in Donahue, T.M.; Smith, P.A.; Thomas, L.: *Space Research*, Vol. X. COSPAR, Paris, 1969; Lebedinsky, Alexander, *et al.*: Spectrum of Earth's heat radiation according to Cosmos 45, 65 and 92, in Smith Rose, R.L., ed.: *Space Research*. COSPAR, Paris, 1966; Lebedinsky, Alexander, *et al.*: Earth's ultraviolet spectrum according to Cosmos 65, in Smith Rose, R.L., ed.: *Space Research*. COSPAR, Paris, 1966; Bashulin, P.A., *et al.*: Cosmos 45 measurement of angular and spatial distribution of Earth's infrared radiation. *Cosmic Research*, Vol. 4, 1966; Marov, Mikhail, *et al.*: Water vapour in the mesosphere according to measurements from rockets and artificial Earth satellites Cosmos 45 and 65. *Cosmic Research*, Vol. 7, 1969. Krasnopolsky, Vladimir; Lebedinsky, Alexander: Measurements of night airglow by Cosmos 92, in Smith Rose, R.L., ed.: *Space Research*. COSPAR, Paris, 1966; Krasnopolsky, Vladimir: Features of ultraviolet night airglow according to Cosmos 92, in Donahue, T.M.; Smith, P.A.; Thomas, L.: *Space Research*, Vol. X. COSPAR, Paris, 1969.

[29] Babkov O.I., *et al.*: Measurement of ion density in the Earth's atmosphere from 200km to 6,000km, and Borisov, S.I.; Nikolayev, V.D.: Ion density fluctuations at heights of 200–1,300km in the ionosphere. *Cosmic Research*, Vol. 10 (4–6), 1972.

[30] Powell, Joel W.: Nauka modules. *Journal of the British Interplanetary Society*, Vol. 41, No. 3, March 1988; Clark, Phillip S.: Classes of Soviet/Russian reconnaissance satellites. Paper presented to British Interplanetary Society, 2 June 2001.

[31] Bratolubova–Tzulukidze, L.S., *et al.*: Measurement of high energy gamma ray intensity in primary cosmic rays on Cosmos 208, in Kondratyev, Kirill; Mycroft, M.J.; Sagan, Carl: *Space Research*. COSPAR, Paris, 1970; Anisimov, M.M., *et al.*: X-ray and gamma ray observations from artificial Earth satellites, in Vernov, Sergei; Kocharev, G.E. eds: *Proceedings of the VIth winter school in*

space physics, Apatity, 18 March–1 April 1969, Part 1.

[32] Radio Moscow, 16th March 1972; Basharinov, Anatoli: Some results of microwave sounding of the atmosphere and ocean from Cosmos 243, in Kondratyev, Kirill; Mycroft, M.J.; Sagan, Carl: *Space Research.* COSPAR, Paris, 1970; Basharinov, Anatoli, *et al.*: *Radio Remission of the Earth as a Planet.* NASA, TTF 16,078, 1975; Basharinov, Anatoli, *et al.*: *Radiation Temperatures of the Earth's Mantle in the Superhigh Frequency and Infrared Regions Based on Data from an Experiment on Cosmos 384.* NASA, TTF 14,734, 1973; Akvilonova, A.N., *et al.*: *Study of Cloudiness Parameters Based on Measurements from the Cosmos 384 Satellite.* NASA, TTF 14,735, 1973; Basharinov, Anatoli, *et al.*: *Microwave Radiation Characteristics of Dry and Moisty Ground Covers.* NASA, TTF 44,975, 1973; Malkevich, M.S.; Gurvich, A.S.: *Estimation of the Atmospheric Effects in the Problem of Cosmic Study of the Natural Resources of the Earth by Emission Measurements from Cosmos Satellites.* NASA, TTF 14,977.

[33] Salamonovich, A.E., *et al.*: Satellite measurements of submillimetre radiation of Earth's atmosphere, in Mycroft, M.J., ed.: *Space Research*, Vol. XVI. COSPAR, Paris, 1975.

[34] Gregoryan, O.R., *et al.*: Precipitating particles and electric fields in the polar ionosphere, in Mycroft, M.J.: *Space Research*, Vol. XVI. COSPAR, Paris, 1975; Blokhintsev, I.D., *et al.*: Cosmos 856 and 914 satellite measurements of high energy γ rays. *Cosmic Research*, Vol. 20, 1982; Kuznetsov, S.N., *et al.*: Pulsations of quasi-captured electrons. *Cosmic Research*, Vol. 19, 1981; Grigoryan, O.R., *et al.*: Outpouring of energetic particles and phenomenon in peripheral regions of the magnetosphere. *Cosmic Research*, Vol. 16, 1978; Kuznetsov, S.N.; Stolpovsky, V.G.: Relativistic electrons in the auroral zone. *Cosmic Research*, Vol. 16, 1978.

[35] Panasyuk, Mikhail: *Cosmic Journeys over 50 Years.* Moscow State University, Moscow, 2007.

[36] Ivanov, G.V., *et al.*: Investigations of variations of ion composition and dynamics of the topside ionosphere from Meteor, in Rycroft, M.J., ed.: *Space Research.* COSPAR, Paris, 1978; Kovalskaya, I.Y., *et al.*: Features of intensity variation and spectrums of low energy protons in the polar cusp, in Rycroft, M.J., ed.: *Space Research.* COSPAR, Paris, 1974; Grafodatsky, O.S., *et al.*: Observation of a diffusion wave of energetic electrons in the outer radiation belt in June 1986. *Geomagnetism and Aeronomy*, Vol. 31, No. 6, 1991.

[37] Shevchenko, V.: Moon research for half a century, in Zakutnyaya, Olga; Odintsova, D., eds: *Fifty Years of Space Research.* Institute for Space Research, Moscow, 2009.

[38] Panasyuk, Mikhail: *Cosmic Journeys over 50 Years.* Moscow State University, Moscow, 2007.

[39] Monatersky, R.: New radiation belt spotted around the Earth. *Science News*, June 1993; Mikhailov, Vladimir: Experiment Mariya 2. Unpublished presentation.

[40] Sharma, Jagannath: *Space Research in India.* Physical Research Laboratory,

Ahmedabad, 1961.

[41] Central Intelligence Agency (CIA): Scientific intelligence report, long range capabilities of the Soviet Union in major scientific fields 1957–67, Monograph I, *Summary Estimate*. Washington, DC, 1959.

[42] Vereschetin, V.; Rimsha, M.: Intercosmos – twenty years on. *Science in the USSR*, 1987, No. 6, November–December 1987; Massevich, A.G.; Hawal, M.J.: Intercosmos laser-ranging stations, in Rycroft, M.J., ed.: *Space Research*. COSPAR, Paris, 1976.

[43] Blamont, Jacques: From Sputnik 1 – to where?, in Zakutnyaya, Olga, ed., *Space, the First Step*. IKI, Moscow, 2007.

[44] Clark, Phillip S.: The Skean program. *Spaceflight*, Vol. 20, No. 8, August 1978.

[45] Wernik A., *et al.*: Anistropy and energy spectra of superthermal elections at 214–650km, in Kondratyev, Kirill; Mycroft, M.J.; Sagan, Carl: *Space Research*. COSPAR, Paris, 1970; Shtern, M.I.: *Investigations of the Upper Atmosphere and Outer Space Conducted in 1970 in the USSR*. Report to the 14th meeting of COSPAR. NASA, TTF 666, 1971.

[46] Sheldon, Charles S., II: *Interkosmos*, published by globalsecurity.org, 2009.

[47] Bankov, L.; Dachev, T.: Longitudinal specifics of irregularity distribution in equatorial ionosphere, in Rycroft, M.J.: *Space Research*. COSPAR, Paris, 1977. The solar series of Intercosmos missions is reviewed in Sylwester, Janusz: *Impact of the Intercosmos Program on the Past, Present and Future of Solar Space Research*. Space Research Centre, Polish Academy of Sciences, Wrocklaw, Poland, 2001.

[48] Kutiev, I., *et al.*: NO+ ions in equatorial orbit, in Rycroft, M.J., ed.: *Space Research*, COSPAR, Paris, 1977.

[49] Bano, M., *et al.*: Spatial distribution of charged particles beneath the radiation belts as measured on board Intercosmos 3, in Rycroft, M.J.; Reasenberg, R.D.: *Space Research*. COSPAR, Paris, 1973; Lichter, Yuri., *et al.*: Early results from the Intercosmos 5 VLF experiment, in Rycroft, M.J.; Runcorn, S.K., eds: *Space Research*. COSPAR, Paris, 1972; Moore, T.E.; Wilczynski, J.: *Technology in Comecon*. McMillan, London, 1974; Best, A., *et al.*: Interpretation of coordinated electron density, temperature and very long frequencies on Intercosmos 10, in Rycroft, M.J., ed.: *Space Research*. COSPAR, Paris, 1978; Lichter, Yuri., *et al.*: ELF-VLF emissions, ion density, fluctuations and electron temperature in the ionospheric trough, in Rycroft, M.J., ed.: *Space Research*. COSPAR, Paris, 1978; Lichter, Yuri., *et al.*: Proton whistlers in the equatorial F region, in Rycroft, M.J.; Reasenberg, R.D.: *Space Research*. COSPAR, Paris, 1973; Kudela, Karel; Matisin, Jan: Outer zone electron boundary according to Intercosmos 13 measurement. *Studia Geog & Geod.*, Vol. 26, 1982; Larkina, V.I.: Low frequency emissions on board Intercosmos satellites related to ring current variations. Paper presented at Wrocaw, Poland, 23–25 June 1998; Zacharov, I., *et al.*: Results of investigating meteoric matter on the Intercosmos 14 satellite. *Astronomical Institutes of Czechoslovakia Bulletin*, Vol. 32, No. 4, 1981. *Kopernik* results come from Hanasz, J., *et al.*: *Low Frequency Solar Radio Bursts Observed with the Intercosmos-Kopernik 500 Satellite*. Polish Academy of

Sciences, Torun and Institute of Radio Engineering and Electronics, Moscow; Gdalevich, Gennadiy, *et al.*: *Studies of the Equatorial Anomaly in the F Region and the Upper Atmosphere with Spherical Ion Traps on the Intercosmos 2 Satellite.* NASA, TTF 16153.

[50] Siddiqi, Asif: A secret uncovered – the Soviet decision to land cosmonauts on the Moon. *Spaceflight*, Vol. 46, No. 5, May 2004.

[51] Lardier, Christian: Les 50 ans de Youjnoe a Dnepropetrovsk, *Air & Cosmos*, No. 1987, 10 June 2005; Lardier, Christian: Les 1039 satellites de la NPO PM de Krasnoiarsk, *Air & Cosmos*, No. 1713, 27 August 1999.

[52] Misyura, V.A., *et al.*: Ionospheric electron content and its horizontal gradients at middle and high latitudes, in Rycroft, M.J.; Reasenberg, R.D.: *Space Research*. COSPAR, Paris, 1973; Misyura, V.A., *et al.*: Complex ionospheric observations, in Mycroft, M.J.: *Space Research*, Vol. XVI. COSPAR, Paris, 1975; Benediktov, E.A.: Reception of signals from Cosmos 381 from magnetically conjugate regions. *Cosmic Research*, Vol. 10 (1–3), 1972; Getmantsev, G.G., *et al.*: Measurement of imhomogeneity parameters in ionospheric electron concentrations by Cosmos 381. *Cosmic Research*, Vol. 11 (1), 1974; Shtern, M.I.: *Investigations of the Upper Atmosphere and Outer Space Conducted in 1970 in the USSR.* Report to the 14th meeting of COSPAR. NASA, TTF 666, 1971; Avakyan, S.A.: Doubly charged ions of atomic oxygen in the disturbed ionosphere. *Cosmic Research*, Vol. 16, 1978; Komrakov, G.P., *et al.*: Results of measurements of large-scale electron concentrated irregularities in magnetically quiet and disturbed conditions by Cosmos 381. *Cosmic Research*, Vol. 15, 1977; Komrakov, G.P., *et al.*: Electron densities observed at 1,000km by Cosmos 381 during calm and disturbed conditions. *Cosmic Research*, Vol. 14, 1976.

[53] Karpachev, A.T.; Demimova, G.F.; Pulinets, Sergei: Ionospheric changes in response to IMF variation. *Journal of Atmospheric & Terrestrial Physics*, Vol. 57, No. 12, 1995; Ponomarov, V.N., *et al.*: Velocity of ordered motion of ions at 600km. *Cosmic Research*, Vol. 8 (4–6), 1970.

[54] Shuiskaya, F.K.; Mularchik, Tatiana: Superthermal electrons in the polar ionosphere, in Mycroft, M.J.: *Space Research*, Vol. XVI. COSPAR, Paris, 1975; Gringauz, Konstantin; Gdalevich, Gennadiy: Investigations in the ionosphere on Cosmos 378. *Geomagneticsm & Aeronomy*, No. 14, 1974; Ozerov, V.D.: Properties of high latitude irregularities based on Cosmos 378, in Mycroft, M.J.: *Space Research*, Vol. XVI. COSPAR, Paris, 1975; Afonin, V.V., *et al.*: Some peculiarities of the ionospheric plasma in the southern polar cusp, in Mycroft, M.J.: *Space Research*, Vol. XVI. COSPAR, Paris, 1975; Gringauz, Konstantin: Observation of electron fluxes and related variations of ionospheric plasma parameters in south polar cusp, in Rycroft, M.J.; Runcorn, S.K., eds: *Space Research*. COSPAR, Paris, 1972; Shtern, M.I.: *Investigations of the Upper Atmosphere and Outer Space Conducted in 1970 in the USSR.* Report to the 14th meeting of COSPAR. NASA, TTF 666, 1971.

[55] For accounts of the series and its background, see Shelton, William: *Soviet Space Exploration – the first decade.* Arthur Baker, London, 1969; Siddiqi, Asif

A.: Korolev, Sputnik and the IGY, in Launius, Roger D.; Logsdon, John; Smith, Robert, eds: *Reconsidering Sputnik – forty years since the Soviet satellite.* Harwoood, Amsterdam, 2000; Panasyuk, M.I.: Cosmic rays are wanderers of the universe, in Zakutnyaya, Olga; Odintsova, D., eds: *Fifty Years of Space Research.* Institute for Space Research, Moscow, 2009; Grigorov, Naum: Problems and perspectives in cosmic ray research. *Cosmic Research,* Vol. 2, 1964; Volodichev, N.N.: Solar cosmic burst of 7th July 1966 and its measurement by Proton 3. *Cosmic Research,* Vol. 10 (4–6), 1972.

[56] Grigoriev, Naum, *et al.*: Study of energy spectra of primary cosmic rays at very high energies in the Proton series of satellites, in Bowhill, S.A.; Jaffe, L.D.; Rycroft, M.J.: *Space Research.* COSPAR, Paris, 1971; Mikhailov, Vladimir: Experiment Mariya 2. Unpublished presentation, INCOS/MEPHI; Shtern, M.I.: *Investigations of the Upper Atmosphere and Outer Space Conducted in 1970 in the USSR.* Report to 14th meeting of COSPAR. NASA, TTF 666, 1971; Grigorov, Naum: Electrons with energies of hundreds of megaelectron volts in near Earth space. *Cosmic Research,* Vol. 20, 1982; Logachev, Yuri: *Exploration of Space – the first 50 years of the cosmic era.* Moscow, 2007 (in Russian).

[57] For background and analysis, see Clark, P.S.: The Soviet space year of 1983. *Journal of the British Interplanetary Society,* Vol. 38, No. 1, January 1985. Clark, P.S.: The Soviet space year of 1984. *Journal of the British Interplanetary Society,* Vol. 38, No. 8, August 1985; Clark, Phillip S.: Classes of Soviet/Russian reconnaissance satellites. Paper presented to British Interplanetary Society, 2 June 2001.

[58] Wolczek, Olgierd: Poland in the Intercosmos program. *Spaceflight,* Vol. 22, No. 5, May 1980; Logachev, Yuri: The beginning of the space era at the Skolbeltsyn Institute of Nuclear Physics, in Zakutnyaya, Olga, ed., *Space, the First Step.* IKI, Moscow, 2007; Logachev, Yuri I.: *40th Anniversary of the Space Age in the Research Institute of Nuclear Physics of Moscow University.* Moscow State University, Moscow, 2009; Nymmik, Rikho: Detection of electrons of over 10^{12}eV in primary cosmic rays. *Cosmic Research,* Vol. 129, 1981.

[59] Panasyuk, Mikhail: *Cosmic Journeys over 50 Years.* Moscow State University, Moscow, 2007.

[60] Vaisberg, Oleg: Sputnik 1 and something else, in Zakutnyaya, Olga, ed., *Space, the First Step.* IKI, Moscow, 2007.

[61] Kopik, A.: Research data from the Interball projects are still requested today. *Novosti Kosmonautiki,* No. 10, 285, 2006.

[62] Cambou, F., *et al.*: Characteristics of interplanetary plasma near Earth observed during the solar events of August 1972, in Rycroft, M.J., ed.: *Space Research.* COSPAR, Paris, 1974; Sagdeev, Roald Z.: The principal phases of space research in the USSR, in USSR Academy of Sciences, History of the USSR, New Research, 5, *Yuri Gagarin – to mark the 25th anniversary of the first manned spaceflight.* Social Sciences Editorial Board, Moscow, 1986.

[63] Galeev, Albert; Tamkovich, G.M., eds: *35th Anniversary of the Institute of Space Research of the Russian Academy of Sciences.* Author, Moscow, 1999; Sylwester, Janusz: *Impact of the Intercosmos Program on the Past, Present and Future of*

Solar Space Research. Space Research Centre, Polish Academy of Sciences, Wrocklaw, Poland, 2001; Burgin, M.S., *et al.*: Helium atoms in interstellar and interplanetary medium. *Cosmic Research*, Vol. 21, 1983.

[64] Archangelsky, V.V., *et al.*: Observation of solar x-radiation on Prognoz 6, in Rycroft, M.J., ed.: *Space Research.* COSPAR, Paris, 1978; Kurt, Vladimir: Analysis of energetic particle events following solar flares, in Rycroft, M.J., ed.: *Space Research.* COSPAR, Paris, 1978; Zvereva, A., *et al.*: Prognoz 6 data about ultraviolet sky background in dark and Milky Way regions, in Hudson, H.S., ed.: *Advances in Space Research*, Vol. 1, No. 13, Paris, 1981.

[65] Belotserkovsky, O.: Window on the universe. *Soviet Weekly*, 15 January 1983; Shea, M.A.; Smith, E.J.: The international heliospheric study. *Advances in Space Research*, Vol. 9, No. 4, 1989; Yermolaev, Y.: Variations of solar wind, photon and alpha particle hydrodynamic parameters in Prognoz 7 observations; Zastenker Georgi, *et al.*: Large and middle scale phenomenon in the interplanetary medium; Kovtunenko, Vyacheslav: Radiation factors in space-flight – a system for their monitoring, in Russell, C.T., ed.: The magnetosheath. *Advances in Space Research*, Vol. 14, No. 7, July 1994; Yermolaev, Yuri I.; Zastenker, Georgi: Differential flow between protons and alphas in the solar wind: Prognoz 7 observations. *Journal of Geophysical Research*, Vol. 99, No. 12; Yermolaev, Yuri I.; Stupin, Vitaly V.: Helium abundance and dynamics in different types of solar wind streams: Prognoz 7 observations. *Journal of Geophysical Research*, Vol. 102, No. 2; Badalyan, O.G., *et al.*: Study of helium abundance on low-speed solar wind streams from the data of Prognoz 7 and 8 satellites. *Cosmic Research*, Vol. 37, No. 2; Omelchenko, A.N.; Vaisberg, Oleg; Rassel, K.T.: Further analysis of plasma bursts in Earth boundary later at high altitude. *Cosmic Research*, Vol. 21, 1983; Vaisberg, Oleg, *et al.*: Prognoz 7 observations of heavy ions in the solar wind. *Cosmic Research*, Vol. 18, 1980.

[66] Borodkova, N.L.: Interplanetary shock waves in post solar maximum, in de Jager, C.; Svestka, Z. eds.: *Advances in Space Research*, Vol. 6, No. 6, *Physics of Solar Flares.* COSPAR, Paris; Klimov, S.I., *et al.*: Comparative study of plasma wave activity in plasma sheet boundary and near-Earth plasma sheet, in Russell, C.T., ed.: *Solar Wind Interactions, Advances in Space Research*, Vol. 6, No. 1, 1986, Paris; Smirnov, V.N.; Vaisberg, O.L.: Further analysis of non-linear density of fluctuations in the foot of quasi-perpendicular shock, in Russell, C.T., ed.: Physics of collisionless shocks. *Advances in Space Research*, Vol. 15, No. 8–9, 1995; Popielawska, B., *et al.*: On low energy electrons in the magnetotail, in Sandahl, A.; Saunders, M.A., eds: Auroral and related phenomena, *Advances in Space Research*, Vol. 13, No. 4, April 1993; Zastenker Georgi, *et al.*: Bow shock motion in two-point observations, in Russell, C.T., ed.: Multipoint magneto-spheric measurement, *Advances in Space Research*, Vol. 8, No. 9–10, 1988.

[67] Strekhov, I., *et al.*: CMB anistropy testing from small satellites, in Trümper, J.; Cesarsky, C.; Palumbo, G.G.C.; Bignanmi, G.F., eds: Space astronomy, *Advances in Space Research*, Vol. 13, No. 12, December 1993; Simpson, Clive: Discovering the universe. *Spaceflight*, Vol. 50, No. 7, July 2008; Strukov, I.A.: Relikt experiment. *Earth & the Universe*, 4/84.

[68] Boer, M., *et al.*: Signe II gamma ray burst exposure on Prognoz 9, in Hurley, K.; Vedrenne, G., eds: Gamma ray astronomy, *Advances in Space Research*, Vol. 6, No. 4, 1986; Galeev, Albert; Tamkovich, G.M., eds: *35th Anniversary of the Institute of Space Research of the Russian Academy of Sciences*. Author, Moscow, 1999; Atteia, J.L., *et al.*: Discovery of a source of repeated soft short gamma bursts in Sagittarius. *Letters in Astronomy Journal*, Vol. 13, November 1987, summarized in NASA Technical Reports.

[69] Geranios, A., *et al.*: Energetic particle events near Earth after the flare of 24th April 1985, in Shea, M.A.; Smith, E.J.: The international heliospheric study. *Advances in Space Research*, Vol. 9, No. 4, 1989; Galeev, Albert: *Project Intershock – complex analysis of bow shock crossing of 7th May 1985*, in Russell, C.T., ed.: *Solar Wind Interactions, Advances in Space Research*, Vol. 6, No. 1, 1986.

[70] *Soviet Space Achievements*. Novosti, Moscow, 1965; *Soviet Space Studies*. Novosti, Moscow, 1983.

RUSSIAN-LANGUAGE REFERENCES

1. Агапов В. К запуску первого ИСЗ серии «ДС». // Новости космонавтики, №06, 1997.
2. Зайцев Ю.И. Спутники «Космос» – М., Наука, 1975.
3. Логачев Ю.И. Исследования космоса в НИИЯФ МГУ: Первые 50 лет космической эры. / Ю.И. Логачев; под ред. проф. М.И. Панасюка – Москва, 2007.
4. Губарев Владимир. Окна из будущего. Судьба науки и ученых в России – М.: ИКЦ «Академкнига», 2007.
5. Шкловский И.С. Эшелон – М., 1991.
6. Скулачев Дмитрий, Они были первыми – Наука и жизнь, №6, 2009.
7. Струков И.А., Кремнев Р.С., Смирнов А.И. Взгляд в прошлое Вселенной – Наука в СССР, №4, 1992.
8. Застенкер Г.Н. О проекте «Интершок» – Обратный отсчет времени, М., 2006.

3

Revealing the Moon

Beyond the Earth and its environment, the Moon was the next target of Soviet space science. As far back as 1954, when he wrote *Report on an Artificial Satellite of the Earth*, Mikhail Tikhonravov outlined how a rocket could send a satellite not just into Earth orbit, but on its way to the Moon. In February 1956, the State University of Leningrad convened a national conference to review the state of knowledge of the Moon and the planets. In April 1956, at the Academy of Sciences conference that reviewed the sounding rocket program ("On Rocket Research into the Upper Layers of the Atmosphere"), Sergei Korolev outlined how it would be possible to fly a rocket to the Moon in the "not too distant future". In January 1958, Korolev and Tikhonravov formally sent proposals to the government for the construction of the first spacecraft to be sent to the Moon ("On the Launches of Rockets to the Moon") and, with the United States also announcing its intention to send rockets to the Moon, this was quickly approved. The first Soviet attempts to launch Moon rockets failed, success coming on the fourth attempt, in January 1959. Leaving out technical challenges, the Moon program was a serious scientific undertaking. First, lunar probes would be the first artificial objects to leave near-Earth space and enter interplanetary space, the parameters of which had hitherto been assessed only theoretically. Second, the Moon itself was an intriguing object, for its composition might be a clue to the origin of the Moon and the Earth–Moon system, not to mention the far side of the Moon, which had never been observed before.[1]

FIRST COSMIC SHIP: DISCOVERY OF THE SOLAR WIND

The first successful Moon probe, called the First Cosmic Ship at the time but later renamed Luna 1, was the same spherical shape as Sputnik, but much larger, with a

---

[1] This chapter focuses, as do the other chapters, on the scientific outcomes. Readers interested in the technical aspects of the missions should read this author's *Soviet and Russian Lunar Exploration*, while readers interested to follow the technical aspects of planetary missions should read *Russian Planetary Exploration* (Praxis/Springer, 2007).

Table 3.1. First and Second Cosmic Ship instruments.

Gas component detector
Magnetometer (fields of Earth and Moon)
Meteoroid detector
Cosmic ray detector
Ion traps (four)
1 kg of sodium vapor

diameter of 80 cm (compared to 56 cm). The final payload, dispatched in its cone to the Moon on 2nd January 1959, weighed 361 kg, but the actual Moon probe was 156 kg. The sphere was pressurized with four antennae and scientific instruments on top.

The First Cosmic Ship spacecraft carried instruments for measuring radiation, magnetic fields, and meteorites (Table 3.1). The magnetometer was only the second carried by a Soviet spaceship, following Sputnik 3, but installed this time to detect magnetic fields around the Moon. The magnetometer was called a triaxial fluxgate magnetometer, with three sub-instruments and sensors with a range of −3,000 to +3,000 γ. The ion traps and cosmic ray detectors were the descendants of those flown on the balloons in the 1930s and the *Akademik* flights in the 1940s. The meteoroid detector was developed by Tatiana Nazarova of the Vernadsky Institute. Essentially, it comprised a metal plate on springs to record any impact, however tiny. But nobody thought that what the spacecraft would discover on the way out to the Moon was more important than when it reached its destination.

But would the probe even hit the Moon? As it headed deeper into space, it was possible to use the radio signals to make very precise measurements of its direction and velocity. Astronomer Iosif Shklovsky had, the previous year, suggested to Sergei Korolev that an artificial comet could give a more precise idea. So, an R-5 rocket was fired from Kapustin Yar on 19th September 1958 to 430 km, where it released a sodium cloud, which expanded as a bright orange light over 20° of the sky over the following half-hour, scattering solar radiation in the wavelengths 5890–5896 Å. On 3rd January, some 113,000 km out from Earth, the First Cosmic Ship released a golden-orange cloud of sodium gas so that astronomers could track it. The cloud was visible in the sky over the Indian Ocean and it confirmed that the probe would come quite close to the Moon – but would miss it [1]. As a tracking mechanism, it worked better than expected and Shklovsky was later awarded the Lenin prize for the idea. This experiment provided the basic data for Vladimir Kurt's candidate dissertation, in which he calculated the atmosphere's density on the basis of sodium vapor diffusion.

On 4th January, the First Cosmic Ship passed by the Moon at a distance of 5,965 km some 34 hr after leaving the ground. It went on into orbit around the Sun between the Earth and Mars between 146.4 million and 197.2 million km. There, it was renamed *Mechta* (Russian for "dream"). Signals were picked up for 62 hr, after which the battery presumably gave out, at which point the probe was 600,000 km away. Scientifically, this was a bonus, because it enabled signals to be sent and received for much longer than had the First Cosmic Ship actually hit the Moon.

Konstantin Gringauz, discoverer of the solar wind

The results of the mission of the First Cosmic Ship were published by mission scientists Sergei Vernov and Alexander Chudakov in *Pravda* on 6th March 1959, followed by more information from the President of the Academy of Sciences, Alexander Nesmyanov, opening the Academy's annual general meeting that spring, which ran from 26th to 28th March. First, no magnetic field was detected near the Moon, but scientists were aware that the spacecraft was possibly too far out to detect one. Instead, the magnetometer noted surprising fluctuations in the Earth's magnetic field as the First Cosmic Ship accelerated away into deep space. A full contour map of the Earth's radiation belts was published, showing them peak at 800 γ some 24,000 km out and then fall away to a low level some 50,000 km out (Figure 3.1). The ion trap showed a sharp drop-off some 20,000 km out and a fall in plasma levels 30,000 km out. Second, the meteoroid detector, which

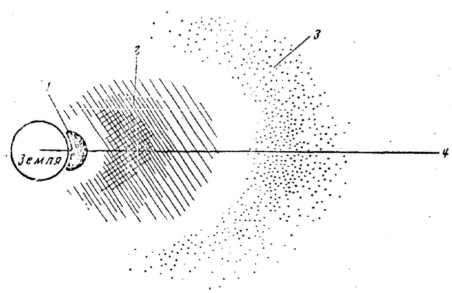

Figure 3.1. New map of Earth's radiation belt, based on passage of the First Cosmic Ship.

was calibrated to detect dust of a billionth of a gramme, suggested that the chance of being hit by dust on the way out to the Moon or back was minimal. Third, in the biggest finding, Konstantin Gringauz's ion traps detected how the Sun emitted strong flows of ionized plasma. This flow of particles was weak, at about 2 particles/cm^2/sec, because the Sun was at the low point in its cycle, but the ion traps had determined the existence of a "solar wind". This was one of the great discoveries of the space age and Gringauz estimated that the wind blew at 400 km/sec.

A map was later published of the journey of the First Cosmic Ship, past the Moon and into solar orbit. The First Cosmic Ship was able to measure Earth's gases far out, measuring the hydrogen geocorona as far out as 20,000 km. Then, it found the edge of the outer radiation belts, measuring its total electron flux at 3×10 elec/cm^2/sec. Now, on the sunny side of Earth, it flew directly into the solar wind, measured the gas component of interplanetary space, and detected Sun-ejected corpuscular fluxes. The temperature of plasma was measured at 10,000 K some 4 Earth radii out [2].

The contour of Earth's radiation led to great controversy. Gringauz identified, 30,000 km out, a sharp drop in the magnetospheric plasma density distribution that later received the name plasmapause. Over time, Soviet scientists began to characterize what became known as the plasmasphere: the doughnut-shaped cold plasma cloud that is full of ions and electrons trapped along the magnetic field lines. Gringauz's analysis challenged everything that had been known at that time and controversy erupted when he submitted his findings for publication in the *Proceedings of the Academy of Sciences* the following year. Gringauz was not well known and had only three papers published before, all from the *Akademik* series of sounding rockets. Normally, publication in *Proceedings* was automatic, but a procedure was in place whereby the President of the Academy, Mstislav Keldysh, would review space-related publications to ensure that there was no error that would embarrass the Soviet Union internationally. Keldysh knew nothing about plasma physics, so he sent the paper for review to those who did: Peter Kapitsa, Yuri Alpert, Sergei Vernov, and Iosif Shklovsky. Vernov thought the findings must be erroneous while Alpert said that what Gringauz claimed to have found just couldn't happen. Only Shklovsky voted yes.

Faced with a two-to-one negative, Keldysh decided against publication, but he reckoned without the persistence of A.L. Mintz, director of the Radio Technical Institute, who took the moral high ground and argued that the prohibition was a denial of academic freedom. After some negotiation, which delayed publication by months, a formula was agreed whereby if the authors were found to be in the wrong, they would be disowned publicly by the Academy. Publication went ahead in April 1960 and Gringauz brought his findings to the COSPAR meeting in Florence, Italy, in May 1961 with the suitably sonorous title of "Structure of the Earth's Ionized Gas Envelope based on Local Charged Particle Concentrations Measured in the USSR". Granted the significance of his findings, they generated remarkably little reaction, his colleagues possibly underestimating their importance. One who did appreciate them was an American scientist at Stanford University, Don Carpenter, and they began a lifetime's collaboration on the issue.

Konstantin Gringauz (1918–1993) was, along with Sergei Vernov, one of the great scientists of the early period of space exploration [3]. He was born in Tula on 5th July 1918 but his family moved to Samara when he was three. At school, he became an early radio amateur, so it was natural for him to study radio at the Electrotechnical Institute in Leningrad, where he acquired a diploma in frequency modulation in 1941. He was evacuated from there when the Germans approached the city, being relocated to Belovo, where he made tank transmitters. In 1947, he joined Korolev's design bureau, OKB-1. His first experiment flew on an R-1 rocket in 1948 and his second one, on a sounding rocket on 26th June 1954, found that electron density did not decrease above the F layer, as had been expected. He confirmed this with further sounding rockets three years later, on 16th May, 25th August, and 9th September 1957. The following year, further rocket tests probed the F layer (21st February, 27th August, and 31st October 1958). His instruments on the First Cosmic Ship, while primitive by today's standards, found the solar wind, the plasmasphere, the plasmapause, the inner plasma sheet, and magnetosheath plasmas. The discovery of the solar wind, the permanent flux of solar plasma, meant that an elusive "agent" in Sun–Earth relations had been found at last. It had long been suspected that such medium existed, providing the "energy link" between the Sun and the Earth, but only indirect evidence of such a medium had been observed before the space era.

SECOND COSMIC SHIP: THE LUNAR ENVIRONMENT

The Second Cosmic Ship, launched on 12th September 1959, carried an identical suite of scientific instruments, although Shmaia Dolginov's magnetometer had been modified to reduce the range of measurement to between -750 and $+750$ γ, where a response was considered more likely. Some 156,000 km out, the Second Cosmic Ship released sodium vapor, expanding into a 650-km-diameter cloud spotted by observatories in Alma Ata, Byurakan, Abastuma, Tbilisi, and Stalinabad. This time, observers there confirmed that it was dead on course. The Second Cosmic Ship took measurements every minute during the flight out, creating 14 km of teletype. The ship flew a different course compared to its predecessor, the side away from the Sun, and found negative currents, the plasmasheath.

As the probe neared the Moon, the instruments were working perfectly and were searching for lunar magnetic and radiation fields (none was found when the last measurement came in, 55 km out). The Second Cosmic Ship encountered and measured the solar wind met by its predecessor. Other instruments measured alpha particles (nuclei of carbon, nitrogen), X-rays, gamma rays, high- and low-energy electrons, and high-energy particles. Some 34 hr into its mission, the Second Cosmic Ship crashed into the center of the Moon at great speed east of *Mare Serenitatis* near the craters Aristides, Archimedes, and Autolycus, the sudden end of transmissions signaling its successful arrival.

The scientific results of the mission of the Second Cosmic Ship were published the following spring. Neither a magnetic field nor a radiation belt was found around the

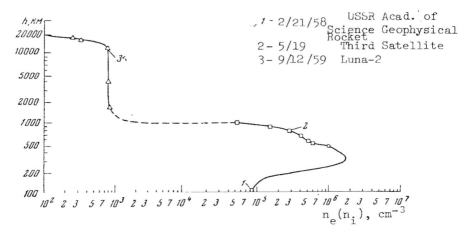

Figure 3.2. The passage of the Second Cosmic Ship (Luna 2) through the radiation belts.

Moon (the instrument was sensitive to a field only a ten-thousandth that of Earth, so if there was one, it must be weaker than that). The Moon did have, though, a rarified gas envelope. Again, the discoveries on the way out were as important. The outer belt of electrons in the Earth's charged particles reached out as far as 50,000 km. As the Second Cosmic Ship flew through the radiation belts, the ion traps detected electrons of varying intensities, peaking at 500 km and then falling back at 1,000 km, steady to 15,000 km, and falling off again at 20,000 km, with the last traces at 40,000 km, a curved shape that became known as "Alpert's curve". The four ion traps on the outside measured the flows of the currents of ion particles all the way out to the Moon. Their concentration varied, sometimes at less than 100 particles/ cm^3. But 8,000 km out from the Moon, current intensities increased, suggesting the existence of a shell of a lunar ionosphere. The data from the outward passage were later compared to those of earlier missions (Figure 3.2).

There were three main results from the mission. The first was the discovery of the limit of Earth's plasmasphere, the ionized gas envelope of the Earth stretching in low latitudes up to 4 Earth radii, with a rather sharp boundary – the plasmapause. The second result was that the amount of electron fluxes in the outer radiation belt was significantly (by three orders of magnitude) diminished compared to the results of Van Allen. The third was the observation of solar wind outside the Earth's geomagnetic field.

MAPPING THE FAR SIDE OF THE MOON

The third Moon probe, at the time called the Automatic Interplanetary Station (AIS, or AMS in Russian) but retrospectively Luna 3, began the project of mapping the Moon, especially the far side not visible from Earth. Scientific director of the mission was Yuri Lipsky (1909–1978) of the Sternberg Astronomical Institute, one of the least known personalities in Russian space science. Yuri Naumovich Lipsky was

Yuri Lipsky with his lunar globes

born in Dubrovno, Belarus, on 22nd November 1909. He entered a school for "working class youth", trained as an electrician for the railway industry, but managed to enter Moscow University to study physics in 1932, graduating and going on to head Kuchinskoi Astrophysical Observatory in 1941, part of the Sternberg State Astronomical Institute of the State University. During the purges, several of his colleagues there had been denounced, but he courageously stood by them (including Iosif Shklovsky) and, quite unusually, the denunciation failed. He served in the army from 1942 until his demobilization as major in 1945, having been wounded three times. Yuri Lipsky returned to the Sternberg Institute, being awarded a candidate degree in 1948, "Evaluation of the Mass of the Lunar Atmosphere from Polarization Studies of its Surface". His controversial polarimetric observations led him to the conclusion that the Moon possessed a significant gaseous envelope, contradicting contemporary views. The issue was not settled until the American Surveyor missions in the 1960s, which proved the polarization effect but found a different explanation: tiny dust particles suspended above the surface due to static electrical charges.

Photographing and mapping the lunar far side was a daring and enormously complex operation for its day: such a mission must take place when the far side was lit up by the Sun and bring the probe on a trajectory back to the Earth high over the Soviet Union so that it could transmit back the pictures. Optimum conditions were met infrequently: in October 1959 (photography after approaching the Moon) and April 1960 (photography while approaching the Moon). The first opportunity would bring the AIS out over the *eastern* side of the Moon as seen from Earth (in uninverted telescopes).

Table 3.2. Automatic Interplanetary Station instruments.

Camera photography system, 200 and 500 mm
Cosmic ray detectors
Micrometeoroid detector

The Automatic Interplanetary Station was probably the most complicated spacecraft of its day, with a three-axis stabilization system, cameras, photographic development system, and transmitters. It looked quite different from any Soviet spacecraft previously built, a 278-kg cylindrical canister with solar cells, 1.3 m tall, 1.2 m in diameter at the widest but 95 cm for most of its body. The canister was sealed and pressurized at 0.23 atmospheres. Shutters opened and closed to regulate the temperature, being set to open if it rose above 25°C. Four antennae poked out through the top of the spacecraft, two more from the bottom. The cameras were set in the top and the other scientific instruments were mounted on other parts of the outside. In addition to the cameras, the main payload, the spacecraft carried a cosmic ray detector and micrometeoroid detector (Table 3.2).

Launched on 4th October 1959, the Automatic Interplanetary Station swung over the sunlit far side of the Moon three days later in a long, elliptical Earth orbit that took it far out beyond the Moon. In the course of a 40-min session, the cameras took 29 pictures of no less than 70% of the far side (the first image was taken at a distance of 63,500 km after passing the Moon, looking back at the sunlit far side, while the last image was taken 40 min later from 66,700 km). The photographs were then developed, spooled, dried, and scanned at 1,000 lines by a cathode ray television system and then, when the station's orbit took it back to Earth, transmitted down to the ground, all at a time before digital photography and commercial faxes had been invented. Seventeen useable pictures were received. The spacecraft eventually burned up in the Earth's atmosphere on 29th March 1960 after 11 revolutions of the Earth–Moon system.

Yuri Lipsky later presented the images to the Sternberg Astronomical Institute. He became one of the head scientists to study the photographs of the far side and developed new methods to obtain the information from the images. As a result, around 500 new features were identified on a previously unknown lunar surface. Although the initial run of the pictures was hazy and fuzzy, it gave a bird's-eye view of the Moon's hidden side, the first time that a space probe had ever revealed spectacles invisible from Earth. The far side was found to be mainly cratered highlands and was quite different from the near side. In the tradition of exploration, to the finder fell the privilege of naming the new-found lands. There was one huge crater, with a remarkable central peak, which the Russians duly called Tsiolkovsky, and two seas, which they duly named the Moscow Sea (*Mare Moskvi* or, Latin, *Mare Moscoviense*) and the Sea of Dreams, *More Mechty* (Sea of Mechta, to mark the 1959 *Mechta* mission, or, Latin, *Mare Desiderii*) but later called *Mare Ingenii* by the International Astronomical Union. A first lunar map and then globe based on the AIS images were published the following year. A geological reconstruction and

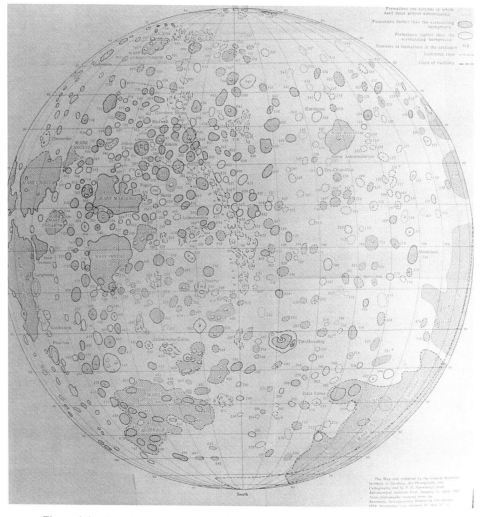

Figure 3.3. The Moon map compiled by the Automatic Interplanetary Station.

cross-section was made of the Moscow Sea. Although the published pictures appeared at first sight to be fuzzy, interpretation of the enhanced images enabled a map of extraordinary detail to be compiled (Figure 3.3).

The spacecraft identified in the south-eastern part of the far side the slopes of what turned out to be the largest ring impact structure in the solar system, 2,500 km across and 12 km deep, the *Mare Ingenii*, some four billion years old. It was a huge area, 3.8 million km² across, with 1,538 craters over 10 km across, rivaling the Ocean of Storms in scale. Many years later, it became the candidate for the subsequently abandoned American project to return to the Moon in 2020, but it went through something of a renaming process in the meantime. Most of the current nomenclature of the Moon derives from the extensive American Lunar Orbiter mapping project of

1966–1968. The Lunar Orbiters did not characterize this part of the Moon, so the *Mare Ingenii* was relegated to a small, dark structure, only 270 km across. Years later, the area was rediscovered by the American Galileo, Clementine, and Lunar Prospector orbiters and the original *Mare Ingenii* acquired the new name of the South Pole Aitken basin, Aitken being a crater marking its outer periphery. During the period when the Americans planned their return there, Russian scientists reconstructed the geology of the basin, going back to the 1959 pictures. It was of great scientific interest, for it was probably caused not by a meteorite, but by a comet from the Edgeworth–Kuiper belt making a low-density but wide oval-shaped impact during the period of what was called the Late Heavy Bombardment (LHB). Meantime, the station's cosmic ray detectors found that once the probe had cleared Earth's radiation belts, the intensity of cosmic rays was very steady (6.79 pulses/cm^2/sec) [4].

During the following window of opportunity to map the far side of the Moon, in April 1960, two more Moon probes were launched, but both failed. For a variety of reasons, the Russians were not able to return to the Moon-mapping project until summer 1965, when an unlaunched Mars probe became available: Zond 3. This was a more capable spacecraft, weighing in at 950 kg, 1.1 m in diameter, with 4-m-wide solar panels, telemetry systems, 2-m transmission dish, an engine for mid-course maneuver, and an equipment section, in this case for cameras. The camera system worked according to similar principles as the Automatic Interplanetary Station, but was redesigned by Arnold Selivanov, with significant improvements. It could hold a similar number of images (40), but could transmit them at low or high resolution (either 550 lines taking 125 sec (quick-look) or 1,100 lines taking 34 min), use infrared and ultraviolet filters, and retransmit from far away from Earth, up to 30 million km distant. Its trajectory enabled it to swing around the other side from that covered in 1959 (Figure 3.4). Zond 3 carried more extensive instrumentation (Table 3.3).

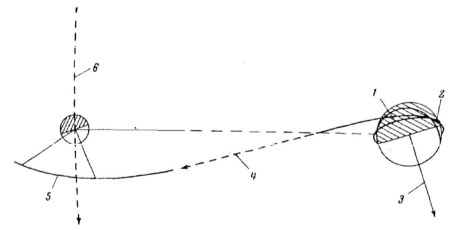

Figure 3.4. Zond 3 trajectory, showing period of photography of western limb of the Moon.

Table 3.3. Zond 3 scientific instruments.

Two cameras
Infrared and ultraviolet spectrometer
Magnetometer
Cosmic ray detector
Solar particle detector
Meteoroid detector

In July 1965, Zond 3 passed 9,219 km over the *Mare Orientale* on the *western* limb of the Moon and duly took 25 wide-angle views and three ultraviolet scans from distances of 11,570–9,960 km over a period of 68 min. The photos covered 19 million km^2 of the lunar surface (Figure 3.5). As Zond swung around the Moon's leading edge, whole new mountain ranges, continents, and hundreds of craters swept into view, including the mountainous outer rings of the *More Mechty*. Whereas the near side was dominated by seas (*maria*), mountain ranges, and large craters, the far side was a vast continent with hardly any *maria* but pockmarked with small craters. The images also showed distinct features of a large ring-shaped formation near the terminator at the western outskirts. It was not until years later that Canadian Phil Stooke combined two images of Luna 3 and Zond 3 to find the shape of the largest ring structure in the solar system, the South Pole – Aitken Basin. Height profiles were made of the far-side craters, finding them to be similar in structure to those on the near side. Two years later, a second Sternberg atlas (*Atlas of the Lunar Farside, Part II*), with interpretative analysis and commentary, was published, again edited by Yuri Lipsky, achieving this time complete coverage of the far side (Figures 3.6 and 3.7). The first *Complete Map of the Moon* and a new globe were issued in 1967. Geologists concluded that the far side was a single highland plate with just small depressions and craters filled with lava.

One of the craters on the far side discovered during Zond 3 observations was named Korolev. Sergei Korolev, known to the people by his title "the chief designer of Soviet cosmonautics", passed away on 14th January 1966, while Zond 3 data were still processed. Zond 3 was his last experiment in deep space exploration. Unfortunately, he did not see the success of Luna 9 soft landing three weeks later (see below). Unlike the AIS, Zond 3 continued into heliocentric orbit, measuring cosmic rays until 3rd December, when it was 39 million km away. Zond 3 identified the gradual rise in the solar cycle over the period, and detected solar flares on the 2nd and 7th October and low-energy proton intensity bursts. Zond 3 readings were compared to Venera 2, enabling Sergei Vernov to get simultaneous measurements from both inside Earth's orbit and outside. This led to a puzzling result, for protons of around 1 MeV seemed to be of solar origin, but they seemed to form a ridge of increased intensity some 1.5 AU away from the Sun. At the end of the mission, a chart was published of cosmic ray levels dating back to 1959, linking the levels of the solar cycle [5].

The discoveries of the first Moon probes are summarized in Table 3.4 and the series in Table 3.5.

Zond 3 speeds past the far side of the Moon

Table 3.4. First Moon-probe discoveries.

Failure to detect magnetic field, radiation belts around Moon
Lunar far side had few *mare*; mainly chaotic upland, one large crater
Finding and measurement of the solar wind

Table 3.5. The first Moon probes.

2 Jan 1959	First Cosmic Ship	Missed Moon, entered solar orbit
12 Sep 1959	Second Cosmic Ship	First spaceship to reach Moon
4 Oct 1959	AIS	First photography of far side
18 July 1965	Zond 3	Completed photography of far side

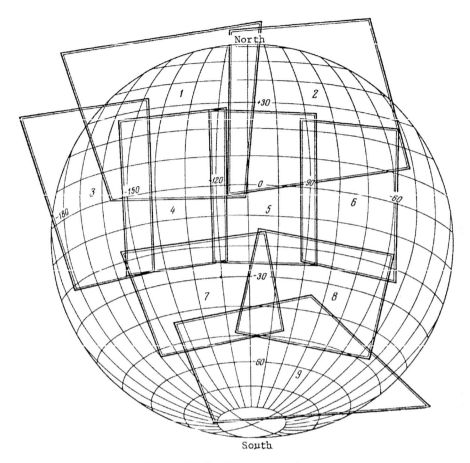

Figure 3.5. Zond 3 mapping frames.

SOFT-LANDING ON THE MOON

Following the successful missions in 1959, the Soviet lunar program moved on to attempt the first soft-landing on the Moon, primarily so as to pave the way for a manned landing, plans for which were also put in progress. There had been much debate among scientists as to how solid was the lunar surface (some predicted that it would be so dusty that spacecraft would sink). At one meeting, Korolev became impatient and concluded the discussion by writing down on his pad "The Moon is solid" and signing it. The landers were small spacecraft, only 100 kg in weight and egg-shaped. The much larger main spacecraft would brake the speed of descent to the Moon to close to zero and eject the lander, which would bounce before coming to a halt. Then, petal-shaped covers would unfurl and it would begin its scientific program. The landers had enough battery for about four or five days. The cameras would characterize the surface, a radiometer would measure radiation levels, while a

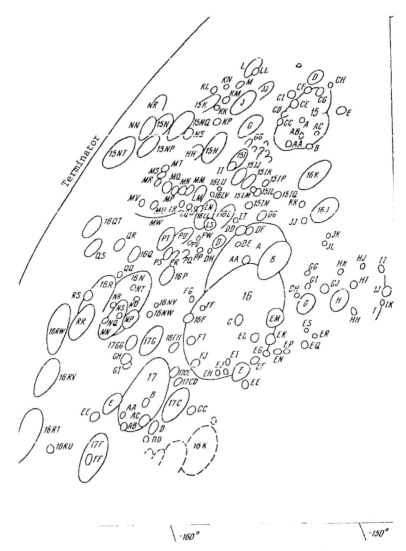

Figure 3.6. Turning the photographs into maps, Zond 3.

thermometer would take the temperature (important for the lunar lander and cosmonaut spacesuits). The later model also carried, on two 1.5-m-long extensible arms, a penetrometer to test the density of the soil (important for designing the landing legs of the manned lunar lander). The soft-lander instruments are listed in Table 3.6.

Achieving a soft-landing proved to be excruciatingly difficult and the first successful landing was not made until the tenth attempt. Useful data were returned from only two of the unsuccessful missions: Luna 4, which missed the Moon in April 1963, and Luna 6, likewise in June 1965. Luna 4 transmitted from 2nd to 15th April,

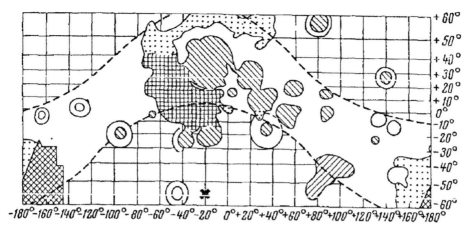

Figure 3.7. Zond 3 interpretative map.

Table 3.6. Luna soft-lander instruments.

Camera
Radiometer
Dynamograph/penetrometer ("gruntmeter")
Thermometer
Cosmic ray detector

found an average radiation level of 19.16 pulses/sec, but dipping to 18.5 on 7th April and rising to 19.5 on 12th April, leading Sergei Vernov to postulate a 14-day sub-cycle. Luna 4 found that parts of Earth's geomagnetic tail could seep out as far as the Moon's orbit (in fact, it stretches even further). Soviet scientists marked Luna 4 by bringing together four years of data on the solar wind, 1959–1963, showing the range of its speed: 230 km/sec (First and Second Cosmic Ships), doubling to 445 km/sec (Luna 4) [6]. Additional measurements from Luna 6 and Zond 3 enabled a longitudinal study of cosmic rays (Figure 3.8).

The first success was achieved by Luna 9, in January 1966, followed by Luna 13 that December, both coming down in the Ocean of Storms. Luna 9 transmitted for 8 hr 5 min in the course of a week, sending back eight panoramic pictures of craters, rocks, stones, and hollows. From them, the Academy of Sciences was able to compile a topographical atlas, published the following year as *First Panoramas of the Lunar Landscape* (*Pervi panorami lunnoi poverkhnost*). This provided a detailed character-ization of the two landing grounds, up to 1.4 km distant from the landers, marking the maximum distance seen by the cameras. The cosmic rays experiment led by NIIYaF found that the cosmic ray flux at the lunar surface was only 1.6 times lower than in open space, while the detector field of view was two times smaller due to the Moon shielding. The difference was explained by the lunar surface radioactivity and lunar albedo particles. Thus, it was possible to estimate the radioactivity of the lunar surface, which was close to that of Earth soil.

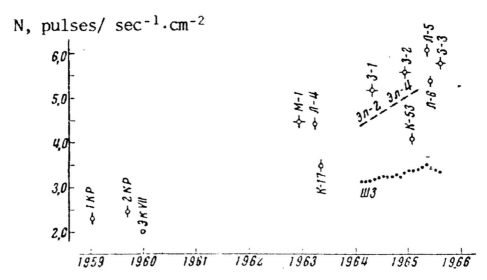

Figure 3.8. Luna 6 contributed, along with other spacecraft, to a picture of the strength of cosmic rays over 1959–1965.

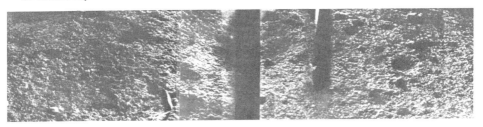

Luna 9 close-up of Moon soil

Luna 13 deployed its arm with a mechanical soil meter and thumper that tested the density of the soil to the depth of 4.5 cm with an impulse of 7 kg and force of 23.3 kg/m^2. The conclusion was that Moon soil was similar to medium-density Earth soil (0.8 g/cm^3, to be precise) and reasonably solid, but with a dusty layer on its surface that could be as much as 20–30 cm deep. Luna 13 determined that the soil had a bearing capacity of 0.68 kg/cm^2 and scientists described it as "porous, weakly bonded and grainy". At the end of the other arm was a radiation density meter that found that levels of radiation on the Moon were modest and would be tolerable for humans. Luna 13 found that the Moon absorbed about 75% of cosmic ray particles reaching the surface, reflecting the balance, 25%, back into space, but that radioactivity levels in the soil were low. Solar radiation was measured by an instrument developed by Viktoria Kurt and estimated as ranging from 40 to 300 keV. The daytime temperature of the lunar surface was measured at 143°C. It transmitted five panoramas at different sun angles. The principal outcome of the missions was confirmation that the Moon's surface was sufficiently hard for a spacecraft to safely land [7]. The series is summarized in Table 3.7.

Luna 13 penetrometer at work

Table 3.7. Luna soft-landing missions.

Luna 9	31 Jan 1966
Luna 13	21 Dec 1966

ORBITING THE MOON

The next step in lunar exploration was to characterize the environment of the Moon through means of an orbiter. Although a purpose-built orbiter had been designed, it was not ready in time, so a unique design was built in only 25 days in the Lavochkin bureau (the first attempt, Cosmos 111, failed). It was equipped with seven scientific instruments, including Gringauz's charged particle trap, Vernov's Geiger counter, and Dolginov's magnetometer on a long boom. Scientists would also measure gases in the lunar environment by examining signal strengths as the probe appeared and reappeared behind the lunar limb, and watch for changes in the orbit due to the lunar gravitational field (Table 3.8).

The 245-kg Luna 10 was fired from Earth on 31st March 1966. Three days later, 8,000 km from the Moon, its engines were turned on to knock 0.64 km/sec off its speed, just enough to let it be captured by the Moon's gravity field. The boiler-shaped instrument cabin separated on schedule 20 sec later. Luna 10 was pulled into a high inclination orbit of 349–1,015 km, 71.9°, 2 hr 58 min, and became the first spacecraft to orbit the Moon. Luna 10's mission lasted way into the summer and did not end until 30th May, after 56 days, 460 lunar revolutions, and 219 communication sessions. Data were transmitted on 183-MHz aerials and also on smaller 922-MHz aerials.

Luna 10 in lunar orbit

Table 3.8. Luna 10 instruments.

Meteorite particles recorder
Gamma spectrometer 0.3–4 MeV
Magnetometer with three channels
Solar plasma experiment
Infrared recorder
Radiation detector
Charged particle detector

A stream of data was sent back, the main outcomes being given at a press conference on 16th April. The most startling finding, coming from the sweeps of the magnetometer (Figure 3.9), was that there were anomalous zones of mass concentrations below the lunar surface disturbing the lunar orbit (mascons). Mascons were probably the most important discovery by Luna 10. Mascons were

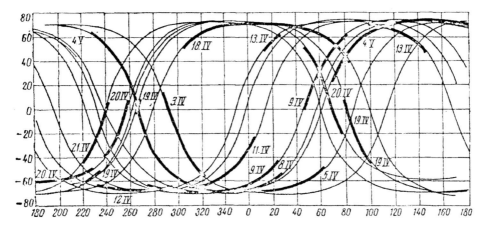

Figure 3.9. Luna 10 magnetometer measurements.

worrying, for they pulled orbiting spacecraft out of their predicted orbit by several kilometers. These distortions could make all the difference to where a spacecraft was targeted for landing and to the success of subsequent link-ups in lunar orbit. We can only presume that Soviet mission planners took careful note of the mascons in plotting future lunar trajectories. Luna 10 found that mascons disturbed an orbit by about 750 m per orbit, causing acceleration while traveling over the lunar seas. Later that year *Determination of the Lunar Gravitational Field* was published, with a three-dimensional map of the distortions of the field, showing it to be pear-shaped, flattened at the pole and pulled out on the far side.

Using its gamma-ray spectrometer, Luna 10 began the first initial survey of the chemistry of the Moon, enabling a preliminary chemical map to be compiled. Six spectra were taken in the 0.3–3.1-MeV band on 5th, 8–9th, 18th, and 21st April. Lunar rocks gave a composition signature broadly similar to basalt, but other important clues to its composition were picked out, such as the identification of uranium, thorium, and potassium 40. Their concentrations were twice that of Earth's rocks, suggesting that the Earth and Moon had different origins. Four forms were suggested: granites, basic rocks, chondrites, and ultra-basic rocks. The radiation on the surface was measured at between 20 and 30 microröntgens, but there were significant variations in radiation levels on the Moon, being higher in the Sea of Clouds, for example, and generally higher in the western maria. Incoming cosmic radiation was measured at 5 particles/cm^2/sec. Between 13 and 26% of the radiation falling on the lunar surface was reflected back. As for meteorites and dust in the lunar environment, Luna 10 took 247 micrometeorite hits, including 50 meteorite hits in one orbit, and scientists determined that meteorite hits were higher in Moon orbit than in the flight to the Moon. Intriguingly, some were impacts from the Moon that had bounced back and hit the spacecraft.

Luna 10's magnetometer was put on the end of a 1.5-m boom and took measurements every 128 sec for two months. Designer Shmaia Dolginov – who had built the original magnetometer on the First Cosmic Ship – was able to refine the

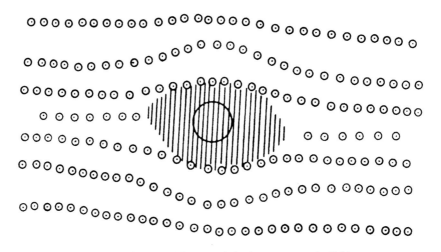

Figure 3.10. Luna 10 map of the lunar magnetic field.

range to between –50 and + 50 γ. It found a very weak magnetic field around the Moon, 0.001% that of Earth. Here, the Moon's magnetic field was measured at 15–40 γ, but it was really an area of trapped particles rather than an intrinsic, organic magnetic field. There was no north–south dipole to the field. During its mission, Luna 10 swept through the Earth's hypothesized magnetic tail four times. When it did so, solar wind protons fell off rapidly, unable to penetrate the tail of Earth's magnetic field. Luna 10 found a shell of a lunar ionosphere at an altitude of 350–1,000 km with a concentration of particles of 300/cm³. A basic outline of the lunar magnetic environment was published (Figure 3.10). The cosmic ray counter measured cosmic rays coming in at 11.2 pulses/sec, but varying according to the height of orbit, day and night, and its flux at $4.7/cm^2/sec$ [8].

In the popular mind, Luna 9's soft-landing was a more impressive mission, largely because of the pictures sent back, but the scientific haul from Luna 10 was significant and enduring (Table 3.9).

Table 3.9. Luna 10 discoveries.

- Weak magnetic field around the Moon, 0.001% Earth, but not an organic field
- Moon sweeps through Earth's magnetic tail
- No lunar magnetic poles
- Cosmic radiation in lunar orbit, variable radiation absorption by surface
- Meteoroid impacts, more in lunar orbit than in the flight to the Moon
- Meteorites bounce off the lunar surface to reach orbiting spacecraft
- No gaseous atmosphere, but a limited ionosphere up to 1,000 km
- Mascons
- Surface composition: basalt

Table 3.10. Luna 11 and 12 instruments.

Magnetometer
Gamma-ray spectrometer
Gas discharge counters
Electrode ion traps
Meteoroid particle detector
Infrared radiometer
Low energy X-ray photon counter
Cameras

MAPPING THE MOON IN DETAIL: LUNA AND ZOND

The success of Luna 10 paved the way for five more lunar orbiting missions and four circumlunar flights, with a focus on compiling detailed surface maps for later manned and unmanned missions. Luna 11 was the mission originally intended to be first to orbit the Moon and carried the same camera system as that flown on Zond 3, but the images would obviously be taken from a much closer vantage point, being expected to cover 25 km^2 each, with a resolution of 15–20 m. Once taken, the photographs would be developed, dried, and scanned by a television system on board and transmitted to Earth. Besides the camera system, seven scientific instruments were carried – the same as Luna 10 (Table 3.10). The spacecraft was three times heavier, at around 1,620 kg.

Luna 11 arrived in a middle-inclination 27° lunar orbit on 27th August 1966, but a foreign object stuck in the thrusters, making it impossible to point the cameras properly. The other instruments continued to operate until the batteries were exhausted on 1st October after 38 days, 277 revolutions, and 137 communications sessions. Luna 11 radar mapping enabled a fresh calculation of surface density to be made (the range was 1.14–1.28 gm/cm^3) and it made some useful solar and deep-space observations. Here, Luna 11 recorded powerful solar eruptions on 28th and 29th August and 2nd September, with electron energies of up to 30 eV and protons up to 50 eV. Three active regions crossed the Sun and Luna 11 detected the subsequent changes in proton intensities, solar flares, magnetic storms, and neutrons. Long-wave radio bursts were detected on 28th and 29th August and 8th September, the last of 7 min, scientists concluding that they were interplanetary because the Sun was below the horizon at the time (see Figure 3.11) [9].

As a result of the camera failure, the mission had to be re-flown. Luna 12 entered a 15° equatorial lunar orbit on 25th October: this time, thrusters were used to point Luna 12 towards a series of planned equatorial landing sites, transmitting the full batch of its images within 24 hr, so quickly as to prevent their downloading by American and British trackers. On the second day in Moon orbit, the spacecraft was put into a slow roll so as to accomplish the other scientific objectives of its mission.

The Luna 12 mission lasted three months and ended on 19th January 1967 after a much longer period: 85 days, 602 orbits, and 302 communications sessions. It

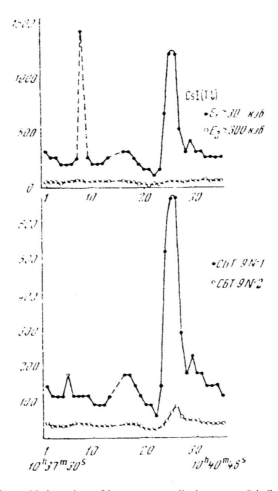

Figure 3.11. Luna 11 detection of long-range radio bursts on 8th September 1966.

appears to have achieved its primary purpose, for not long afterwards, the landing sites for the first landing by Soviet cosmonauts were selected: Ocean of Storms (*Oceanus Procellarum*), the Central Bay (*Sinus Medii*), and the Sea of Tranquility (*Mare Tranquillitatis*) (not the Apollo 11 site). Although 40 images were taken, only a small selection of Luna 12 pictures was published, the quality of reproduction being poor. The originals may have been better, for the history of the Sternberg Institute credits Luna 12 as a major contribution to the detailed lunar maps drawn up at this period. Specifically, it is stated that, as a result of Luna 12, Yuri Lipsky was able to present detailed schemes of the prospective landing sites, with terrain descriptions, flight profiles in and out, micro-relief of craters, stones and slopes, and information on soil strengths. This information was then programmed into the Soviet lunar lander (LK) computer system to guide the descent to the surface. Quick-read navigation charts were prepared for the LK cosmonauts so they could spot the various features during descent. As for the scientific aspects of the mission, assessments were made of the reflectivity of the lunar surface to infer its density, which was measured at 1.4 g/cm^3. Lunas 11 and 12 between them made 150 radar images of the surface so as to determine its relief. The gamma spectrometer suggested that the solar wind made the surface radioactive down to a depth of 35 cm. Luna 12 was hit by meteorite showers that were heavier at the end of the year. X-rays reaching lunar orbit were measured, most coming from the Sun. Luna 12 detected a burst of protons on 13th December and, not long after, transited through a region of magnetized plasma of low-energy protons of 0.5–10 MeV. Radio levels rose as Luna 12 swept into Earth's magnetospheric tail [10].

Luna 14 flew out to the Moon on 7th April 1968 (there were two failures in this series). It was primarily a detailed study of the Moon's gravitation field so as to measure mascons more precisely and a test of communication systems for the manned lunar landing, but it also carried a gamma spectrometer and instruments to measure charged solar particles, cosmic rays, and ions in the lunar environment. Its 170-cm radar was used to make density measurements of the lunar surface, finding overall densities of 1.18–1.28 gm/cm^3 across the maria and highlands, but individual regions varied much more widely, from 0.62 at a low point to as high as 2.20 gm/cm^3. The regolith was estimated to have an average depth of 9 m. Luna 14 enabled a determination of the respective mass ratio of Earth and the Moon. Transmissions lasted until 11th June [11].

The project to map the Moon was continued through the Zond program over 1968–1970. Zonds 4–8 comprised a program to pave the way for the first manned flight around the Moon – a mission that was never actually flown – by circling the Moon and returning to Earth at high speed (except Zond 4, which was orbited a similar distance but away from the Moon and not recovered). The absence of a human crew on the precursor missions provided a weight margin enabling much larger cameras to be carried. Granted that the cabins were to be recovered, there was no need to install an onboard development, relay, or transmission system, so the photographs could be developed afterwards using traditional techniques that provided higher-quality imaging than television scanning. The cameras flown were either 400 mm (black and white, Zond 6, 8) or 300 mm (color, Zond 5, 7), taking 13 × 18-cm frames and were built by the Moscow State Institute for Geodesy and Cartography. Zonds 4 and 5 carried detectors to measure solar proton fluxes, measuring them at 35–50 sterads in the 10–50-MeV range for Zond 5, lower for Zond 4.

Due to an orientation failure, no pictures of the Moon were received from Zond 5. It did take some photographs of the departing Earth from 90,000 km out, detailed versions of which were used to redraw the geological and tectonic map of Arabia and the continent of Africa (Figure 3.12). Zond 5 carried out an important biological mission that is reviewed in Chapter 6.

Its successor, though, Zond 6, was much more successful, taking 169 images of its close approach, crater floors, Earthrise, and a receding full Moon from distances ranging from 9,290 to 2,430 km. The Zond cabin depressurized during the return to Earth due to premature parachute deployment and crashed at some speed, but the

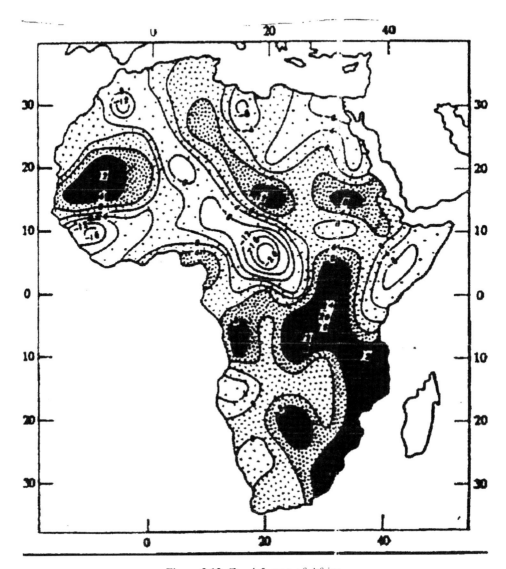

Figure 3.12. Zond 5 map of Africa.

institute was able to recover 52 images from the wreckage, although some of these were fogged or torn. Zond 6's trajectory took it around the south-western corner of the Moon, where the topography was still quite uncertain. The southern polar regions had a giant depression some 5–7 km below the median (*Mare Australae*), while some rims had elevations up to 10 km above the median. Zond 7 took 35 images as it flew 2,200 km past the Ocean of Storms, over the Leibnitz Mountains on the western limb, rounding the Moon, snapping remarkable pictures of a full Earth Earthrise. Zond 8 took 98 high-density 6,000–8,000-pixel images, some of the best photographs of the Moon ever taken, many from its close approach of 1,350 km,

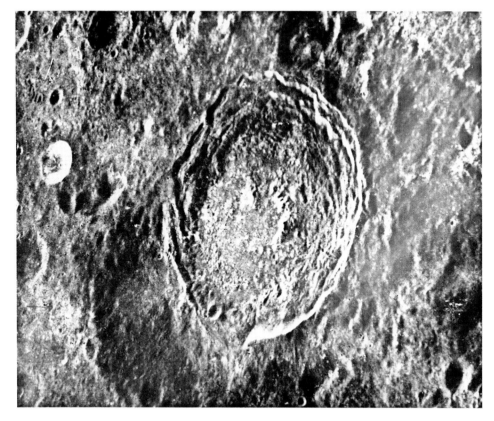

Craters from Zond 6

including a direct pass over the *Orientale* basin, showing its 960-km ring structure, cinder cones, domes, furrows, fissures, and dark material, explicable by vulcanism after the impact basin was created but before it was filled by *mare* flooding. There were few fresh craters. The sea was flooded by lava over a prolonged period of 1.5 billion years from 3.7 to 2.3 billion years ago. It was depressed as deep as 4.7 km, but with ridges as high as 10 km. Analysts made a relief map of the "vast lowland" around the south pole, now known as the South Pole Aitken Basin [12].

Two final lunar orbiting missions were flown in the Soviet Moon program: Luna 19 and Luna 22, both very similar. These were multi-purpose missions, designed to map the Moon and continue the work of characterizing the lunar environment begun by Luna 10. They carried both panoramic cameras and close-look cameras, both spacecraft maneuvering over the lunar surface to inspect particular targets of interest. There was supposed to be a third lunar orbiter, to go into polar orbit, but it was cancelled.

Luna 19's mission began when it entered lunar orbit on 2nd October 1971 and lasted a full year to 3rd October 1972 and 1,000 communication sessions were held in the course of 4,000 orbits. Luna 19 reported back on magnetic fields, mascons,

Zond 7 crosses boundary of lunar night and day

Zond 8 sweeps around the edge of the Moon

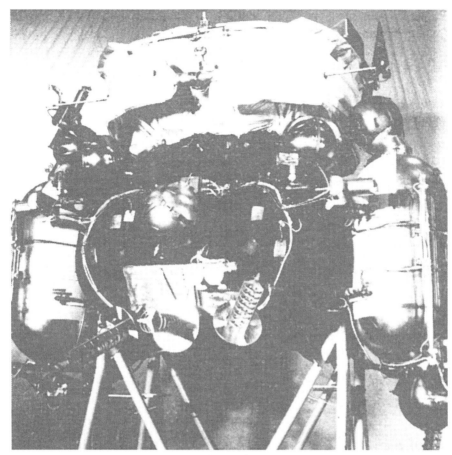

Luna 19

cosmic rays, the lunar gravity field, and meteoroids, and sent back televised pictures of an area 30–60°S and 20–30°E, the quality of publication much improved compared to Luna 12 in 1966. In February 1972, it swept over the Torrid Gulf near the crater Eratosthenes (11°W, 15°N), and filmed rock-strewn plains above which reared a volcanic-like summit. Studies were made of craters Godin and Agrippa at 10°E, 3°N and of Ryumker crater. Radar maps were made of two high northerly regions from 40° 51′N to 54° 45′N, the Bay of Dew, *Sinus Roris*. It had been intended that Luna 19 descend to an altitude of only 16 km, but due to an engine fault, this part of the program, which involved mapping and altitude measurement, was canceled and the perilune remained 126 km.

Some science reports were issued, noting how Luna 19 had measured solar flares and plasma, mascons, the lunar surface, and the composition of its soil. The magnetometer measured magnetic fields as the Moon moved in and out of the Earth's long magnetic tail: its strengths on the near and far sides of the Moon were compared, being stronger on the daylight side. A radio occultation experiment was carried out in May 1972 and this found a thin layer (10^3/cm^3) of plasma near the sunlit lunar surface up to an altitude of 10 km above the surface. The level of meteorites was measured in an experiment using similar detectors to those installed on Cosmos 213, 470, and 502, and the orbital station Salyut. Accounts give figures of between six and 10 impacts, but low in any case [13]. Cosmic rays were studied: averaged data on solar protons fluxes were obtained, with proton energies from 1 to 40 MeV. During most of the mission, the Sun was highly active and Luna 19 obtained results on the impact of solar activity on solar cosmic rays, such as decreases and increases in the proton fluxes due to solar flares.

Luna 22 entered lunar orbit on 8th June 1974. Its scientific program was a little different, with some technological experiments. It repeatedly altered orbit, swooping down to as low as 24 km, returning to an operational orbit at 200 km and then raising its far point to as much as 1,437 km out from the Moon. Some individual regions were mapped in detail (e.g. craters Debes, Tralles, and Cleomedes). In between the high-altitude panoramic photography and the low-altitude swoops, there were two extended periods in which no maneuvers were made, presumably so as to give time for uninterrupted measurement of changes to its path arising from the distortions in the Moon's gravitational field. Once the mission was over, a new model was published of the Moon's gravitation field. The new model, called M8 (the eighth so devised, including American models in the series), was based on 465,000 measurements, taken from not just Luna 22, but all the previous orbiters.

Few scientific results were released from the mission, although they should have been substantial, as evidenced by the heavy radio traffic to and from the probe over the 18 months of its operation. Lunar topography was mapped carefully through the use of an altimeter and a gamma-ray spectrometer analyzed the composition of the surface. Further investigation was made of the lunar plasma found at 10 km by Luna 19. A dual-frequency dispersion interferometer was turned on as the station passed behind the Moon (LOS, or Loss of Signal) and again as it came round again (AOS, or Acquisition of Signal), this being the moment to detect it against the blackness of space. The new probe indicated that the sheath of ionized gas formed 8 km over the lunar surface during sunlight spread out over 2–10 km and reached as high as 50 km. A chart was published showing densities, the maximum being between 450 and 1,000 electrons/cm^3. Tables were published comparing the levels of protons over 30 MeV and β particles over 2 MeV between translunar coast, high circular orbit, and swoop orbits. Luna 22 recorded 44 micrometeoroid impacts between 1st June 1974 and 28th January 1975 – a much lower rate because of its higher orbit. There were two highs: 14 in June, during the *Arietid* meteor shower, and 11 in December from the *Geminids* and *Ursids* [14].

The Luna 19 and 22 photographs brought to a conclusion the lunar mapping project initiated by Yuri Lipsky 20 years earlier. In March 1976, the Sternberg Institute published the *Atlas of the Reverse Side of the Moon, Part III* – a compilation of all the images of the AIS, Zond 3, and Zonds 6–8, Lunas 12, 19, and 22, with reliefs and interpretive tables. This was accompanied by the *Complete Map of the Moon* (1977), covering 99.5% of the surface and a new globe (1979). Geomorphological interpretations accompanied the Zond 8 images.

Table 3.11. The lunar orbiting missions, with orbiting parameters.

Luna 10	31 Mar 1966	349–1,015 km, 71.9°, 2 hr 58 min
Luna 11	24 Aug 1966	159–1,193 km, 27°, 2 hr 58 min
Luna 12	22 Oct 1966	100–1,737 km, 15°, 3 hr 25 min
Luna 14	7 Apr 1968	160–870 km, 42°, 2 hr 40 min
Luna 19	28 Sep 1971	140–148 km, 40.58°, 2 hr 04 min
Luna 22	2 Jun 1974	219–221 km, 19.6°, 2 hr 10 min

Table 3.12. The circumlunar missions, with distance from Moon.

Zond 5	15 Sep 1968	1,950 km
Zond 6	10 Nov 1968	2,420 km
Zond 7	8 Aug 1969	2,000 km
Zond 8	20 Oct 1970	1,100 km

The promoter of this 20-year project, Yuri Lipsky, died on 24th January 1978, with his lifetime's project essentially complete. Apart from the great atlases, he published over 100 works and it was fitting that a crater on the far side was named after him. The work of lunar mapping was continued by Vladislav Shevchenko (b. 1940), who took over his post in the Institute (one that he still holds) and Zhanna Rodionova. The outcome of the previous 20 years' research was published as Vladislav Shevchenko's *Modern Selenography of the Moon* (Nauka Press, 1980). The final set of missions is summarized in Tables 3.11 and 3.12.

RECOVERING LUNAR SAMPLES

When the chance of landing a cosmonaut on the Moon began to recede, the Soviet Union devised an alternate program to retrieve samples from the Moon by automatic spacecraft. This program suffered a high failure rate, with several spacecraft being lost en route to the Moon and three more lost during their attempts to land on the Moon (Lunas 15, 18, and 23). Despite this, samples were recovered from three sites by Lunas 16, 20, and 24, the last being a core sample. Lunas 16 and 24 were *mare* samples, while Luna 20's were taken from the lunar uplands. Although the sample sizes were small, electron microscopes meant that they could be examined in minute detail. Samples were later distributed to planetary geologists further afield, principally in the United States, India, and Britain (Newcastle, Cambridge).

The sample return spacecraft was 5 tonnes in weight at launch and 3.96 m tall. The normal profile of the mission was to spend four days en route to the Moon, four days in lunar orbit, and then make a stop-and-drop burn to the landing site, small engines cushioning the final fall to the Moon. With its fuel depleted, the lander weighed 1.8 tonnes, comprising a landing stage, robotic arm, ascent rocket, and spherical cabin. In the course of a day on the Moon, the spacecraft would photograph the landing site, following which the robotic arm would drill rock out of the surface and place it in the return cabin, which would be sealed for the return to Earth. The landing site was carefully chosen so that the ascending rocket could make a single burn that would bring it on a direct course to a landing site in the Soviet Union without the need for a course correction. The small return cabin, only 50 cm across and 40 kg in weight, would descend under parachute and be brought back to Moscow for analysis. The lander also carried a cosmic ray counter that could transmit data for up to three days after landing, until the battery ran out.

Luna 16 landed smoothly on the shoreline of the Sea of Fertility on 20th September 1970. The flat and stony ground was marked only by a few small craters. The drill head bored into the lunar surface at 500 rpm using electric motors for 7 min and then scooped the grains of soil down to its maximum depth of 35 cm. There, it began to hit hard rock and rather than risk damaging the drill, the boring was terminated and the sample collected and put into the container attached to the drill head. Like a robot in a backyard assembly shop, the drill head jerked upwards, brought itself alongside the small spherical recovery capsule, turned it round, and pressed the grains into the sealed cabin, which was then slapped shut. After a day on the Moon, the upper-stage motor fired, sending the cabin back to Earth for a touchdown in the normal landing ground in Kazakhstan.

Brought under police escort through a nighttime Moscow to the Vernadsky Institute, the gray grains of Moon dust poured out – loose lumps of dark-gray, blackish powder like very dark, wet, beach sand. They were sent for analysis by a team of young scientists led by Alexander Vinogradov, the academician, and Kirill Florensky (1915–1982), whose father, Pavel Florensky, was a religious philosopher. Kirill had earlier led expeditions to the remains of the Siberian impact of 1908.

The sample, although small (105 g), provided a considerable amount of scientific information. It was described as "blackish powder in friable lumps" with molten, rounded, and angular grains averaging 0.1 mm in size, getting larger further down. The density of the soil was calculated at 1.17 g/cm^3, with grain sizes of 60–120 μm, bigger with depth. The solar wind was found to act on the surface particles down to a depth of 5 cm to a great extent, but to a much diminished extent below than. Overall, the rocks were not that different from the American Apollo 11 and 12 samples. They were basalt, called feldspathic basalt, as expected, but with much more aluminum (10–19%), iron, and magnesium. The following were the main features of the sample:

- it was a uniform, unstratified sample, but with thee main parts (Figure 3.13);
- age of the oldest sample was estimated at 4.455 billion years;
- basalts, the main element, made of plagioclase, pyroxene, ileminite, and olivine;
- there were melted vitrified, glassy, and metallic particles;
- the main elements were breccias, gabbro, and anorthosite;
- seventy different chemical elements were identified;
- the sample comprised a mixture of powder, fine, and coarse grains;
- it had good cohesive qualities, like damp sand, but poor bearing capacity;
- samples comprised *mare*, quickly cooled *mare*, and ejected highland rock;
- the samples had absorbed quantities of solar wind.

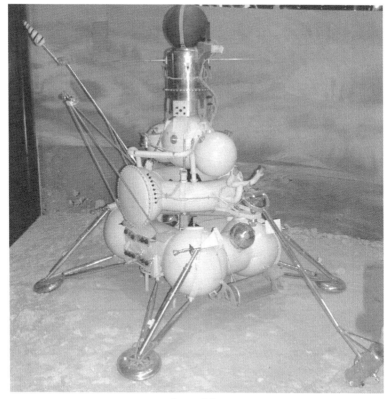

Luna 16

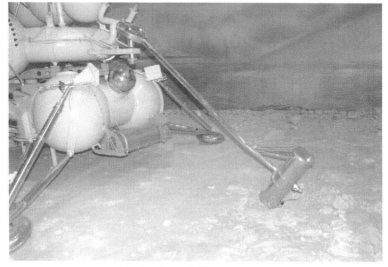

Details of the drilling system

Samples from Luna 16

The sample thus comprised the original lava of the Moon, the basalt, but also breccias, caked rock fragments formed during impacts. The composition was, in the basaltic material, silicon 43.8%, iron 19.35%, aluminum 13.65%, calcium 10.4%, magnesium 7.05%, and titanium 4.9%, with slightly different proportions for the fine-grained material. The rare gases found in the regolith (the lunar surface layer, including impact material and dust) included Argon 36, Argon 40, krypton, and xenon. The proportion of glassy fragments was a surprise: the glass particles comprised a small volume of the lunar rock, at only 0.01%, but they came in strange forms – spheres, cylinders, globules, and dumbbell shapes.

Judging by the basaltic fragments, the Sea of Fertility appeared to have formed from at least two distinct volcanic events between 3.347 and 3.421 billion years ago, the main date range of the samples. Agewise, this placed it somewhere between the two other sets of samples collected at that point: Apollo 11 and 12. The regolith was thin with a young blanket of ejecta (rocks gouged out during a crater impact). Later analysis showed that 25% of Luna 16's rock was actually highland rock, ejected and scattered there by impacts elsewhere. Some of these rocks, older and younger, were mixed in with the main, local sample. The intense heat of impact caused some rock to form into spherical particles and basalt glasses – what are called impact melts. A footnote on the lunar soil density measured by Luna 16 was that it was possible to extract, from the radar measurements of all the landers, surface density data that were later found to be well calibrated with the real density measured from returned Moon rock. The next probe was Luna 18, which made the first attempt to land in the lunar highlands. It had the misfortune to land on a steep slope and was wrecked, but mission scientists were able to recover soil density data from the altimeter readings and match them against earlier missions. These are given in Table 3.13, along with other missions [15].

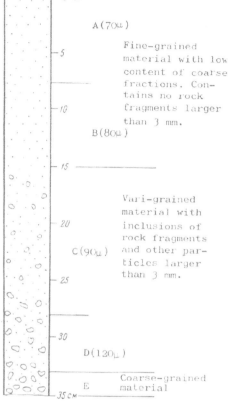

Figure 3.13. Luna 16 particle analysis.

Table 3.13. Soil density measurements, later Lunas.

Probe	Density g/cm^3
Luna 11	1.14–1.28
Luna 12	0.72–2.55
Luna 14	1.07
Luna 16	0.74
Luna 17	0.80
Luna 18	0.85
Luna 19	1.48

LUNA 20: INTO THE LUNAR HIGHLANDS

Luna 20, in February 1972, came down in 1,500-m-high mountains north of the Sea of Fertility just 120 km away from Luna 16. It touched down in a plateau between two peaks – a risky landing site (Luna 18 had crashed there earlier), on a continental region reckoned to have been formed long before the Sea of Fertility was flooded. Safely on the lunar surface, the drilling operation began, rotating anti-clockwise at 500 rpm, cutting away with sharp teeth and then putting material into a holding tube. The drill had two engines: one for the main drilling, but a second to take over if it faltered. The drill was kept sealed until the moment of drilling began, for it was important to keep it lubricated right up to the moment of operation. If it were exposed to a vacuum too early, there was the danger that the lubricant would evaporate. The drilling operation took 40 min and was photographed throughout. The rig encountered stiff resistance at 10 cm and operations had to stop three times, lest it overheat. When it reached 25 cm, the samples were scooped into the return capsule for the long journey home. The recovery on Earth was probably the most difficult of all the sample recovery missions, for the cabin landed on an island in a river during a snowstorm. The conditions were undoubtedly tough and the sample probably much smaller than hoped for, at only 30–50 g.

But it was Moondust all the same and light ash-gray dust was calculated to be older, at 3.9 billion years. The color was lighter and had more particles than the previous sample. Soil densities were very similar to Luna 16: 1 g/cm^3 for loose material and 1.8 g/cm^3 for packed material, with an average density of 1.15 g/cm^3. Its composition, though, marked it as different from both the Apollo and Luna *mare* samples. The sample consisted mainly of anorthosite débris, with olivine, pyroxene, and ileminite, which scientists called "highland basalt". High-quality non-rusting iron was found – one of the most interesting findings. Luna 20's samples had the highest content of aluminum and calcium oxides of all the Moon samples. Iron and titanium were low. There were fewer glasses than Luna 16, but the rocks had as many as 300 crystals of feldspar, olivine, and pyroxene, with some tiny brown and green globules. There were no *mare* basaltic fragments. Lunas 16 and 20 found three minerals – chromium, titanium, and zirconium – that are rare on Earth, but with very little lead, sodium, or potassium and only minute traces of gold or silver. Some Luna 20 troctolites and gabbros were dated to 4.19 billion years and the troctolites were the most primitive ever found – one of the oldest parts of the Moon recovered. The Americans came to characterize these highland rocks as ANT, or Anorthosite, Norite, and Troctolite. Seven Rare Earth Elements (as in KREEP) were identified.

As was the case with Luna 16, particle size was bigger the deeper one drilled. According to analyst V. Gromov, "the samples consist of small mineral particles that differ in shape. The particles easily stick to each other to form separate clots and aggregates. In its granulometric composition, lunar soil resembles dusty sand". Luna 16 particles averaged a size of 0.0303 mm, Luna 20 0.0132 mm (American highland particles were also smaller). One of the most intriguing samples was a complex pyroxene grain classified as pigeonite K4, which came from slowly cooling magma – but was closely similar to a meteor impact in Moore County in the United States, presumably from a lunar impact of sufficient force to send a fragment crashing into Earth. Another sample, a phosphorous metal particle 2004-016, was like an iron meteorite, but seemed to be so old that it might have come from an early period of planetary formation.

The top layer of the regolith was estimated to be 2–12 m deep, the bottom layer 18–38 m. Luna 20 found that solar rays had penetrated down to 20 cm. Not only that, but cosmic rays had worn away the crystal lattice of the aluminum rocks, transforming them into metal aluminum particles. Some of the sample appeared to be ejecta from the nearby Apollonius C crater and seemed to have been well irradiated by solar cosmic rays. Luna 20's rocks, at 3.9 billion years old, were not that different from the highland rock samples collected by America's Apollo 16 later that year, but with less aluminum and more anorthosite. The point of 3.9 billion years was a significant one in the evolution of the Moon, for it marked the formation of the large basins, the light-colored plains, and the end of what was the Late Heavy Bombardment. One sample, dated as 4.51 billion years old, showed evidence of lunar cataclysms at 3.9 billion and again between 4.17 and 4.2 billion years ago – either large-impact events or intensive bombardment. Later that year, 2 g of Luna 20 samples were exchanged with American Apollo 15 samples [16].

LUNA 24: THE LONG VIEW

The final mission, marking the end of the Luna series, was Luna 24, which landed in the flat Sea of Crisis in August 1976. This time, the drill had been adapted to penetrate far into the lunar soil and extract a long core sample, 2.5 m down. As the rotary percussion rig drilled into the soil at an angle of 30°, the sample was stored in a rubber pipe in such a way as to prevent clogging and compression and minimize the grains falling off. The little return capsule landed off course in Siberia, but was safely retrieved with a 170-g sample, the largest from the series. Luna 24's landing site was in the middle of the Sea of Crisis, 40 km from its rim of 4-km-high mountains, between crater Fahrenheit and Hill 5408. The *mare* is depressed about 5 km lower than surrounding highlands and comprises dark *mantle* deposits and ridges, the lava coming from at least one wave of lava flooding.

Some 20 different layers could be identified in the core sample, but there were four main parts. The upper part (10 cm) had large, loose fragments, while the middle upper part (60 cm) had smaller fragments rich in iron, the lower middle (30 cm) coarse fragments, and the lowest part (50 cm) was light-gray regolith with few rock fragments and little iron. The sample had 1,937 loose particles.

Pravda, on 5th September, related how 60 different chemical elements had been found, including calcium, in rocks that were dark-gray to brown in color. They appeared to be laid down in layers. Luna 24's sample comprised breccias (31.8%), basalt (21.1%), various minerals (21.2%), and glasses (6.1%), and analysts concluded that *Mare Crisium* was probably the second *mare* area to be filled with basalts. The regolith was deeper and more developed than Luna 16. For his 70th birthday that December, Soviet President Leonid Brezhnev was given a beautifully sculpted steel model of Luna 24, with three tiny samples of its Moon soil in trays with a magnifying glass over them so he could see them. The model and samples were subsequently minded by his widow Viktoria and then his daughter Galina, who later sold them to a space collector.

Little attention was paid internationally to Luna 24 at the time and the Soviet press gave it cursory coverage, for the lunar program was winding down at that stage. Luna 24 was the last of the American or Russian missions that brought back lunar rock over 1969–1976, was the last lunar mission for any country until the 1990s, and they remain the last Moon samples brought back to Earth. The lack of attention given to Luna 24 was disappointing, for it was much the most productive of the Soviet lunar missions. That was certainly what the Americans thought as well, for NASA hosted a three-day conference on Luna 24 in Houston, Texas, in December 1977. The results were so substantial that a subsequent 500-page book of conference papers was published: *The View from Luna 24*. New papers on the Luna 24 rocks were still being published more than 30 years later. Strangely enough, Luna 24 was back in the news then: European SMART orbiter scientists were surprised when SMART "discovered" calcium in the lunar surface when it flew over both *Mare Crisium* and then the lunar highlands east of the Sea of Fertility – until someone drew their attention to the Luna 24 and 20 samples, which had found plenty of calcium so many years earlier.

Part of the significance of Luna 24 lay in its landing site. Russian determination to land in the Sea of Crisis was evident, for it had been the target for Lunas 15 and 23 earlier (indeed, their wreckage cannot have been far away). Now, the explanation became clear: the Sea of Crises had a number of craters whose ejecta could tell us quite a lot about the history and age of the Moon. Most of Luna 24's rocks were the local rock, a basaltic flow formed when the *mare* was formed (their age was pinpointed at 3.273 billion years), with "melt inclusions" or rocks that had solidified at high temperatures when the *Mare Crisium* filled with lava, thus preserving some of the history of the melting process. But there were interesting layers on top, mainly ejecta from *mare* rocks and a little from the highlands.

Luna 24 landed 18.4 km from Fahrenheit, a 6.4-km-wide crater with a 335-m rim. Fahrenheit is estimated to have been caused by a meteorite that ploughed into the lunar surface 2 billion years ago to a depth of 100 m and threw up so much ejecta (rocks up to 20 m across) at a speed of up to 230 m/sec so that it comprised a significant part of the Luna sample. There was evidence of one smaller local cratering event since then, probably 500 million years ago. The rest of the sample came from primitive magma, some from deep down in the lunar interior, what was called a primary melt, and from craters and their secondary impact further afield. Impacts from two other *Mare Crisium* craters were also apparent in the Luna 24 sample: Picard and Pierce, both young craters formed after the *mare* filled. The Luna 24 samples contained high levels of magnesium-full minerals and glasses from a *mare* where such substances are otherwise absent – yet Apollo 15 data indicated that Picard and Pierce were high-magnesium craters, so this is most likely where the material came from. The second type of rock in the Luna 24 sample is estimated to be 500 million years old, so this could indicate the age of impact of these craters. Some Luna 24 material came from further afield still, judging by their rays, from craters Langrenus, Taruntius, Proclus, Swift, Greaves, Thales, and Bruno, the last being a far-side crater, so Luna 24 may have obtained what in effect are far-side samples. Some metallic fragments found in the sample were even more exotic and were meteor remains from deep space.

From this, scientists were able to reconstruct the history of the area. The *mare* was probably once a lava lake formed during a single extrusion in the period 3.27–3.65 billion years ago, much later than the Apollo 17 site. The basalts melted over the course of the period 3.36–3.14 billion years ago, about the same time as the Ocean of Storms, where they were compared against the Apollo 12 sample. This made it some of the youngest lunar rock, confirmed by low levels of argon 36, argon 40, and helium from the Sun. The surface appeared to have been undisturbed for the last 300 million years, indicating very little bombardment during this period. Indian scientists who analyzed the samples confirmed this, for nitrogen comprised 55 ppm, mainly coming from the solar wind – quite a low value compared to some Apollo samples and indicating that the soil was relatively undisturbed in recent geologic years.

Turning from the history to the detailed composition, there was much of interest. Luna 24's rocks were quite distinctive. The range of elements found in the rocks was a long one: olivine crystals, a little nitrogen, pyroxene, plagioclase, iron, magnesium, silicon, and in many forms, such as glass crystal powders, metal particles, basalt, glass fragments, troctolite, agglutinates, clods, melted rock, basaltic glass, and norites. Some basaltic fragments had within them olivine glass. There were 80 grains of olivine. Rhönite was found in 8-μm tan to dark-brown grains, looking like silica. This was the first such discovery on the Moon, rhönite being an unusual combination of titanium and pyroxene. Lithium was found but at quite a low level, at 8–10 ppm, indicating low KREEP. Analysis by Indian scientists found olivine, plagioclase, pigeonite, enstatite, native iron, and metals. Titanium levels at the Luna 24 site were the lowest on the Moon (1.1%) to the point that the Americans called the rock VLT, or Very Low Titanium, matched only by their own nearby Apollo 17. This VLT feature was the most distinctive aspect of the Luna 24 sample, effectively a new type of basic rock of low titanium and high magnesium, possibly thrown up when the Fahrenheit meteorite burrowed deep into the lunar magma. In effect, Luna 24 brought back rock that came from deep down within the Moon. VLT was matched by high levels of aluminum (19%) and iron (20%).

Years later, using Luna and Apollo data, scientists put together a titanium map of the Moon and, as a result, imagined how lava had mixed during the early years of the Moon. The Luna 24 glasses and glass beads were black, brown, tea-colored, and green, and scientists speculated that the brown ones came from came from lunar magma and the green ones from meteorite impacts fusing with the regolith. Some of the beads were largely hollow or with big interior voids and some showed microcraters within them.

Information was provided on the depth to which cosmic rays penetrated the surface. The abundant presence of helium and argon in the soil was evidence of solar wind absorption and enabled scientists to range the sample to between 3.24 and 3.30

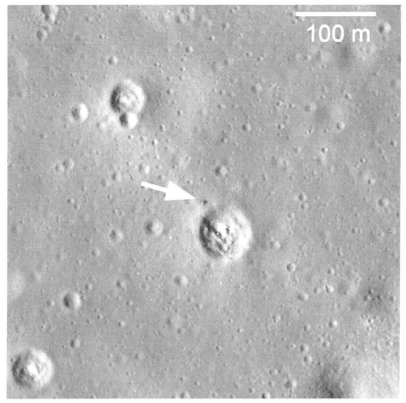

Luna 24 on the lunar surface (Credit: NASA)

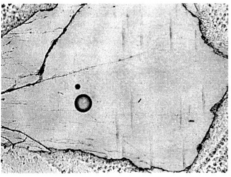

Luna 24 samples under the microscope
(Credit: Natasha Khisina)

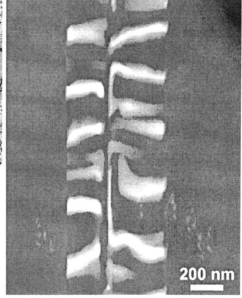

Table 3.14. Sample return missions.

Luna 16	12 Sep 1970	Sea of Fertility
Luna 20	14 Feb 1972	Apollonius
Luna 24	9 Aug 1976	Sea of Crises

billion years. Cosmic ray tracks were found in the samples and some galactic cosmic rays (mainly thorium uranium) reached lunar crystals as far down as 8 cm. In one olivine crystal, eight solar cosmic ray tracks were found, ranging from quite recent to 2.5 million years old, but providing a record of the changing composition of cosmic rays over time. One of the more controversial findings was made by British scientists in Newcastle on Tyne. They found evidence of prior magnetism and suggested that the Moon may once have had a strong surface magnetic field.

By the end of the series, the Soviet Union had collected three sets of samples: two were *mare* samples (Luna 16, 24), one highland (the anorthosite débris of Luna 20). Sample dates crossed an eon (averages of 3.27 for Luna 24, 3.41 for Luna 16, 3.84 for Luna 20). As one American observer, Don Williams, pointed out, though, "every spot was different" [17]. Generalizations about lunar composition and formation became difficult to make. Mingled in with each sample were the remains of impacts from other points on the Moon, near and far. As the great Apollo geologist instructor, Farouk al Baz, pointed out to his astronaut team, the point of collecting rocks was not so much to ask *What is it made of?* but *How did it get to be there?* The three Lunas answered a somewhat different question: *What can we learn from its past history and evolution of the rock?* The three landing sites, between them, provided an excellent cross-section of the eastern side of the Moon.

By the end of the series, Soviet scientists were able to publish a set of comparative tables of the composition of lunar rock, matching Lunas 16, 20, and 24 against the Apollo samples, according to the main chemical elements, trace, and rare elements. It was possible to devise a model for the evolution of the Moon, from its formation 4.6 billion years ago to the creating of the *maria* (4–3.8 billion years), the upwelling of lava to form *mare* (3.8–3.1 billion years), the end of the great bombardment to the relatively quiet period from 3 billion years ago. The missions are summarized in Table 3.14.

ROVING THE MOON

The final part of the Soviet Moon program was the lunar rover. The idea of a lunar roving vehicle went far back in the Moon program, the first such ideas being sketched in 1957. Approval for a lunar roving vehicle was given in the government decision to send cosmonauts to the Moon in August 1964. It was intended to land rovers on the Moon *before* cosmonauts arrived, where they would test out the suitability of landing grounds.

In the event, the rovers became, like the soil sample missions, part of the program of automatic exploration of the Moon when the possibilities of achieving a manned lunar landing receded. Because of their ability to travel far across the lunar surface, the Moon rovers came to provide a substantial volume of science about the Moon. The vehicle, to be called "Lunokhod" or "moon walker" in Russian, weighed 756 kg and was 4.42 m long (lid open), 2.15 m in diameter, and 1.92 m high. Its wheel base was 2.22 × 1.6 m. The main container was a pressurized vehicle, looking like an upside-down bathtub, carrying cameras, transmitters, and scientific instruments. It was kept warm by a small decaying radioisotope of 11 kg of polonium 210. The eight 51-cm-diameter wheels were made of metal with a mesh covering.

Lunokhod, to be delivered to the Moon on a landing stage similar to that of the sample return missions, would be steered by a five-person ground crew in a mission control: commander, driver, navigator, engineer, and radio operator. They used four 1.3-kg panoramic cameras able to scan 360° around the rover and two television cameras to scan forward, with a 50° field of view and 1/25-sec speed. Lunokhod would work 14 days at a time, during the lunar day, and then be closed down for the subsequent two weeks of lunar night. The control center was in the Crimea. Lunokhod carried several instruments (Table 3.15): a RIFMA X-ray fluorescent spectrometer to determine the composition of the Moon rock; PROP (*Pribor Otsenki Prokhodimosti*, literally Instrument for Passability Estimation), which was a penetrometer and detector to measure the distance traveled; an X-ray telescope; a cosmic ray telescope; an energetic particles detector; and a French-built 3.7-kg laser reflector, designed to measure the precise distance between Earth and the rover.

Lunokhod and its laser

Lunokhod wheel system

In the event, the Lunokhod mission attempted to combine as extensive an exploration as possible over the lunar surface, with stops to investigate rocks of interest. Scientists were most interested in the Lunokhod stopping to sample rocks with the RIFMA and test the density of the soil with the penetrometer. Engineers, though, were interested in seeing how far Lunokhod could travel and the press, especially *Pravda*, would telephone each day to ask "How far did Lunokhod travel today?" There were tensions between the engineers who wanted to travel and the scientists who wanted to investigate, the former once famously quarreling with the scientists, reminding them that the mission was "Lunokhod, not Lunostop!".

The first Lunokhod was delivered by Luna 17 to the Sea of Rains in November 1970. On the way out to the Moon, Luna 17 recorded fluxes of protons in the 1-5-MeV range, matched against Earth-orbiting satellites and a spacecraft then on the way to Venus, Venera 7. The 3-cm radar, which brought Luna 17 in to land, was also used to measure the density of the lunar soil – 0.8 gm/cm^3, slightly harder than that

Table 3.15. Lunokhod instruments.

Laser reflector
RIFMA (Röntgen Isotopic Fluorescent Method of Analysis) X-ray fluorescent spectrometer
Extra-galactic X-ray telescope
Cosmic ray background radiation detector
PrOP (*PRibor Otsenki Prokhodimosti*) penetrometer.

of Luna 16 (0.74 gm/cm^3). The exploration area was chosen to be a quiet area free of obstacles on the north-west of the Moon and Lunokhod landed some 3.5 km from the nearest identifiable crater, Heraclides F. The area was a plain interspersed with ridges and craters.

Locating its precise position, though, took a little time and here the laser came into its own. Bad weather prevented the use of the laser until early December. With the help of Nikolai Basov of the *Molekularni* missions, a search area of 10-km diameter was established, which was bombarded with 170 laser pulses over 40 min sent both from the Crimea and the French Pic du Midi observatory. Lunokhod was located about 5 km west of the center of the search area (35° 53′W, 38° 17′N).

Lunokhod was one of the sensations of the space age, attracting much admiration and favorable press comment. On the first lunar day, it traveled only 197 m, testing out its systems. On the second lunar day, it made a 1,502-m journey to the south-east, returning to the landing site in January 1971 on its third lunar day. On the fourth lunar day (February 1971), it began a long journey to the north, achieving a record 2,004 m. The following lunar day, it became trapped in a crater, but resumed its journey the following lunar day. In July 1971, power began to fail, so it traveled much less, stopping more frequently (every 65 m) and carrying out mainly static experiments. Energy supplies finally ran out in October, by which time this superbly built robot had traveled 10.54 km, covered an area of 80,000 m^2, sent back 20,000 pictures including 200 panoramas, and X-rayed the soil at 25 locations. It lasted three times longer than planned.

The first point of scientific interest was the regolith over which Lunokhod traveled. Television images showed that the surface had been bombarded by numerous small particles that made shallow pits between 5 and 10 cm deep. The soil analyzer RIFMA bombarded the surface with X-rays and enabled ground control to read back the chemical composition of the basalt-type soil: the rocks were abundant in aluminum, calcium, silicon, iron, magnesium, and titanium. The lunar regolith was found to be crushed basalt, the effusion of basaltic lava coming out of the *Mare Imbrium*. There was little variation at the many stations sampled in the course of Lunokhod's 11-km journey. The surface was dusty but cohesive: as Lunokhod drove, the soil disintegrated into lumps of fine-grained material, forming steep walls that did not crumble. Lunokhod's wheel typically made a tread of 0.5–1 cm. At one stage, Lunokhod was deliberately run over the edge of an elongated rock in order to break it, which it did, splitting off a fragment. The regolith was estimated to be between 2 and 6 m deep. The chemical composition changed little over its route, averaging 21% silicon, 10% iron, 8% aluminum, typical of *mare* basalt.

From time to time, the PROP mechanical rod jabbed into the soil to test its strength and this was an important aspect of the work of Lunokhod. The vehicle would stop and PROP would be lowered at the back of the vehicle, coned at an angle of 60°. First, it was forced into the surface to a depth of up to 5 cm to test force. Some pressure was applied, equivalent to a sixth of the weight of the rover. Then, vanes inside the cone were rotated 90° for torque, again another measurement of surface strength. This was done 500 times. The results of penetration and trafficability tests were published in detail, finding that Lunokhod operated on

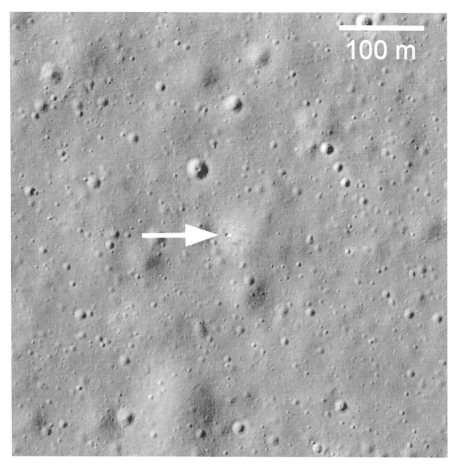

Lunokhod's final resting place; its last wheel marks are still visible (Credit: NASA)

surfaces that were much weaker than the lunar roving vehicles driven by the Apollo 15–17 astronauts. Its bearing capacity was estimated at ranging from 0.2 to 1.1 kg/cm.

Although most of Lunokhod's work focused on the Moon, it was also used to measure solar cosmic ray bursts and these were matched with measurements made by Venera 7, which received them about 17 hr earlier. There were several increases in solar cosmic rays (5–15th November, 11th–21st December), with the two spacecraft charting increases, peaks up to 30 MeV, and decreases, matching them against flares from the center and east of the Sun and subsequent magnetic storms. Solar flares were detected on 18th and 19th November and there was a sudden burst of cosmic rays, between 12th and 22nd December. There was a 60-sec solar X-ray burst on 10th December and Lunokhod charted the rise of the burst (8 sec), folding (15 sec), and then fizzling. The photon telescope found two strong sources of cosmic X-rays.

Even when Lunokhod finished active operations, it continued to function, for it had been parked in such a position as to aim its laser toward Earth, although the main laser-ranging experiments were not conducted until 1974–1975. It was used to fix the Earth–Moon distance with great accuracy and also to pinpoint Lunokhod's own final position (for the record, 38.3689°N, 35.1537°W). To raise funds for the ailing space program in the 1990s, the rover was sold at auction to a private collector, so this information is important so as to arrange collection. The story had a sequel, for in spring 2010, Lunokhod was relocated on the Moon's surface by the powerful cameras of the American Lunar Reconnaissance Orbiter. In April, astronomers at Apache Point Observatory in New Mexico re-activated the laser with their 3.5-m telescope, getting a surprisingly strong signal of 2,000 photons, making possible the first ever daytime ranging. It was expected to track the way in which the Moon's orbit was continuing to recede from the Earth [18].

LUNOKHOD 2: ALONG THE RIM OF LE MONNIER

Off the edge of the Sea of Serenity, the now eroded remains of the Le Monnier crater cut into the edge of the rocky Taurus Mountains.[2] Le Monnier was a 55-km-wide crater filled by *mare* rather than a pure *mare*: it was a flooded rim, rather than a sharply defined crater, formed around the time of the Imbrium Basin. Moreover, it marked the transition between the low *mare* and the upland continental area. There, the second Lunokhod landed in January 1973. Lunokhod 2 was a distinct improvement over its predecessor, 100 kg heavier at 840 kg. It could travel at twice the speed, up to 2 km/hr, and had higher and faster cameras to facilitate the drivers. There were new scientific instruments, most notably a photodetector called *Rubin* to detect ultraviolet light sources in our galaxy and the level of Earthglow on the nighttime Moon (Table 3.16).

The first part of Lunokhod's mission was to head south to the southern end of Le Monnier crater and explore its edge. On its first day, Lunokhod 2 crossed the gentle undulations of the crater floor, peppered by small crates, the regolith being estimated as up to 6 m deep. Initial chemical analysis gave an aluminum content of 9% and iron 6%, suggesting a thin supply of highland material mixed with the upper layer of the *mare* floor. Lunokhod duly reached the crater edge after a 9.8-km journey in the

Table 3.16. Lunokhod 2 instruments.

Soil mechanics tester
Solar X-ray detector
Magnetometer
Photodetector (*Rubin*)
Laser reflector (France)

---

[2] Some journals use "Lemonnier" but "Le Monnier" is the standard.

course of its second lunar day (February 1973). The nature of the ground varied from soft and loose to hard and firm. It ascended the edge of a 2-km-wide crater at the Vstrechnye Hills (Russian for "oncoming"). Here, the regolith was 10 m deep. The highland nature of the rock was evident from declining iron content (4.9%) and rising aluminum (11.5%). Landslide terraces of up to 15 m were seen, with gradients up to 20°. The panoramic pictures showed, beckoning in the distance, the lunar mountains to the east of the Le Monnier rim. Lunokhod 2 had to continuously avoid small boulders, analysis of which suggested that they were the fragments of cratering impacts.

Now, Lunokhod set out on its greatest adventure. At the end of the previous year, America's Apollo 17 astronauts, the final Moonlanding, had taken images of a hitherto unphotographed rille (tectonic breach) on the eastern edge of Le Monnier, 6 km long, 60 m deep, and up to 500 m wide, called *Fossa recta* (Latin) or *Borozh Pryamaya* (Russian). Lunokhod 2 set out there in March, in the course of which it traversed a record 16.5 km. It traveled along the *mare* side of the rim, the iron level rising back to 6%. Depth of the surface was only a few centimeters, but the wheels would sink in up to 20 cm in soft soil at the foot of slopes, making ruts like a tractor ploughing a field.

Then, it began the exploration of both sides of the rille – a dangerous undertaking, with many risks of hitting boulders, becoming trapped in craters at the edge, or even falling in. The rille was 80 m deep in places, with a slope of 35° and cluttered with rubble. The rille was full of interesting lunar geology and outcrops of bedrock. It cut below the regolith, exposing pristine basaltic magmatic bedrock. It was a tectonic fault, formed long after the *mare* itself. Iron content was typical of *mare*, at 8%. Although there were large stones at the rille's edge, Lunokhod 2 managed to navigate gingerly to the very edge and analyze the surface there. Pictures showed the winding rille against lunar hills in the far distance. Its rocky edges showed how the rille had cut into the regolith to the point that it almost disappeared at its rocky border.

Unfortunately, disaster struck the following month, when Lunokhod brushed against a crater wall that deposited clods of lunar soil on the insulating system, causing it to overheat. By the time the mission was over in June, Lunokhod 2 had traveled 37 km, sent back 86 panoramic pictures and 80,000 television pictures, and had covered four times the area of its predecessor. Up to 1,000 penetrations of the

Lunokhod 2 in Le Monnier bay, surrounded by mountains

Lunokhod 2 looking down into the steep rille Borozh Pryamaya

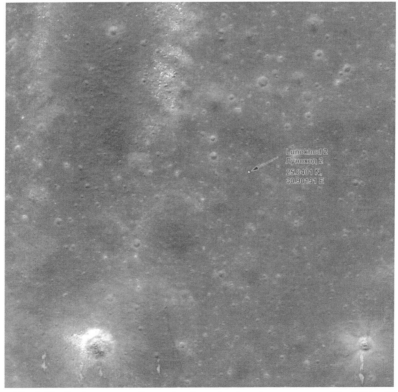

Lunokhod 2's final stop; track marks from its last journey are still visible (Credit: NASA)

lunar surface were made, while the laser fired 4,000 times to determine the Earth–Moon distance to an accuracy of 25 cm. When Lunokhod 2 failed, it was assumed that the dust blanket had also covered the laser, but the laser signal was recovered by the Pic du Midi observatory in France on 17th December 1973 and by the McDonald observatory in the United States the following year, enabling its final resting place to be measured: 30.8549°E, 25.842°N.

Again, the first point of scientific interest was the lunar surface. Detailed tables of the composition of the lunar rocks were published, comparing those sampled by Lunas 16 and 20, Lunokhod and Lunokhod 2. In the case of Lunokhod 2, it was possible this time to make comparisons between the composition of the *mare* and continental rocks, the proportions of aluminum, silicon, potassium, and iron being different. The most striking feature of the regolith was that it was a high alumina basalt. Chemical analysis of the surface was made at eight points, from the landing stage to the foothills, along the crater floor and to the rille. At the landing site, the RIFMA-M X-ray fluorescent spectrometer measured the soil near the lander as 24% silicon, 8% calcium, 6% iron, and 9% aluminum but at the edge of the Taurus Hills, there was a sharp rise in the iron content. The choice of route was vindicated, for the content readings were quite different at the boundary between the crater and the mountains. Aluminium content rose from 8.8 to 11.6%, calcium from 8 to 9%. Iron content had marked variations from 4% in the mountains to 6% in the crater and the highest levels were over 8% along the eastern wall of the rille. The proportion of iron became a good test of the age of rocks, as the last stage of lava outwelling from the core of the Moon onto the *mare* had the highest iron content. Eventually, a detailed geological and chemical map was published (for an outline, see Figure 3.14).

Back in Moscow, mission scientists made a geological map of Le Monnier bay, complete with slices of the surface, bedrock, and underlying strata. Lunokhod 2 made possible a definitive depth of the regolith. In Le Monnier, it was found to be 1 m at its minimum, 10 m at maximum, but the normal depth to be between 4 and 6 m (the deepest ever found was 25 m, at the American Surveyor 6 landing site). In traveling over the surface, Lunokhod 2 found that its wheels easily destroyed what appeared to be stones, but they were really large clods of fine-grained material stuck together. Typically, the top, dusty layer of the regolith comprised powdery granules of particles in the range 0.001–0.01 mm. Rock fragments were most plentiful on the outside of craters, notably those of 20-m diameter or more and, after that, the larger the crater, the more stones around it.

Following Luna 19's work in trying to measure local magnetic fields, one of Lunokhod 2's tasks was to measure the magnetic field along its journey. The average intensity was 20–30 γ, but fell to 10–15 γ in craters more than 50 m wide, suggesting a connection between impacts and magnetic anomalies. These findings were later matched against Apollo data to compile local lunar magnetic maps. Lunokhod 2's magnetometer determined that there was a very weak permanent magnetic field around the Moon. Data from Lunokhod 2's magnetometer, combined with that of Apollo 16, triangulated underneath the Sea of Serenity a large electrical anomaly, attributable to an upswelling of magma to the surface enriched in radioactive potassium, uranium, and thorium.

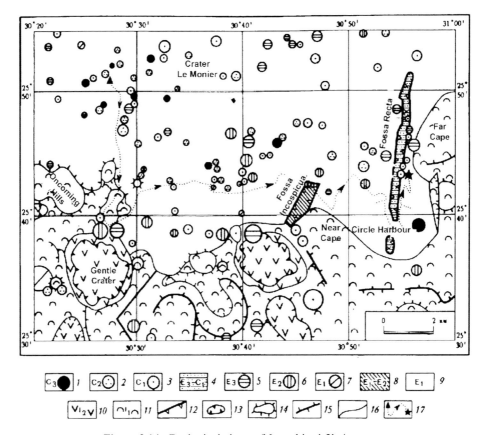

Figure 3.14. Geological chart of Lunokhod 2's journey.

One of its most interesting findings actually had nothing to do with the lunar surface, but the suitability of the Moon as a base for observing the sky. Whilst it would be excellent during the lunar night, during the daytime, the lunar sky was surrounded by a swarm of dust particles – a kind of atmosphere that would make telescopic observations very difficult. The astrophotometer determined that the lunar night sky was, in Earthglow, 15 times brighter than Earth's night sky in Moonlight. The night sky was unexpectedly bright in ultraviolet and visible light, with a twilight for 30 hr after the rover stopped for the night. The temperature of the lunar night was measured at –183°C. Lunokhod 2 continued the measurements of solar and stellar events begun by Lunokhod. For example, Lunokhod 2 detected a 29-keV X-ray burst on 17th January 1973, possibly from the Sun, although it was during a period of quiet Sun: it lasted 10 min, with a rise time of 1 min, then falling away [19]. The two rover missions are summarized in Table 3.17.

Table 3.17. The lunar rovers.

Lunokhod	10 Nov 1970	Sea of Rains
Lunokhod 2	8 Jan 1973	Le Monnier

The American Lunar Reconnaissance Orbiter (LRO) also relocated Lunokhod 2. Pulses were sent to its laser reflector, but it still had dust on it from its mishap, the signal return being only 750 photons. LRO also found Luna 20 in the highlands of the Moon and in *Mare Crisium*, both Luna 24 and, a mere 2.4 km away, the damaged Luna 23 with its unfired upper stage.

REVEALING THE MOON: WHAT WAS LEARNED?

The Soviet lunar program concluded in 1976. For the Russians, the principal disappointment must be that they had neither flown cosmonauts around the Moon nor landed them there, although we know that they had the technical capacity to do so. Nevertheless, there were substantial scientific outcomes for the missions conducted over the years 1959–1976. The following is a broad outline of the scientific outcome:

- the Moon was mapped and atlases made (Automatic Interplanetary Station, Zond 3, and the Zond 5–8 missions); mapping of selected areas of the near side was carried out by Lunas 12, 19, and 22;
- the environment of near Moon space was characterized by the orbiting missions: Lunas 10–12, 14, 19, and 22; data were obtained on the levels of solar and cosmic radiation, cosmic and solar particles, and gravitational fields;
- the Moon's surface was studied and characterized from above by instruments on Lunas 11–14 and 18;
- the Moon's gravitational field was first mapped by Luna 10, then in detail by Luna 14 and refined by Lunas 19 and 22; attempts to identify and then measure the Moon's magnetic field were made by the First and Second Cosmic Ships; a determination was made by Lunokhod;
- the chemical characteristics, composition, and density of the Moon rock were determined *in situ* by Lunokhod and Lunokhod 2 and through samples brought back to Earth by Lunas 16, 20, and 24; the lunar samples, although small, were shared internationally;
- a model of the regolith was established;
- the physical properties of the surface were determined by Luna 13 and the two Lunokhods (RIFMA);
- precise distances between Earth and Moon were measured by laser reflectors on Lunokhods 1 and 2;
- radiation levels and temperatures on the surface of the Moon were measured by Lunas 9, 16, 20, 23, and 24;

- the nature of the lunar micro-atmosphere was measured by the Lunokhods.

Neither the Apollo nor the Luna missions unambiguously answered the question as to the origin of the Moon. The Moon rocks had certain similarities with rocks from deep inside Earth's mantle – but not enough similarities – to suggest the two had exactly the same origin. They did, though, point the way over the following 40 years to a new understanding (e.g. the Nice Model, based on the city in southern France) of how the solar system formed – one that included a catastrophic impact of a planetary-sized body (almost as large as Mars) with Earth, which subsequently formed the Moon from elements of both, more so the impactor. Ring basins, for example, are far from unique to the Moon, suggesting common origins and systems of formation.

Substantial scientific results were achieved, starting with the First Cosmic Ship. These were the main scientific outcomes and the missions responsible:

- the solar wind was found (First Cosmic Ship);
- the Moon has no significant indigenous magnetic field (Second Cosmic Ship), but a trapped field that is a function of being in Earth's magnetic tail; in effect, this constitutes a permanent, but extremely weak, magnetic field (Luna 19, Lunokhod); it had some striking local features (Lunokhod 2);
- the far side of the Moon comprises largely highland areas, with one *mare* (the Moscow Sea), a *mare*-type floor of a large crater (Tsiolkovsky), and a huge basin towards the south pole (*More Mecht*, Sea of Dreams, or *Mare Igenii*) (Automatic Interplanetary Station, Zond 3);
- the Moon's surface is hard and sufficiently strong to take spaceships and vehicles from Earth (Luna 9, 13); the surface, though, is covered in a layer of dust for which it is important that vehicles have sufficient traction to cross. The surface varies in strength from one part of the Moon to another and can be soft and loose in places (Lunokhod);
- the lunar environment has extreme temperatures; radiation levels are within acceptable limits for humans, except during unusual events (e.g. solar flares) (Luna 9, 13);
- the Moon has mass concentrations (mascons) that disturb the paths of orbiting spaceships (Luna 10, 14);
- the basic composition of the Moon's rock is basalt (Luna 10); it comprises a mixture of powder, coarse, and fine grains, but is cohesive, rather like damp sand; glassy particles are embedded in the rock; it has absorbed cosmic rays over the years (Luna 16), numerous chemical elements can be found, 60–70 being typical (Luna 16, 20, 24); the highland samples had a high level of metallic elements, anorthosite, olivine, pyroxene, ileminite, and non-rusting iron, with high levels of aluminum and calcium (Luna 20); the color of the Moon rock was a mixture of gray and brown, with the soil set down in layers (Luna 24); the main elements in the basaltic rock are aluminum, silicon, potassium, calcium, and iron, but their proportions differ from place to place (Lunokhod 2); the age of the rocks ranged from 3 billion years (Luna 20) to 3.3 billion (Luna 24), but some fragments were found as old as 4.19 billion years (Luna 20);

- the Moon is surrounded by dust particles, sufficient to interfere with astronomical observations; the surface of the Moon in Earthlight is 15 times greater than the nighttime Earth in Moonlight (Lunokhod 2);
- the Moon is slowly moving away from the Earth, 38 mm a year (Lunokhod combined with Apollo data);
- the lunar environment is affected by the meteor showers that pass the Earth– Moon system and spacecraft may also be affected by impacts bouncing off the surface and reaching orbit.

REFERENCES

[1] Cherepaschuk, Anatoly: Sternberg Astronomical Institute and the beginning of the space exploration era, in Zakutnyaya, Olga, ed.: *Space, the First Step*. IKI, Moscow, 2007.

[2] Lemaire, J.F.; Gringauz, Konstantin: *The Earth's Plasmasphere*. Cambridge University Press, Cambridge, 1998; *Soviet Space Achievements*. Novosti, Moscow, 1965; Axford, Ian: The beginning, in Zakutnyaya, Olga, ed.: *Space, the First Step*. IKI, Moscow, 2007.

[3] Axford, W.I.; Verigin, M.: Tribute to Konstantin Gringauz (1918–1993), in Coates, A.J., ed.: Comparative studies of magnetospheric phenomena. *Advances in Space Research*, Vol. 16, No. 4, 1995.

[4] Barabashev, N.P.; Mikhailov, A.A.; Lipsky, Yuri: *Atlas of the Far Side of the Moon*. Academy of Sciences, Moscow, 1960; Shevchenko, Vladislav: Moon research for half a century, in Zakutnyaya, Olga; Odintsova, D., eds: *Fifty Years of Space Research*. Institute for Space Research, Moscow, 2009; Chikmachev, A.I., et al.: *Generalized Topography of the Lunar South Pole Aitken Basin*. Sternberg Institute, Moscow, 2009; Chikmachev, A.I.: *On the Discovery of the South Pole Aitken Basin*. Sternberg Institute, Moscow, 2009; Rodionova, Zhanna; Kozlova, E.A.: Morphological analysis of the cratering of the South Pole Aitken Basin. *Solar System Research*, Vol. 34, No. 5, 2000. For discussion on the role of the Late Heavy Bombardment (LHB), see Doressoundiram, Alain; Lellouch, Emmanuel: *At the Edge of the Solar System – icy new worlds unveiled*. Praxis/Springer, Berlin, 2010; Vernov, Sergei, et al.: *Radiation Measurements during the Flight of the Third Soviet Cosmic Rocket*. NASA, TTF 8,033; Dolginov, Shmaia, et al.: *Experiments on a Program of World Magnetic Survey*. NASA, TTF 14,407, 1972.

[5] Vernov, Sergei, et al.: Measurement of cosmic ray intensity by Zond 3, in Smith Rose, R.L., ed.: *Space Research*. COSPAR, Paris, 1966; Mirovna, M.N.: Structural characteristics of some craters on the far side of the Moon and Height profiles of 29 craters on the far side of the Moon, in Koval, I.K.: *Physics of the Moon and Planets*. NASA, TTF 566; Vernov, Sergei, et al.: *Variations of Cosmic Rays according to the Data of Interplanetary Probes Zond 3 and Venera 2*. NASA, Goddard Space Flight Centre, ST CR PF IS 10,550; Lyubimov, G.P.: *Measurement of the Intensity of Cosmic Radiation during the Flights of Automatic*

Interplanetary Stations Zond 1,2,3 and Luna 5,6. NASA, Goddard Space Flight Centre, ST CR IS 10,655.

[6]   Vernov, Sergei: State of prospects for studying the radiation zones of Earth, in Skuridin, G.A., *et al.*, eds: *Space Physics*, papers from conference held in Moscow, 10–16 June 1965; Galeev, Albert; Tamkovich, G.M., eds: *35th Anniversary of the Institute of Space Research of the Russian Academy of Sciences*. Author, Moscow, 1999; Vakulov, P.N., *et al.*: Investigation of cosmic rays, in Muller, P., ed.: *Space Research*, Vol. IV. COSPAR, Paris, 1963; Vernov, Sergei: Radiation investigations during the flight of interplanetary stations Mars 1 and Luna 4. *Cosmic Research*, Vol. 2, 1964.

[7]   Minchin, S.N., *et al.: Earth – Space – Moon*. NASA, TTF 800, undated.

[8]   Nazarova, Tatiana, *et al.*: Investigation of solid interplanetary matter in the vicinity of the Moon, in Mitra, A.P.; Jacchia, L.G.; Newman, W.S., eds: *Space Research*, Vol. VIII. COSPAR, Paris, 1967. Shelton, William: *Soviet Space Exploration – the first decade*. Arthur Baker, London, 1969; Shevchenko, Vladislav: Moon research for half a century, in Zakutnyaya, Olga; Odintsova, D., eds: *Fifty Years of Space Research*. Institute for Space Research, Moscow, 2009; Akim, E.L., *et al.*: Determination of the gravity field of the Moon from the motion of the artificial satellite Luna 10. *Cosmic Research*, Vol. 4, 1966; Mandelstam, S.L., *et al.*: Investigation of lunar X-ray emission with the help of the lunar satellite Luna 10. *Cosmic Research*, Vol. 4, 1966; Gringauz, Konstantin, *et al.*: Study of plasma in neighbourhood of the Moon using charged particle traps on the first lunar artificial satellite. *Cosmic Research*, Vol. 4, 1966; Vinogradov, Alexander, *et al.*: Mass gamma radiation of the lunar surface by Luna 10. *Cosmic Research*, Vol. 4, 1966; Zhuzgov, L.N., *et al.*: Investigation of the magnetic field by Luna 10. *Cosmic Research*, Vol. 4, 1966; Krupenio, N.N.: *Radar Studies of the Moon*. NASA, TTF 18,847; *Press Conference: Major Landmark in the Investigation of Outer Space*. NASA, Goddard Space Flight Centre, ST PR LPS 10479; Grigorov, Naum, *et al.: Study of Corpuscular Radiation on Spacecraft Luna 10*. NASA, Goddard Space Flight Centre, ST PF CR LPS 10,553, and *Investigation of Cosmic Radiation on the AMS Luna 10*. NASA, Goddard Space Flight Centre, ST LPS CR 10,526. Dolginov, Shmaia, *et al.: Possible Interpretation of the Results of Measurements of Luna 10*. NASA, Goddard Space Flight Centre, ST LPS NLS 10,618.

[9]   Grigoriev, N.L.: Solar cosmic radiation from Luna 11, in Mitra, A.P.; Jacchia, L.G.; Newman, W.S., eds: *Space Research*, Vol. VIII. COSPAR, Paris, 1967; Slysh, V.I.: Long wavelength solar radio emissions observed by lunar satellites Luna 11 and 12. *Cosmic Research*, Vol. 5, 1967.

[10] Sotnikov, B.I.; Baidal, G.M.; Sizentsev, G.A.: *The Part Played by the Department of Lunar and Planetary Studies of the Sternberg Astronomical Institute of Moscow State University in Lunar Exploration by Means of Rocket and Space Technology*. Sternberg Institute, Moscow, 2007; Surkov, Yuri, *et al.*: Gamma spectrometer analysis of lunar samples from Luna 16, in Rycroft, M.J.; Runcorn, S.K., eds: *Space Research*. COSPAR Paris; Mandelstam, S.L., *et al.*: Lunar X-ray and cosmic X-ray background measured by Luna 12. *Cosmic*

Research, Vol. 5, 1967; Grigorov, Naum, *et al.*: Flux of low energy protons near the Moon on 13th December 1966. *Cosmic Research*, Vol. 7, 1969; Grigoreva, V.P.; Slysh, V.A.: Long wave cosmic radio radiation in circumlunar space. *Cosmic Research*, Vol. 8 (1–3), 1970.

[11] Krupenio, N.N.: Lunar surface layer density measurement from spacecraft radar, in Kondratyev, Kirill; Mycroft, M.J.; Sagan, Carl: *Space Research.* COSPAR, Paris, 1970; Yakovlev, O.I., *et al.*: Scattering of meter radio waves on the lunar surface. *Cosmic Research*, Vol. 5, 1967; Kolosov, M.A., *et al.*: Measurement of characterization of ionosphere using radio signals from Luna 14. *Cosmic Research*, Vol. 8 (4–6), 1970.

[12] Grigoriev, A.A.: Study of characteristics of Earth's surface and some meteorological elements in global pictures from Zond 7, in Kondratyev, Kirill; Mycroft, M.J.; Sagan, Carl: *Space Research*. COSPAR, Paris, 1970; Grigoriev, A.A.: Study of lineaments from a Zond 5 picture of northern Africa, in Rycroft, M.J.; Runcorn, S.K., eds: *Space Research*. COSPAR, Paris, 1972. For Zond 8 photography analysis, see Basilevsky, Alexander; Ronca, L.B.; Ivanov, B.A.: *On the Rate of Ancient Meteoroidal Flux – studies of pre mare crater populations on the Moon.* NASA, LPI, undated; Florensky, Kirill; Basilevsky, Alexander; Grebennik, N.N.: *On the Dependence of Morphology of Lunar Craters from their Sizes.* NASA, LPI, undated; Sukhanov, A.L.: *Volcanism in Mare Orientale.* NASA, LPI, undated; Rodionova, Zhanna; Skobeleva, T.P.: *Multi-Ring Basins on the Moon, Mars and Mercury.* NASA, LPI, undated; Vinogradov, V.; Kondratyev, Kirill: Global photography of Earth and the possibility of data interpretation, in Lunc, Michal, ed.: *Proceedings of the XX International Astronautical Conference*, Mar del Plata, 1969; Rodionov, B.N., *et al.*: New data on the Moon's figure and relief based on results from Zond 6. *Cosmic Research*, Vol. 9 (1), 1970; Bredov, M.M., *et al.*: Investigation of proton fluxes in the 1.50 - 50 MeV range with Zond 4 and 5 interplanetary stations, in Vernov, Sergei, *et al.* eds: Physics of solar activity and dynamic processes in interplanetary space and in the Earth's magnetosphere, unreferenced papers; Kadel, S.D., *et al.*: *History of Mare Vulcanism in the Orientale Basin – mare deposit ages, compositions and morphologies.* NASA, XXIV Lunar Planetary Science Conference; Rodionov, B.N.: Relief of the Moon's reverse side according to Zond 8 photographs. *Cosmic Research*, Vol. 15, 1977.

[13] Nazarova, Tatiana: Investigations of meteoric matter in vicinity of Earth and Moon from orbital station Salyut and Moon satellite Luna 19, in Rycroft, M.J.; Runcorn, S.K., eds: *Space Research*. COSPAR, Paris, 1972. Krupenio, N.N.: Results of radar experiments performed on Luna 19 and 20, in Rycroft, M.J., ed.: *Space Research*. COSPAR, Paris, 1974; Savich, N.A.: Cislunar plasma model, in Mycroft, M.J.: *Space Research*, Vol. XVI. COSPAR, Paris, 1975; Nazarova, Tatiana: Measurement of meteoric material from spacecraft. *Cosmic Research*, Vol. 14, 1976.

[14] Vyshlov, A.S.: Preliminary results of cislunar plasma research by Luna 22, in Mycroft, M.J.: *Space Research*, Vol. XVI. COSPAR, Paris, 1975; Nazarova, Tatiana; Rybakov, A.K.: Investigation of meteoric matter by Luna 22 and Mars

7, in Mycroft, M.J.: *Space Research*, Vol. XVI. COSPAR, Paris, 1975; Akim, E.L.; Vlasova, Z.P.: Research on the Moon's gravitation field from measurement data on trajectories of Soviet artificial satellites of the Moon. *Cosmic Research*, Vol. 20, 1982; Vyshlov, V.S.: Observation of radio source occultations by the Moon and the nature of plasma near the Moon. *Cosmic Research*, Vol. 16, 1978; Chuchkov, E.A.; Tulipov, V.I.: Corpuscular albedo of the Moon based on data of the Luna 22 artificial lunar satellite. *Cosmic Research*, Vol. 16, 1978; Pavelyev, A.G., *et al.*: Bistatic planetary radar scattering characteristics. *Cosmic Research*, Vol. 16, 1978.

[15] Florensky, Kirill, *et al.*: Morphology, types and distribution of sizes of regolith particles in the Sea of Fertility, in Bowhill, S.A.; Jaffe, L.D.; Rycroft, M.J.: *Space Research*. COSPAR, Paris, 1971; Vinogradov, Alexander, *et al.*, eds: *Current Concepts Regarding the Moon*. NASA, technical translations, undated; Vinogradov, Alexander: Rare gases in the regolith from the Sea of Fertility, in Bowhill, S.A.; Jaffe, L.D.; Rycroft, M.J.: *Space Research*. COSPAR, Paris, 1971; Farrand, William: Highland contamination and minimum basalt thickness in northern Mare Fecunditatis, in Ryder, Graham, ed.: *Proceedings of the 18th Lunar and Planetary Science Conference*. Cambridge University Press, Cambridge, 1988; Cherkasov, I.I.; Shvarev, V.V.: *Lunar Soil Science*. Keterpress, Jerusalem, 1975; Ivanov, A.V., *et al.*: *Chondrule-Like Particles from Luna 16 and Luna 20* and *On the Intensity of Sodium Vaporization from the Mare Regolith*. NASA, Lunar and Planetary Institute (LPI henceforth); Kurat, G.; Kracher, A.: *Luna 16 Revisited – a progress report*. NASA, LPI; Mason, Brian, *et al.*: *Composition of Luna 16 and Apollo 14 Fines*. NASA, LPI; Reid, Arch M.: *Luna 16 Revisited – the case for aluminous mare basalts*. NASA, LPI; Khisina, Natasha, *et al.*: *Luna 16 and Luna 20 Exsolved Clinopyroxenes – estimation of subsolid cooling rates*. NASA, LPI; Albee, A.L., *et al.*: *Mineralogy, Petrology and Chemistry of Luna 16*. NASA, LPI; Schnetzler, J.A., *et al.*: *Chemical Composition of Apollo 14, Apollo 15 and Luna 16 Material*. NASA, LPI; Krupenio, N.N.: Radar measurements on automated spacecraft Luna 23. *Cosmic Research*, Vol. 16, 1978; Krupenio, N.N.: Lunar soil density measurements based on direct and indirect measurements. *Cosmic Research*, Vol. 15, 1977; Krupenio, N.N.: Radar measurements in automated lunar stations. *Cosmic Research*, Vol. 14, 1976.

[16] Kashkarov, L.L.: Track investigations of lunar soil returned by Luna 20, in Rycroft, M.J., ed.: *Space Research*. COSPAR, Paris, 1974; Kashkarov, L.L.: *Chemical Modification in the Lunar Olivine Microcrystals*. Vernadsky Institute Moscow; Leonovich, A.K.: Luna 16 and 20 investigations of physical and mechanical properties of lunar soil, in Rycroft, M.J., ed.: *Space Research*. COSPAR, Paris, 1974; Gromov, V.: *Physical and Mechanical Properties of Lunar Soil*. VNII Transmash, St Petersburg, undated; Ashikhmina, N.A., *et al.*: *The First Finding of Metal Aluminium Particles in Lunar Soil*. NASA, LPI; Ghose, Subrata, *et al.*: *Luna 20 Pyroxenes – exsolution and phase transformation as indicators of petrologic history*. NASA, LPI; Huneke, J.C.; Wasserburg, G.J.: *Evidence from Luna 20 Rocks from Lunar Differentiation prior to 4.51 AE Ago*.

NASA, LPI; Meyer, Henry: *Luna 20 – mineralogy and petrology of fragments less than 125mm size.* NASA, LPI; Nazarov, M.A., *et al.*: *Luna 20 P Rich Metal Particle Bulk Composition and Trace Element Particle.* NASA, LPI; Ridley, W.I., *et al.*: *Major Element Composition of Glasses in Two Apollo 16 Soils and a Comparison with Luna 20 Glasses.* NASA, LPI; Albee, A.L., *et al.*: *Luna 20 Samples 22012 and 22013.* NASA, LPI; Nazarov, M.A.: *Luna 20 – model mineralogy of highland rocks.* NASA, LPI; Philippot, J.C.; Jérome, D.Y.: *Elemental Composition of Luna 20 Regolith and Highland Basalt.* NASA, LPI; Walton, James: *Evidence of Solar Cosmic Ray Proton Production of Neon and Argon in Luna 20 Single Particles.* NASA, LPI; Nguyen, L.D., *et al.*: *Rare Earth Elements in Luna 20 Soils.* NASA, LPI.

[17] Kurat, G.; Krachev, A.: Luna 24 – a second case for early magnesian mare filling. Lunar and Planetary Science Conference, 17–21 March 1980, NASA; Assonov S.S., *et al.*: *Noble Gases in Agglutinates, Breccias and Fines from Luna 24 Regolith Sample.* Houston, LPI, 1978; Padia, J.T., *et al.*: *Total Nitrogen in Luna 24 Samples.* Kanpur, Indian Institute of Technology, 1979; Cohen, A., *et al.*: *Argon-40–Argon-39 Chronology and Petrogenesis along the Eastern Limb of the Moon from Luna 16, 20 and 24 Samples.* LPI, Houston; Deshpande, V.V.: *Thermogravimetric and X-ray Diffraction Analysis of Luna 24 Samples.* Bhabba Atomic Research Centre, Bombay, 1979; Murty, S.V.S., *et al.*: *Lithium in Luna 24 Samples.* Indian Institute of Technology, Kanpur, 1979; Graham, A.L.; Hutchison, R.: Mineralogy and petrology from Luna 24 core, in MPI für Chemie; Treimanh, Allan H.: Rhönite in Luna 24 pyroxenes and implications for volatiles in planetary magmas. *The American Mineralogist*, Vol. 93, No. 2, 2008; Weiblen, Paul W.: *Petrologic Peculiarities in Melt Inclusions in Luna 24 Samples.* University of Minneapolis, Minnesota. Hennessy, J.; Turner, G.: 40Ar–39Ar Ages and irradiation history of Luna 24 Basalts. *Philosophical Transactions of the Royal Society of London*, Vol. 297, No. 1428, 17 June 1980. Maxwell, Ted A.; Baz, Faroukh El: Sources of highland material in the Luna 24 regolith. Paper presented to conference on Luna 24 in Houston, Texas, 1–3 December 1977; Gillis, J.J.; Jolliff, B.L.: *Bimordial TiO$_2$ Content in Mare Basalts.* LPI, Houston, 2000; Stephenson, A.; Collinson, D.W.; Runcorn, S.K.: Rock magnetic and palaeomagnetic studies on Luna 24 samples, in Merrill, R.B.; Papike, J.J., eds: *Mare Crisium – the view from Luna 24.* Pergamon, New York, 1978; Perelygin, V.P., *et al.*: *Study of Fossil Tracks due to VH and VVH Groups of Cosmic Ray Nuclei in Olivine Crystals from Luna-16 and Luna-24.* Dubna, Poland and Bombay, India, 1979; European Moon probe founds calcium on lunar surface. *Science Daily*, 16 August 2006; Khisina, Natasha, *et al.*: *Lamellar Chromite Diopside Symplectites in Olivine from the Luna 24 Regolith.* Vernadsky Institute, Moscow, 2009; Williams, Don: *To a Rocky Moon – a geologist's history of lunar exploration.* University of Arizona Press, Tucson, 1993; Carr, Michael, *et al.*: *The Geology of the Terrestrial Planets.* NASA, Washington, DC, 1982, SP 469; Butler, P.; Morrison, D.A.: *Geology of the Luna 24 Landing Site.* NASA, LPI; Barsukov, V.L.; Florensky, Kirill: *The Lunar Soil from Mare Crisium – preliminary data.* NASA, LPI; Barsukov, V.L., *et al.*: *Preliminary*

Description of the Regolith Core from Mare Crisium. NASA, LPI; Smith, J.V., *et al.: Luna 24 – implications for remote sampling of unweathered regolith.* NASA, LPI; Kashkarov, L.L., *et al.: Track Studies in Four Samples of Luna 24 Core.* NASA, LPI; Grove, T.L.: *Cooling Histories of Luna 24 Low Ti Ferrobasalts and Ferrogabbros.* NASA, LPI; Basu, Abhijit, *et al.: Impact Melt Origin of Luna 24 Olivine.* NASA, LPI; Bell, P.M.; Mao, H.K.: *Crystal Field Spectra of Luna 24 Glass.* NASA, LPI; Norman, Marc: *Considerations of an Olivine Vitrophyre as the Parent Magma at Luna 24.* NASA, LPI; Norman, Marc; Taylor, Lawrence: *Chemistry and Petrology of Luna 24 Glass Chips.* NASA, LPI; Rao, M.N.; Venkatesan, T.R.: *Noble Gas Studies of Luna 24 Drill Core Samples.* NASA, LPI; Schaeffer, O.A., *et al.: Argon and Petrologic Study of Mare Crisium – age and petrology of Luna 24 samples.* NASA, LPI; Settle, M.J., *et al.: Mode of Emplacement of Fahrenheit Ejecta at the Luna 24 Site.* NASA, LPI; Tarasov, L.S., *et al.: Petrological Peculiarities of Basaltic Rocks from Mare Crisium* and *Geochemical Features of Lunar Rocks from Mare Crisium – peculiarities of the distribution of rare Earth elements and the problem of generation of magmas.* NASA, LPI; Vilas, Faith: *Mare Crisium from a New Angle – chemical composition from low altitude X-ray fluorescence data.* NASA, LPI; Allison, R.J.; McDonnell, J.A.M.: *Microscale Accretionary Particles and Impact Pits on Luna 24 Core Spherules.* NASA, LPI; Goswami, J.N.; Lal, D.: *Particle Tracks and Microcraters in Luna 24 Drill Core Sample.* NASA, LPI.

[18] Vernov, Sergei: Solar cosmic ray bursts in November–December 1970 from Venera 7 and Lunokhod, in Bowhill, S.A.; Jaffe, L.D.; Rycroft, M.J.: *Space Research.* COSPAR, Paris, 1971; Krupenio, N.N.: Results of radar experiments performed on Luna 16 and 17, in Rycroft, M.J.; Runcorn, S.K., eds: *Space Research.* COSPAR, Paris, 1972; Kocharev, G.E.: Investigation of solar X-rays from lunar surface by Lunokhod 2, in Rycroft, M.J., ed.: *Space Research.* COSPAR, Paris, 1974; Florensky, Kirill, *et al.:* Geomorphological analysis of area of Mare Imbrium explored by Lunokhod, in Bowhill, S.A.; Jaffe, L.D.; Rycroft, M.J.: *Space Research.* COSPAR, Paris, 1971; Calame, O.: Location of Lunokhod and determination of Crimea–McDonald chord length from lunar laser range measurements, in Mycroft, M.J.: *Space Research,* Vol. XVI. COSPAR, Paris, 1975; Kocharev, G.E.; Victorov, S.V.: Experimental data on the investigation of lunar surface chemical composition, in Rycroft, M.J.; Runcorn, S.K., eds: *Space Research.* COSPAR, Paris, 1972; Abdrakhimov, A.M.: *Reexamining Lunokhod Sites – old and new chemical data.* Vernadsky Institute, Moscow, 2009; Kokurin, Yuri, *et al.:* Laser radar location of light reflector on Lunokhod. *Cosmic Research,* Vol. 9 (4–6), 1971.

[19] Florensky, Kirill, *et al.:* The floor of Le Monnier crater – a study of Lunokhod 2 data, in *Proceedings of 9th Lunar and Planetary Science Conference,* 1978. NASA, 1979; Severney, A.: Preliminary results obtained with an astrophotometer on Lunokhod 2, in Rycroft, M.J.; Reasenberg, R.D.: *Space Research.* COSPAR, Paris, 1973; Barker, E.S.: Improved coordinates for Lunokhod 2 based on laser observations from McDonald observatory, in Rycroft, M.J., ed.: *Space Research.* COSPAR, Paris, 1974. Kocharev, G.E., *et al.:* Chemical

composition variations of the lunar surface in mare-highland contact zone, in Rycroft, M.J., ed.: *Space Research*. COSPAR, Paris, 1974; Dolginov, S.S.: Magnetic field in Le Monnier, in Pomeroy, John; Hubbard, Norman, eds: *Soviet American Conference on the Cosmochemistry of the Moon and Planets*. NASA, 1974; Shevchenko, Vladislav: Moon research for half a century, in Zakutnyaya, Olga; Odintsova, D., eds: *Fifty Years of Space Research*. Institute for Space Research, Moscow, 2009; Malenkov, M.I.; Stephanov, V.V.: Russian technologies of planetary rover locomotion systems, in Zakutnyaya, Olga; Odintsova, D., eds: *Fifty Years of Space Research*. Institute for Space Research, Moscow, 2009. For a discussion on the role of iron, see Lu, Yaoxiaoyi: *Iron Abundances in Regolith as Cosmogonical Indicator of the Evolution of the Moon*. Vernadsky Institute, Moscow, 2009; Abdrakhimov, A.M.; Shashkina, V.P.: *Lunokhod 2 Site – panoramas digitizing and new geochemical remote data*. Vernadsky Institute, Moscow, 2009; Dyal, P., *et al.*: *Mare Serenitatis Conductivity Anomaly Detected by Apollo 16 and Lunokhod 2 Magnetometers*. NASA, LPI, undated. For the story of the Lunokhod laser reflectors, see LRO teams helps track laser signals to Russian rover mirror. *Space Daily*, 27 April 2010, and Old Moon rover beams surprising laser flashes to Earth, *Space Daily*, 4 June 2010; Gurshteyn, A.A.: *Lunokhod 2 on the Lunar Continent*. NASA, TTF 15,401.

RUSSIAN-LANGUAGE REFERENCES

1. Голованов Я. Королев: факты и мифы – М.: Наука, 1994. – 800 с.
2. Логачев Ю.И. Исследования космоса в НИИЯФ МГУ – Москва, 2007.
3. Энциклопедия «Космонавтика».
4. В.С. Авдуевский, В.П. Сенкевич. Автоматические межпланетные станции для исследования Луны, Марса, Венеры и межпланетного пространства, созданные в СССР до 1982 г – М. ИИЕТ РАН. 1999 // Из истории ракетно-космической науки и техники. Вып. 3. стр. 48–90.
5. Соболев И. LRO – новые находки на Луне. // Новости космонавтики.
6. Шевченко В.В. Юрий Наумович Липский (к 100-летию со дня рождения) – Земля и Вселенная, №6, 2009, 48–58.

4

Unveiling Venus

Soviet space science achieved its greatest successes with the planet Venus, revealing Venus to be a very different world from what anyone had anticipated. In the 1950s, before the space age, there was a consensus that Venus was a benign world with a surface temperature of 60–75°C, watery oceans, and exotic plants like water lilies. Using his filters in Pulkovo University, astronomer Gavril Tikhov guessed that the plants might be blue in color. A few observers predicted that Venus might be a hot, dry world, but no one wanted to listen to their gloomy prognostications.

FIRST VENUS PROBES

Sergei Korolev and Mikhail Tikhonravov first proposed to the government in summer 1958 the sending of spacecraft to Venus and Mars. Calculations of the trajectories for the Mars and Venus windows were performed by Mstislav Keldysh and his applied mathematics institute. The orbits of the planets were such that opportunities arose to send spacecraft to Venus only every 18 months and Mars every 25 months. Originally, Korolev entertained hopes of getting a probe ready for the June 1959 window to Venus, but this turned out to be unrealistic, so he had to wait until February 1961.

Exploring a planet was a difficult proposition when the conditions at arrival were so unknown. Unlike the Americans, whose first Venus probes were designed to fly past the planet, Korolev aimed to reach the surface at the first attempt. Accordingly, the 1VA spacecraft was built in early 1961 – a cylinder with a dome on top, large parabolic aerial at the side, and other scientific instruments protruding. The underlying assumption was that the 1VA would, on arrival at Venus, meet a pressure of 1.5–5 atmospheres, a temperature of up to 75°C, and a composition of carbon dioxide and nitrogen. The dome was a thermal cover to protect the spacecraft during descent and buoyant so as to help the probe to float once it splashed down onto Venus's oceans. The dome included what was called a phase state sensor – a little like a builder's spirit level, designed to tell whether the spacecraft was stable or, if it moved, floating and which would be activated by a sugar lock when Venus's water

Table 4.1. First Venus probe instruments.

Cosmic rays and gamma fields detector
Counter to measure charged particles of interplanetary gas and corpuscular streams from Sun
Micrometeoroid detector
Magnetometer

---

melted the sugar. 1VA was 2.035 m high, 1.050 m in diameter, weighed 644 kg with cylindrical body and two solar panels with an area of 2 m^2 to soak in sunlight and turn it into electricity. It had instruments to send back information on radiation, micrometeorites, and charged particles, but there was insufficient weight to carry cameras (Table 4.1). There was no time to design and fit a mid-course correction engine, so the launch had to be precise.

The first attempt to send a probe to Venus, on 4th February 1961, failed, but the second probe was launched successfully on 12th February, being originally but confusingly called the "Automatic Interplanetary Station", the same as the lunar probe that flew around the Moon in 1959 (later, it was retrospectively named Venera 1). It was the first ever launching towards another planet, adding to the ever lengthening lists of Soviet space "firsts". As it sped away from Earth, its instruments detected Earth's magnetosheath. The magnetometer recorded a faint interplanetary magnetic field of 3–4 nT (nanoTesla) (Earth's is between 24,000 and 66,000 nT), so this was close to zero. The solar wind detected by the First Cosmic Ship to the Moon in January 1959 was confirmed. Konstantin Gringauz presented the results at a conference in Florence, Italy, later that year.

Regrettably, this was the sum total of useful information received, for contact with the probe was lost after five days – a failure of the thermal regulation system causing the orientation system to break down, meaning that signals could no longer be directed towards Earth. Calculations of its path suggest that it would have missed Venus by 100,000 km, so its buoyancy on the Venusian oceans never came to be tested.

The Venus probes were redesigned, with a standard design for Venus and Mars probes called the 2MV and, following this, the 3MV, with a parallel series to test new Venus and Mars probes called Zond. These comprised a main module and below it either an instrument compartment for photographing the planet on flyby or, alternatively, a landing module with a parachute. These were larger spacecraft, at 3.6 m tall, 1.1 m in diameter, weighing up to 1 tonne, with solar panels giving them a span of 4 m, and they were also equipped with a mid-course correction engine.

The first attempts to launch the new type of spacecraft failed in August 1962 and the first successful launch of the new spacecraft to Venus did not take place until 2nd April 1964, called Zond and carrying a lander. The increased weight enabled a much wider range of scientific instruments to be carried (Table 4.2).

Zond was the first Russian interplanetary spacecraft to carry a Lyman-alpha (α) photometer to detect cosmic rays. The lander had a counter to measure gamma rays from the surface of the planet, while the gas analyzer would measure the chemical

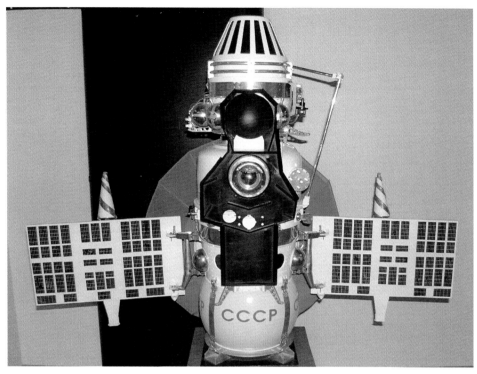

Venera 2MV and 3MV series

Table 4.2. Zond instruments.

Main spacecraft
Radiation detector
Cosmic ray detector
Magnetometer
Ion detector
Atomic hydrogen detector
Charged particles detector
Micrometeroid detector

Zond instruments: lander
Temperature, density and pressure sensors
Barometer
Thermometer
Radiation detector
Microorganism detector
Atmospheric composition detector
Acidity measurement detector
Electrical activity detector
Luminosity detector
Photometer

Table 4.3. Venera 2 and 3 instruments.

In-flight instruments
Magnetometer
Cosmic ray detector
Low-energy charged particle detector
Solar plasma detector
Micrometeorite detector (not Venera 3)
Hydrogen and oxygen spectrometer
Radio detectors in 150-m, 1,500-m, and 15-km bands

Venera 2 instruments
200-mm camera
Infrared spectrometer
Spectograph
Spectrometer

Venera 3 instruments: lander
Temperature, density, and pressure sensors
Gas analyzer
Photometer
Movement detector
Gamma-ray counter

composition of the atmosphere once it reached the surface. Although there was no camera on the lander, a photometer was installed to measure the light levels, but because the cabin was set to land on the nighttime side of Venus, this was made sensitive to the lowest possible level of light (0.001 lux) right up to 10,000 lux.

Like some deep-space probes that preceded it, the most valuable information came from the early part of its flight, as Earth receded in the distance. Zond took four photometer images of Earth at distances of 37,000, 47,000, and 15 million km in an attempt to measure Earth's hydrogen field – it found that the hydrogen corona extended out to 250,000 km, but was at its most intense from 1,600 to 20,000 km and that its intensity was down 25% compared to Mars 1 and Luna 4 figures, reflecting a decline of solar activity. The intensity of cosmic rays was measured at 2.1 particles/cm^2/sec. But Zond was soon in trouble – a weld break causing the cabin pressure to fall to zero, but communications were maintained until 24th May, a distance of 14 million km [1].

The next in this MV series were launched in November 1965, called Venera 2 (flyby) and Venera 3 (lander). The instruments described in Table 4.3 were installed.

Venera 2 passed Venus at 23,950 km on 27th February, but a thermal failure caused communications to be lost as it approached the planet. Communications with Venera 3 were lost even earlier, on 19th February, but the probe is calculated to have impacted on the planet on 1st March, the first spacecraft to hit another world. Denied information at Venus itself, Soviet scientists had again to content themselves with data from the in-flight instruments. The solar wind detectors followed the ups

Table 4.4. The first Venus probes.

Automatic Interplanetary Station	12 Feb 1961
Zond	2 Apr 1964
Venera 2	12 Nov 1965
Venera 3	16 Nov 1965

Missions failing to leave Earth not included

and downs of solar energy, including a spike on 15th December. They calculated that the solar wind took 7 hr to reach Earth. Meteorite measurements were also made and these overlapped with the flight of the Moon probe Zond 3. Together, they gave more than half a year of meteorite data across the solar system, both inwards and outwards of the Sun. Detectors showed Zond 3 passing through the *Perseids* and *Aquarids*, while Venera 2 passed through the *Taurids* – but there was almost entirely empty space otherwise. By the mid 1960s, it was possible to conclude that interplanetary space comprised these irregular streams of meteors concentrated in dense formations – but that there was almost nothing in between. Later, theories of planetary formation came to recognize the capacity of the gravitational dynamics of the solar system to sweep space into streams, rings, disks, moons, and planets.

By this stage, the Americans had received information from Mariner 2 as it passed Venus long before, in December 1962. Dreams of lily ponds were now fading away and when President of the Academy of Sciences Mstislav Keldysh wrote an article in *Pravda* on 6th March 1966 to mark the Venus landing, he expressed his personal belief that surface temperatures were in the order of 300–400°C and that the planet had succumbed at some earlier stage in its history to the "greenhouse effect", one of the first times this now common term had been used (the phrase had been invented and popularized by Carl Sagan in 1960, its main Russian promoter being Leningrad atmosphere and climate expert Kirill Kondratyev). As a precaution, future missions would be sent to come down at nighttime, when conditions would be as cool as possible. Table 4.4 summarizes the first probes.

AT LAST, SIGNALS FROM VENUS

Despite whatever Keldysh might say, those who believed in Venusian oceans had still not given up. The next probe, Venera 4, had an improved design and the cabin was, like its predecessors, designed to float. Weighing 1,106 kg, Venera 4 stood 3.5 m high, its solar panels spanned 4 m, and had an area of 2.5 m², while the high-gain antenna was 2.3 m in diameter. The descent cabin weighed 383 kg, comprising instruments, altimeter, thermal control, battery, and two transmitters in a pressure shell, topped by a lid and parachute. Venera 4 carried the same gas analysis set as had been installed on Zond, but also a hydrometer to measure the amount of water vapor in the atmosphere (Table 4.5). Despite its buoyancy, few thought the cabin would end up bobbing on a Venusian ocean wave, so the movement detector and its sugar lock were taken off to save weight.

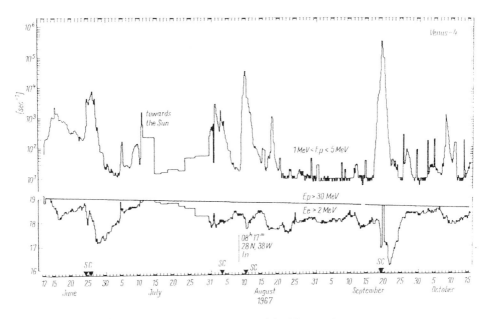

Figure 4.1. Radiation measured by Venera 4 en route.

In-flight measurements were again made (Figure 4.1). Galactic cosmic rays were counted throughout the route from Earth to Venus, with a record of the levels covering every day of the mission, showing quiet conditions (early September) and peaks (20th September). Venera 4 measured protons of 1–5 MeV, finding that their intensities en route were up 100-fold on 1965 data from Venera 2: there were radiation bursts during 50% of the journey, compared to 20% earlier. Venera 4 data were compared with an identical instrument on the failed Moon probe, Cosmos 159, stuck in high Earth orbit (400–60,000 km). This presented the unexpected result that Earth's magnetosphere was normally transparent, with just as many protons detected by Cosmos 159 as Venera 4. This was not the case during magnetic storms, when such protons might be captured and directed into the radiation belts. As it closed in on its destination, Venera 4 found an increase in solar radiation levels compared to Veneras 2 and 3, explicable by the now rising stage of the solar cycle.

Venera 4 arrived at Venus on 18th October 1967. Viktor Kerzhanovich (b. 1938) had built a Doppler tracking system, designed to follow probes closely during their descent. It duly began making ever more frequent clicking sounds as the Venera accelerated into Venus's atmosphere. The approach to the planet was perfect, communications were steady, and the lander was deployed, to descend under parachute to the surface. Signals from the main spacecraft ceased on schedule, presumably as it burned up, and then signals began to come through from the lander.

At 7.44 am that early autumn morning, telemetry operators exclaimed "We have a signal!" – the first time this had been received from a probe descending into the atmosphere of another world. The atmosphere was expected to activate the

Table 4.5. Venera 4 instruments.

Main spacecraft
Magnetometer
Cosmic ray detector
Ion detector
Spectrometer

Venera 4 instruments: lander
Thermometer, barometer
Radio altimeter
Gas analyzers

parachute when it reached a certain density – a point expected at about 26 km above the surface. The telemetry operators called out the readings every few minutes. The first was "pressure 960 mb, temperature 78°, altitude 28 km". A descent profile was published accordingly (Figure 4.2). When signals ceased 93 min later, the density of the atmosphere had risen to 22 atmospheres but recording had stopped there because it was off-scale (the expectation was 5 atmospheres) and the temperature was 277°C. The chemical analysis came in immediately, one scientist yelling out breathlessly "It's carbon dioxide! Oxygen 1%, no nitrogen! No water!". Presumably, at this point, the spacecraft was now on the surface, but no one was sure because the altimeter had failed.

An American spacecraft, Mariner 5, had flown past Venus at the same time, but it

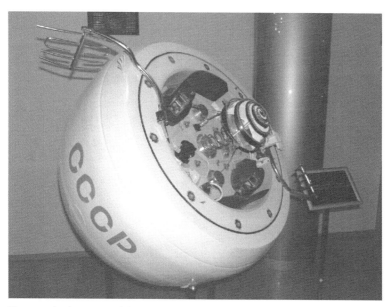

The Venera 4 descent cabin; it was built to withstand extreme pressures and temperatures, the instruments being exposed on the top

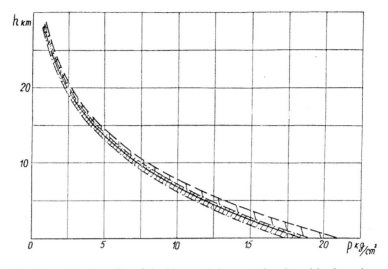

Figure 4.2. The original profile of the Venera 4 descent, showing altitude and pressure.

indicated that surface temperatures and pressures were even higher. Over the next year, Russian and American scientists met at a number of international events and attempted to reconcile their differing data (one logical but improbable explanation was that Venera 4 had landed right on the top of a very high mountain) (Figure 4.3). What actually happened was this. The Venus atmosphere was sufficiently dense to trigger the altimeter at 52 km, twice the height expected (or indicated), and the lander had penetrated to somewhere in the 22–26-km range, at which stage it was simply crushed. The final readings therefore represented conditions far above the surface. Recalculation of the descent profile hinted at a steady wind of up to 50 km/sec at around 50-km altitude, dying out at 40 km, though not much attention was paid to these findings at the time. But it left the designers with the unpalatable question: if this is what Venus is like at this altitude, what must the surface be like further down? At least it won't be a manned flight, commented chief designer Georgi Babakin.

Despite this, Venera 4 was the first planetary probe to return data from within the atmosphere of another planet (Table 4.6). The most striking findings came, of course, from the descent, where the instruments had already found high pressures and temperatures. The atmospheric composition was mainly carbon dioxide, not the nitrogen most planetologists had expected, with a single-digit proportion of nitrogen. Strangely, but as it turned out correctly, the hydrometer found the atmosphere to be almost dry – an unlikely finding granted its thick clouds! Anatoli Basharinov, who had led the Cosmos 243 mission to measure Earth's water vapor levels, was quick to draw attention to how dry Venus was, at less than 0.7% water vapor. This was evidence of water having leaked out of Venus's atmosphere a long time ago. Either way, the findings had the proponents of a watery Venus in headlong retreat. With Venera 4, scientists could begin to publish data, interpretation, analysis, and modeling of the planet – amongst the first to do so being Mikhail Marov, one of the most prolific writers about Soviet exploration of the planets.

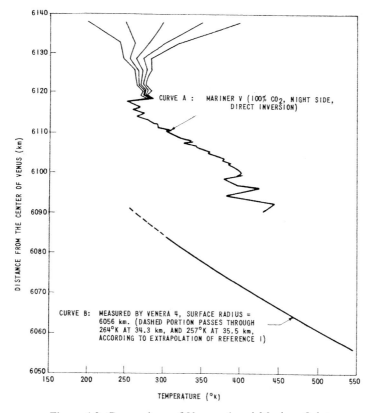

Figure 4.3. Comparison of Venera 4 and Mariner 5 data.

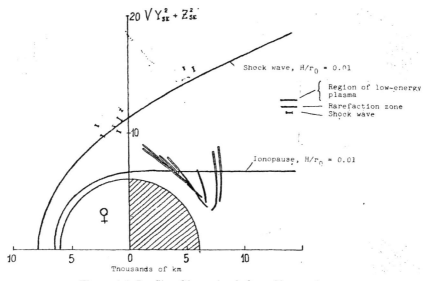

Figure 4.4. Profile of bow shock from Venera 4.

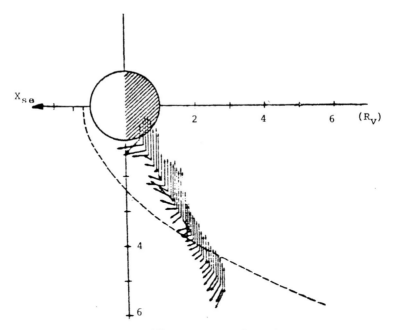

Figure 4.5. Profile of Venus magnetic field after Venera 4.

The ultraviolet spectrometer build by Vladimir Kurt (b. 1933) detected a weak hydrogen corona over Venus, 1/1,000th that of Earth some 10,000 km out from the planet. Venera 4 was important for confirming Vladimir Kurt's theory that the hydrogen atoms came not just from the Sun, but also from an interstellar wind that penetrated into the solar system, and Venera 4 found that they had reached into Venus's upper atmosphere. Venera 4 found a magnetic shock wave 19,000 km from Venus (Figure 4.4). This automatically suggested some form of magnetic field, but the instruments found it to be very weak: 18 γ at 38,000 km, falling to 14 γ at 28,000 km, but then rising abruptly to 32 γ at 23,000 km, falling back to 30 γ at 15,000 km and then back to 18 γ. It was therefore not a field like Earth's (Figure 4.5) [2].

Table 4.6. Venera 4 science.

Carbon dioxide: 90–95%
Nitrogen: not more than 7%
Molecular oxygen: 0.4–0.8%
Water vapor: 0.1–1.6%, dry atmosphere
Wind of 50 km/sec at 50 km, dying out at 40 km
Temperature: 270–280°C at point of crush
Pressure: 20 kg/cm^2 or 15–22 atmospheres at point of crush
Bow shock but very weak magnetic field
Rising radiation levels in interplanetary space, matching the solar cycle
Weak hydrogen corona
Interstellar wind of hydrogen atoms

DESCENT INTO THE CLOUDS

The experience of Venera 4 challenged the designers to toughen the landers so that next time they would actually reach the surface. Veneras 5 and 6 were duly launched in January 1969 and targeted to come in 300 km apart. The parachute was much smaller, at only 15 m across, so as to speed up the descent from 3 to 6 m/sec. A better calibrated altimeter was fitted, so there could be no dispute as to whether the probe reached the surface or not and how close it got. A photometer was added to the instrument suite. En route, on 21st and 30th March, Venera 6 picked up solar bursts, with proton energies up to 30 MeV, and on 11th April a flare speeding out of the eastern side of the Sun, compiling a detailed record of the event. Veneras 5 and 6 carried photon counters that measured the light intensity of the Milky Way and found it surrounded by an extended hydrogen envelope [3].

They reached the planet a day apart in May 1969. Venera 5 penetrated 36.7 km into the clouds, getting down to around between 26 and 16-km altitude, and transmitted for 53 min. Full chemical sampling took place at 0.6 and 5 atmospheres. Venera 5 strained, groaned, and eventually broke up when pressures rose above 27 atmospheres and temperatures above 320°C. During the descent, the first attempt was made to gauge the light level, the photometer taking a single reading just 4 min before destruction and finding that there was a light level of 250 W/m^2.

Three hundred kilometers across the planet, Venera 6 plunged into the atmosphere, descending between 34.1 and 37.8 km into the clouds and transmitting for 51 min before being likewise crushed, but it seems to have got much further down, possibly between 10 and 12 km over the surface. The atmosphere was sampled at 2 and 10 atmospheres. Between the two, they were able to pinpoint the precise proportions of gases in the atmosphere, all the more important in the context of Venera 4's unexpected findings (Table 4.7).

By this stage, the debate on life on Venus was coming to a close. At the time of Venera 4, it was accepted that animal life must now be ruled out and that the carboniferous swamps had probably boiled off – but some primitive, carbon dioxide-breathing forms of vegetable life might still persist. Even this now seemed unlikely and *Pravda* ran an article in which the anonymous chief designer of the Venera probes admitted that the planet was not very suitable for life and it was unlikely that humans could ever land there. A few die-hards clung to the idea that some silicon-based bacteria could still hang on, even in such hostile conditions, but they were bucking the trend.

Table 4.7. Venera 5 and 6 science.

Carbon dioxide: 93–97%
Inert gasses: 2–5%
Nitrogen, oxygen: less than 0.4%
Water vapor content: traces of 4–11 g/liter
Temperature: 327°C (at 18 km, Venera 5), 294°C (at 22 km, Venera 6)
Pressure: 27.5 atmospheres (Venera 5), 19.8 (Venera 6)

Table 4.8. Descents into Venus's atmosphere (launch dates).

Venera 4	12 Jun 1967
Venera 5	5 Jan 1969
Venera 6	10 Jan 1969

By the end of 1969, Soviet scientists were able to draw up a profile of the atmosphere of Venus, based on readings taken from 55 down to 32 km above the surface. They presented the results at the conference of the International Astronautical Federation in the lakeside town of Konstanz in the south of Germany in 1970. Carbon dioxide was clearly the main chemical component of the atmosphere, up to 97%, oxygen up to 1.5%, nitrogen fluctuating a lot according to altitude but around 5%, with some limited evidence of water. The scientists gave an imputed surface temperature of 412–499°C and a pressure of 102–106 atmospheres. Already, they had come to the hypothesis that Venus had suffered a runaway greenhouse effect. Table 4.8 summarizes the probes that entered the Venusian atmosphere.

AT LAST, THE SURFACE OF VENUS

Even submarine designers were consulted about how to improve the next Venus probe in such a way that it might reach the surface. New materials, such as titanium, were used; the parachutes were made even smaller and deployed later; the jettisoning of the lander was delayed to a later stage; and the cabin was pumped with coolant just before the descent. On the downside, though, few instruments were carried (Table 4.9).

Venera 7 was launched in August 1970, arriving on 15th December. Ground controllers followed the lander through the descent, but no signals were received from the surface. At least, that was what they thought, until a month later, a radio technician went through old signal tapes once again and managed to pick out a later, stable, 23-min transmission with a steady temperature reading that could only be coming from the surface. Venera 7 had made it to the surface after all. The parachute came out at 0802, fully extended at 0813. Subsequent analysis of the data suggested that it may have been lucky to get there at all, for the lander hit air pockets during

Table 4.9. Venera 7 instruments.

Mother craft
Cosmic ray detector
Solar wind detector

Lander
Thermometer
Pressure meter

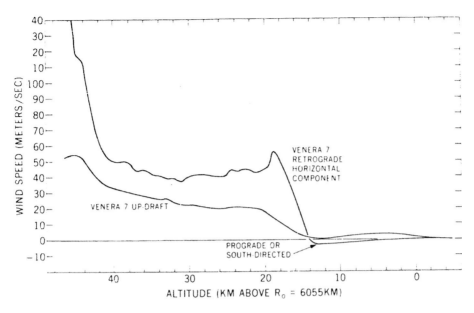

Figure 4.6. Profile of wind during Venera 7 descent.

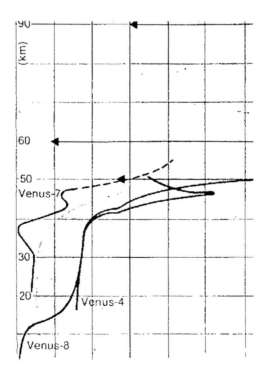

Figure 4.7. Wind profile of Veneras 4, 7, and 8.

Table 4.10. Venera 7 science.

Pressure: 92 atmospheres
Temperature: 475°C
Surface wind: 2.5 m/sec
Violent high atmosphere
Soils ranging from light sand (first impact) to volcanic rock (second)

---

the descent (0819), the parachute tore off at 3 km, the cabin bouncing on landing, hitting the surface hard at 17 m/sec, leaving the transmitter pointing in the wrong direction (0837), hence the problem with the signal (Figure 4.6). Venera 7 was able to give a temperature, pressure, and wind readout, but not a chemical analysis of the atmosphere or surface (Table 4.10). A descent profile was compiled, showing a steady rise in temperature from 300 K at the start of the descent to 730 K at the landing point. Sideways wind velocities were as much as 14 m/sec at 48 km, falling to 7 m/sec at 45 km, much quieter below 38 km, but picking up again for the last 3,400 m. Later, re-analysis of Venera 4 data showed how it, too, had been blown sideways during its descent, as subsequent probes would be (Figure 4.7). Analysis of Venera 7's two impacts led to an imputed calculation that one surface was light sand, the other a volcanic outcrop.

With a landing achieved at last, after almost 10 years, reaching a planet radically different from what had been expected at the start of the 1960s, the next probe could be more confidently designed both to reach the surface and achieve a higher scientific return. Venera 8 carried an ammonia detector (ammonia is an indicator of vulcanism), speed indicator, and gamma-ray spectrometer modeled on that flown by Luna 10 to test the composition of the surface (Table 4.11). Because of new instrumentation, the lander weight was up again, to 495 kg. For the first time, Venera 8 was designed to come down in daylight, its light indicator designed to pave the way for later spacecraft with cameras. An omni-directional antenna was fitted, so that regardless of which way it was pointing, the signals would still reach Earth.

Venera 8 began to provide data from 50 km above the surface, when its

Table 4.11. Venera 8 instruments.

Mother spacecraft
Solar wind detector
Cosmic ray detector
Ultraviolet spectrometer

Lander
Temperature and pressure sensors
Anemometer
Photometer
Gamma-ray spectrometer
Gas analyzer
Altimeter

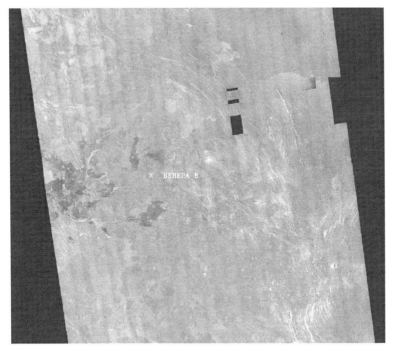

Venera 8 landing site (Credit: NASA)

instruments were first turned on, at 1137. The first radar reading came in at 45 km, the last at 900 m, and 35 data points were received. As it descended, the radar detected two hills (one 1,000 m high, another 2,000 m), a downward slope, a hollow 2,000 m deep, and then a gentle slope upwards to the landing spot – a rugged profile according to the radar analysts. The winds were again strong and the capsule was blown 60 km sideways during its 53-min descent, or slightly more than 1 km a minute (over 60 km/hr). There was a stiff wind blowing the cabin sideways down to about 15 km. Venera 8's instruments to determine its chemical composition were turned on 33 km up. Composition of the upper atmosphere was 94% carbon dioxide, 2% nitrogen, and 0.1% oxygen. Ammonia was a small (0.1%) but important element. Surface pressure was in line with Venera 7 at 90 atmospheres, temperature 475°C.

Venera 8 reached the surface at 1217 and signals came back for 63 min from the surface before the craft was overcome by the heat. Daytime temperatures at this landing site were little different from the nighttime Venera 7 landing point. A huge volume of information came in about the clouds. After traveling 10 km through cloud, Venera 8 seemed to come out of the cloud layer at 35 km. Venera 5 had used its photometer only once during its nighttime descent, but Venera 8 measured light levels at 27 points during the descent. Light decreased steadily from 50 to 35 km as the spacecraft went through the cloud layers. Then, from 35 km downwards, out of the cloud layer, the level decreased more quickly and would have been like Earth on a

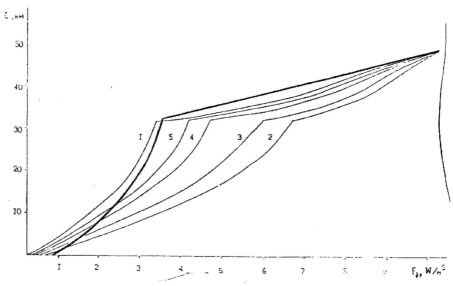

Figure 4.8. Light levels during the Venera 8 descent; note how they drop rapidly under 32 km.

cloudy day. The photometer measured the surface brightness as poor, similar to Earth's at sunrise, giving visibility to the distance of about 1 km. Only 1% of sunlight reached the surface – but a finding that had to be treated cautiously, for the Sun was only 5° over the horizon at the time. The atmosphere below 32 km was essentially transparent, this altitude marking an important point of transition (Figure 4.8). And on the question of the clouds, the gas analyzer gave the first hints to explain the mystery of the dry clouds, finding sulfuric acid. Now, we know that the upper clouds are a fine sulfuric haze. Although the clouds are otherwise dry, they had a high density of sulfuric acid droplets, making for a steady drizzle. A fresh descent profile was published (Figure 4.9).

From this, it was possible to draw a profile of the cloud layer, concluding that it was more like a fog layer and that it had two main levels: thick cloud at the top layer (65–50 km), then haze between 35 and 48 km, but clear below this level. The cloud – or fog – had high winds at parts of the upper level, but the winds were more gentle on the surface. Venera 8 found something extraordinary. Below 100 km, the clouds begin to rotate faster than the planet itself, 40% faster, at 100 m/sec, falling to 60 m/sec at 50 km but decreasing to zero near the surface. Viktor Kerzhanovich's finding was called "super-rotation" and the precise reason is still not understood (on Earth, there are narrow bands of super-rotation, the jet-streams).

No less interesting were the findings from the surface. The gamma-ray spectrometer made measurements during the descent and took two readings on the surface itself, giving a reading of potassium 4%, 2.2 ppm (parts per million) uranium, and 6.5 ppm thorium, indicating something like alkaline basalt. The soil had a density of 1.5 g/cm^3 and appeared to be loose, crushed rock. The radar

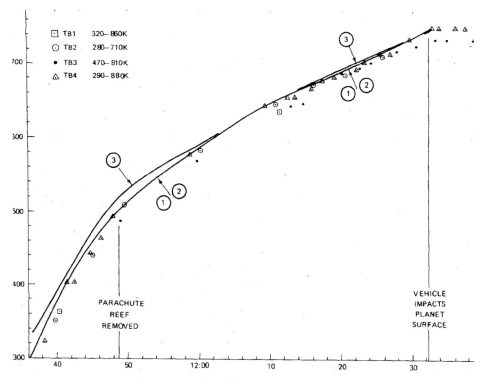

Figure 4.9. Descent profile of Venera 8, temperature on left axis.

altimeter had the ability, during the descent, to estimate the density of the surface layer, calculated at 800 g/m^3, which would make it like granite. The landing area was upland plain, reckoned to be typical of 65% of the planet. The finding of a granite-like rock by Venera 8's instruments was contested, even ridiculed, as most of the surface was supposed to be basaltic. Subsequent mapping by Europe's Venus Express in the new century shows that the instrument was correct and that Venus comprised a mixture of granite highlands and basaltic lowlands, the granite indicating the planet's watery past. More detailed geological analysis suggested that the Venera had come down in a zone of granite, flowing lava, and small volcanoes, fortuitously one of the most unusual sites of all the landings [4]. The outcomes are summarized in Table 4.12.

PHOTOGRAPHING THE SURFACE

Venera 8 was the first spacecraft to provide baseline data from the surface of another planetary world and marked a significant achievement in Russian space science. The next step was to take images of the surface, which would now be possible because the lighting conditions on Venus were known. Venera 8 marked the end of that particular series of spacecraft, the MV. Using the UR-500 Proton launcher, it was

Table 4.12. Venera 8 science.

Temperature: 475°C.

Pressure: 93 atmospheres

Soil: loosely composed, uncompacted, alkaline basalt like granite: 4% potassium, 2 ppm uranium, and 6.5 ppm thorium

Wind: 100 m/sec above 48 km, between 40 and 70 m/sec between 42 and 48 km, but only 1 m/sec for the last 10 km down to touchdown point. Surface wind: 1 m/sec

Atmospheric composition: 97% carbon dioxide, 2% nitrogen, 0.9% water vapor, 0.15% oxygen, ammonia of 0.01–0.1% between 32 and 44 km

Surface density: 1.5 gm/cm^3

Cloud layer: base of cloud layer: 35 km, with thin haze to 48 km and then a thick layer to 65 km. Skies more like fog than Earthly cloud. Super-rotation

Little or no dust in atmosphere

---

now possible to send much larger spacecraft to Venus, weighing over 5 tonnes, with much more capable instrumentation (Table 4.13). The mother ships were large, 2.8 m tall, with a solar panel span of no less than 6.7 m, and were designed to drop the landers either before orbiting Venus or flying past into solar orbit.

Venera 9–14 series

Table 4.13. Venera 9 and 10 instruments.

Landers
Panoramic telephotometer
Photometer to measure chemical composition of atmosphere
Instrument to measure radiation in atmosphere from 63 to 18 km
Temperature and pressure sensors
Accelerometer to measure G forces during descent
Anemometer to measure wind
Gamma-ray spectrometer to measure radioactive elements in rocks
Radiation densitometer
Mass spectrometer

Orbiters
Panoramic camera
Infrared spectrometer
Infrared radiometer to measure cloud temperatures
Photopolarometer
Spectrometer
Magnetometer
Plasma electrostatic spectrometer
Trap for charged particles
Ultraviolet imaging spectrometer
32-cm radio radar

The new lander looked like a mixture of a pressure cooker and a kettle with a metallic ring about it – this was a titanium aerobrake or disk brake designed to slow the spacecraft as it descended. Before, the parachute had been kept on until landing; now, the parachute would be severed at 50 km and the disk brake would guide the cabin to the surface as fast as possible, but not so fast as not to survive touchdown. For photography, two downward-facing cameras with goldfish-bowl lenses were located just under the disk brake, one on each side and angled in such a way that each would take a single 180° panorama, two completing 360° through high-pressure windows made of quartz 1 cm thick.

Veneras 9 and 10 left for Venus on 8th and 14th June 1975, arriving on 22nd and 25th October, respectively. Once they had landed, the surface instrumentation swung into action and the race was on to return the data while transmissions still lasted. A disappointment was that one of the cameras failed on each spacecraft, so only a 180° scan was received.

The first pictures were received on Earth an hour later and they surprised the scientists, who had expected to see only the rocks in the immediate foreground. Instead, the horizon could be seen 300 m away and the immediate area was full of round and curved rocks on a dark surface. Venera 9 came down in a young mountainscape 2,500 m above the datum or reference line, the equivalent of sea level (datum is defined as a radius of 6,051 km of Venus) on the side of a hill or volcano. Although the overall landing area was flat, Venera 9 came down on a steep fault line

Table 4.14. Venera 9 lander science.

Sufficient natural light (10,000 lux) on the surface to take pictures without floodlights
Wind: 0.4–0.7 m/sec (less than 10 km/hr)
Temperature: 480°C
Soil density: 2.7 g–2.9 g/cm^3
Pressure: 90 atmospheres

Table 4.15. Venera 10 lander science.

Wind: 0.8–1.3 m/sec
Dust stirred up on landing
Temperature: 465°C
Pressure: 92 atmospheres
Surface density: 2.7 g/cm^3

with a slope of 30°, the angle of a sand dune on Earth, scattered with sharp and rounded stones and slabs from a recent eruption. There was no dust and rocks could be identified up to 100 m away. It was possible to make out boulders of some 70 cm across and 20 cm tall with fine soil between them. Some sunshine was reaching the surface, for some rocks had shadows. The rocks were not eroded by wind, so either they were young or there was little wind erosion on Venus. The radioactive chemical composition of the rocks was 0.3% potassium, 0.0002% thorium, and 0.0001% uranium, with a rock density of between 2.7 and 2.9 g/cm^3, more like ordinary basalt than the granite indicated by Venera 8. The intensity of solar radiation on the surface was measured. The outcomes are summarized in Table 4.14.

Venera 10 instead came down on a rolling plain with outcrops of 1–3 m of hard but layered and well weathered rocks, later called "pancake rocks" because of their shape. The differences between the two landscapes provided rich discussion for geologists. Venera 10 seemed to have landed on exposed bedrock. The lander was leaning backward on a 3-m slab, giving a more distant horizon for the single panorama. Subsequent photo analysis found that 13 min after landing, a 30-m-high dust cloud of 25-mm particles blew up and hovered for 3 min in the dense atmosphere, possibly delayed-action wake turbulence as a result of the impact of the touchdown on the surface. It was eventually dispersed by the surface wind of 1 m/sec. The penetrometer measured the density of the soil, finding it to be 2.7 g/cm^3, reflecting its rockiness. Later, geological interpretation maps were issued of the surface area where the probes landed. The results are summarized in Table 4.15.

Meantime, the two mother ships entered orbit around Venus, carrying out a program of observations until 22nd March 1976 (preliminary results of the orbiting mission were published in *Pravda* earlier, on 21st February). The orbiters carried panoramic cameras similar to those used on Mars 4 and 5, which enabled scientists to profile the cloud system. Imaging took place at a distance of 7,000–35,000 km out in the course of 17 passes on nine dates from 26th October to 25th December. Although the clouds appeared to be uniform to the gaze of Earthly astronomers, the

Venera pictures showed them to be swirling and full of features, with bands of darker and darker clouds, belts, and possible storms, with at least three distinct layers. Observers identified Y- and V-shaped formations, cellular structures, filament bands, eddies, spirals, dark bands at the equator, and changes on contrast over the period. They found the upper clouds to have a temperature of around –35°C and nighttime clouds were about 10°C warmer than daytime clouds. The maximum altitude of the atmosphere was 88 km. Veneras 9 and 10 used a bistatic radar to make the first, basic radar maps of the planet, focusing on the southern hemisphere. Its relief was found to be irregular, alternating between rough and smooth surfaces, with altitude gradients of up to 3,000 m. Fifty radio probes were made up to March 1976 to determine the structure of the atmosphere, temperatures, and pressure, and to paint a gradient accordingly.

The Venera 9 and 10 orbiters indicated a surface density from 2.4 g/cm^3 in granite regions to 3.2 g/cm^3 in basaltic regions and indicated that there were mountains 1.5– 2.5 km high. The first geological radio imaging maps of Venus were published, showing relatively smooth surfaces and gradients at latitudes 27–28°S and longitudes 155–175°. Another study, at 217–219° longitude and 24.8–25.3°S, found the surface to be "extremely smooth". Interpretative analyses were issued (Figure 4.10).

The issue of Venus's magnetic field was one that perplexed scientists for many years, though it was not as controversial as the Martian field (see Chapter 5). The orbiters found that Venus, although a non-magnetic planet, nevertheless had a magnetic plasma tail. The planet's ionosphere formed a magnetic barrier preventing the solar wind from entering the planet's atmosphere and causing an impact wave on the Sun's side and a tail on the other. Shmaia Dolginov found a cavity of magnetic

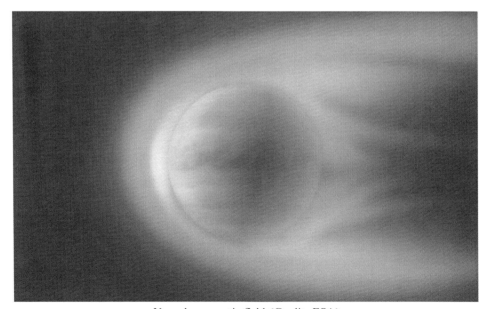

Venus's magnetic field (Credit: ESA)

energy on the night side, stretching out to 2.5 Venus radii away from the Sun, most at 10–15 γ, but up to 20 γ close in. Konstantin Gringauz eventually concluded that Venus could not be proven to have a traditional, intrinsic magnetic field, but a complicated magnetosphere and tail with a magnetospheric current – weak but sufficient to deflect solar wind. Venus's magnetosphere had similarities to that of Mars, but the tail was not as sharp.

Veneras 9 and 10 followed plasma in this tail up to 12 radii downstream, noting its falling density, temperature, and speed. Veneras 9 and 10 found nighttime ionospheric plasma and made 13 electron density profiles of the daytime ionosphere, finding it to lie between 120 and 160-km altitude. Studies of the Venus magnetosphere led Oleg Vaisberg and Lev Zelenyi to present in 1984 what is called the model of the "induced accretion magnetosphere" whereby protons, reaching a gaseous mass, were replaced by photons and escaped along draped flux tubes. They believed that such a model could be applied to all such bodies in the solar system.

Nighttime oxygen airglow was discovered and measured by Vladimir Krasnopolsky. Finally, although not much attention was paid to the discovery at the time,

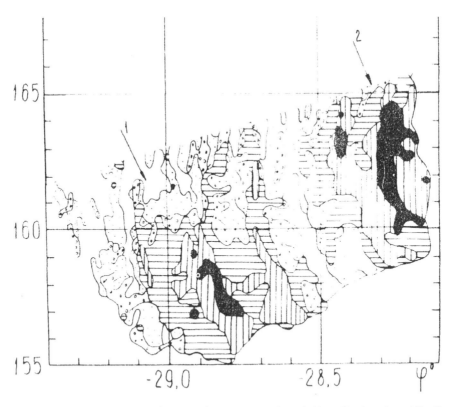

Figure 4.10. Surface map of Venus from Venera 9 and 10 radio imaging. (Credit: COSPAR)

Table 4.16. Venera 9 and 10 orbiter science.

Three cloud layers
Cloud base of 30–35 km, layered to 64 km
Cloud layers glow at night.
Nighttime oxygen airglow
High levels of corrosion due to sulfuric acid, especially upper atmosphere
Bromide and iodine vapor in lower clouds
First cloud maps: clouds denser at the equator, spiral towards the poles
Cloud temperatures: 35°C light side, 45°C dark side
Highest cloud temperatures 40–50 km above surface
Temperature before cloud entry: –35°C
Carbon dioxide to water ratio at 38 km: 1000:1
Non-traditional magnetosphere with tail and current: induced accretion model
Smooth surfaces in some southern latitudes
Measurements of surface densities (3.2 g/cm^3)
Detection of lightning

---

lightning was detected in the nighttime atmosphere of the planet. Typical flashes lasted 0.25 sec, as on Earth, and they had a power in the order of 10^8 joules. Krasnopolski said that at one level, this should not be surprising, for Earth's atmosphere generated 2,000 thunderstorms at any one time, with 100 strikes a second [5]. The outcomes from the orbiter are summarized in Table 4.16.

STRANGE ATMOSPHERE

Veneras 9 and 10 were a substantial advance over Venera 8 and showed what was possible with larger and better-instrumented spacecraft. Although the instruments of both Veneras 9 and 10 had been able to measure the composition of the soil indirectly, more valuable would be direct laboratory chemical sampling. Accordingly, the next two Veneras were designed to carry a chemical laboratory for *in situ* analysis. The cameras were also improved, so as to provide color imaging with a much improved rate of transmission. Extensible arms would deploy a penetrator to measure the strength of the soil, called PROP-V (*PRibori Otchenki Prokhodimosti Venera*). Instruments to sample the atmosphere were improved, with a gas chromatograph to take samples at nine points between 42 km and touchdown, drawing them into porous materials where their elements could be measured. Gas chromatograph experiments were built by Lev Mukhin (b. 1933) to provide the most detailed ever chemical analysis of the atmosphere on the way down. The instrument suite is summarized in Table 4.17.

Veneras 11 and 12 were launched late in their launch windows, which had the practical consequence that there was insufficient fuel for the energy required for entry into Venus orbit, so the probes were dropped off as the mother craft flew by.

Table 4.17. Venera 11 and 12 instruments.

Mother craft
Omnidirectional gamma and X-ray radiation detector (France)
Konus cosmic ray detector
KV-77, for high-energy particles
Plasma spectrometer
Ultraviolet spectrometer
Magnetometer
Solar wind detector

Landers
Panoramic color camera
Gas chromatograph
Mass spectrometer
Gamma-ray spectrometer
Lightning detector
Temperature and pressure sensors
Anemometer
Nephelometer
Optical spectrophotometer
X-ray fluorescent spectrometer
Accelerometer
Soil penetrator (PROP-V)

Veneras 11 and 12 arrived at the planet in reverse order, Venera 12 on the 21st December and Venera 11 on the 25th. As it touched down, Venera 12 churned up dust, which took 25 sec to settle and blew away under the 1-m/sec surface winds, but Venera 11 made a dustless landing. A strange thing happened to both probes during the descent when at between 12.5 and 8-km altitude above the surface: all the instrument readings went off-scale and there was a sudden electrical discharge from the spacecraft. This was the first intimation of what became a disconcerting, frustrating period for mission control. Once Venera 12 reached the surface, the lens caps stuck on the cameras, melted in place, and would not fall off. The PROP-V system was not strong enough to survive the hardness of the landing. Then, the seals in the chemical laboratory broke. Worse still, when Venera 11 came in four days later, it went through an identical experience. There were neither pictures nor soil analysis.

Meantime, both mother ships made deflection maneuvers and continued on their journey in the solar system. As they flew by, they detected a helium envelope around the planet at 150 km and hydrogen envelope extending from 250 to 50,000 km. The Venera 12 mother ship was the first Soviet spacecraft to observe a comet, taking images of comet Bradfield with its ultraviolet spectrometer. Perhaps the most exciting new instrument on board was *Konus*, developed by E.P. Masets in the Joffe institute. Its objective was to locate the source of and characterize gamma-ray bursts. Gamma-ray bursts had first been discovered in the 1960s – accidentally, in the event

– by the American Vela spacecraft. The first Soviet spacecraft to detect gamma-ray bursts and background was Cosmos 461 (up to 80 MeV), but *Konus* was the first systematic attempt to record and characterize them. Gamma-ray bursts were processes of powerful energy release from distant areas of the universe that could explain important questions about the formation of the universe. Veneras 11 and 12 captured 143 gamma bursts, of which 11 were long (over 66 sec).

Where were the gamma bursts coming from? Their sources were triangulated between the two Veneras and a Prognoz orbiting solar observatory. The work of *Konus* continued after the flyby and, in the end, 175 radiation sources were mapped. A map was compiled of all the gamma-ray bursts, called the "Mazets catalog". On 5th March 1979, both *Konus* detected a gamma burst and what was first considered to be an X-ray pulsar but what was later identified as a rotating neutron star with a strong magnetic field in a binary system, with the esoteric name of Dorado and the less glamorous catalog name of FXP 0520-66. The instrument also tried to ascertain the nature of high-energy bursts from Pisces, thought to be some form of pulsar first

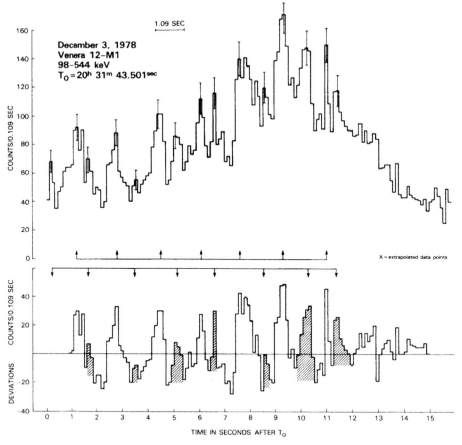

Figure 4.11. Flare record by Venera 12, 3rd December 1978.

detected by the Soviet Earth orbiting observatory Prognoz 7 in March 1979. *Konus* identified 120 solar flares and 20 X-ray eruptions from the Sun, one powerful burst of solar flares lasting 2 hr. Intensities and duration were measured (Figure 4.11).

Data continued to be collected by Venera 11 until 1st February 1980 and Venera 12 until 1st April 1980. *Konus* was later fitted to Veneras 13 and 14 and, many years later, to the naval electronic intelligence satellite Cosmos 2326. Its type of low orbit, coupled with its precise pointing, made it an ideal platform to pinpoint gamma-ray bursts and it detected and localized 15 in its first two months in orbit. Many years later, the Laboratory for Experimental Astrophysics of the Ioffe Physical Technical Institute published a *Konus* database covering 1978–2000.

Despite the problems with the landings, both probes transmitted for an hour and a half from the surface and there was a significant scientific return, even if it lacked photographic backing. The Russians devoted an entire issue of *Kosmicheskiye Issledovaniya* (*Cosmic Research*) to Veneras 11 and 12 (the September–October 1979 issue). The principal gains were in our knowledge of the atmosphere taken from instruments during the descent. Photometers were switched on 63 km above the surface. Chromatographs took the most precise measurement of Venus's atmosphere to date: 176 spectra, from 23 km down to 1.5 km, with 54 chromatographs from 42 km down to the surface, a chemical breakdown of each one being relayed as they descended, giving detailed readouts of the proportions, with every sample measured several times, after which the container was pumped out to a vacuum before the next one was taken in. The main element was, of course, carbon dioxide, at 97%, with nitrogen at 2%, but Venera 12 was able to measure some very small elements as well. Here, argon was detected – important because it is a decay product of potassium, which, in turn, is an indirect indicator of volcanism. Nitrogen was measured at between 2.5 and 4.5%, with small amounts for water vapor (0.01%), chlorine (2.8%), argon (0.4%), sulfur (1.3%), chlorine, neon, and krypton. Veneras 11 and 12 confirmed sulfuric acid droplets in the atmosphere, first meeting them at altitudes of between 54 and 47 km. Over the course of time, the importance of sulfuric acid as the key atmospheric chemical came to be appreciated.

As for the clouds, the nephelometers found that the final cloud layer was between 48 and 51 km, with mist below. Two aerosol layers were found, one at 30 km and a thin one of 0.1-μm particles at 48 km. Cloud effectively ended at the 48-km mark. The level of dryness was confirmed: Venera 11's instruments found only 0.5% water vapor at 44 km, falling to 0.1% at 24 km, while Venera 12's found none at all. The spectrophotometer found that the clouds were relatively transparent: although 3–6% of sunlight reached ground level, the Sun was so dispersed that you could never see it directly from the surface. The level of solar radiation reaching the surface was measured at 78 w/m^2.

Perhaps the most intriguing experiment was the attempt to find lightning on Venus. The *Groza* instrument was carried to measure the sounds of Venus – wind, thunder, and lightning. The lightning detector, an external loop antenna, was turned on at 60 km. It brought to prominence one of Russia's great planetary experts, astronomer Leonid Ksanformaliti (b. 1932). He certainly found lightning, all the way down from 32 to 2-km altitude. Venera 11 recorded five bursts during its descent.

Table 4.18. Venera 11 and 12 science.

Surface pressure: 88 atmospheres (Venera 11) and 80 atmospheres (Venera 12)
Surface temperature of 446°C (Venera 11) and 500°C (Venera 12)
Presence of argon 36–40 and 56 in the atmosphere, 1/200 times smaller than Earth
Role of sulfuric acid
Large-scale, intense thunder and lightning
Mazets catalog of gamma-ray bursts

Atmosphere
Major elements
Carbon dioxide   97%
Nitrogen   2–3%
Minor elements
Water   700–5,000 millionths
Argon   110 millionths
Ne 12 millionths
Kr 0.3–0.8 millionths
Oxygen   18 millionths
Sulfur dioxide   130 millionths
Carbon monoxide   28 millionths

Actual discharges originated from an altitude of 71 km down, but tended to die out at 3 km. But they were intense: Venera 11 once picked up 1,000 almost simultaneous discharges at 30 km. On the surface, Venera 11 counted up to 25 lightning strikes a second during one burst and Venera 12 detected a total of 1,200 strikes. After it landed, a massive thunderclap reverberated around the Venera 12 site for 15 min, with sufficient force to be heard 3,000 km distant. One thunderstorm occupied an area of 150 km horizontally down and 2 km vertically. He reckoned that the lightning did not take place in the clouds, but below them – what, on Earth, we call "clear sky" lightning. Years later, a sound version of the instruments' recording was released. The find of lightning on Venus was contested internationally, but came to be accepted as a breakthrough [6]. The outcomes are summarized in Table 4.18.

CHEMICAL LABORATORY ON THE SURFACE

The next two missions were effectively repeats of Veneras 11 and 12. Flying a landing path in from the north-east of the planet, Venera 13 stirred up a cloud of volcanic dust as it landed on hilly rolling land 2 km above the mean level – in fact, it bounced from its 7.5-m/sec first impact on the crumbly rock and then settled back again. The temperature outside was 457°C and the pressure 89 atmospheres, with a wind speed of 0.3–1 m/sec. This time, the camera worked perfectly, revealing a scene of stony desert, with an outcropping of bedrock and depressions in between with loose, fine-grained soil.

Now came the time for the chemical laboratory to begin work. First, the empty chemical laboratory on board was scanned, so as to set a zero baseline reading. A mechanical ladder straightway extended onto the surface and began to drill the rock using screw drills 30 mm into the surface, extracting a 2-cm^3 sample. The samples were carried in and blown through three locked chambers of decreasing pressure and temperature (the last was at 30°C), analyzed by an X-ray fluorescent device, irradiated by plutonium U-235 and iron-55, and scanned 38 times. The system had to work against time, ever conscious that the probe would soon yield to pressure, temperature, and sulfuric rain. The first drilling and scanning were performed within 4 min and the overall retrieving and measuring of the sample took 32 min – the period of life for which the lander was guaranteed to operate. The pressure inside the chamber was 1/2,000th that of outside.

The rock analysis was 45% silica oxide, 4% potassium, 7% calcium oxide, with a general composition of basalt, basaltic rocks, and alkaline potassic salts. Venera 13's rocks were of the type found in the world's oceans and reckoned to be typical of two-thirds of Venus, not the granite that Venera 8 had led them to expect. Scientists believed that what they saw at the Venera 13 site was the planet's old, badly eroded crust, lava flows covered with a layer of crushed fine-grained material and rubble.

Venera 14 came down on a low-lying basaltic basin 500 m above sea level, with a pressure of 93 atmospheres, temperature 465°C. Both probes experienced the same electrical anomaly as their predecessors. Venera 14 picked up large electric pulses at 20 km and between 6 and 12 km. The pictures were different, looking more like icing on a baked caked surface or, as Moscow radio put it more scientifically, "wrinkled brownish slate-like sandstone". It seemed to be a harsher, rockier, stonier, more weathered plain with fine-grained sedimentary layered rock, no loose surface soil, and a continuous rocky outcrop stretching towards the horizon. Subsequent analysis suggested that the surface was entirely made of exposed bedrock. There was little fine dust on the surface here. The drilling arm reached down, scooped the rock into a hermetically sealed chamber, and likewise put it through X-ray and fluorescent analysis, taking 20 spectra. The lander took in a 1-cm^3 soil sample drilled from a depth of 30 mm. Each lander carried a seismometer: Venera 13 detected nothing, but Venera 14 two events. One was at 950 sec after landing, the other at 1,361 sec, but they were weak and could even have been wind. The *Groza* microphone was even able to detect the lens caps falling off the cameras, the drilling sound (quite loud), and the laboratory at work and, in the background, the wind, blowing lightly, at 30–50 cm/sec (Venera 13) and 30–35 cm/sec (Venera 14). The PROP V penetrometers on Venera 13 worked but Venera 14's had the misfortune of deploying right on top of the ejected camera lens, which it analyzed instead.

Eight images were taken by each probe, the ground being orange–brown in color and showing dust being blown onto the lander at up to 1–2 km/hr. Venera 13 sent back separate panoramas in red, green, and blue showing a stony rolling plateau and the curved horizon in the distance with stones, pebbles, and flat rocks scattered all over it. There was close-grained soil studded with stones up to 5 cm in diameter. Above was an orange sky. Due to the effects of the atmosphere, which absorbed the blue part of the spectrum, the sky appeared to be orange while the lava, the stones, and the sand were greenish yellow. Because the lander operated for 127 min, there

Venera 14 image of the surface; the original images used a curving goldfish lens but photographs were soon issued with a straightened-out horizon

was sufficient time for successive pictures to show wind gently blowing fine-grained sand off the Venera 13 lander at a speed of 0.3–0.6 m/sec. Winds were generally light and not enough to cause significant surface erosion.

The Venera 14 landing site showed that the ground comprised plates in five separate layers without any marked elevations and a fairly straight horizon. The smooth flat rock-strewn plain suggested a long, continuous sedimentary process. It was stratified and little weathered, meaning that the structure was young. The sulfur content was 0.3%, suggesting it was the youngest Venus rock yet found. The main chemical elements of the rocks were determined to be magnesium, aluminum, silica, potassium, calcium, manganese, and iron. A table was compiled of the respective chemistry of the two landing sites, with some differences between the two (Table 4.19 and Figure 4.12).

Veneras 13 and 14 built on the knowledge of the atmosphere developed by Veneras 11 and 12. No fewer than 6,000 spectrographs were taken during the descent. When they first hit the uppermost fringes of the atmosphere at 90 km, there was a pressure of 0.0005 atmospheres and a temperature of –100°C. By the time they were down to 75 km above the surface, the atmosphere had grown to 0.15 atmospheres and the temperature had risen to –51°C. The probes solved the problem of discrepancies in the measurement of water vapor by previous probes. Veneras 13 and 14 found that water vapor, small though it could be, was uneven, being most concentrated between 40 and 60 km (peak was 0.2% at an altitude of 48 km), lowest in the very high atmosphere and immediately above the surface.

During the descent, the gas chromatograph made precise measurements of the level of water (700 ppm), oxygen (4 ppm), hydrogen (25 ppm), hydrogen sulfide

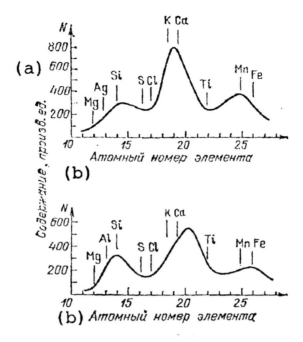

Figure 4.12. Telemetry readout of chemical composition, Veneras 13 and 14.

Table 4.19. Venera 13 and 14 landing sites chemistry.

	Venera 13	Venera 14
Magnesium	11.4%	8.1%
Aluminum	15.8%	17.9%
Silicon	45.1%	48.7%
Potassium	4%	0.2%
Calcium	7.1%	10.3%
Titanium	1.59%	1.25%
Iron	9.3%	8.8%

(80 ppm), and carbonyl sulphide (40 ppm). Chlorine and sulfur were detected. Confirming previous descents but in more detail, the nephelometers identified, at this stage, three distinct layers of cloud:

- dense clouds at 57 km and above;
- a transparent mid-layer at 50–57 km;
- a final denser layer at 48–50 km.

The amount of sunlight reaching the surface was 2.4% (Venera 13) and 3.5% (Venera 14). The outcomes of the mission were reported in *Planet Venus* by Leonid Ksinformaliti in 1982 and are summarized in Table 4.20.

Table 4.20. Venera 13 and 14 discoveries.

Chemical composition of the Venus rock
Two minor seismic events
Characterization of two landing sites
The blowing away of dust by wind
Strength of the soil
Temperature, chemical, moisture of the atmosphere
Dispersal of water vapor in the atmosphere but concentrated at 40–60 km
Nature of cloud layers

Finally, the Venera 13 and 14 mother ships, which flew past the planet, measured the chemical composition of the atmosphere and found a new chemical compound called SF_6. They again found the atmosphere was now highly dehydrated, but with the suggestion that water had in the past played an important role in the formation of the cloud layer. *Konus* detected 89 gamma bursts and 300 solar flares in the course of the orbital mission, from November 1981 to 1st June 1982. The gamma bursts were strange. Most lasted a few seconds. At the two extremes, 10% lasted less than a second, but were still intense. At the other, six lasted over 100 sec, one 140 sec. There was still no clarity as to whether the origin of the gamma bursts was galactic or extra-galactic. The mother ships continued to send back data until May [7].

MAPPING THE SURFACE OF VENUS

Just as the mapping of the Moon had been a priority within the lunar program, so, too, was the mapping of Venus a scientific priority within the Venus program. Veneras 9 and 10 had carried a 32-cm radio altimeter to make the first outline radar map of the surface, but to achieve the detail required, a dedicated radar mission was organized: the double Veneras 15 and 16. These used the same basic spacecraft as Veneras 9 and 10, 11 and 12, and 13 and 14, but instead of the landing craft, there was a large radar system. The main promoter of the project was Academician Vladimir Kotelnikov (1908–2005), director of the Institute of Radio Engineering and Electronics, whose institute had the instruments to make such a mission possible.

Its heart was the *Polyus* radar dish, 6 m from end to end and 1.4 m wide, designed by Russia's leading expert in radar astronomy, Oleg Rzhiga (b. 1930). The radar was built to "paint" the surface of Venus with microwaves for 3.9 milliseconds every 0.3 sec, making 3,200 overlapping images with a resolution of 1 m on each pass, then transmitted back to Earth and fed into a SPF-SM super-computer. The radar required considerable electrical energy, so the solar panels were almost doubled in area. Transmitter capacity was increased 30 times, with an information flow of 108 kbytes/second. There was also a radio altimeter with an accuracy of 50 m. Apart from radar mapping, the objective was to compile a temperature profile of the atmosphere, determining how the planet conserved and distributed its heat: Veneras 15 and 16 were equipped with a 35-kg Fourier infrared radiometer made in

Table 4.21. Venera 15 and 16 instruments.

Polyus radar
Omega radiometric system
Fourier infrared spectrometer (GDR)
Radio occultation device *Dispersion*
Cosmic ray detectors
Solar wind detector

the GDR to remotely sense the atmosphere, cloud layers, and thermal radiation, and to measure surface temperatures at about 60 measurement points on every pass (one failed after two months, but not before returning substantial volumes of data). The instruments are listed in Table 4.21.

Veneras 15 and 16 entered polar orbit around Venus in October 1983. The objective was to compile a radar map of the northern half of the planet, fuel being insufficient to do both halves and barely enough for one (so as to carry extra fuel, the spacecraft was already 1 m taller than the previous Veneras). The northern was considered the most geologically interesting hemisphere, for it contained the two upland areas of the planet, *Maxwell Montes* and *Ishtar Terra*. On each northbound pass, the spacecraft would turn on its radar for 15–16 min, covering 1,000,000 km², beginning to image at 80°N, going over the pole, down to 30–35°N on the other side, compiling 8,000-km-long radar strips of 120 × 160-km width. At the same time, the narrow beam altimeter would acquire height data to an accuracy of 50 m and, altogether, 415,000 readings would be made. The orbiting spacecraft would then move 155 km along the equator each revolution, or about 1.48° longitude each day, requiring eight months to map the planet's northern region.

Before they ran out of fuel in July 1984, Veneras 15 and 16 mapped over 40% of the planet's surface. The maps were published by the Soviet Academy of Sciences (*Atlas Poverkhnosti Veneri*, edited by Vladimir Kotelnikov) – a 27-mosaic radar, relief, and geological atlas of the planet in 1987, comprising 115 million km² of all latitudes north of 30°N. The Venus maps were remarkable not just for the individual features discovered, but because they provided such a rich level of detail as to permit planetary geologists to make strides in developing a considered interpretation of Venus's past history and its present state of development and evolution. Later, contour maps were published (Figure 4.13) with analysis (Figure 4.14). A heat atlas, arising from the GDR infrared spectrometer, was also published.

French scientist Jacques Blamont, who participated in the later Soviet Venus missions, commented that the Venera 15 and 16 landscapes presented "a morphology encountered nowhere else in the solar system". The radars found some extraordinary features:

- dome-shaped hills rising above the plains, sometimes a couple of kilometers across;
- long, linear ridges of 100–200 km in length, sometimes with parallel ridges between 8 and 14 km alongside;

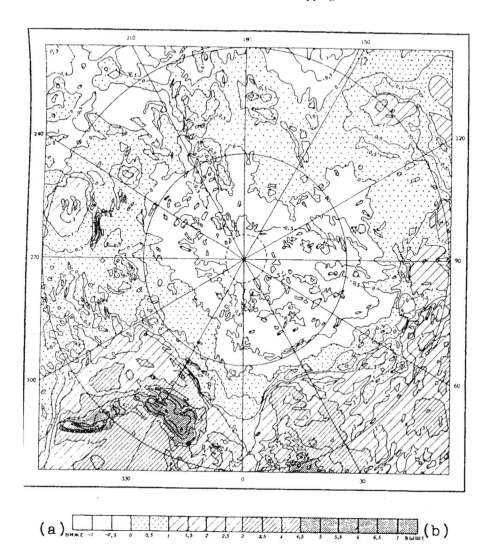

Figure 4.13. Map of the northern latitudes of Venus from Veneras 15 and 16.

- impact craters, from 8-km diameter upward, the largest being Klenova, 144 km across; Venus's thick atmosphere was able to burn up small meteorites, but anything above 1 km across would get through; Alexander Basilevsky estimated that most impact craters were about 1 billion years old;
- about 30 coronæ (e.g. *Bachue, Anahait, Pomona*), with jumbled relief in the middle and ridges around the edge, possibly the outpourings from volcanic hot spots; these could have been domes that bubbled up and subsequently collapsed;
- large volcanic calderic circular depressions up to 280 km across (e.g. *Collette, Sacajawea*);

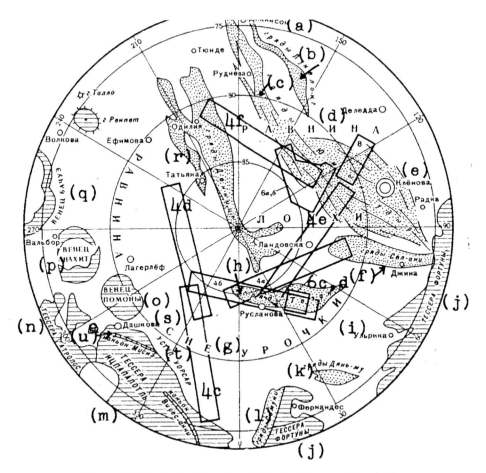

Figure 4.14. Analysis of northern latitudes, Veneras 15 and 16.

- at 100° longitude, 64°N, 200 hills in the shape of shield volcanoes;
- a large depression at 70°N, subdivided into two regions of undulating plain: *Snegurochka planitia* and a flat and ridge belt plain, *Louhi planitia*;
- grooves (e.g. *Laksmi Planum*) and elliptical structures of ridges and grooves measuring 300 × 500 km (e.g. *Tethus*);
- massive mountain peaks of 8, 9, and even 12 km (e.g. *Vesta Rupes*);
- ridges and grooves like parquet flooring (*tessera*);
- haloes around volcanoes;
- lava flows, seen as streaks on mountain slopes (e.g. *Theia, Rhea Montes*);
- troughs (*graben*) in upland areas (e.g. *Beta Regio*);
- thousands of probably extinct volcanoes, most small, but one 95 km across.

Both large-scale and more detailed maps were issued of these remarkable features. (See Figures 4.15–4.18 for a sense of the analysis of Venus made possible by the Venera 15 and 16 radar imaging missions.) The most remarkable discoveries were

ВЕНЕРА   ЗЕМЛЯ ИШТАР   ПЛАТО ЛАКШМИ

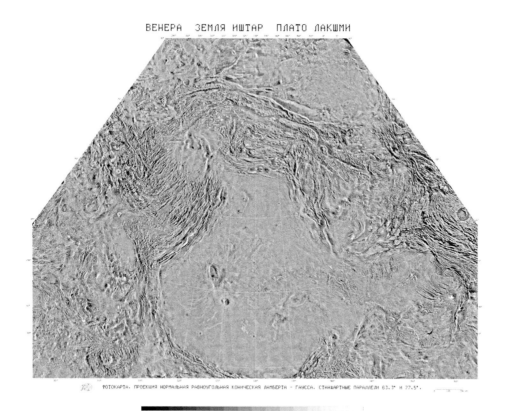

ФОТОКАРТА. ПРОЕКЦИЯ НОРМАЛЬНАЯ РАВНОУГОЛЬНАЯ КОНИЧЕСКАЯ ЛАМБЕРТА - ГАУССА. СТАНДАРТНЫЕ ПАРАЛЛЕЛИ 63.3° И 77.5°.

Venera 15 map of *Ishtar Laksmi*

those of large annular structures 400 km in diameter, called *coronæ*, which had no direct comparators to Earthly or other planetary geology. American planetary scientists Rosaly Lopez and Michael Carroll describe them as oblong or circular fractures, wrinkles or blemishes on the crusts, hundreds of kilometers long, often surrounded by a trough, probably volcanic domes formed by upwelling magma. The dome spreads, collapses, flattens, cools, sinks, and cracks. They had a sub-category of pancake coronæ, possibly formed from thicker lava, now estimated to number 1,250. Also new were the arachnoids, domes of concentric fractures and crests 50–230 km across – but crisscrossed by spidery lines that may be cracks or ridges from upswelling lava.

Veneras 15 and 16 found 146 impact craters and a 400,000-km^2 lava flood plain, as well as plateaus, depressions, shields, plains, and valleys. Some features were adjudged to be asteroid craters up to 3 billion years old. Images showed craters, ancient ring structures, elliptical craters, and volcanic domes. Finding craters at all was a breakthrough, for lunar scientists had used the density of craters to measure the age of the Moon's surface (the more craters, the older it was). The small number of craters was significant, for they contrasted with Earth's battered Moon. On Venus, volcanism and movements of the crust had obliterated

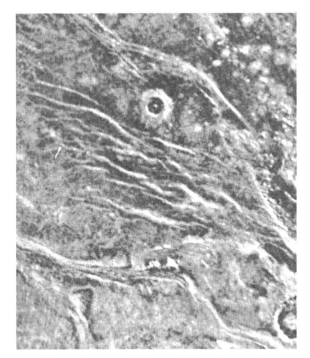

Venera images of ridges and irregular terrain

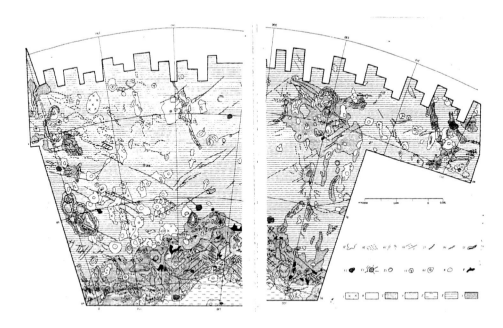

Figure 4.15. Large-scale geological features.

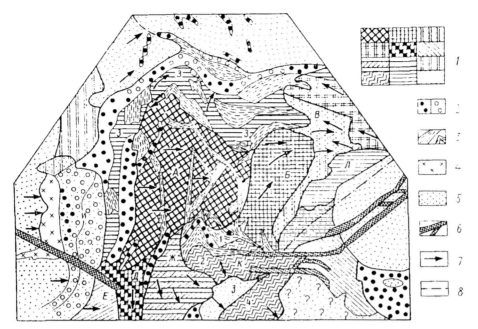

Figure 4.16. *Tessera Fortuna.*

Figure 4.17. *Parquet.*

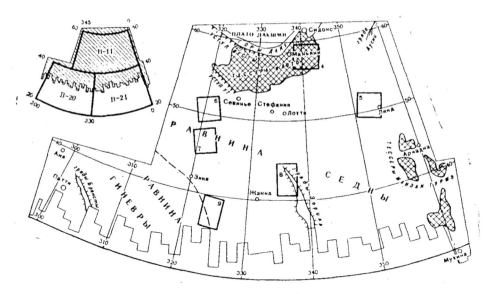

Figure 4.18. *Ravina Sedna.*

traces of most smaller craters older than 700 million years, making the surface of Venus one of the youngest in the solar system. American lunar scientists developed various crater density models and from them, Soviet planetologists calculated the age of Venus's surface as between 300 million and 1 billion years, depending on the model adopted.

The other factor forming Venus's surface was tectonism, the way in which the crust pushed into other parts of the crust (on Earth, this results in mountain formation as one continent pushes against another). Tectonism appeared to have created two sets of uplands on Venus: ridges and grooves, and smooth domes. The largest uplands were belts (e.g. *Maxwell Montes*, standing 4,000–10,000 m above their surrounding plains) and ridges and grooves, with ridges often intersecting at right angles, causing what was originally called parquet terrain but later tiled terrain or *tessera*, the Greek word. Smooth domes, such as *Beta Regio*, were sometimes 1,000–2,000 km across, standing 3–5 km over their surrounding plains, with gentle slopes and elevation. Venera radar images suggested that Venus experienced a prolonged period of basaltic vulcanism and intensive tectonism – a bit like Earth, but without the benefit of oceans, water, or weathering. Venus showed an astonishing absence of erosion, probably due to the constant nature of its dense atmosphere. Venus gave the impression of being more active than Mars, but less active than Earth. It was not always possible to tell the difference between volcanic or impact craters and the Veneras were not able to detect active vulcanism. Analysts concluded that Venus's vulcanism was much less vigorous than Earth's, but that it was probably still going on. Venera 15 and 16 data suggested that the surface of Venus was not unlike the volcanic plains of the Moon and Mercury: primarily basaltic lava, with rock eroded by wind.

The Academy of Sciences map divided the landscape into flat plain, undulating plain, plain with mounds and ridges, parquet, ridge belts, ridges, hills, craters, valleys, and ledges. Venus was characterized as mainly rolling highlands (almost three-quarters), smooth lowlands (almost a quarter), and high plateau (the balance).

The rolling uplands are now thought to be the preserved old crust of Venus, covered by crater and reforming features, while the lowlands, because they have no impact craters, are thought to be young. Areas were considered upland if more than 500–2,000 m above the datum, reference level. Much of the northern hemisphere comprised flat terrain, similar to the lunar *maria*, the plains of Mars, or the ocean floors of many parts of Earth, but these plains were interspersed with substantial upland areas, like *Ishtar Terra*, rising 4,000 m high above the plains, and *Maxwell Montes*. On the plains, the radars detected what appeared to be outflows of basaltic lava, which, even in Venus's high temperature, quickly formed a cooling crust as they vented. In *Ishtar Terra*, the size of Australia, one mountain was found to be over 13 km high – higher than Mount Everest. The northern pole was found to be relatively depressed, with most below the datum line.

The average surface temperature was measured at 500°C, but the infrared spectrometer on Venera 15 (16's failed) found hot spots ("localized thermal anomalies") with temperatures of 700°C, possibly volcanoes. Several localized thermal anomalies were also found. One was in the *Beta Regio* from 281 to 288° longitude and from 17 to 32°N; the second thermal anomaly was at 0–15°N latitude from 60 to 70°E, in *Maxwell Montes*.

Turning to the clouds, the other instruments were able to draw up a temperature–altitude–pressure profile from between 60 and 95 km in the cloud layer. This showed that the cloud system was both dynamic and complex. First, the probes found thermal variations between latitudes: for example, there was a warmer belt at 79–80°N. Cloud temperatures were lowest in the morning and rose in the course of the day. There was a cold layer, an inversion of dense clouds, between 60 and 70 km, where the temperatures were 40°C lower. Above 60°N, the atmosphere was 20–25°C cooler. The hottest clouds were the upper layers at the equator and in a belt from 40°S to 40°N at altitudes of 58–65 km. The top cloud layer began at 67–68 km, but was lower at the pole, 60–62 km, descending to 56–58 km, with a hole over the pole. The main observable strata of cloud were observed from orbit to lie from 47 to 70 km, broadly matching the earlier descent data.

A jet-stream was found. Venera 15 discovered a mid-latitude jet-stream of wind of 100 m/sec at an altitude of 70 km and that deep and dense polar clouds moved down to the equator, where they warmed up. The jet-stream circled the planet at 50–55°N, separating the equatorial atmosphere from the northern atmosphere. Finally, chemical data were refined. Venera 15 detected sulfur dioxide at 60–65 km at the rate of 1–4 ppm, but the levels doubled at night. The upper clouds were mainly composed of sulfuric acid. The level of water vapor was 20 ppm at 58 km.

The mapping of Venus was a technological triumph and a scientific success (Table 4.22). The members of the team responsible for Veneras 15 and 16 were rightly awarded state prizes [8].

Table 4.22. Venera 15 and 16 achievements and discoveries.

Mapping of 40% of surface (northern and polar regions)
Thermal characterization of atmosphere: identification of anomalous cold spots, hot spots
Typology of planetary surface (mountain plateaus, lowlands, rolling uplands)
Typology of individual and unusual surface features
Characterization of a dynamic and complex atmosphere
Discovery of jet-stream

BALLOONS INTO THE ATMOSPHERE

By this stage, Venus had been mapped and there had been eight landings on the planet.

Over 1981–1982, a successor mission was studied: a long-duration lander powered by a nuclear isotope, placing a spacecraft on the surface for a month to test for Earthquakes (or Venusquakes). This was abandoned in favor of a once-in-a lifetime opportunity to send a balloon to Venus and then fly on to comet Halley.

Originally, only a balloon mission was planned. The balloon project had a convoluted origin in discussions with France over deploying a large balloon in the Venus atmosphere. The French had suggested such a mission to the Academy of Sciences as far back as 1967 and pressed the point again, suggesting that such a mission would mark the 200th anniversary of the first ever balloon flight, in France, in 1783.

These discussions coincided with the imminent return, to the inner solar system, of the comet Halley, which orbited the Sun every 176 years. Calculations by the Institute for Space Research showed that it would be possible to send a spacecraft to Venus to drop a balloon, alter course, and then intercept comet Halley. Vladimir Kurt made the formal proposal for the Venus–Halley route. The dual objectives, though, meant that the balloon would not be the large one the French had hoped for.

The project became the largest collaborative international endeavor organized by the Institute for Space Research and Intercosmos, involving 13 countries, with 120 kg available for international experiments. It was named the VEGA project, VEGA standing for Venus and Halley ("G" and "H" are the same in Russian), the spacecraft called VEGA 1 and VEGA 2 (rather than Veneras 17 and 18). Instead of foreign scientists handing over their instruments and never seeing them again, the mission and instrumentation were planned jointly, the Institute for Space Research hosting a growing series of international visitors who had open access to the building.

The VEGA probes were of the standard Venera type, but much more complex because of the balloons, an extra fuel load, wider solar panels, a protective dust screen, and, for the first time, a movable scanning platform to film the interception of comet Halley. The closing speed with the comet would be 80 km/sec – so fast that a moving platform was essential to enable the cameras to swivel and track the comet.

It was equipped with color filters to detect the key chemical and thermal elements of the comet. In the battle over the instruments to be carried for the interception, the plasma scientists lost out to the dust lobby led by the great dust authority Eberhard Grin (but they included one surreptitiously attached to another experiment). VEGA carried a fast data rate transmission system, able to function at 65 kbytes/sec. As for the balloons, these were 3.4 m in diameter, 21 kg in weight, with 2 kg of helium to inflate them. Underneath, on a 12-m-long cable, was a 1.2-m-long gondola, 14 cm in diameter, weighing 6.9 kg with a 4.5-W transmitter designed to take readings every 75 sec during the expected 50-hr journey of each balloon. Chief designer was Vyacheslav Linkin (b. 1937). The instrument package is summarized in Table 4.23.

The two VEGAs were launched in December 1984. VEGA 1 arrived at Venus on 9th June 1985, deploying the lander and balloon. The VEGA 1 lander's mass spectrometer was turned on at 64 km when the parachute was opened and it at once began to detect cloud particles, sulfur dioxide, and chlorine. The X-ray fluorescent spectrometer was switched on 25 km up. Then things began to go wrong. At 18 km over the *Mermaid Plain*, the electrical anomaly experienced by previous

Table 4.23. VEGA experiments.

Mother craft
Television system
Three-channel spectrometer, to analyze the chemical composition of the comet's matter in
    visible, ultraviolet, and infrared bands
Particle impact mass spectrometer, to examine mass of solid dust particles
Neutral gas mass spectrometer, to determine molecular composition of comet's gas
Dust particle detector, to measure the intensity of the dust particles
Charged particle analyzer, to characterize electrons and ions in the comet's plasma
Plasma wave analyzer, for the near comet plasma
Magnetometer to measure comet's own magnetic field

Lander
Meteo, to measure weather, pressure, and temperature below 110 km
IFP, to measure aerosols between 50 and 40 km
Nephelometer/scatterometer/spectrometer to measure composition of atmospheric gases
    50 km to surface
Mass spectrometer to measure composition of atmosphere at 40–50 km
Gas chromatograph to measure composition of gases 35–50 km
Hydrometer, to measure water vapor between 35 and 50 km
X-ray fluorescent spectrometer to detect uranium, iron, magnesium, and silicon
Gamma-ray spectrometer to detect uranium, thorium, and potassium on surface

Balloon
Pressure sensor
Thermometer
Wind meter
Nephelometer to measure clouds and aerosols
Photometer to measure light and lightning
Position and drift indicator

---

landers hit VEGA 1, prematurely triggering the drilling and chemical analysis program while it was still airborne. VEGA 1 did touch down, but there was no chemical laboratory analysis, which, considering that this was a lowland area, was a big loss. Surface temperature was inferred at 468°C and pressure 95 atmospheres.

On the way down, VEGA 1 went through two cloud layers: the first from 50 to 58 km, then the lower layer to 35 km. The level of sulfur in the upper clouds was 5.8 mg/m^3; chlorine 4.1 mg/m^3 and, for the first time, phosphorous, 7.7 mg/m^3. Cloud particles were measured as smaller than 1 µm, like an Earthly fog, except more corrosive. An improved hydrometer was carried to refine the water vapor data from Venera 4, finding that water vapor was only 0.15% at upper altitudes and much less lower down.

VEGA 2's landing was entirely successful. During its descent, VEGA 2 found not only sulfur, chlorine, and iron, but also, in the lower cloud layer, phosphorous again (6 mg/m^3). VEGA 2 was targeted at an upland slope at an elevation of 1.8 km. Surface temperature was 460°C and pressure 90 atmospheres. The surface rocks were duly drilled and chemically analyzed. The VEGA 2 rocks were found to be like Earthly basalt, not unlike the northern Apennines in Italy, rich in aluminum and silicon, but poor in iron and magnesium. The rocks were like the anorthosite samples found on the Moon. There was evidence that water had been present in the rocks when they melted on the Venus mantle, many geological years ago. Sulfur at the VEGA 2 landing site was higher, at between 2 and 5%, much more than Venera 13's 0.65%, suggesting that the VEGA 2 rocks were the oldest yet found. VEGA 2 was the 10th and last lander on Venus. The series is summarized in Table 4.24.

Meantime, both VEGAs had released their balloons high in the atmosphere (Figure 4.19). The snow-white VEGA 1 balloon deployed on the night side of Venus, at an altitude of 54 km, so selected because the cloud was thickest there. VEGA 2's balloon was deployed a similar distance south of the equator. To be precise, the VEGA 1 balloon deployed at 7° 11′N, longitude 177° 48′, VEGA 2 at 6°S, longitude 181° 31′ and they then began their journeys around the planet. Both flew for about 48 hr between 50 and 55 km (Figure 4.20).

Table 4.24. Venus lander summary.

Lander	Launch date	Location	Transmission time
Venera 7	17 Aug 1970	*Navka Planitia*	23 min
Venera 8	27 Mar 1972	*Navka Planitia*	63 min
Venera 9	8 Jun 1975	*Beta Regio*	56 min
Venera 10	14 Jun 1975	*Beta Regio*	66 min
Venera 11	9 Sep 1978	*Navka Planitia*	95 min
Venera 12	14 Sep 1978	*Navka Planitia*	110 min
Venera 13	30 Oct 1982	*Navka Planitia*	127 min
Venera 14	4 Nov 1982	*Navka Planitia*	57 min
VEGA 1	15 Dec 1984	*Mermaid Plain*	56 min
VEGA 2	21 Dec 1984	*Aphrodite*	57 min

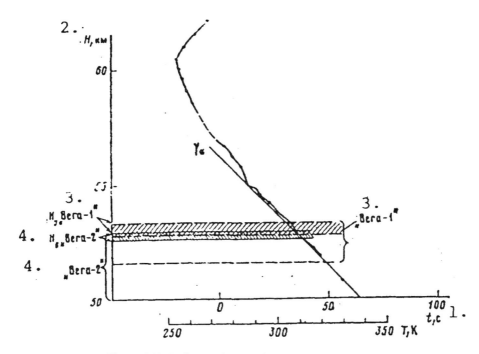

Figure 4.19. Balloon release points, VEGAs 1 and 2.

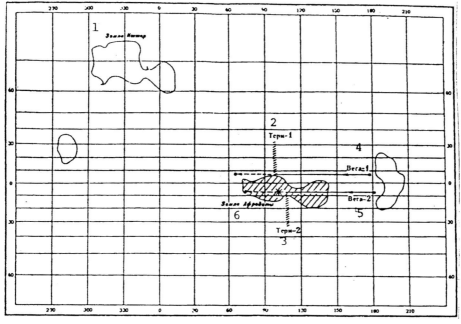

Figure 4.20. Journey of VEGA 1 and 2 balloons.

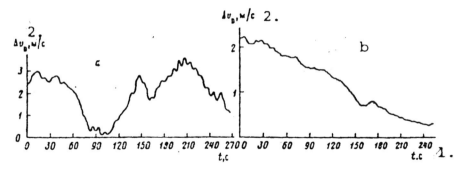

Figure 4.21. VEGA 1 balloon speeds.

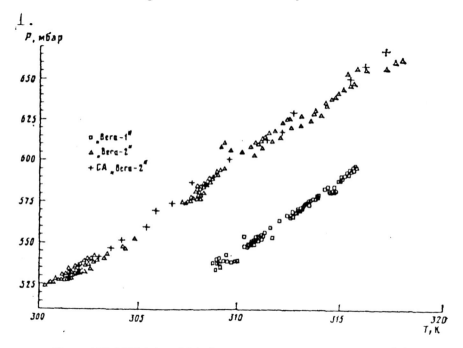

Figure 4.22. VEGA 1 and 2 balloons: pressure, temperatures recorded.

The gondola instruments reported back on temperature, winds, light, and the composition of the atmosphere. The horizontal wind speeds averaged a stormy 69 m/sec (VEGA 1) and 66 m/sec (VEGA 2) (Figure 4.21). VEGA 1, which traveled altogether 11,000 km, was blown south by a northerly wind, while VEGA 2 was blown 500 km northward towards the equator. At 53.5 km, pressure was typically 530 mb, temperature 300K, but rising in places to 900 mb and 340K (Figure 4.22).

The main surprise was the strength of the vertical winds, with periods of turbulence. The balloons had a bumpy journey as they bobbed in the planet's

Table 4.25. Venus atmospheric science from the VEGAs.

The clouds of Venus are a thin fog of sulfuric acid
Phosphorous found
Density of sulfuric acid is 1 µg/m^3 in the altitudes 48–63 km
Cloud particles small, like Earthly fog
Surface like Earthly basalt
Evidence of water some time ago
Water content was between 0.01 and 0.02% at 25–30-km and between 0.1 and 0.2% at 55–60-km altitude
Atmosphere is turbulent, even violent in places

atmosphere. Both balloons found themselves regularly bouncing 200 and 300 m up and down, with downgusts of 1 m/sec, sometimes reaching 3 m/sec. The VEGA 1 balloon was soon caught in turbulent air currents, sped up to 240 km/hr, and was tossed up and down in air currents. After a 46-hr journey of 9,000 km through the darkness, the VEGA 1 aerostat reached the daylight side of the planet, where the Sun heated its envelope and it burst. VEGA 2's journey started quietly but became ever more unsettled, was whirled about through hurricanes, at one stage fell 3 km in 30 min, and eventually burst. Over the *Aphrodite* mountains, where the strongest currents were encountered, the balloon plunged 2,400 m in an air pocket. The balloons found that the clouds traveled in gales at up to 60–70 m/sec. No one had expected the atmosphere to be so rough. The balloons enabled a remodeling of the Venus atmosphere. Before, it had been thought to be calm, with a uniform cloud deck; now, it was an atmosphere of dramatic change and vigorous convection. The outcomes are summarized in Table 4.25.

FINALLY, TO HALLEY

The climax of Soviet Venus exploration was the mission to comet Halley. After dropping the landers and balloons at Venus, the two mother craft altered course to intercept the comet. On 6th March 1986, VEGA 1 closed in at an interception speed of 79.2 km/sec, 100 times faster than a bullet. VEGA 1 passed the comet at a distance of 8,890 km, taking over 500 images in 3 hr through different color filters of the comet's oblong core and dust jets spewing material out. VEGA 1 was damaged during flyby, losing 40% power on its solar panels. VEGA 2 passed the comet at a distance of 8,030 km at 76.8 km/sec, taking images of the south pole of the comet's nucleus. Imaging began on 7th March, interception took place on the 9th, and the comet was a receding blob in the distance in the last pictures on the 11th.

All the foreign scientists were invited to mission control for a live broadcast of the interception, each experimenter being given a desk. It was a dramatic and emotional moment when the oblong-shaped comet came up on screen, surrounded by the psychedelic false colors of the imaging system. The dust-recording instrument made a first "pip" as it encountered the first dust particle and then the "pips" started pinging in ever faster as they collided at 78 km/sec.

Between them, the two VEGAs sent back over 1,500 pictures of the comet's glowing gasses, the multispectral images giving the blob of the comet a red-and-yellow gassy surround against a blue and purple sky. It turned out that comet Halley was an oblong monolith of black clathrate carbon dioxide ice from which millions of tonnes of vapors were escaping, with a surface temperature of between 300 and 400 K but an internal one of 100,000 K. The main chemicals of the comet were found to be water vapor and carbon dioxide, with atomic hydrogen, oxygen, carbon, molecular carbon monoxide, dioxide, hydroxide, and cyan.

The comet was covered with a porous crust and found to contain elements familiar in some stony or metallic meteorites. The VEGAs took a chemical analysis of the dust particles blown away, some of them hundredths of a micrometer in size. These particles contained mainly carbon, but also sodium, magnesium, calcium, iron, metals, nitrogen, silicates, oxygen, and hydrogen. As many as 2,000 particles were studied by the dust-impact mass spectrometer. From all of this, the scientists concluded that cometry nuclei were probably formed near the Sun between Jupiter and Neptune and subsequently hurled away. These dust particle studies were unique and unmatched until the American probe Stardust returned to Earth with samples in 2006. They provided unique data on composition, size, and density.

The mass of the comet was estimated at 300 billion tonnes. A million tonnes escaped every 24 hr at perihelion passage, with a depth of material lost in each solar orbit of 6 m or 40 tonnes. Halley had a dark surface, reflecting only 5% of light falling on it, possibly with a black porous skin 1 cm thick. Halley was a solid, irregular object, $15.3 \times 7.2 \times 7.2$ km, density of 0.6 g/cm^3, with a volume of 400 km^3 and surface area of 300 km^2. The short axis rotated every 2.2 days ± 0.5 and the long axis rotated every 7.4 days ± 0.5. The comet emitted two jets that spewed gases as far as 19,000 km out, one towards the Sun, one upwards. Active regions comprised 10–15% of the surface.

As a result of the VEGA interceptions, Russian scientists were now able to model the comet as a conglomerate of ice with a self-renewing thin layer of black porous matter of low heat conductivity. From time to time, the inside of the comet would evaporate and burst open on its surface, creating active zones. Carbon dioxide and ice would break through and solid particles would be carried away, the comet losing tonnes of material a day, but quickly rebuilding itself. As gas escaped from the comet at 1 km/sec, the solar wind would ionize the evaporated gases, causing the long, luminous cometry tails and a plasma cloud 1 million km wide and 25 times greater than Earth's magnetosphere. VEGA confirmed the snowball model of cometary development, finding water ice, volatiles, non-volatiles, and organic elements. The comet had weak but cohesive tensile strength, with a two-stage crust. It was formed by soft accretion in the outer solar system, aggregating interstellar dust and small particles. It was in pristine condition and had changed little since the start of the solar system.

As predicted by the Mars scientists, VEGA 1 and 2 found a bow shock around

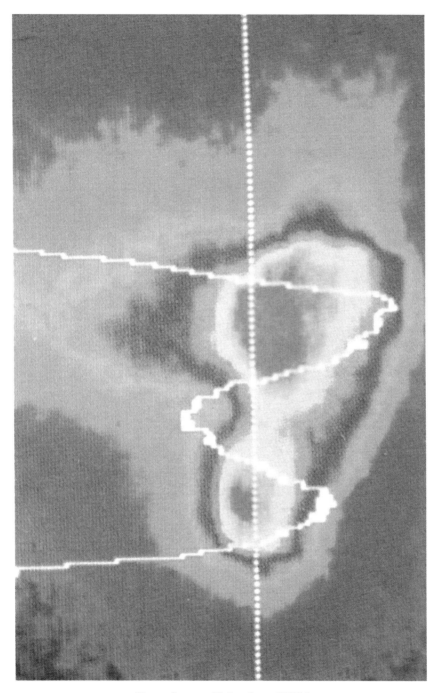

Core of comet Halley from VEGA

Halley, the "cometopause", a sharp boundary separating it from the solar wind at a distance of about 1.6 million km. Behind it, ions were found and the solar wind cooled. It had two bow shock boundaries: the first at 800,000 km, comprising hot electrons, 0.8–3.6 keV, with high helium density, and the second at 330,000–370,000 km. Protons around the comet started to rise 150,000 km out. The comet had its own magnetic field of 75 nT [9].

UNVEILING VENUS: WHAT WAS LEARNED?

The knowledge of Venus had improved out of all proportion compared to the 1950s, when astronomers had superimposed their speculations of lily ponds on top of a necessarily extremely limited base of knowledge. The many failures before Venera 4 illustrated just how difficult it would be to obtain a working knowledge of Venus. Now, with ten landings, three descents into the atmosphere, and a radar map from orbiters, the knowledge of Venus had advanced unimaginably. After the first shock of the nature of Venus – the boiling temperatures, the sulfuric rain, the crushing pressures – the later missions brought fresh rounds of surprises: the extraordinary terrain, unique in the solar system; the layering of the clouds, super-rotation and the jet-stream; the dynamic, complex, and violent atmosphere; and the lightning.

The first learning point was a knowledge of the Venus surface. The broad selection of landing sites meant that a representative cross-section of Venus was studied (Table 4.26).

Across the Veneras, surface temperature ranged from 445 to almost 500°C. Pressure varied from 80 to 95 atmospheres, largely a function of where the probes landed with respect to the reference level. Chemical analysis of the rocks revealed the main elements to be silicon (typically around 45%), aluminum (about 16%), magnesium (11%), and iron (8%). This was the overall picture, but there were variations between sites. Venera 8's was quite different from the others, reflecting its landing on a granite outcrop, with 4% potassium (less than 0.45% elsewhere), high uranium (2.2 ppm, less than 0.69 ppm elsewhere), and high thorium (6.5 ppm, less than 3.65 ppm elsewhere). Venera 8 landed some distance from the other landers, but analysis of all the sites suggested that there could be quite large local variations, even in geologically homogeneous areas. Some of the lowlands had quite rough local areas (e.g. Venera 9). The variety of sites landed provided the contrast necessary for interpretation – for example, Venera 14's surface was entirely exposed bedrock, at Venera 9 it not been exposed at all, and the others were in between. The surfaces had

Table 4.26. Typology of landing sites.

Old upland rolling plains	Venera 8, 13
Flat lowlands	Venera 14
Young volcanic structures	Venera 9, 10
High mountain massif slopes	VEGA 2
(Lowland plain	VEGA 1)

Table 4.27. Venus: cloud layers.

Top layer	64 km
High density	50–60 km
Moderate density	32–49 km
Mild	18–31 km
None	Below 18 km

the characteristic appearance of cooled lava flows that had been subsequently eroded. For the lowland landers, the rocks were rounded, the most likely reason being wind erosion, confirmed by the wind speed measurements of 0.4–1.3 m/sec. The strength of Venusian rock varied, being weak at the Venera 9 site, strongest at Venera 10. Venera 13's had the consistency of heavy clay combined with compacted fine sand, while Venera 14 was more like foam concrete. The probes identified volcanism as the main feature shaping the surface.

The second learning point was the atmosphere and its clouds. Knowledge of the Venusian atmosphere was obtained by a mixture of instruments on the descending landers or the orbiters. The first discovery was that the clouds could be broken into a number of layers, though below 50 km they were really more like mist or haze than our clouds (Table 4.27).

The later Veneras showed that the clouds were not only layered, but had complex systems of currents, up and down draughts, a jet-stream, and variations in temperature.

Venera 12 measured the atmosphere's composition in some detail and is probably the best guide, finding the main elements to be carbon dioxide at 97%, with nitrogen 2–3%, and then a series of small elements. The composition of the atmosphere varied according to height: for example, sulfur and chlorine were most evident from 46 to 63 km. Most solar energy was absorbed at high altitudes, over 50 km. Only 1% of solar energy reached the surface, but this was sufficient to maintain high temperatures in the low atmosphere and contribute to a greenhouse effect.

With this hard knowledge, the debate on Venus could move on to interpretation. The nature of this debate was reflected in 1998, when one of the scientists most involved in the Soviet Venus program, Mikhail Marov, wrote with American colleague David Grinspoon a summary of everything that had been assembled by the American and Russian space programs to date [10]. The question was no longer: *What is Venus like?* but *How did Venus become like this?* The Venera missions did not, could not, tell us how Venus evolved, how it became so hot, or what had happened to its atmosphere, although there was now an abundant bed of knowledge upon which to construct different theories, possibilities, and explanations. Before the space age, it had been assumed that planets of similar size, formed in a similar point in space with a similar chemical composition and early history, should follow a similar evolutionary path – hence the images of swamps on Venus. Although both planets had significant water at an early stage, their evolutionary paths had diverged, possibly as long as 4 billion years ago. Venus appeared to have undergone what was called the "runaway greenhouse effect" in which a change in the atmosphere, possibly carbon, led to an unstoppable rise in temperature that plateaued at 600 K. This, in turn, served as a grim warning as to what uncontrollable carbon release could do to Earth's atmosphere.

There was one consolation, though: those artists who drew those moist pictures of Venus swamps, jungle canopies, and blue plants in the 1950s were right after all – but they had drawn pictures of what Venus *had* been like, several billion years earlier.

REFERENCES

[1]   Kurt, Vladimir: Observations of scattered Lα radiation out of geocorona, in King-Hele, D.G.; Muller, P.; Rinhini, G., eds: *Space Research*, Vol. V. COSPAR, Paris, 1965; Avdyushin, S.I., *et al.*: Radiation intensity based on measurements on Zond 1, in Skuridin, G.A., *et al.*, eds: *Space Physics*, papers from conference held in Moscow, 10–16 June 1965.

[2]   Vernov, Sergei, *et al.*: Study of solar and cosmic radiation from Venera 4, in Champion, K.S.W.; Smith, P.A.; Smith-Rose, R.L., eds: *Space Research*, Vol. IX. COSPAR, Paris, 1969; Logachev, Yuri I.: *40th Anniversary of the Space Age in the Research Institute of Nuclear Physics of Moscow University*. Moscow State University, Moscow, 2009. For research into meteor concentrations, see Nazarova, Tatiana: Investigation of meteoric dust by rockets and satellites. *Cosmic Research*, Vol. 4, 1966; Podgornyi, I.M.: Space physical investigations in the laboratory, in Sergei Vernov, *et al.*, eds: *Physics of Solar Activity and Dynamic Processes in Interplanetary Space and in the Earth's Magnetosphere*, unreferenced papers; Basharinov, Anatoli; Kutuza, B.G.: *Estimate of Water Content in the Atmosphere of Venus According to Data from Radio Astronomical Measurements and Space Probes*. NASA, TTF 15,428.

[3]   Vernov, Sergei, *et al.*: Solar cosmic burst with proton energies up to 30MeV observed by Venera 6, and Kurt, V.G.; Smirnov, A.G.: Ultraviolet radiation in space and the Venus atmosphere, in Kondratyev, Kirill; Mycroft, M.J.; Sagan, Carl: *Space Research*. COSPAR, Paris, 1970.

[4]   Marov, Mikhail: *Venus*. NASA, TTF 16,210, 1975; Marov, Mikhail; Davydov, V.D.: *Earth Type Planets – Mercury, Venus, Mars*. NASA, TTF 15,198, 1973; Marov, Mikhail: *The Planets of the Solar System*. NASA, TM 88015, 1986; Kerzhanovich, Viktor: Wind velocity and certain characteristics of the surface of Venus derived from Venus 7. *Cosmic Research*, Vol. 10 (103), 1972; Surkov, Yuri, *et al.*: Composition and structure of the cloud layer of Venus, in Rycroft, M.J.; Reasenberg, R.D.: *Space Research*. COSPAR, Paris, 1973; Blamont, Jacques: Exploration of the planetary atmospheres, in Zakutnyaya, Olga; Odintsova, D., eds: *Fifty Years of Space Research*. Institute for Space Research, Moscow, 2009. For the Venus Express analysis, see European Space Agency: *New Map Hints at Venus' Wet, Volcanic Past*. ESA, Paris, press release, 14 July 2009; Bashmashnikov, M.V., *et al.*: Characteristics of the surface of Venus by radio altimeter in the landing area of the Venera 8 descent vehicle. *Cosmic Research*, Vol. 14, 1976.

[5] Vaisberg, Oleg, *et al.*: Ion populations in the tail of Venus, in Coates, A.J., ed.: Comparative studies of magnetospheric phenomena. *Advances in Space Research*, Vol. 16, No. 4, 1995; Pavelyev, A.G.: Radio image of unexplored regions of Venus from bistatic experiments, in Rycroft, M.J.: *Space Research*. COSPAR, Paris, 1979; Lemaire, J.F.; Gringauz, Konstantin: Comparison of the magnetosphere of Mars, Venus and Earth. *Advances in Space Research*, Vol. VI, No. 1, 1981; Dolginov, Shmaia, *et al.*: *The Magnetosphere of Venus.* NASA, LPI, undated; Florensky, Kirill, *et al.*: *First Panoramas of the Venusian Surface.* NASA, LPI, undated; Milekhin, O.E., *et al.*: Roughness of the surface of Venus from bistatic radar data. *Cosmic Research*, Vol. 21, 1983; Krasnopolsky, Vladimir: Lightning on Venus according to information from Venera 9 and 10. *Cosmic Research*, Vol. 18, 1980; Kerzhanovich, Viktor: The structure of the Venusian cloud layer according to Venera 9 televised pictures. *Cosmic Research*, Vol. 17, 1979; Moshkin, B.E., *et al.*: Dust on the surface of Venus. *Cosmic Research*, Vol. 17, 1979; Selivanov, Arnold: Survey of Venus' cloud layer from Venera 9 orbiter. *Cosmic Research*, Vol. 16, 1978; Yavkovlev, O.I., *et al.*: Radioscopy of the night time atmosphere of Venus by probes Venera 9 and 10. *Cosmic Research*, Vol. 16, 1978.

[6] Mitrofanov, Igor: On the nature of a recurrent source of gamma and X-ray bursts in Dorado, in Hudson, H.S., ed.: *Advances in Space Research*, Vol. 1, No. 13, Paris, 1981; Barsukov, V.L., *et al.*: Geochemical model of troposphere and lithosphere of Venus based on new data, in Rycroft, M.J.: *Space Research*. COSPAR, Paris, 1979; Istomin, Vadim, *et al.*: Mass spectrometer measurements of composition of lower atmosphere of Venus, in Rycroft, M.J.: *Space Research*. COSPAR, Paris, 1979; Gelman, B.G., *et al.*: Gas chromatograph analysis of chemical composition of Venus atmosphere, in Rycroft, M.J.: *Space Research*. COSPAR, Paris, 1979; Ksinformaliti, Leonid, *et al.*: Electrical discharges in Venus atmosphere, in Rycroft, M.J.: *Space Research*. COSPAR, Paris, 1979; Bertaux, J.L.; Chassefierre, E.; Kurt, Vladimir: Venus extreme ultraviolet measurements of hydrogen and helium from Venera 11 and 12, in Keating, G.M.; Kliore, A.J.; Moroz, Vasili, eds: *The Atmosphere of Venus – recent findings*, *Advances in Space Research*, Vol. 5, No. 9, 1986; Beli, B.M.: Different temporal behaviours in gamma ray busts, in Bassani, L., Palumbo, G.G.C.; Vedrene, G., eds: *Recent Results and Perspectives in Instrument Development in X- and Gamma-Ray Astronomy*, *Advances in Space Research*, Vol. 11, No. 8, 1991; Zelenyi, Lev: Fifty years to change our views of the world, in Zakutnyaya, Olga; Odintsova, D., eds: *Fifty Years of Space Research*. Institute for Space Research, Moscow, 2009; Gavrik, A.L., *et al.*: Formation of daytime Venusian ionosphere – results of dual frequency occultation experiment, in Rycroft, M.J.: *Space Research*. COSPAR, Paris, 1979; Barsukov, V.L., *et al.*: *The Metal Chloride Sulfur Condensates in the Venusian Troposphere.* NASA, LPI, undated; Garvin, J.B.; Head, J.W.: *Venera Lander Site Geologic Characteristics from Pioneer Venus Radar Measurements.* NASA, LPI, undated; Garvin, J.B.: *Dust Cloud Observed in Venera 10 Panorama of Venusian Surface.* NASA, LPI, undated; Garvin, J.B.; Head, J.W.: *Venus – the nature of the surface from Venera*

panoramas. NASA, LPI, undated; Masursky, Harold: *Venus – tectonics and volcanism based on latest Pioneer Venus and Venera data.* NASA, LPI, undated; Golovin, Yuri, *et al.*: Atmosphere of Venus from optical measurements by Venera 11 and 12. *Cosmic Research,* Vol. 20, 1982; Ksanformaliti, Leonid: Electrical activity in the atmosphere of Venus. *Cosmic Research,* Vol. 21, 1983; Golovin, Yuri; Ustinov, E.A.: Aerosol below clouds in the atmosphere of Venus. *Cosmic Research,* Vol. 20, 1982; Golovin, Yuri: Aerosol component of the atmosphere of Venus according to spectrometer measurements on Venera 11 and 12. *Cosmic Research,* Vol. 19, 1981.

[7]   Beli, B.M.: Temporal structures in the Signe GRBs, in Gehrels, N., ed.: *Gamma Ray Astronomy, Advances in Space Research,* Vol. 15, No. 5, May 1995; Andreev, O.N., *et al.*: Observation of cosmic γ bursts by Konus experiment on Venera 13, 14. *Cosmic Research,* Vol. 21, 1983.

[8]   Blamont, Jacques: Exploration of the planetary atmospheres, in Zakutnyaya, Olga; Odintsova, D., eds: *Fifty Years of Space Research.* Institute for Space Research, Moscow, 2009; Schäfer, K., *et al.*: Infrared Fourier spectrometer experiment from Venera 15, in Keating, G.M., ed.: *The Venus Atmosphere, Advances in Space Research,* Vol. 10, No. 5, 1990; Zasova, L.V.; Moroz, V.I.: Latitude structure of upper clouds of Venus, in Shortill, R.W., *et al.*: *Recent Results from Mars and Venus, Advances in Space Research,* Vol. 12, No. 9, September 1992; Titov, D.: Radioactive balance in the mesosphere of Venus from Venera 15 infrared spectrometer, in Keating, G.M., ed.: *Exploration of Venus and Mars Atmospheres, Advances in Space Research,* Vol. 15, No. 4, April 1995; Blamont, J., *et al.*: VEGA balloon meteorological measurements, in Keating, G.M., ed.: *The Venus Atmosphere, Advances in Space Research,* Vol. 10, No. 5, 1990; Zasova, L.V.: Structure of the Venusian atmosphere at high latitudes, in Taylor, F.W., ed.: *Atmospheres of Venus and Mars, Advances in Space Research,* Vol. 16, No. 6, 1995; Schäfer, K., *et al.*: Venus mesopshere radiative transfer on the basis of Venera 15, in Keating, G.M., ed.: *Exploration of Venus and Mars Atmospheres, Advances in Space Research,* Vol. 15, No. 4, April 1995; Lopes, Rosaly; Carroll, Michael: *Alien Volcanoes.* Johns Hopkins University, Baltimore, 2008; Wood, Charles A.; Coombs, Cassandra: The three ages of Venus. NASA, undated paper.

[9]   Gombosi, T.T., *et al.*: Cometary environments. *Advances in Space Research,* Vol. 9, No. 3, 1989; Sagdeev, Roald, *et al.*: Comet Halley – nucleus and jets (results of VEGA mission), in Grün, E., ed.: *Comets Halley and Giacobini Zinner, Advances in Space Research,* Vol. 5, No. 12. COSPAR, Paris, 1986; Stoeva, P.V., *et al.*: Space distribution of the dust continuum around Halley comet nucleus obtained by three-channel spectrometer on board Vega 2, in Zelnyi, Lev; Geller, M.A.; Allen, J.H., eds: *Auroral Phenomena and Solar Terrestrial Relations.* Proceedings from a conference held in memory of Yuri Galperin, 3–7 February 2003. Troitsk, IZMIRAN with Scientific Committee for Terrestrial Physics.

[10] Marov, Mikhail Y.; Grinspoon, David H.: *The Planet Venus.* Yale University Press, New Haven and London, 1998.

5

The path to Mars

The person who shaped Russia's scientific approach to Mars was Gavril Tikhov, the balloonist astronomer and the man who had earlier studied Venus. He had been observing Mars from Pulkovo observatory in Leningrad since the 1930s. The colors of Mars – red, brown, green – encouraged him to compare them with how the colors of Earth would look from space. Mars, he figured, displayed the same colors as the Soviet Arctic, with its snows, mountains, soils, and vegetation. *Ergo*, Mars might be similar to Earth's northern latitudes. He undertook a spectral analysis of seasonal changes on Mars and in the mid 1950s, towards the end of his long life, he was invited to give public lectures on the theme *Is there life on Mars?* It was no surprise that they were well attended.

Gavril Tikhov at Pulkovo observatory

Tikhov may be ridiculed for his Venus plants now, but his general approach was less wide of the mark than he was given credit for. He took the view, unfashionable at the time, that life could survive in conditions that were not necessarily Earthlike. Even on Earth, there were plants that survived without oxygen, preferring ammonia. He once wrote that life was a phenomenon that was "extraordinarily tenacious and stubborn. It can exist in conditions differing greatly from those on Earth". He argued that life could

Gavril Tikhov in the 1950s

be possible on Mars, just as life could be found on cold, high plateaus on Earth. Tikhov traveled thousands of kilometers across southern Russia to the Caucasus and to the Siberian arctic to show the extreme conditions in which life could survive and thrive, bringing 15 expeditions there (Moscow State University continues these studies to the present day). Accordingly, finding life on the planet became a dominant, early theme in Russian Mars exploration. Unfortunately, he did not live to see the first probe to the planet, for although Tikhov lived to 85, he died in Alma Ata, Kazakhstan, on 25th January 1960, just before the Soviet Union launched its first Mars probe.

Tikhov contributed to the shaping of a view that Mars was a relatively benign world, with an atmospheric pressure estimated to be in the order of 80–120 mb. Although no one expected to find waterways, most astronomers agreed that the polar caps melted every Martian year, releasing water that encouraged the seasonal growth of vegetation. When Korolev first designed his Mars probes, he worked on the assumption that the temperature of Mars was in the order of $-70°C$ to $+20°C$, with a thin but useful atmosphere comprising mainly nitrogen and some oxygen. He originally entertained the hope of sending a first probe to Mars in August 1958, but he did not have an upper stage ready in time.

From early 1958 to August 1959, Mstislav Keldysh's Applied Mathematics Division of the Mathematical Institute of the Academy of Sciences carried out the calculations necessary for a flight to Mars in October 1960. The first mission was designed to photograph the surface of Mars with 750-mm cameras to a detail of 3–6 km, with picture sizes of 50 × 150 mm, with infrared instruments to detect plant life or other organic compounds, while a small lander would be sent down to the surface. The first probes weighed 500 kg and were built for eight instruments, including a magnetometer, ion traps, cosmic ray detectors, micrometeorite detectors, radiometer, charged particle detector, spectroreflectometer, spectrometers, and cameras. Unhappily, when launched in October 1960, neither even reached the Earth parking orbit from which they would be sent on their way to Mars.

Mars 1; its communications systems of dishes and domes enabled it to send back information on interplanetary space up to a record distance of 106 million km

FIRST MARS MISSION: MARS 1

For the next Mars window, in autumn 1962, a new type of Mars probe was developed: the 2MV series, soon replaced by the 3MV (see Chapter 4). These were sophisticated spacecraft for their day, carrying large and heavy cameras and many other instruments (Table 5.1). At one stage, it was even intended to put a small micro-biological laboratory on the lander, which would parachute down to the surface. Of the three probes launched in 1962, only one successfully left Earth – Mars 1, a flyby mission, on 1st November 1962.

After only five days, most of the nitrogen in the attitude control system had leaked out, meaning that it would be impossible to orientate the spacecraft for pictures at Mars. Remarkably, the rest of the spacecraft remained in good working order and communications were maintained through the semi-directional antenna until the following 21st March, 106 million km away from Earth – a record, but long before passing Mars the following June. As was the case with the early Venus probes, the value of the first Mars probe was in the early stages of its flight as it left Earth (Table 5.2). Mars 1 left Earth at night in the shadow of the solar wind and detected low-energy fluxes at high latitudes over the Earth out to 18,000 km. The Mars and

Table 5.1. Mars 1 instruments.

Television system for close approach to Mars
Spectro-reflexometer to detect organic compounds on the planet's surface
Spectrograph to measure the Martian atmosphere and ozone
Magnetometer to detect Martian magnetic field
Gas discharge scintillation counter to detect radiation belts around Mars and cosmic radiation
Cosmic radiation counter
Radio telescope for cosmic waves in 150–1,500-m band
Charged particle trap for low-energy protons and electrons and ions
Micrometeoroid detectors for cosmic dust

Venus probes enabled comparison between day and night radiation during the fast passage out of the Earth's orbit, Venus missions heading out on the daytime side into the Sun, Mars on the night side outward-bound. A new map of Earth's environment was published (Figure 5.1), building on the trajectories of the early probes.

Mars 1 soon began to sense the solar wind and detected the weak interplanetary magnetic field detected by Venera 1. Mars 1 noted how background radiation had risen by an average of 60% since 1959. The meteoroid detector picked up a high rate of hits when it met dust clouds and when it went through the *Taurid* meteor shower on the first day of its mission, there was a hit every 2 min between 6,000 and 40,000 km distance, peaking at 21,000 km and again when it met the same stream again between 20 million and 40 million km out. In between was empty space. Then, from 31st December to 30th January, at a distance of 23–45 million km, it took 104 hits in 4 hr 13 min from an unknown stream of meteors that did not cross Earth's orbit, with a similar density to the *Taurids*. Tatiana Nazarova had designed the

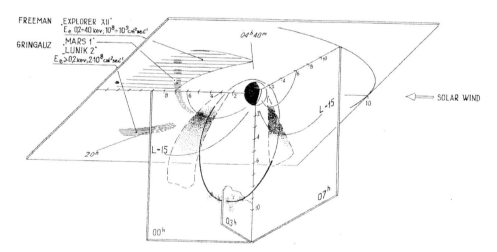

Figure 5.1. Earth's environment, as mapped from the outward trajectory of Mars 1 and other early space probes. (Credit: COSPAR)

Table 5.2. Mars 1 science.

Measurement of the distribution of charged particles in Earth's plasma envelope as the probe
  receded from Earth

Measurement of the speed of solar wind

Detection of solar storm of 600 million particles/cm^2/sec on 30th November 1962

Intensity of the background cosmic radiation up 50 and 70% above level detected by the First
  Cosmic Ship in 1959

High level of collision with micrometeoroid dust near Earth, decreasing rapidly as craft
  proceeded out from Earth's orbit, increasing intermittently as the craft passed through the
  remains of meteorite showers

---

micrometeorite sensors as far back as Sputnik 3, but with Mars 1, their value became
really apparent. On 30th November 1962, Mars 1's instruments picked up a solar
storm and measured the particles as flying at 600 m/sec. Mars 1 detected cosmic rays
at a record distance from Earth [1].

The story of the next Mars probe, Zond 2, launched in November 1964, was
similar. There was considerable difficulty controlling the probe and communications
broke down at 150 million km, some time before the probe reached Mars. Zond 2
achieved a technical first by using pulsed plasma engines for 70 min some 5 million
km from Earth. There were some scientific outcomes from Zond 2, for it measured
atomic hydrogen as it left Earth. It carried Gringauz's integral particle trap and
modulation charged particle trap and compiled a map of electron fluxes as it sped
away from Earth. Heading outbound from the Sun, Zond 2 was launched at night,
shooting right through the radiation belts (radiation peaked at 25,500 km). Zond 2
found low-energy electrons near the equator during its nighttime passage but a new
zone of charged particles when it came out on the sunny side. The solar plasma flux
was initially very quiet, but there was a sudden 21-hr storm on 7th December,
reaching the outbound Zond 2 some time after Earth [2].

The principal scientific discoveries of this period were made by Zond 2's rival on
the way to Mars, America's Mariner 4, which flew close past Mars in July 1965 and
found craters and a thin atmosphere, only 8 mb. This discovery meant that one
could not rely on parachutes alone to land on Mars, for engines would also have to
be used. This, in turn, required a complete redesign, so the 3MV design was
abandoned as suitable for Mars. Zond 2 had carried a lander, but we know now that
the air was too thick for its parachute to work, so it would certainly have crashed.

With the availability of the new UR-500 launcher (see Proton, in Chapter 2),
much larger spacecraft could be sent to Mars, weighing up to 5 tonnes. For the 1969
window to Mars, two orbiters were built to carry out a three-month mapping
mission from an altitude of 500–700 km. Although the UR-500 had a successful
introduction into service, it then went through a long period of difficulty and both
orbiters were lost in launch failures. For the 1971 window to Mars, it was planned to
send an orbiter (Mars 71S, "S" for "Sputnik") to Mars to get precise measurements
of the planet's orbit to pave the way for two subsequent combined orbiter/landers.
Trying to calculate the right trajectory was quite a challenge and an entire floor of

the Institute for Space Research was given over to a computer built for the purpose, the *Ural 14*, which covered a floor area of 50 m^2.

A first priority was the mapping of the planet. The orbiters were designed to map as much of Mars as possible, using narrow and wide-angle cameras called Zufar (350 mm) and Vega (52 mm), with film able to take up to 480 images. It was intended that most of the imaging be done during the first 40 days in Mars orbit during the points of closest approach, with sets of 12 images taken each time, 35–40 sec apart.

The exact nature of the instruments to be carried was the subject of a lively debate. The relatively young, 36-year-old Oleg Vaisberg made a pitch for the orbiters to carry plasma instruments to detect the interaction of Mars's atmosphere with the solar wind – an unwelcome distraction for the planetary scientists, who accused him of being "opinionated". But the design conference was chaired by Mstislav Keldysh, who insisted the plasma instruments fly.

The landers followed similar design principles to the Luna 9 and 13 landers, which had reached the surface of the Moon (see Chapter 3). They were egg-shaped, with their batteries designed to last a number of days while key information on weather, wind, and pressure could be relayed, as well as images. Landing would be carried out by a combination of engines and parachute. The landers carried a small walking robot or skid rover, called PROP-M *Pribori Otchenki Prokhodimosti-Mars* (literally "instrument for evaluating cross-country movement") with a mass of 4.5 kg and tethered to the craft for communications. The skid rover was a squat box 250 × 200 × 40 mm, with a dynamic penetrometer and radiation densitometer, designed to walk on skids up to 15 m, the limit of the cable. It was programmed to stop to make measurements every 1.5 m. The skid rover had built-in artificial intelligence: when it met an obstacle, it was programmed to reverse and use the skids on alternating sides to walk around the obstacle.

The orbiter failed to leave Earth orbit, but the two subsequent, identical probes left Earth successfully as Mars 2 and 3 in May 1971. These carried the largest assembly of instrumentation of any Mars probes launched up to that time (Table 5.3). These were large spacecraft, 4.2 m high, 2 m in base diameter, and 5.9 m wide with solar panels extended.

Leaving Earth, the plasma detectors were turned on and they found traces of the tail of Earth's magnetosphere as far as 3 million km from Earth. As it sped outward, Mars 3 recorded hydrogen particles in the solar wind traveling at between 300 and 600 km/sec (for example, on 29th June, a typical day, 414 km/sec). The strength of the interplanetary magnetic field was measured at a baseline 5 γ. Alarmingly, though, Earth-based astronomers and an approaching, rival American probe, Mariner 9, found that the Martian atmosphere had now been enveloped in a huge dust storm, the biggest in almost 20 years.

The loss of the 71S orbiter meant that the deployment of the lander and the entry to Mars orbit must be carried out without the benefit of the precise measurements of the path of Mars expected from the orbiter. When Mars 2 arrived over Mars on 27th November, the lander was released using insufficiently accurate calculations about the planet's orbit. The lander entered the atmosphere at too shallow an angle

Table 5.3. Mars 2 and 3 instruments.

Orbiter

Infrared radiometer in 8–40-micron range, to study temperatures to –100°C
Photometer to analyze water vapor concentrations
Infrared photometer
Ultraviolet photometer to detect hydrogen, oxygen, argon
Lyman-α sensor to detect hydrogen in atmosphere
Visible-range photometer to study reflectivity of surface and atmosphere
Radio telescope operating a 3.4-cm radiometer 60 cm tall to determine reflectivity of surface
  and atmosphere, giving temperatures just below the surface (50 cm)
Infrared spectrometer to measure carbon dioxide
Photoelevision unit with one narrow-range 350-mm camera (Zufar) and one 52-mm wide-
  angle camera (Vega) with red, green–blue, and ultraviolet color filters with on-board
  scanning system on 1,000 lines and 10-m resolution
Radio occultation experiment for information on structure of atmosphere
Spectrometer to find water vapor and carbon dioxide
Magnetometer

In-flight experiments

Instrument to measure galactic cosmic rays and solar corpuscular radiation
Eight electrostatic plasma sensors to determine speed, temperature, and composition of solar
  wind in 3,000–10,000 eV
Three-axis magnetometer to measure interplanetary, Martian field on boom from solar panel
Fluxgate magnetometer
Cosmic ray particle detector
Electron and photon charged-particles spectrometer

Lander

Mass spectrometer
Temperature and pressure sensors
Anemometer
Cameras
Soil analyzer
Skid rovers PROP-M

and crashed, but nevertheless became the first spacecraft to reach the surface of the planet. It had been intended to land in Hellas, one of the smoothest areas identified by American Mariner 6 and 7 photographs in 1969.

Mars 3 arrived five days later, on 2nd December 1971. The lander was released 4.5 hr before closest approach and the Mars 3 lander began its descent in the southern hemisphere in the light-colored regions called *Electris* and *Phaetonis*. It took only 3 min for Mars 3 to go through the atmosphere and for the parachute to open. An accelerometer triggered a pilot parachute and then a reefed main parachute, still at supersonic speed. At mach 1, the main parachute was inflated. The heat shield, glowing red hot, was dropped. The altimeter turned on. The descending cabin must have been buffeted by the ferocious winds then blowing across the dusty red landscape. Twenty-five meters above the surface, a tiny rocket above the cabin

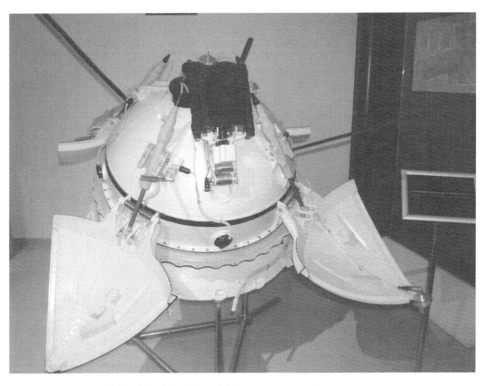

Mars 3 soft-lander, with instrument package on top

blasted the parachute free so that it would not fall on top of the 450-kg landing cabin and another rocket fired briefly to slow the final landing speed. This was a sophisticated combination of maneuvers designed to minimize the shock of descent and landing, while keeping the weight down.

The soft landing was achieved at 16.50.35 on 2nd December 1971. Four petals opened and the domed shape of the capsule rested there on the sands of Mars, the first ever soft-landing. Antennae popped out, aerials searched skywards, and TV cameras began scanning. Video transmission via the mother craft began 90 sec later at 16.52.05 on two independent channels at 72 kilobytes per second (kbps) – and then, exasperatingly, fizzled after only 14.5 sec, for reasons that are still not known, but probably connected to the dust storm then raging. The picture received has been disputed ever since. Some people claim that the picture shows a flat, even horizon some distance away and rocks in the foreground, at least the beginnings of a real image, but other photographic experts consider it to be only "noise" and not a true image.

THE ORBITAL SCIENCE MISSIONS OF MARS 2 AND 3

In the meantime, the Mars 2 and 3 mother ships both entered Martian orbit. There were two immediate problems. First, both spacecraft were scheduled to enter orbits of 25 hr, but the problems of defining Mars orbit meant that orbital insertion maneuvers were imprecise. Mars 2 entered a close, 18-hr orbit, 1,380–25,000 km, at 48.9° inclination. Mars 3 had a similar periaxis (1,500 km) but flew much further out, to between 190,000 and 209,000 km, taking no less than 12 days 19 hr to complete an orbit at 60°. Mars 2's orbit was more suitable, but its telemetry return was poor and more data came from Mars 3 in a less convenient orbit. Second, Mars was enveloped in the dust storm for several months, which was quite a problem because both probes were designed to carry out their operational program in the first three months after arrival. Despite this, there was a substantial scientific return.

Mapping Mars, the first objective was delayed due to the sandstorm and the first useable pictures were not received until 22nd January 1972. The normal procedure was to turn on the main scanning instrumentation for about 30 min at the point of closest approach to the planet while it passed over the planet at 4 km/sec, scanning the surface in sections of 6 × 50 km. For visual mapping, the spacecraft concentrated on the regions from 30°N to 60°S. Mars 3's main batches of photographs were taken in four sessions: on 10th and 12th December, 28th February, and 12th March.

About 60 useable pictures were sent back. They illustrated the dynamic nature of the planet's atmosphere, with localized dust storms, clouds, winds, and streaks.

The images were less than satisfactory, being keyed on too light a setting, giving them an over-exposed appearance and due to telemetry problems could only be relayed at the lowest number of lines (250, not the intended 1,100). The imaging combined a mixture of distance scans of northern latitudes (e.g. *Albera Patera*) and some long, thin strips (e.g. *Elysium*). The first pictures showed cratered regions, mountains up to 15 km high, and craterless plains (the main one being the *Hellas* basin). The next images showed mountains of up to 3,100 m and depressions of 1,200 m below the reference line. They picked up volcanic peaks up to 22 km above the datum line (including massive *Olympus Mons*, now estimated at 24 km high). Mars 3 took some striking pictures of the thin crescent of Mars beckoning in the distance from the apoaxis of its orbit. One of the first pictures showed the limb of the atmosphere with the dust swirling within the thin air.

On 16th February, Mars 2 flew over the *Hellespont*, *Iapygia*, and *Syrtis Major*, marking elevations up to 12 and 15 km high, measuring pressure (5.5–6 mb) and temperatures. The highest surface pressure ever measured was 6.7 mb by Mars 2. Mars 3 indicated the presence of water, but the proportion of moisture or water vapor was tiny, "only the thickness of a human hair", with a level of water concentrations 1/2,000–1/5,000 that of Earth, varying from one region to another (Figure 5.2) but most as it passed over the equatorial regions. Water vapor content of the atmosphere averaged only 6 μm, but rose to 20 μm in March 1972. Humidity fell abruptly above 50° latitude, presumably due to freezing. If all the water in the atmosphere were to be condensed, it would cover the surface with a mere 15 microns. Bright clouds were spotted at latitude 35–40°, their particles being very small (1 mm).

The probes were able to make a profile of the composition of the atmosphere: carbon dioxide to 100 km, carbon monoxide breaking up into hydrogen to 400 km,

Mars 3 image of Mars from a
far point of its elliptical orbit

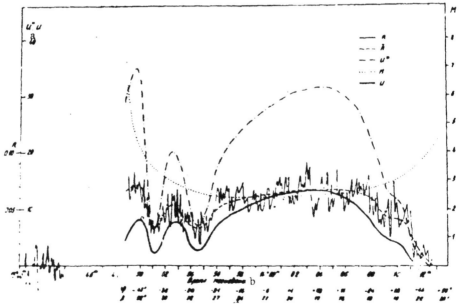

Figure 5.2. Mars 2 and 3 instrument readout of water in the atmosphere.

lighter hydrogen above that to 10,000 km. Traces of oxygen could be found as high as 800 km and a hydrogen corona was found 10,000–20,000 km above the planet. The ionosphere was measured below 330 km and could be divided into three layers: below 210 km, below 57 km, and below 36 km. The ionospheric maximum ranged from an altitude of 110–145 km, with a maximum of 100,000 particles/cm^3. There was still sufficient atmosphere to present a perceptible airglow 200 km behind the terminator at night. The atmosphere had a temperature ranging from 230 to 325 K, mainly carbon dioxide with 2% atomic oxygen.

Mars 2 and 3 found that the dust storms reached up to 15 km high in the atmosphere and that cloud particles reached up to an altitude of 40 km. High in the atmosphere, long, thin clouds were found in the ultraviolet filter. In their report on the great dust storm of 1971, Soviet scientists concluded that:

- the storm lasted three months, well into January 1972, and was highly irregular in nature, being intense in some parts of the planet and less so in others;
- the storm comprised an active early phase in the course of which dust was lifted off the highlands, with a long settling phase with occasional surges of activity;
- the particles, being mainly silicate, were very light and took months to settle down;
- while the storm raged, surface temperatures fell by about 25°C while the atmosphere instead absorbed solar radiation. Vasili Moroz called this a form of "anti-greenhouse effect";
- the volume of dust was measured at 10^9 tonnes;
- dust particles were in the range 0.5–1 µm, occasionally as big as 10 µm.

The radio telescope, developed by Arkady Kuzmin (b. 1923), was switched on during the 1,500–5,000-km phase of its elongated orbit and covered 200 km^2 at a time, the main sessions being on 15th and 27th December 1971, 9th January 1972, and 16th and 28th February 1972. The radio telescope could measure surface temperature down to 50 cm. The thermal map compiled by Mars 2 and 3 found that:

- surface temperatures ranged from –13°C to –93°C in the southern hemisphere, where summer was ending; dark regions were 10–15°C warmer than bright regions;
- temperatures fell to –110°C at the north pole in the midst of its winter;
- soil temperatures were unchanged during the day, but the surface cooled very quickly after sunset, suggesting a low level of conductivity and dry sandy soil;
- the Martian dark areas or seas cooled more slowly than the light areas or continents;
- there were thermal "hot spots" where temperatures were much higher (up to 10°C higher) than surrounding areas;
- temperatures 0.5 m below the surface never rose above –40°C;
- soil temperatures in the equatorial belts averaged –40°C, but by 60°S latitude, they were down to –70°C, irrespective of day or night.

As for the surface, it was possible to estimate soil density at between 1.1 and 1.9 g/cm^3. But there were variations. In the *Cerberus* region, one of the warmer regions, density was calculated at 2.4 g/cm^3 and there were some places where there were concentrations of 3.5 g/cm^3. The surface was believed to be covered with dust made mainly of silicon oxide but the average depth of dust was only 1 mm. The surface had low conductivity, like dry dust. The altimeter measured surface height variations of up to 8 km, from the lowlands of *Hellas* to the uplands of the *Syrtis Major*.

An early discovery was that of orbital anomalies, apparent by the first week of January 1972, akin to lunar mascons and an observable flattening of the Martian poles in the order of 35 km. Mars 2's perigee was bent from 1,250 to 1,100 km in one month, and by April, the scale of the anomaly was measured to average 150 km.

The most controversial aspect of the mission was Mars's magnetic field – a story that was to run and run until the next century. Russian scientists were unsure whether this was a real, indigenous magnetic field or particles assembling in the shadow of the planet away from the Sun and giving the appearance of a magnetic field, called an induced field. The nature of the Martian magnetosphere – the initial Russian view favored an indigenous field – became an unusual point of contention between American and Soviet scientists in the 1970s, the main challenger being C.T. Russell of University of California Los Angeles. Scientific papers from the period referred politely to "discussions still continuing" on the issue, while others, of a more inflammatory disposition, deplored "dangerous" speculations. Konstantin Gringauz once wrote a polite but lengthy rebuttal of the American position, arguing that the instrumentation on the Mars probes was consistent, properly calibrated, and gave unambiguous measurements. This was an important issue in several respects, for the stronger a magnetic field, the greater the likelihood that Mars was protected from the most severe effects of radiation and that life existed, either now or in the past.

Part of the problem was that the American Mariner 4 had not found a magnetic field – an outcome its scientists were inclined to defend (though it did find an increase of 5 γ on flyby at 13,200 km). By contrast, Mars 2 and 3 instruments gave a figure of up to 60 γ, being consistent from both probes and across all their instrumentation. Mars appeared to form a shock wave to resist the solar wind, for the probes entered and exited from areas of charged particles. As it arrived, Mars 2 found a bow shock beginning 1,400 km out and a subsequent magnetosphere. The main bow shock was found to be 500–1,000 km out, stretching out 2.8 Mars radii, with an electron gas tail out to 200,000 km (Figure 5.3). Mars 2's initial measurement of the intensity of the magnetic field was 30 γ, with its dipole along the Martian equator. On 21st January 1972, Mars 2 measured a magnetic field of up to 170 eV. Mars 2 and 3 made multiple crossings of the bow shock and outer magnetosphere and found an outer boundary later and compression of the magnetotail. There was a thick tail downstream, where there was a flow of planetary plasma with a sharp boundary. Electron temperatures and ion currents rose, indicating it was a far-from-dead electrical environment (Figure 5.4). A polar cusp may have been identified. There was corroborating evidence in that whereas the density of the electrons of the solar wind was 0.4/cm^3, near the planet it was up to

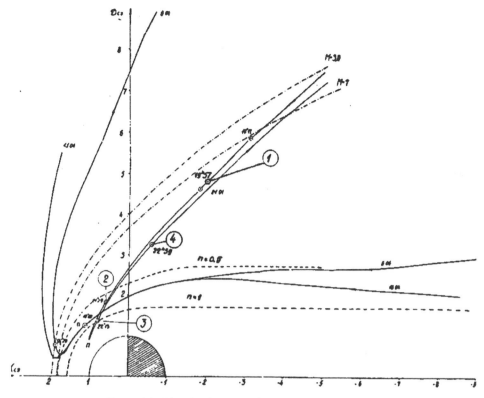

Figure 5.3. First Soviet map of Martian magnetic field.

$0.8/cm^3$, suggesting that they had been caught and trapped. The Mars 2 and 3 figures were remarkably consistent and precise and quite a challenge to the original Mariner findings. The magnetometer suggested a very weak magnetic field and Konstantin Gringauz ventured his own measurement of the magnetic field (for the record, $m = 2 \times 10^{22}$ gsxcm^3 or 0.015% that of Earth). Scientists were able to construct a model of the Martian magnetosphere – one that still largely holds today – and predicted that there would be a similar boundary around other solar system objects, from planets like Venus to small bodies like comets.

The final scientific outcome concerned not the planet, but its moons, for Mars 3 data indicated that the moon Deimos was out-gassing and perturbing the solar wind and could be forming a dust ring around Mars.

The success of Mars 2 and 3 was eclipsed in the public mind in the West by that of America's Mariner 9. Although Mariner 9 was a small spacecraft compared to the Russian ones – it weighed only 1,031 kg, including fuel – it carried a full suite of scientific instruments and these matched Mars 2 and 3 findings in respect of the atmosphere, temperatures, pressures, and environment around the planet. Mariner 9 mapped 70% of the planet, taking 7,000 images, providing a level of detail much deeper than the bleak images relayed during the Mariner 4, 6, and 7 flybys.

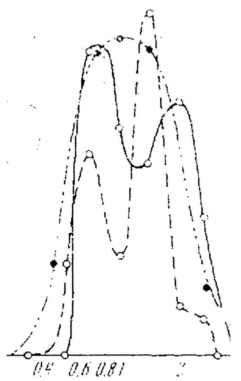

Vasili Moroz, who speculated in the 1970s
that Mars had a watery past

Figure 5.4. Mars 2 and 3 instruments registering
disturbed ions.

Mariner 9 had scientists gasping as it sent back spellbinding pictures of craters,
volcanoes, river beds, canyons, and sand dunes, with more than a hint that the planet
had a watery past. Mars had suddenly become an interesting place to explore.

The scientific results from Mars 2 and 3 were published in papers over the
following years, some focusing on the surface results, some on the atmosphere,
others on the space environment around Mars [3]. One of the main interpreters of
Mars, based on the data from the two probes, was Altma Ata observatory
astronomer Vasili Moroz (1932–2004). Presenting a paper at the International
Astronautical Federation in Baku, Azerbaijan, the following autumn (1973), Vasili
Moroz argued that although dry now, Mars could have supported large reserves of
water – as recently as 25,000 years ago. It is ironic that this was presented as a
sudden "finding" from American Mars probes early the following century, but the
Mars 2 and 3 scientists had reached a similar conclusion 30 years earlier. In
summary, Mars 2 and 3 had accomplished much to expand our knowledge of Mars,
as Table 5.4 indicates.

Table 5.4. Mars 2 and 3 science.

The main discoveries
Mascons affecting orbital path up to 150 km
Flattened poles by 35 km
Dust storms 10 km high
Shock wave resisting solar wind
Surface temperature, pressure, density
Composition of lower and upper atmosphere
Carbon dioxide decomposes at 100 km
Profile of surface temperatures by day and night
Probable presence of large volumes of water

Mars surface
Temperature: $-110°C$ to $+13°C$
Pressure: 5.5–6 mb, 1/200 that of Earth
Surface density: between 1.2 and 1.6 g/cm^3, in places up to 3.5 g/cm^3
Composition: dust mainly of silicon oxide, depth 1 mm

Atmosphere
Composition: carbon dioxide, 90%; nitrogen, 0.027%; oxygen, 0.02%
Water vapor: 10–20 μm, 1/2,000–1/5,000 of Earth
Argon 40: 0.016%

Environment
Magnetosphere, bow shock, and tail, with very weak magnetic field, probably intrinsic (60 γ)
Base of the ionosphere: 80–110 km
Oxygen 800 km out, hydrogen up to 10,000 km
Hydrogen corona between 10,000 and 20,000 km out

DESCENT THROUGH MARS'S ATMOSPHERE

For the next, 1973 window to Mars, the Soviet Union assembled no fewer than four spacecraft of the same type as Mars 2 and 3, the overall title being Mars 73. Similar instruments were carried (Table 5.5). Planetary trajectories were less favorable this time, which meant that it was no longer possible to send combined orbiter/landers. Instead, two spacecraft were kitted out as orbiters (Mars 4 and 5) and two as landers (Mars 6 and 7). All were successfully launched in July–August 1973. A significant problem for the mission was that a faulty computer chip was identified during a late stage of the preparations, but rather than postpone the missions for two years or risk their cancellation, it was decided to go ahead despite the risk of failure of particular components.

There were some significant improvements compared to the 1971 probes. More film was carried than ever before, no less than 20 m in length, with pictures set at either $\frac{1}{50}$ or $\frac{1}{150}$ of a second. Instead of just slow and fast scanning methods, no fewer than ten scanning rates could be chosen, though, in practice, three were favored: preview (220 × 235 pixels), normal (880 × 940 pixels), and high-resolution

Table 5.5. Mars 73 experiments.

Mars 4 and 5 orbiters
Magnetometer
Infrared radiometer to study surface temperature
Plasma ion traps
Radio occultation device to profile density of atmosphere
Radio telescope polarimeter to probe below surface
Two polarimeters to characterize surface texture
Spectrometer to study upper-atmosphere emissions
Narrow angle electrostatic plasma sensor to study solar wind
Lyman-α photometer to search for hydrogen in the upper atmosphere
Three cameras: Vega camera (52 mm), Zufar (350 mm), panoramic
Carbon dioxide photometers
Water vapor photometer to detect water in atmosphere
Ultraviolet photometer to measure ozone
Stereo 2, to study solar emissions (France)
Zhemo: solar protons and electrons (France)
Photometers (four)
Polarimeters

Mars 6 and 7 flyby modules
Telephotometer
Lyman-α sensor to detect hydrogen in upper atmosphere
Magnetometer
Ion trap
Narrow-angle electrostatic plasma sensor to study solar wind
Solar cosmic ray sensor
Charged-particle detector
Micrometeorite sensor
Solar radiometer (France)
Radio occultation device

Mars 6 and 7 landers
Cameras and telephotometer
Thermometers
Pressure, density, wind sensors
Accelerometer
Atmospheric density meter
Mass spectrometer for atmospheric composition
Activation analysis experiment for soil
Mechanical properties soil sensor

(1,760 × 1,880 pixels). There were two panoramic cameras to assemble an image that would cross 30° of the planet from one horizon to another, scanning at four lines per second. In an important change in the transmission arrangements, it was decided that the landers should send some basic data during descent, rather than wait for a successful landing for the subsequent transmission of recorded descent data.

The feared computer problem duly materialized. Mars 6 ceased transmission at the end of September, though ground controllers held out the forlorn hope that it might still carry out its program automatically. The four Mars probes reached their destination in February–March 1974. Mars 4 was unable to make its second course correction, so it could not reach the right point for a burn into orbit and sailed past the planet at 1,844 km. The Mars 7 lander's computer belayed the orbiter computer command for ignition and failed to carry out its descent burn.

Despite this, ground controllers managed to retrieve something from these disappointments. Mars 4 carried out a full mapping program during the 6 min of closest approach, taking 12 Vega pictures and two panoramas, sent back radio occultation data, and continued to transmit from solar orbit. The Mars 4 imaging covered a swathe of the planet from the meridian to 90°W and 20–50°S, sandwiched between the Mars 6 and 7 landing sites. Mars 7 continued to transmit until at least September 1974, the last of the four to fall silent. Instruments on Mars 7 provided a year of solar data, recording a 1-MeV increase in solar protons in September 1973, and characterized individual solar flares, all the more intriguing granted its vantage point so far out from the Sun (1.5 AU). Solar radiation reached Mars 7 some 90 min after Earth (where it was measured by Prognoz 3). At one stage, it recorded how solar wind rose from its normal velocity of 300 km/sec to 700 km/sec. One of the most interesting outcomes was a chart of electron fluxes of 40 keV to 6 MeV between August 1973 and March 1974, which went up and down irrespective of the solar cycle and during a period of solar calm. The conclusion? They were coming from the other side of the spacecraft from Jupiter! They were Jovian electrons. This was not a fresh discovery, for this had already been suggested by American probes Pioneers 10 and 11. Mars 7 must have traveled through fairly empty space, for its meteorite recorder noted only one impact in the course of a whole year.

Ground controllers had almost given up on the Mars 6 spacecraft, due to arrive on 12th March: nothing had been heard since the previous September. Although they did not know it, Mars 6 had, despite its transmitter failure, continued to respond to ground commands and to operate autonomously. The computer had orientated the spacecraft correctly over Mars and adjusted the trajectory in the way intended by Mars 2 and 3. The lander was duly separated at 55,000 km/hr and the mother ship flew past the planet at a distance of 1,600 km, cruised on its solar orbit, and orientated itself back towards the lander act as a retransmission relay. It carried out a radio occultation experiment whose results matched those of Mars 4 and 5 and these found a nightside ionosphere with an electron density of 4,600 elements/cm^3.

The descent transmission channel installed on the lander during the descent now proved its worth. Mars 6 barreled in at an angle of between 11.5 and 12°. The first ground control knew that the spacecraft was functioning was when a Doppler signal started streaming in from Mars 6 at 11.39 am some 4,800 km out as the spaceship dived towards Mars at an angle of 11.7°. These were the first signals in six months, with the cabin coming down to a landing spot at 23° 54′S, 19° 25′W longitude, in the *Margaritifer Sinus*.

Entry into the Mars atmosphere took place at 11.53.38 on 12th March. Braking slowed the descent module to 600 m/sec at 20-km altitude. Prompted by the high-altitude radar, the parachute was reefed and then fully opened above the 645-kg lander. Now, the scientific instruments came fully on, transmitting for 149 sec, but were lost as the spaceship hit the surface at 61 m/sec, its standard speed of descent, but much higher than the intended touchdown speed of 6.5 m/sec. The signal fizzled over a period of 1.3 sec and cut out. The time was now 12.04 pm. Signals were not recovered after landing – we still do not know why – but at least the descent data had been retrieved. This was the first time data had been received from a spacecraft descending through the Martian atmosphere to the Martian surface.

Live transmissions from the descent did broadcast data on pressure, temperature, and atmospheric composition and, as a result, it was possible to compile a profile of the atmosphere (Figure 5.5 and Table 5.6). The pressure readings show the pressure climbing from 2 mb high up in the atmosphere to 3 mb at time of parachute deployment to 5.45 mb at time of landing. Temperature readings show a rise in temperature from between –131 and –109°C at 29 km to between –63 and –58°C at 12 km and –27°C at touchdown. Wind measurements were taken from 7.3-km altitude down to 200 m and they fluctuated between 12 and 15 m/sec. There were contradictory reports on the chemical composition of the atmosphere during the descent, determining it to be mainly carbon dioxide, but also finding argon as high as 25–45% in places. The real rate is now known to be about 1.6%, so it is probable that the instrumentation was faulty. Unlike 1971, visibility during the descent was good and a photometer on board took color filter images of the atmosphere during the descent.

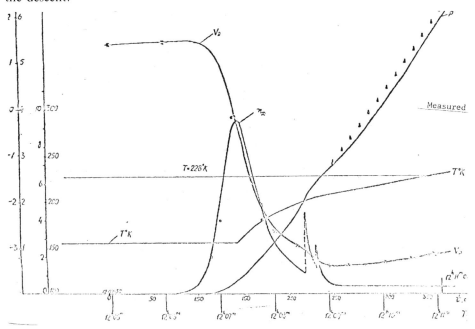

Figure 5.5. This complex diagram reads out the Mars 6 data for pressure, temperature, and altitude.

Table 5.6. Mars 6 science.

Descent
Composition of atmosphere: mainly carbon dioxide
Pressure: 2 mb, climbing to over 5.4 mb
Temperature: $-131°C$, rising to $-27°C$
Wind: between 12 and 15 m/sec

Science from Mars 6: near the surface
Wind: 8–12 m/sec
Surface temperature: $-27°C$ (246 K)
Surface pressure: 5.45 mb

ORBITAL SCIENCE: OBSERVATIONS FROM MARS 5

The other, albeit short, success of the great Mars fleet was Mars 5, which, on 12th February 1974, entered Mars orbit of 5,154–35,980 km, period 24 hr 52 min 30 sec, inclination 35° 19′ 17″ precisely. No sooner had it arrived in Mars orbit than its house-keeping instruments reported a slow leak out of the pressurized, sealed instrument compartment. As a result, its planned three-month scientific program was abruptly compressed into three weeks. Mars 5 lasted 22 orbits until 28th February, when it eventually depressurized. This was less a problem than it suggests, for, like the Cosmos series, Russian spaceships were designed to collect substantial information intensively over a short period, rather than the longer-term approach of the Americans.

The cameras took about 12 pictures during each close approach, the main sweeps being on 17th, 21st, 23rd, 25th, and 26th February 1974. The five panoramic pictures scanned a 30° swath with up to 512 pixels per scanline, taking panoramas covering swaths from 5°N to 20°S, 130–330°W, which included the *Vallis Marineris*. One hundred and eight photographs were taken, of which 43 were of reasonable quality, and five panoramas were assembled. They showed volcanoes, dried-up river beds, tectonic faults, sandy-bottomed craters, and erosion of the landscape. Soviet scientists carried out an examination of the features of Mars photographed by Mars 4 and 5, finding what were interpreted as former river valleys, glaciated regions, and rivers, with the inference that in the past, there were many rivers on Mars and the climate was once much warmer. Detailed descriptions were published of individual regions. For example, a sweep from the *Memmonia Fossae* to the *Margaritifer Sinus* plotted a relief map as the spacecraft flew over hills some 5–8 km over the Martian mean at *Memmonia Fossae*, across to sets of hills, down to the 1 km below the mean in the basin of *Argyre* and then back up to 3–4 km above the mean in *Margariter Sinus*. Clouds formed over the mountain ridges. The surface pressure averaged 6.1 mb. The terrain had abundant craters, indicating it was old and the soil was very fine-grained. Interpretation of some features suggested ancient natural water courses and meandering river channels. A selection of Martian terrains from Mars 5 is shown in the photos.

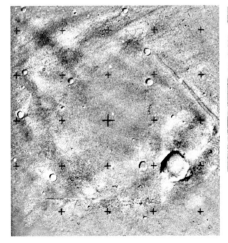

Plains with craters and linear
features, from Mars 5

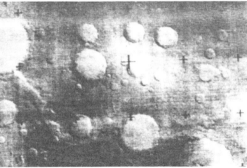

Crater impacts, from Mars 5

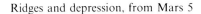

Ridges and depression, from Mars 5

Larger crater with dark flat floor,
from Mars 5

Mars 5 was the first planetary probe to carry a photopolarimeter, designed to measure haze in the atmosphere and dust and sand grains on the surface. The atmosphere was found to have an aerosol layer of water condensate and dust at 20 km, which rose up to 40 km at night. Analysis suggested that *Mare Erythraum* was like a dusty lunar *mare*, while other regions had what seemed to be dunes (*Claritas Fossae*, *Thaumasia Fossae*, and *Ogygis Rupes*). Within the craters *Lampland* and *Bond*, there were large boulders, but the wind had blown the dust off them. During perigee passes, surface temperatures were measured, from –1°C daytime to –73°C by night. Some surface pressures – presumably at lower locations than Mars 6's landing

Table 5.7. Mars 5 discoveries: composition of Mars rock.

O	44%
Si	17%
Al + Fe	19%
K	0.3%
U	0.6 ppm
Thorium	2.1 ppm

Source: Surkov, 1997

site – reached 6.7 mb, in line with Mars 2 earlier. A survey across the southern equator from 140°W to 330°E gave an average particle size in the region of 0.05–0.5 mm. Surface density was measured at 1.57 g/cm^{-3}, compared to 1.37 for Mars 3, but varying according to region and soil depth.

Mars 5's 256-channel gamma-ray spectrometer made six 9-hr-long datasets from Mars orbit. The gamma-ray spectrometer switched on during seven low passes over *Thaumasia, Argyre, Coprates, Lacus Phoenicis, Sinus Sheba, Tharsis,* and *Aria Mons* volcano, a big swathe of the southern hemisphere from 20°N to 50°S from *Amazonis* across to *Mare Erytheraeum,* an area of 400,000 km^2. The uranium, thorium, and potassium content of the surface corresponded to basic igneous rocks on Earth, but proportions varied, depending on the ages of the rocks concerned and whether they were old highland formations or younger volcanic formations. The remarkable thing, though, was just how abundant were oxygen, silicon, iron, uranium, thorium, and potassium. Estimates of the size of the rock analyzed ranged from grains of from 0.04 mm in sandy wind-blown areas to 0.5 mm in areas of tectonic activity where the grains were coarser. Mars 5 found the content of Martian rock shown in Table 5.7.

Mars 5 made important discoveries about the atmosphere of Mars. Mars 5 found an ozone layer 30 km out (but 40 km above the equator) and reconfirmed the earlier finding of atomic hydrogen 20,000 km out, the concentration being 1/1,000 that of Earth. Mars 4 and 5 found that there was a nighttime ionosphere, undetected before, of 4.6×10^{-3} electrons from 110 to 130 km up. The identification of some argon was confirmed, but in small proportions. Sandstorms blew up from time to time. Cirrus-like clouds were identified in the high atmosphere, as were yellowy clouds with fine dust particles. This time, the atmosphere was largely dust-free, but there were occasional local dust storms and veils of white clouds. The temperature in the exosphere was 325 K.

Mars 5 returned to the issue of water identified as crucial by Vasili Moroz. Its instruments found five-fold fluctuations in atmospheric humidity along the flight trajectory and variations from one region to another (Figure 5.6). Water vapor could be found up to 10 km up, but hardly any above 20 km. Mars 5 found that there was water in the atmosphere in the form of vapor and clouds, on the surface as carbon dioxide ice, and under the soil as permafrost. The total water content it still estimated as small, at 10–50 μm, with occasional maxima of up to 100 μm, such as near the Southern Spot (the higher rate was later confirmed by America's Viking,

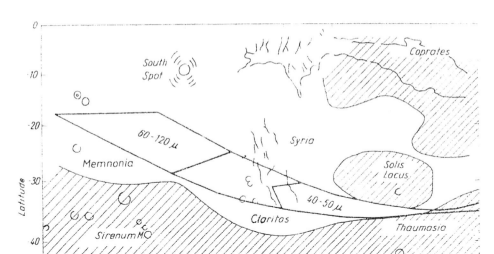

Figure 5.6. The Mars 5 map of water vapor. (Credit: COSPAR)

which found a rate as high as 100 μm over the polar regions). A map was published showing the relative levels of humidity across the planet. Kuzmin came to the conclusion that Mars had considerable sources of water, mainly locked in below the surface, but able to form as ice on the surface above latitude 70°. There had been considerable amounts of liquid on the surface in Mars geological history, he speculated. As they contemplated this, they began to form the theory of what is now called the "warmer, wetter" Mars. Like Earth and Venus, Mars formed oceans in its early history (though Mars was slowest and last). About a billion years ago, Mars was a "paradise of a dense atmosphere, rivers and warm climate", wrote Vasili Moroz and Lev Mukhin. Another commentator, V.D. Davydov, spoke of how Mars 5 had revealed "riverlike relief features" and speculated that there might be subterranean reservoirs under the surface.

The issue of a magnetic field continued to attract attention. Mars 5 confirmed the original data from Mars 2, meeting a magnetosphere at a similar point, between 1,500 and 1,700 km above the surface (400 km when pressed by intense solar wind) and a magnetic tail 900 km behind the planet, with an intensity 1/1,500 that of Earth. Mars 5 went in and out of ions through the transition zone. It could not be proved absolutely that it was an intrinsic magnetic field, but Russian analysts now seemed more certain that there was one there, located 15° off centre, currently in the southern hemisphere. Mars 5 again measured the velocity of solar wind at Mars at mach 7, the bow shock forming 350 km in front of Mars. Mars 5 measurements of 20th February 1974 found that the outer tail was filled with planetary ions and that

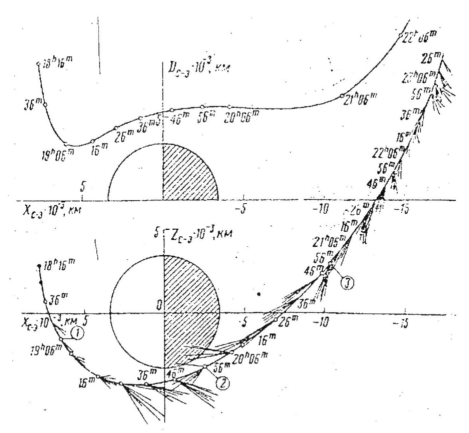

Figure 5.7. Map of the Martian shock wave and magnetic field following the Mars 5 mission.

solar protons did not penetrate the sharp boundary of the tail. Mars 5 data led scientists to draw up a new map of Mars's magnetic field (Figure 5.7). There was a weak but dynamic plasma tail behind Mars of several hundred eV, with three plasma zones: undisturbed solar wind, an area behind the bow shock, and the tail. There was a certain level of consistency between the levels of plasma measured: 30–140 eV (Mars 2), 30 eV–10 keV (Mars 3), and 300 eV–20 keV (Mars 5). Again, the consensus was that the magnetic field was intrinsic, its strength being 2.5×10^{22} gauss/cm^3, about 0.03% that of Earth. Finally, Mars 5 returned to the role of the moon Deimos in the Martian system. On 15th February 1974, Deimos passed in front of Mars 5 and momentarily blocked it, creating what was called a "corpuscular shadow". The scientific outcome of the mission is summarized in Table 5.8.

The Mars fleet of 1973 was reported as a failure in the West and it certainly fell short of its ambitions. For its part, the Russian popular press did little to promote the missions, most likely for fear of drawing attention to the embarrassing problem of the computer circuits. Mission scientists, though, were free to go ahead and

Table 5.8. Mars 5 discoveries.

Composition of Mars rock
Variations in atmospheric humidity
Levels of water vapor 10–50 mm, occasionally 100 mm
Ozone layer at 30 km, aerosol layer 20 km
Existence of argon
Surface temperatures: –44 to –2°C day, –73°C night
Exospheric temperature: 325 K
Small magnetic field, 0.003% Earth, new model
Atmospheric pressure: 6.7 mb (combined with Mars 4 and 6)
Electron density of ionosphere: 4,600/cm^3 at 110 km, nighttime ionosphere
Size of surface grains 0.04–0.5 mm
Evidence of a past warmer, wetter Mars

publicize the scientific outcomes. Impressive, scientifically authoritative and beautifully presented maps were made of Mars using the Mars 4 and 5 photographs – an important element of the project of mapping the three celestial bodies nearest to Earth. A.V. Sidorenko compiled, in *Poverknost Marsa*, a lengthy geographic, geological, geomorphological analysis of the scientific outcomes of the missions. The specific outcomes of Mars 5 mission were published in special issues of *Kosmicheski Issledovaniya* (Vol. 13, No. 1, 1975; Vol. 15, No. 2, 1977), while the outcomes of Mars 4, 5, and 6 were published in book form four years later as *Physics of the Planet Mars*, by Vasili Moroz (Nauka Publishers, Moscow, 1978) followed by Vladimir Krasnopolsky's *Photochemistry of the Martian and Venus Atmospheres* (1982) [4].

MEETING WITH PHOBOS

From a political point of view, the mission of the great Mars fleet was disappointing. It led to what was subsequently called by IKI director Roald Sagdeev "the war of the worlds" – a battle for the future direction of the interplanetary program. Exploration of Venus and Mars had been strongly guided by the desire to be seen to be ahead of the Americans and here the Venus program had been successful. Sagdeev took the view that the Soviet Union should not compete directly with the United States to every space objective – an approach that he considered to be scientifically meritless and economically wasteful. Instead, the Soviet Union should identify areas in which it had unquestionable skills, expertise, and a scientific base on which to build (e.g. Venus) and avoid head-to-head confrontations with the United States in areas in which the latter was likely to get ahead (e.g. Mars). He and his followers were called the "Venusians", while those who wished to compete with the Americans to Mars were called the "Martians". The plan by the Americans to send large orbiters, with landers, to the surface of Mars over 1975–1976 (Viking) forced a crisis, for it obliged the Soviet Union to make a decision as to whether to challenge Viking or not. Viking had been over 10 years in planning, was well funded (each mission cost over $1 billion), and was America's biggest space project since the Moon landings.

Sagdeev, assisted by Keldysh ("the Venusians"), won and there was no Russian Mars fleet to chase the Vikings to Mars, to the disappointment of the "Martians". Instead, the decision was taken to focus on Venus – an approach vindicated by the subsequent results over 1975–1985 (see Chapter 4). Sagdeev called this decision "a real turning point" in Soviet space science – one in which a more rational system of planning overcame short-term political considerations.

When Keldysh stepped down from the presidency of the Academy of Sciences soon after, the vice-presidency went to 80-year-old Alexander Vinogradov, the man who led the successful lunar soil recovery missions, Luna 16, 20, and 24, over 1970–1976 and had been head of the Vernadsky Institute from 1947. His arrival paved the way for a brief and temporarily successful counter-attack by the "Martians". Vinogradov felt that what he had done on the Moon could be matched on Mars and he won governmental approval for what was called Project 5M, to retrieve samples from the surface of Mars, before the Americans, of course. The director of the Lavochkin design bureau, Sergei Kryukov, threw his heart and soul into the project over 1975–1977 and some hardware was even built. But the project became ever more complex and when Vinogradov died, the Institute for Space Research managed to persuade the space minister to cancel the project, restoring the "Venusians" to the ascendant. His place in the Vernadsky Institute was taken by Venus expert, Valeri Barsukov (1928–1992), who led it until his death.

With the success of the Venera and VEGA missions and the near exhaustion of what could be achieved at Venus, it was time to reconsider a return to Mars. The Institute for Space Research was keen to select goals that would not directly challenge the United States, so the small but unexplored Martian moon of Phobos was chosen as a target instead. Phobos had always been of interest and Russian studies dated to 1896, when S.K. Kostinsky had observed the moon from Pulkovo observatory.

Phobos was a tiny, 27-km-across moonlet, originally considered to be a captured asteroid, circling Mars every 7 hr 39 min about 6,000 km distant (the other moon, Deimos, was much further out, at 23,500 km). For the return to Mars, a new spacecraft was designed by Kryukov's successor in the Lavochkin design bureau, Vyacheslav Kovtunenko. This was called the UMVL, or Universal Mars Venus Luna, "universal" because it was designed for a series of possible missions. These were 6 tonnes in weight, with a squat appearance, sitting on a series of large propulsion tanks and engines called the Combined Braking/correction Propulsion System (CBPS). This was a two-part system. The lower part had eight tanks with 3 tonnes of fuel, to be used for mid-course maneuvers, braking into orbit and subsequent maneuvers before being jettisoned. The upper stage, which remained, would be used for attitude control and minor maneuvers near the target.

The Phobos mission, which had a high level of international participation, startled American scientists with its ambition and complexity. The two probes built for the mission had a series of instruments for the in-flight phase, Mars orbit, and the interception of Phobos. For Mars orbit, the most interesting instrument was

Mars UVL spacecraft Phobos

TERMOSCAN, an imager cooled by liquid nitrogen, designed to obtain infrared images so that a thermal map of the Mars surface could be compiled. Most of the public interest centered on the instruments scheduled for the Phobos interception. Here, the mother ship would drop what was called a Long-term Automated Lander (LAL), to transmit for two months, and 100-kg bouncing hopper. Meantime, it would bombard the surface with a laser beam so that its chemical composition could be identified. Its instruments are detailed in Table 5.9.

The missions started brightly, both being launched in July 1988. Phobos 1 left Earth during a magnetic storm, which it was able to observe. During August, its *Terek* telescope transmitted 140 pictures of the Sun, including a flare on 28th August and a picture of a plasma ejection the length of half a solar radius. The solar pictures recorded coronal holes and blobs. The photometer registered 200 bursts of hard gamma radiation and one emission of neutrons of 2.23 MeV was recorded. It was so disappointing that Phobos 1 was lost the following month because of a computer error which mistakenly commanded the systems to shut down.

Despite computer and transmission problems, Phobos 2 made a successful entry into Martian orbit on 29th January 1989. Throughout February, Phobos 2 made a series of complicated maneuvers to match its orbit with the moon Phobos and on the 21st, its cameras took their first, distant pictures of the moon. On 1st March, the platform of instruments for observing Mars's surface was turned on. By 27th March,

Table 5.9. Phobos experiments.

In-flight
Magnetometers
Low-energy electron and ion spectrometer
Solar wind spectrometer
Proton/solar wind spectrometer
Low-energy solar X-ray spectrometer
Plasma wave analyzer
X-ray telescope
X-ray photometer
Solar ultraviolet radiometer
Gamma-ray spectrometer
Solar cosmic ray detector
Solar photometer
Gamma-ray burst detector (same as Veneras 11 and 12)
Energetic charged particle spectrometer, SLED
Solar telescope coronograph

Interception
Remote laser mass spectrometer (LIMA-D)
Remote mass analyzer of secondary ions (DION)

Long-term Autonomous Lander
Television
Seismometer
Spectrometer
Penetrator (*Grunt*)
Telephotometer

Hopper
X-ray fluorescent spectrometer
Magnetometer
Penetrometer (PROP-F)
Dynamograph
Gravimeter

Mars orbit
Scanning infrared radiometer TERMOSCAN
Infrared spectrometer
Gamma spectrometer
Radio sounder
Television unit (FREGAT)
Neutron moisture meter
Atmospheric spectrometer

Phobos 2 had closed to within 191 km of the moon and had started to film. Unhappily, contact with Phobos 2 was lost later that day and never recovered. Although the immediate cause was a signaling, command, and battery problem, the subsequent acrimonious post-mortem revealed a series of management and design errors.

SCIENCE FROM PHOBOS

Although the Phobos mission was, in the public mind, a "failure", the scientific outcomes were far from inconsequential, although they first appeared lost in blood-letting that followed. First results were published in the prestigious English-language science journal *Nature* in 1989 and an international conference was held in Paris in October 1989.

Phobos quickly reopened the controversy of the magnetic field, no wonder, for Phobos 2 provided magnetic field data from 63 circular orbits at 2.8 radii, orbiting every 8 hr 3 min. As soon as Phobos 2 entered Mars orbit, its instruments detected a bow wave 1,000 km above Mars, the weak magnetic field. The bow shock could be observed slowing the solar wind by 100 km/sec. Phobos 2 showed that the solar wind was deflected, formed a boundary, that protons slowed and then reaccelerated, and that electrons were torn away and formed ion clouds. Phobos 2 enabled the compilation of a new map of the Martian magnetic environment, outlining the shape of the magnetopause. Phobos found three different boundaries to the Martian magnetic field, suggesting a weak co-rotating field.

Within the plasma sheet, Phobos 2 found moderately energetic 1-keV heavy ions inside the magnetic cavity, oxygen ions ten times more abundant than protons, and the most oxygen-dominated magnetosphere in the solar system, the magnetotail plasmasheet, was formed mainly by $O+$ ions. The magnetotail electron fluxes were indeed sufficient for formation of a nightside ionosphere. The Irish experiment, the Solar Low Energy Detector (SLED) experiment, found particles of 30–350 eV some 900 km over the planet and 30–200-keV particles in higher orbits. Mars had auroral cold beams, like Earth, which was quite unexpected.

Russian scientists still found the interaction between Mars and its solar wind "difficult to understand". Phobos 2 found that there were multiple plasma boundaries around Mars, making the problem more challenging (compounded by the fact that internationally, several different terms were used to describe the same phenomenon, like magnetopause, planetopause, mass loading boundary). Never-theless, interpretation of the data enabled a much improved characterization of the field, with four features: a magnetopause boundary, shock wave, plasma layer, and tail. The data from Phobos 2 appeared to indicate that Mars's magnetic field echoed that of Venus, being very weak, some 1/10,000 that of Earth. A final comforting finding from the radiation results, which came from SLED, was that the level was well below the threshold that would be dangerous to cosmonauts (30 keV–3.2 MeV range).

One scientist tried to bridge the factions. Zhang took the view, following the way that Phobos 2 swung in and out of the tail, that it was an induced magnetic field with an intrinsic component. In fact, although this was not the eventual explanation, it pointed the way. Indeed, one analysis of the Phobos data had already stumbled on the answer when it noted that magnetotail thickness might be "influenced by magnetization of local crust". The debate raged on for many years and the issue was finally resolved many years later by America's low-orbiting Mars Global Surveyor (MGS), which detected the magnetic field four days after it arrived in Mars orbit in September 1997, about 1,800th that of Earth. What made the difference this time was that it took a much lower orbit, only 400 km above the surface. The surprising result was how the magnetic field measurements rose sharply as MGS made low passes over particular parts of the surface, especially the ancient terrains, highlands, and shield volcanoes of the southern hemisphere. In other words, Mars had magnetic anomalies, like the Earth. Some parts of the surface were intrinsically magnetized, so both sides were right in their own way. Explaining this, analysts believed that 4 billion years ago, Mars had an Earthlike magnetic field, but when the planet cooled, it switched off, remaining only localized where volcanic vents came up to the surface.

Analysis of the ions high over Mars was crucial to the water debate and Mars's past history. The weakness of the magnetic field may have been a determinant of the planet's loss of its water. As noted earlier, Phobos 2 confirmed that the magnetosphere comprised ions of oxygen, which meant that they must have risen there from the surface and from that it was possible to estimate the level of water loss over time. Phobos instruments indicated that Mars was losing its atmosphere at the rate of 1–5 kg/sec, which was significant granted its low density (on Earth, such an amount would be negligible). Detailed calculations of the loss rate were made and they suggested that the rest would eventually be gone in 100 million years. The water vapor content in the atmosphere of Mars between 20 and 60 km was now calculated at 1/10,000 that of Earth and the main component carbon dioxide. There were local variations, too. Phobos 2 found that the Martian regolith released three times more water at noon than the early or late part of the day, but that the rate of release was 30% more from granite rock. But was there more water on or under the surface? Small but variable signs of water were indicated around volcanoes and on the surface around the *Vallis Marineris*. A profile was made of the vertical distribution of water vapor and aerosols in the atmosphere. Phobos found water in high-altitude clouds over *Tharsis*.

As for the atmosphere, Phobos 2 confirmed indications from Mars 5 of ozone in the atmosphere. The infrared spectrometer recorded the signatures of 33 sunrises and sunsets, detecting ozone at an altitude of 12 km. Phobos 2 measured dust and ice particles in the atmosphere. It found that there was constant atmospheric haze all around the planet, mainly at 10–11 km, with little dust above 25 km, but that the haze could be quite quickly triggered into a dust storm. Ice particles were mainly water, but carbon dioxide at higher altitudes. As the infrared spectrometer swiveled towards the limb of the equatorial spring atmosphere, formaldehyde molecules were detected for the first time.

TERMOSCAN now came into play to make a thermal map of the Martian surface. The TERMOSCAN instrument made four passes over Mars, with a maximum resolution of 300 m, and made the first thermal chart of Mars's surface. TERMOSCAN imaged the equatorial zone of Mars in a strip 1,500 km wide with a resolution of 2 km from an altitude of 6,000 km. TERMOSCAN measured surface temperatures ranging from –93 to 7°C, with areas of local variation of 20–30°C. Analysis of the cooling and reheating of the planet's surface as Phobos's shadow crossed Mars provided information on the thermal quality of the planet's surface, finding good insulating material down to 50 microns, but poorer below. Phobos 2 took 40,000 infrared spectra of the surface, focused on the mountains of *Olympus Mons*, *Pavius Mons*, *Avsia Mons*, and *Aseaeus Mons*.

Four photopanoramas were taken, on 11th February, 1st March, and 26th March (daytime and evening). Some showed the small smudge Phobos traversing the rusty Martian landscape below. Major surface features were identified and the height of some of the volcanoes was extrapolated (e.g. *Pavonis Mons*, 5.9 km). Contour maps were made of major geologic features, such as *Tharsis*, *Marineris*, and *Pavonis Mons*. The more detailed pictures were taken by the VSK video spectrometric complex built in Bulgaria, the GDR, and the Soviet Union. Two instruments contributed to an improved knowledge of the composition of Martian rock. The French infrared spectrometer suggested that they had a sedimentary nature. The gamma-ray spectrometer, similar to one carried on Mars 5, was activated four times during the close approaches of the transfer orbit at a distance of 3,000 km over the surface, at all times over the equatorial regions between, where it made 11 tracks between 30 N and 30°S, covering such features as the *Vallis Marineris* and the largest volcanoes. As a result, it was possible to calculate surface composition (Table 5.10).

Broadly, these figures confirmed those of Mars 5. Combined with American Viking results, they also indicated that much of the area concerned was covered with a layer of fine-grained wind-blown material and that these composition figures represented both the surface and the underlying bedrock.

The probe did not come close enough to the moon Phobos to reach many definite conclusions, but its daytime temperature was calculated at 27°C. The cameras took 37 pictures of Phobos itself, covering 80% of its surface, finding six new craters and 11 hollows not previously seen. They were sufficiently detailed for a new globe to

Table 5.10. Phobos 2: composition of Mars rock.

Oxygen	48%
Magnesium	6%
Aluminum	5%
Silicon	19%
Potassium	0.3%
Calcium	6%
Titanium	1%
Iron	9%
Uranium	0.5 ppm
Thorium	1.9 ppm

Source: Surkov, 1997

New globe, issued on the basis of images taken during Phobos photography

be made of Phobos on a scale of 1:100,000. Scanning of Phobos itself revealed a variety of surfaces, suggesting that the moon was a conglomeration of many different materials. Phobos's mass was estimated at between 1.08×10^{16} and 1.08×10^{17} kg. Its density was low, at 1.9 g/cm^3, less than normal carbonacious chondrite, suggesting either that it was made of porous material with many interior voids or had lighter material deep within (e.g. water ice) (or a combination thereof). The infrared spectrometer indicated the former, for it found no sign of hydration. It was sufficiently different from the other carbonaceous chondrites as to create doubt that it really was a captured asteroid. Over time, experts came to believe that Phobos was instead formed from the accretion of material blasted into Mars orbit by some catastrophic event or impact. It was confirmed that Phobos created its own, small dust ring around the planet, proving right the analysis of Mars 3 many years before. One claim was even made, by E. Dubinin, that water was out-gassing from Phobos.

Phobos also carried instruments to measure phenomena beyond the Mars environment. Phobos 1 and 2 between them registered over 100 bursts of X-ray and gamma-ray emissions, one of which, on 24th October 1988, was the most intense ever noted. Phobos 2 recorded six solar flares from Martian orbit in March 1989. Phobos 2 detected 62 gamma-ray bursts. Finally, Phobos 2's parting shot was a remarkable picture of distant Jupiter, the first time it had been seen from Mars orbit [5]. The science outcome of the mission is summarized in Table 5.11.

Table 5.11. Phobos 2 science.

Bow wave 1,000 km above Mars, weak magnetic field
Four features of magnetic field: magnetopause boundary, shock wave, plasma layer, tail
Loss of planetary atmosphere at rate of 1–5 kg/sec
Water vapor content of 1/10,000 Earth
New model of oxygenated magnetopause
Levels of radiation safe for cosmonauts
Temperature of Phobos surface: 27°C
Mapping of surface of Phobos
Density 1.9 g/cm^3
Composition: like carbonaceous chondrite, but lighter or more porous
Creation of small dust ring
Composition of Martian surface rock
Thin layer of blown sandy material covering Martian surface: fine-grain, wind-blown
Surface temperatures of –93 to +7°C

PHOBOS POSTSCRIPT

Phobos 2 was the last mission of the Soviet period. A debate followed as to whether to send the backup craft, Phobos 3, to retry the mission, but this idea was dropped in favor of a large, international project, Mars 94, which, after postponements due to economic difficulties, became Mars 96 and was launched in November 1996. This was the largest interplanetary spacecraft ever launched, with an ambitious program for the in-flight stage and Mars orbit, with the sending down to the surface of two landers and two penetrometers. Despite heroic efforts that led to a successful launch, the upper stage of the UR-500 let them down and Mars 8, as it was called, crashed back to Earth. For Russian planetary scientists, the loss of Mars 96 was the lowest point of their career. Even 15 years later, some are too emotional to talk about it. Various other probes were considered, but they were overwhelmed by the economic collapse that affected the Russian federation for all but the final years of the 1990s.

Something was saved from the ruins, for instruments developed at the time of Mars 96 found their way onto subsequent missions, both European and their old Martian adversaries, the Americans. A good example was the High Energy Neutron Detector (HEND), installed on the Gamma Ray Spectrometer suite on America's *Mars Odyssey* (2001). HEND compiled a summertime and wintertime map of epithermal neutrons on Mars, discovered subsurface water in the polar regions, pointed to arid zones, and modeled the sublimation of the polar caps. Silicon, iron, aluminum, and calcium maps were presented.

Five Mars 96 instruments flew on Europe's *Mars Express* (2003), such as OMEGA, to measure the mineralogical and molecular composition of the surface and atmosphere of Mars. OMEGA made vertical profiles of ozone; charts of oxygen emission and carbon distribution; and geological maps of pyroxene, olivine, and ferric oxide. The SPICAM instrument measured atmospheric density, detected carbon dioxide clouds, mapped aerosols and the hazetop in the atmosphere by longitude and season, and detected the first aurorae (nine logged 2004–2006). Maps were presented of the seasonal distribution of ozone, oxygen, water vapor, dust, and cloud. The planetary Fourier spectrometer made atmospheric maps of aerosols, the levels of water, temperature, and carbon monoxide, and the polar cap. As it passed over *Tharsis* and *Ascraeus mons*, it made a profile of their altitude and the dust and ice on their surface. A "water map" of the planet was compiled, showing its most arid zones and those where groundwater was believed to lie below the surface (*Arabia* and *Amazonis*). Gravity waves were discovered over the north pole. Water clouds were identified over some of the major volcanoes, such as *Olympus Mons*, and ice clouds over *Vallis Marineris*. A black–green–red–blue photo was composed of the ozone layer on the limb.

Gamma-ray spectrometers were carried on the American asteroid missions (NEAR, 1996; Dawn, 2007), MESSENGER to Venus and Mercury (2004), and Moon mission Lunar Prospector (1998). HEND's lunar successor LEND (Lunar Neutron Exploration Detector) flew on the Lunar Reconnaissance Orbiter (2009) to continue the search for lunar hydrogen and water. By the end of 2009, LEND principal scientist Igor Mitrofanov reported that there might be substantial volumes of dirty ice deposited by comets and buried below the surface soil.

America's Pathfinder rover, *Sojourner* (1996), carried an Alpha Proton X-ray Spectrometer, developed by IKI with the Max Planck Institute in Germany, and enabled it to poke at and analyze rocks at close hand. The mission paved the way for America's two extraordinary rovers, *Spirt* and *Opportunity* (2003), which carried a Mössbauer spectrometer developed by IKI with the University of Mainz, and it was the main instrument used for rock, soil, and dust analysis. *Spirit* traveled through olivine basalt at Gusev crater, while in *Meridiani*, *Opportunity* explored basaltic sand and sulfate-rich outcrops.

Europe's *Venus Express* (2005) carried nine instruments with varying levels of Russian involvement. These profiled the atmosphere and clouds, finding many new features such as a double-eyed vortex over the south pole, strong daily variations in the cloud tops, white airglow oxygen clouds over the south pole, and two points at which oxygen was leaking out of the atmosphere (97 km and one nearer to 110 km). These instruments found many unexpected cloud formations (mottled, streaky clouds) and a dark polar cloud contrasted with bright mid-latitude clouds. The cloud tops were found to be very consistent, 65 km at the poles, 70 km at the equator, with local variations of no more than ± 2 km. The instruments enabled scientists to build in unprecedented detail a picture of the clouds and atmosphere, with their composition, temperature, aerosols, and dynamics. Later, Russian instruments will fly on the European Mercury mission, *Bepi Colombo* (spectral imager, ultraviolet spectrometer, planetary ion camera), and the Mars rover mission, Exomars.

THE PATH TO MARS: WHAT WAS LEARNED?

The science gained from Russian Mars missions was comparatively much less than that gained from Venus or the Moon. Mars 1 gained some useful information on conditions in interplanetary space, but it was not until 1971, with Mars 2 and 3, that significant data were collected, principally on the Martian atmosphere. Mars 4 and 5 enabled part of the Martian surface to be mapped and analyzed geologically, while Mars 5 probed the nature of the surface and extended the knowledge of the atmosphere begun by Mars 2 and 3. Mars 6, whose significance was probably greatly underestimated at the time, gave us the first *in situ* measurements of the atmosphere down to its surface and enabled us to have basic information on pressure, density, wind, and composition. Although none of these missions collected as much mapping data as America's remarkable Mariner 9, the knowledge of the atmosphere and environment was at least comparable to the American understanding at the time.

The American Viking missions of 1975–1976 were extraordinarily successful and provided a quantum leap forward in our understanding of Mars, far beyond anything collected by the American Mariners and Russian Mars probes combined to that point. The Phobos 2 mission, although it failed in its primary purpose of landing on Phobos, nevertheless returned important information on Phobos itself, the Martian atmosphere, and magnetic environment. Many of the interpretations of the missions of the early 1970s came to be vindicated as our knowledge of Mars improved in the new century, especially the theory of the warmer, wetter Mars. In summary, the main knowledge gained from Russian Mars missions was as follows:

- interplanetary space:
    nature of Earth's plasmasphere on exit;
    measurement of solar wind;
    measurement of solar storms and other events;
    intermittent meteor showers in trans-Mars trajectory, empty regions between;
- the Martian environment:
    orbital anomalies (mascons);
    characterization, location, orientation of weak magnetic field;
    radiation level tolerable for human expeditions;
    mapping of ionosphere, plasma zones, electron density, bow shock;
- the Martian atmosphere:
    temperatures and pressure from surface through to exosphere;
    levels, variations in atmospheric humidity;
    composition and layers (carbon dioxide, nitrogen, oxygen);
    measurement of levels of ozone, argon, hydrogen;
    water vapor content and variations;
    characterization, frequency, altitude, composition of dust clouds;
    continued evaporation of water vapor, with higher levels in the past;
    clouds: types, altitude, characteristics, nature;
- the Martian surface:
    characterization of key geological features;
    surface temperature (up to $-13°C$) and pressure (up to 6.7 mb);
    surface densities and characteristics;
    wind speeds;
    chemical composition;

photographic and thermal mapping of distinct regions;

nature of fine dust layer;

probability of large volumes of water on the planet in its past;

surface insulation and conductivity;

hot and cold areas, temperature changes according to region, day/night

- the moon Phobos:

characterization of surface;

mass and density;

chemical composition (many elements, mainly carbonaceous); origin as asteroid less certain;

surface temperature.

REFERENCES

[1] *Soviet Space Achievements.* Novosti, Moscow, 1965; Nazarova, Tatiana: Preliminary results of the measurement of meteoritic matter along the trajectory of Mars 1, in Muller, P., ed.: *Space Research*, Vol. IV. COSPAR, Paris, 1963; Chertok, Boris: *Rockets and People*, Vol. III, series editor Asif Siddiqi. NASA, Washington, DC, 2009. Tikhov's quotation is taken from Konstantinov, Boris; Pekelis, V.D., eds: *Inhabited Space*. NASA, TTF 819.

[2] Bezrukih, V.V., *et al.*: Preliminary results of measurements carried out by means of charged particle traps on the interplanetary station Zond 2, in King-Hele, D.G.; Muller, P.; Righini, G., eds: *Space Research*. COSPAR, Paris, 1965; Tikhonravov, Mikhail; Raushenbakh, Boris; Vaisberg, Oleg: Ten years of space research in the USSR. *Kosmicheski Issledovatl*, Vol. 5, No. 5, 1967, NASA translation TTF 11,500; Slysh, V.I.: Measurement of cosmic ray emissions at 210 and 2,200 kc/sec to R_E8 on the automatic interplanetary station Zond 2. *Cosmic Research*, Vol. 3, 1965; Morozov, A.I.; Shubin, A.P.: *Space Electrojet Engines*. NASA, TTF 16,542, 1975; Bezrukh, V.I., *et al.*: *Investigation of Solar Plasma Fluxes on the Interplanetary Station Zond 2*. NASA, Goddard Space Flight Centre, TTF 9,904.

[3] Snyder, Conway; Moroz, Vasili: Spacecraft exploration of Mars, in Kiefer, H.H., *et al.*, eds: *Mars*. University of Arizona Press, Tucson, 1992; Vaisberg, Oleg: Sputnik 1 and something else, in Zakutnyaya, Olga, ed., *Space, the First Step*. IKI, Moscow, 2007; Vaisberg Oleg; Smirnov, V.: Martian magnetotail, in Russell, C.T., ed.: *Solar Wind Interactions, Advances in Space Research*, Vol. 6, No. 1, 1986; Vaisberg, Oleg: Ion flux parameters in the region of the solar wind interaction on Mars according to Mars 4 and 5, in Mycroft, M.J.: *Space Research*, Vol. XVI. COSPAR, Paris, 1975; Gringauz, Konstantin, *et al.*: Measurements of electron and ion plasma components along the Mars 5 satellite orbit, in Mycroft, M.J.: *Space Research*, Vol. XVI. COSPAR, Paris, 1975; Dolginov, Shmaia: *The Magnetic Field in the Very Close Neighbourhood of Mars, According to Data from the Mars 2 and 3 Spacecraft*. NASA, Goddard Space Flight Centre, X 690 72 434, 1972; Moroz, Vasili: *The Mars 2 and 3 Orbital*

Spacecraft – results of the investigation of the Martian surface and atmosphere. NASA, TTF 14,814, 1973; Vaisberg, Oleg: *Measurement of Low-Energy Particles by the Mars 2 and 3 Planetary Probes.* NASA, TTF 15,540, 1974; Ksanformaliti, Leonid, *et al.*: *Mars 3 – pressures and altitudes based on the results of CO_2 altimetry.* NASA, TTF 15,910, 1974; Moroz, Vasili, *et al.*: *Mars 3 – astrophysical study of the lower atmosphere and planet's surface.* NASA, TTF 15,543; Moroz, Vasili, *et al.*: *Dust Storms on Mars According to Photometric Observations Taken On Board the Mars 3 Automatic Interplanetary Station.* NASA, TTF 14,818, 1973; Basharinov, Anatoli, *et al.*: *Results of Radio Emission Observation from the Planet Mars According to Data of the Experiment on the Mars 3 Automatic Space Station.* NASA, TTF 15,542, 1974; Vinogradov, Alexander; Surkov, Yuri: *Current Achievements in Astronautics.* NASA, TTF 14,929; Breuz, T.K.; Gringauz, Konstantin: The nature of obstacles which slow down solar wind near Venus and Mars and the properties of interaction between the solar wind and atmosphere of these planets. *Cosmic Research*, Vol. 18, 1980; Smirnov, V.N., *et al.*: Possible discovery of cusps near Mars. *Cosmic Research*, Vol. 16, 1978.

[4] Kontor, N.N., *et al.*: Solar proton anistropy, in Mycroft, M.J.: *Space Research,* XVI. COSPAR, Paris, 1975; Nazarova, Tatiana; Rybakov, A.K.: Investigation of meteoric matter by Luna 22 and Mars 7, in Mycroft, M.J.: *Space Research,* XVI. COSPAR, Paris, 1975; Marov, Mikhail, *et al.*: Measurements with Mars 6 lander and model of Mars atmosphere, in Mycroft, M.J.: *Space Research,* Vol. XVI. COSPAR, Paris, 1975; Turnill, Reginald: *Observer's Book of Unmanned Spaceflight.* Frederick Warne, London, 1974; Kondratyev, Kirill; Bunakova, AM: *The Meteorology of Mars.* Hydrometeorological Press, Leningrad, 1973, TT F 816; Sagdeev, Roald Z.: The principal phases of space research in the USSR, in USSR Academy of Sciences, History of the USSR, New Research, 5, *Yuri Gagarin – to mark the 25th anniversary of the first manned spaceflight.* Social Sciences Editorial Board, Moscow, 1986; Dollfus, A., *et al.*: Simultaneous polarimetry of Mars from Mars 5 spacecraft and ground-based telescopes, in Rycroft, M.J., ed.: *Space Research,* Vol. XVII, 1976; Kerzhanovich, Viktor V.: Mars 6 – improved analysis of the descent module measurements. *Icarus,* 30, 1977; Lewis, Richard S.: *Illustrated Encyclopedia of Space Exploration – a comprehensive history of space discovery.* Salamander, London, 1983; Kerzhanovich, Viktor; Pikhadze, Konstantin: Soviet Veneras and Mars – first entry probes trajectory reconstruction science. Paper presented to the international workshop on planetary probe atmospheric entry and descent trajectory analysis and science, Lisbon, Portugal, 6–9 October 2003; Mars 5 and 6 flights analyzed. *Flight International,* 4 April 1974; Surkov, Yuri: *Exploration of Terrestrial Planets from Spacecraft – instrumentation, investigation, interpretation,* 2nd edn. Praxis Publishing with John Wiley & Sons, Chichester, 1997; Moroz, Vasili I.; Nadzhip, A.E.: Water vapour in the Mars atmosphere from Mars 5, in Mycroft, M.J.: *Space Research,* Vol. XVI. COSPAR, Paris, 1975; Savich, N.A.; Samovol, V.A.: Nighttime ionosphere of Mars from Mars 4 and 5 dual frequency radio occultation measurements, in Mycroft, M.J.: *Space Research,* Vol. XVI. COSPAR, Paris, 1975; Gringauz, Konstantin, *et al.*: The magnetic field of

Mars, in Pomeroy, John; Hubbard, Norman, eds: *Soviet American Conference on the Cosmochemistry of the Moon and Planets.* NASA, 1974; Kuzmin, R.O., *et al.*: *Morphology of Fresh Martian Craters as an Indicator of the Depth of the Upper Boundary of the Ice-Bearing Permafrost – a photo-geologic study.* NASA, LPI, undated; Kuzmin, R.O.: *On Possible Structure of the Martian Cryolithosphere.* NASA, LPI, undated; Moroz, Vasili; Mukhin, Lev: *About the Internal Evolution of the Atmosphere and Climate of Earth Type Planets.* NASA, LPI, undated; Zasova, L.V., *et al.*: CO_2 altimetry of Memmonia Fossae to Margaritifer Sinus by Mars 5. *Cosmic Research*, Vol. 20, 1982; Alekseev, N.V., *et al.*: Observations of electron fluxes with energies over 40keV unrelated to solar flares by Mars 7. *Cosmic Research*, Vol. 20, 1982; Vakulov, P.V.: Modulation of cosmic rays by high energy streams of solar wind based on measurements by Mars 7 and Prognoz 3 in January 1974. *Cosmic Research*, Vol. 19, 1981; Vdovin, V.V.: Some characteristics of the soil in the equatorial zone of Mars according to thermal radioactivity data from Mars 5. *Cosmic Research*, Vol. 18, 1980; Krasnopolsky, Vladimir, *et al.*: Structure of the lower and middle atmosphere of Mars from ultra-violet photometry from Mars 5. *Cosmic Research*, Vol. 18, 1980; Davydov, V.D.: Ancient riverbeds, ground ices and contemporary sources of water on Mars. *Cosmic Research*, Vol. 18, 1980; Dolginov, Shmaia: The magnetic field of Mars. *Cosmic Research*, Vol. 16, 1978; Krupenio, N.N.: Estimate of the density of Martian soil. *Cosmic Research*, Vol. 16, 1978; Bogdanov, A.V.: Some evidence of a strong interaction of the Martian satellite Deimos with the solar wind. *Cosmic Research*, Vol. 15, 1977; Krasnopolsky, Vladimir; Pashev, V.A.: High altitude profile of water vapour on Mars. *Cosmic Research*, Vol. 15, 1977.; Moroz, Vasili; Kukhin, Lev: Early evolutionary stages in the atmosphere and climate of the terrestrial planets. *Cosmic Research*, Vol. 15, 1977; Vasileva, L.I.: Determination of duration and limits of communications located on entry of the descent vehicle of Mars 6 into the Martian atmosphere. *Cosmic Research*, Vol. 15, 1976.

[5] Sokelman, I.: Diagnostics of the inner corona by XUV imaging of the Sun, in Antonicci, E.; Somov, B.V., eds: *Solar Corona and Solar Wind, Advances in Space Research*, Vol. 11, No. 1, 1991; Möhlmanuel, D., *et al.*: Magnetic field environment of Mars as studied by Phobos 2, in Shortill, R.W., *et al.*, eds: *Recent Results from Mars and Venus, Advances in Space Research*, Vol. 12, No. 9, September 1992; Vereigin, M., *et al.*: Martian atmosphere dissipation problem – Phobos 2 TAUS experiment evidence, in Shortill, R.W., *et al.*, eds: *Recent Results from Mars and Venus, Advances in Space Research*, Vol. 12, No. 9, September 1992; Dubinin, E.; Lundin, R.: Mass loading near Mars, in Coates, A.J., ed.: *Comparative Studies of Magnetospheric Phenomena, Advances in Space Research*, Vol. 16, No. 4, 1995; Zhang, T.-L., *et al.*: Studies of flaring angles of Mars and Earth magnetic tails, in Coates, A.J., ed.: *Comparative Studies of Magnetospheric Phenomena, Advances in Space Research*, Vol. 16, No. 4, 1995; Luhmann, J.G.; Russell, C.T.; Brace, L.H.; Vaisberg, Oleg: The intrinsic magnetic field and solar wind interaction of Mars, in Kiefer, H.H., *et al.*, eds: *Mars.* University of Arizona Press, Tucson, 1992; Moore, Patrick: *On Mars.*

Cassell, London, 1998; Forget, François; Costard, François; Lognonné, Philippe: *Planet Mars – story of another world.* Praxis/Springer, Chichester, 2008; Titov, D.V., *et al.*: Evidence of regolith-atmosphere water exchange on Mars observed from Phobos 2 infrared spectrometers, in Taylor, F.W., ed.: *Atmospheres of Venus and Mars, Advances in Space Research*, Vol. 16, No. 6, 1995; Moroz, Vasili: Experimental data on aerosols on Mars Viking, Phobos and future missions, in Taylor, F.W., ed.: *Atmospheres of Venus and Mars, Advances in Space Research*, Vol. 16, No. 6, 1995; Mitrofanov, Igor, *et al.*: Statistical study of the evolution of gamma ray bursts detected by Phobos and Compton, in Gehrels, N., ed.: *Gamma Ray Astronomy, Advances in Space Research*, Vol. 15, No. 5, May 1995; Moroz Vasili, *et al.*: Characteristics of aerosol phenomenon in the Martian atmosphere, in Shortill, R.W., *et al.*, eds: *Recent Results from Mars and Venus, Advances in Space Research*, Vol. 12, No. 9, September 1992; McKenna Lalor, Susan: Energetic particle studies at Mars by SLED on Phobos 2, in Shortill, R.W., *et al.*, eds: *Recent Results from Mars and Venus, Advances in Space Research*, Vol. 12, No. 9, September 1992; McKenna Lalor, Susan: Planned investigation of energetic particle populations in the close Martian environment, in Keating, G.M., ed.: *Exploration of Venus and Mars Atmospheres, Advances in Space Research*, Vol. 15, No. 4, April 1995; Zakharov, Alexander: Plasma environment of Mars – Phobos mission results, in Shortill, R.W., *et al.*, eds: *Recent Results from Mars and Venus, Advances in Space Research*, Vol. 12, No. 9, September 1992; Goldman, Stuart: The legacy of Phobos 2. *Sky & Telescope*, February 1990; Results of the Phobos project. *Soviet Science & Technology Almanac 90.* Novosti, Moscow, 1990; Zaitsev, Yuri: The successes of Phobos 2. *Spaceflight*, Vol. 31, No. 11, November 1989; Korablev, O.I., *et al.*: Tentative identification of formaldehyde in the Martian atmosphere. *Planetary and Space Science*, 41, No. 6, from NASA Technical Reports.

6

Orbiting space stations

The use of dogs and other animals on sounding rocket flights in the 1950s (Chapter 1), not to mention Sputnik 2, showed how closely linked early Soviet space science was to plans for manned spaceflight. It is no surprise that dogs and other animals continued to be used to anticipate the reactions of humans to manned spaceflight, for not just the early human flights, but also the later, longer-duration missions. Once the first manned flights began in 1961, the reactions of humans to ever longer manned spaceflight became an important discipline within space science. Not only that, but manned spacecraft opened up new opportunities and platforms for the carrying out of scientific experiments in orbit. The establishment of the first orbiting space station, Salyut, in 1971, dramatically expanded the possibilities of installing scientific equipment on orbit, to be operated directly by cosmonauts. Chapter 6 follows Soviet space science arising from the presence of humans in orbit from the first manned flights through to the International Space Station.

BEGINNINGS OF SPACE BIOLOGY: KORABL SPUTNIK

The Soviet Union originally intended to fly its first cosmonaut into orbit before the end of 1960, planning to inaugurate manned space flight with nothing less than a day-long mission. In preparation, the new cabin planned for manned flight, called Korabl Sputnik but based on object D2 of Tikhonravov's much earlier design studies, first flew on 15th May 1960. It carried only one scientific instrument, an air density meter, crucial for determining the degree to which the very limited atmosphere at orbital altitude would slow orbiting spaceships and cause them to re-enter naturally (see Chapter 2). The high orbit of the cabin enabled a four-year density map to be compiled (Figure 6.1).

The second Korabl Sputnik carried a wide range of animals and experiments to anticipate the effects of manned flight. The first such mission was lost on 28th July 1960, when the rocket exploded at 28 sec, killing dogs Bars and Lisichka. The second was successful, called Korabl Sputnik 2, on 19th August 1960. This time, unlike the unfortunate Laika, it was fully intended to recover the crew, which

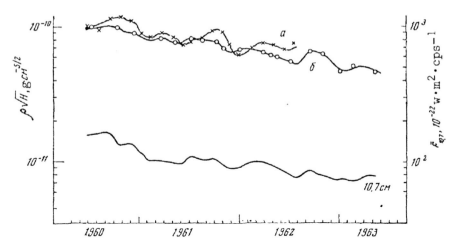

Figure 6.1. Air density map 1960–1963 compiled by the first Korabl Sputnik.

comprised not just one dog, but two (as in the earlier ballistic ascents) and a menagerie of other biological samples. The biological cargo was designed to be fitted onto the cosmonaut's ejector seat in the cabin. Just as the cosmonaut was expected to eject before his parachute landing, so, too, the biological cargo must eject for the final stage of its descent (the main cabin was very heavy and even under a parachute would make a potentially bone-breaking landing). The biological cargo was extensive and went far beyond what a manned precursor test on its own might require. Its ambition reflects the priority given to space biology, which Academician V.A. Engelhardt had made a priority within the Academy. By now, there were no fewer than 14 laboratories dedicated to space biology. Korabl Sputnik 2 was to spend a day in orbit, flying the full profile of the first manned spaceflight.

The two dogs selected were from the Pavlov Institute in Leningrad and had to be of the right size to fit into the two 8-kg dog containers (no longer than 43 cm). One hundred dogs were selected as candidates for the mission and were given two months of training in centrifuges, isolation chambers, and even aircraft flying parabolic curves to simulate weightlessness. The dogs selected were two huskies, Belka and Strelka, meaning "squirrel" and "little arrow", respectively. Also on board were two white rats, 28 white and black mice, guinea pigs, several hundred insects, plants, fungi, bacteria, and seeds (onion, wheat, spiderwort, and maize) as well as cell cultures, enzymes, DNA, and frog eggs. The virgin lands campaign being in full swing, several different types of wheat were flown to test the possibilities of improving crop yield. Three doctors in the Moscow Institute of Experimental Biology, Doctors Rybakov, Novikov, and Kapichnikov, volunteered skin samples from their shoulders and thighs, both for the mission and for a control sample. Some of these samples were, after removal, infected with cancer to see whether radiation would accelerate their cancer growth. The Russians called such a large biological cargo "the ark" and on the ground was a second, control "ark". The ark carried a set of cosmic ray plates: after 10 hr, the plates were split, half being put in a photographic solution for 90 min and then photographed to test for cosmic ray impacts up to that point, the other plates being recovered unopened when the cabin returned.

The Russians followed the mission on live television. Although the Americans did not install live television on their spaceships until 1968 (Apollo 7), the Russians had live television from Korabl Sputnik 2 onwards. As the rocket took off, the animals were clearly affected by the noise and vibration, shifting restlessly around their small cabin, their pulse rates three times normal. For the first 5 min of weightlessness, they hung listless, lifeless, hardly moving, not knowing how to react: they were clearly anxious, for blood pressure, heart, and respiration rates rose, not coming down for 4–6 min. Strelka became distressed, her pulse rose to 180/min, and she at one stage hyperventilated. In addition to television, ground control received data on cabin temperature (18°C), pressure, humidity, and radiation levels. Later, television showed them barking, apparently quite happily. The dogs were twice fed a jelly that combined food and liquid, having been trained to suck on a dispenser when a signal was given (which they did).

On 20th August, after 18 orbits, the retrorocket fired and the cabin began its descent. At 8,000 m, the ejector seat fired the "ark" out of the cabin. Farm workers east of the river Volga spotted the now empty cabin coming down at a fast 10 m/sec, followed by the slower "ark" at a gentler 6 m/sec. Cars were soon bouncing across the fields to release the dogs and the rest of the cargo of the "ark" and bring them back to Moscow. Belka and Strelka were calm after landing and had a good appetite.

The flight of Korabl Sputnik 2 was the world's top news story for several days and the happy dogs were triumphantly presented to a tumultuous press conference on their return. Even as they arrived in an old black car, the dogs were mobbed by journalists and photographers. "We realized something really important had happened", according to Lyudmilla Radkevich, their manager.

Nikita Khrushchev, never one to pass up the opportunity to make a political point, made a gift of one of Belka's subsequent puppies, Pushinka, to Jacqueline Kennedy, a happy addition to the White House (she went through a security check first in case the cunning Russians had fitted listening devices in her). The post-flight press conference marked the first public appearance of the man who became the senior doctor of spaceflight, the thin, moustached military doctor, Oleg Gazenko. Born in Nikolaevka, Stavropol, in 1908, he had worked in the Institute of Aviation Medicine from 1947 and had supervised the sounding rocket flights in the 1950s. At the press conference, he personally lifted aloft the dogs Belka and Strelka, each in a harness, to demonstrate their good health.

The Russians made no attempt to conceal their intentions of soon making a manned spaceflight and the following day's *Pravda* told its readers that the practical conditions for such a mission now existed. The non-infected skin flaps were reattached to their owners, taking longer to heal than control samples kept on the ground for the same period. Korolev decided that there must be two successful dog flights before he would commit to a manned flight.

The big story of Korabl Sputnik did not come out for over 30 years, which was that Belka remained restless for several hours and vomited on her fourth orbit, only

Oleg Gazenko, leader of space medicine for 40 years

calming down afterwards. Belka was the first victim of what is now called "space sickness" or, in American parlance, "space adaptation syndrome". Of all the many ill effects of spaceflight, nausea was low on the list of what was predicted. Essentially, what it boils down to is that one's sense of well-being is determined by one's balance, a function of the inner ear. In weightlessness, one's orientation is affected, the inner ear senses imbalance, and this leads to nausea. About half of all astronauts and cosmonauts suffer from space adaptation syndrome and it is medically impossible to predict in advance for one space traveler from another. Generally, it is a feature of the first two or three days of flight, after which it wears off. It takes different forms with different people, ranging from nausea to vomiting to headaches.

At the time, the Russian doctors probably did not know what to make of Belka's being sick – whether it was an isolated event or not. The medical director of the flight, though, read the event more correctly than he may have realized and recommended that the flight of the first cosmonaut into space be limited to one orbit only, not the full day originally planned [1].

Although most attention was given to the dogs, Korabl Sputnik 2 carried other biological experiments. On board were single-cell green algae (chlorella) to test whether it could absorb carbon acid and generate oxygen – a theory first put forward by Tsiolkovsky in the previous century. Research on the potential of algae to develop closed-cycle ecological systems had begun in January 1958 in the Department of Astrobiology of the Kazakh Academy of Sciences in Alma Ata under the guidance of the world's first astrobiologist, the famous astronomer, Gavril Tikhov, and was given an increase in staff from five to 20 to "develop biology in Sputniks". Another biology experiment was a biodetector, designed to find out whether biological, bacteriological, or bio-microbial specimens could be found in orbit, following earlier scooping devices flown on sounding rockets (for the record, they were not). The interest in flying seeds into space may have surprised Westerners, but it should not have. Leningrad was home to the world's largest seed bank, created by Nikolai Vavilov, where the Soviet Union was world leader [2]. Fruit flies were flown into orbit, modeled on the first experiment carried on a balloon in 1936. There was a slight increase in lethal mutations, but a more pronounced effect in following generations.

One of the most important discoveries of Korabl Sputnik 2 was not in space biology, but in radiation, for, traveling over the South Atlantic, Geiger counters installed by Sergei Vernov found a sudden concentration of protons above 50 MeV, where the inner radiation belt descended far below its normal level. This became known, in time, as the Brazil or South Atlantic Magnetic Anomaly (SAMA). Particles flowing along electric field lines quickly perished there in what became known as a "forbidden region". This information was transmitted back by telemetry during the mission but when Korabl Sputnik's nuclear photo-emulsion traps were examined microscopically on their return, the scientists noted zones of increased radiation not only over the South Atlantic, but also over the Indian Ocean and Cape Town. The Brazil anomaly was subsequently investigated and mapped in detail by counters fitted on Korabl Sputnik 3, and Cosmos 4, 7, 9, 12, and 15 (see Chapter 2). Korabl Sputnik 2 enabled the compilation of a new map of the Earth's lower radiation belt and the first map of the anomaly was published by Sergei Vernov, Ivan Savenko, and Pavel Shavrin in December 1960 (Figure 6.2). Korabl Sputnik 2 made early observations of solar flares. The temperature of the Sun was measured from orbit at 0.9 million K, but during two flares, parts rose to 6.5 million K. The first flare marked a rise in radiation intensity of 63%, the second by 320%. In an exotic side to the cosmic ray experiment, Russia's leading cosmic ray expert, Naum Grigorov, microscopically examined the 489-layer emulsion pile that was used to trap cosmic rays to look for evidence of anti-matter. Although he counted 740 impacts, he found no anti-nuclei, but speculated that Earth might scoop up anti-matter in its path around the solar system – a suitable subject for a future experiment [3].

In advance of the flight of the first cosmonaut, four more dog missions were carried out. The next, Korabl Sputnik 3, was a repeat with two dogs, two guinea pigs, 26 mice, frog ova, DNA, bacteria, viruses, cells, rabbit bone marrow, 21 flasks of flies, with pea, wheat, maize, onion, and beans. It ended tragically: when it appeared that the cabin would come down off course and possibly land outside the USSR, explosives were detonated to destroy the cabin (including dogs Pchelka and Mushka). Before that, though, the radiation counters were used to further map the lower edge of the radiation belt and a new global radiation map was published (Figures 6.3 and 6.4). Korabl Sputnik 3 coincided with a magnetic storm and particles became exceptionally agitated [4]. The next Korabl Sputnik also went amiss, the rocket underperformed, and the cabin came down in wintry Siberia, but despite a long wait in appallingly cold weather, the animals were recovered alive.

The planned manned flight slipped into spring 1961, but two final dress rehearsals were made, each a one-orbit profile. A single dog was carried on each mission, alongside a rubber dummy cosmonaut given the name of "Ivan Ivanovich", who took the place of the second dog. The landing procedure was different: the dummy was ejected in the ejector seat, while the dog rode the cabin to the ground, presumably to test whether it could survive the much faster impact should the ejector system fail. In the event, neither dog appears to have been injured. Chernushka made a single orbit on 9th March, accompanied by 40 white mice, 40 black mice, guinea pigs, reptiles, plants, blood samples, bacteria, and micro-organisms. A similar ark accompanied Zvezdochka into orbit two weeks later, on 25th March 1961. This

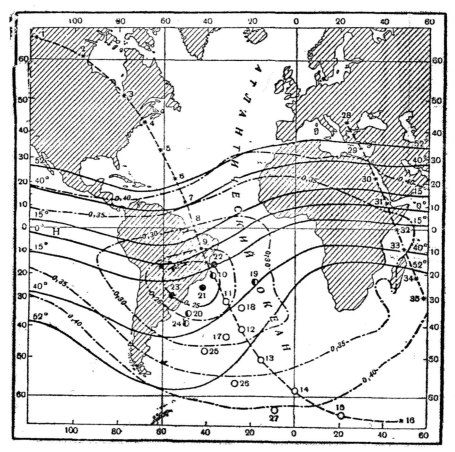

Figure 6.2. Korabl Sputnik 2 map of the South Atlantic magnetic anomaly. (Credit: COSPAR)

marked, for the moment, the end of the dog missions begun almost 10 years earlier in July 1951, in the course of which over 40 dogs had flown into space on more than 30 missions, 18 being lost. At a press conference organized by the Academy of Sciences immediately afterwards, at which Zvezdochka and Chernushka were the stars, a number of veteran space dogs were presented to mark the end of the program. Everyone seemed to know that the next rocket would carry a human.

The Korabl Sputniks were important developments in space biological science and prepared the way for manned spaceflight. Although such missions seem overly cautious from the perspective of 50 years later, this was not how they appeared at the time, where there were genuine concerns about the dangers to humans of the effects of spaceflight. The missions are summarized in Table 6.1 and the mission outcomes in Table 6.2.

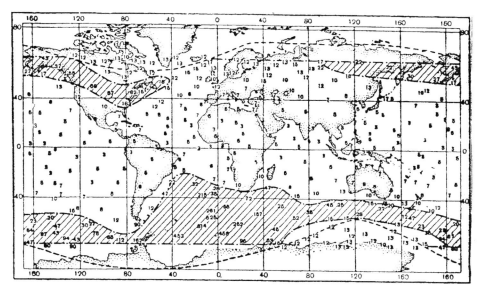

Figure 6.3. Radiation intensity, compiled by Korabl Sputnik 3.

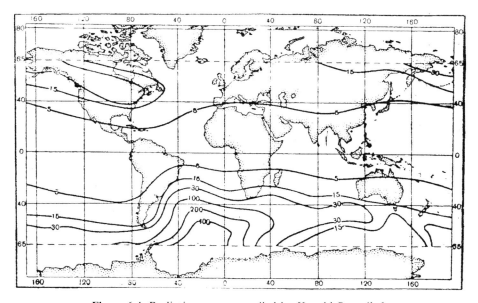

Figure 6.4. Radiation map, compiled by Korabl Sputnik 3.

Table 6.1. Orbited Korabl Sputnik biological missions, with names of dogs.

Korabl Sputnik 2	19 Aug 1960	Belka, Strelka
Korabl Sputnik 3	1 Dec 1960	Pchelka, Mushka (destroyed)
Korabl Sputnik 4	9 Mar 1961	Chernushka
Korabl Sputnik 5	25 Mar 1961	Zvedochka

Table 6.2. Korabl Sputnik mission outcomes.

Sufficiently safe to proceed to manned flight
Significant stress and vibration during launch phase
Period of immediate adaptation to weightlessness: 4–6 min
Animals can function during a day-long mission, recover quickly after return
Radiation levels within limits
Discovery of space sickness
Mapping of South Atlantic magnetic anomaly
Estimates of temperature of the Sun and solar flares

SPACE BIOLOGY BY COSMONAUTS

The first manned space missions were primarily directed around technological, engineering, and indeed political goals: science was not a top priority. Despite this, the early manned missions did have some scientific objectives, especially in the key area of biological science. These have often been overlooked in analyses of these missions.

There was little room for science during Yuri Gagarin's pioneering but short 108-min flight around the Earth, although the cabin included a biological tray. With the second Vostok flight, it was possible to carry out the original 1960 plan of spending a full day in space. Gherman Titov was the pilot of Vostok 2, which entered orbit on 6th August 1961. Using a Konvass camera, he took the first movie images of the Earth from orbit and then still images from his Zritel camera. A small biological payload was carried: fruit flies, dry seeds, and bacteria to test the effects of cosmic radiation on them, using yeast cultures in oleic acid. This was installed in the spacious cabin and did not require his personal supervision.

Scientifically, the principal event took place soon after entering orbit, when Titov was quickly overcome by nausea, felt disorientated, and imagined the control panel floating away from him. He vomited at least once. Only by staying perfectly still could he get the sensations to pass. Titov recovered abruptly on the 12th orbit and was bright and alert when the time came for him to return to Earth. Doctors on the ground were well aware of what was going on and Titov's distress was evident from both telemetry and television. Forewarned by Belka's experience, they were far from surprised and Titov gave a very full account of his experience to the doctors on his return.

Although the Russians were open about the issue, the importance of space sickness was probably not well appreciated in the West at the time. The early American missions were extremely short by comparison – not until 1963 did an American fly for more than a day – and no early American "right stuff" astronaut was likely to admit to vomiting in orbit if he could help it. The automatic inclination was to write off Titov's experience as exceptional.

The subsequent, ever longer flights in the Vostok series provided opportunities for a science program that would push back the frontiers of space biology. The aim was to test the reaction of the cosmonauts to weightlessness, each cosmonaut being equipped with a series of sensors to test for pulse and respiration. In addition, an electro oculogram recorded eye movement while an electrocardiophone noted cardiac muscle. Andrian Nikolayev (Vostok 3, four days) and Pavel Popovich (Vostok 4, three days) orbited at the same time in August 1962 and had a common science program, which involved:

- medical tests: taking regular self-measurements of heartbeats, pulse, temperature, and breathing;
- orientation tests, from mathematical calculations to join-the-dots, to test mental awareness;
- water tests, in which they observed the behaviour of water in weightlessness, both within a flask and in the cabin (Vostok 4 only);
- biology: small biological package of fruit flies, dry seeds, onion, cress, mustard, pine, cabbage, radish, pea, black bean, tomato, carrot, lettuce, sugar beet, bacteria, spores; they crossed packages of male and female fruit flies.

Nikolayev and Popovich landed only a few minutes apart and although their cardiovascular profile took over a week to return to normal, neither had been space sick. Although it took the cardiovascular system 7–12 days after landing to return to pre-flight levels, the principal scientific outcome was that missions of this duration did not pose any particular biological or psychological hazards. Both had an observation program using cameras and binoculars, taking images of Earth, Moon, and stars. The radiation dose to cosmonauts averaged 8 millirads/day, of which 80% came from cosmic rays, 5% from the inner radiation belt, and 15% gamma radiation from radiation belt electrons. The total radiation doses were 63 millirads for Vostok 3 and 50 for Vostok 4. When the seeds were grown back on Earth, it was found that some were inhibited and some were stimulated, which the botanists attributed to weightlessness rather than radiation. Microscopic investigation showed that the spores were affected by chromosomic aberrations. Results were not always consistent: whereas *Drosophila* flies had recessive mutations on the first Vostok, this was not the case on the following three missions. X-ray photographs showed how cosmic ray tracks had scratched the walls of the biological containers [5].

The aim of the next mission, Vostok 5, was to extend the duration record to eight days, but a lower orbit than intended restricted this to five days – still a record. It was a second double mission, Vostok 6 orbiting simultaneously for three days. The two cosmonauts, Valeri Bykovsky and Valentina Tereshkova, had a common scientific program proposed by the President of the Soviet Academy of Sciences a month before the mission, concentrated on Bykovsky's longer Vostok 5 mission. This involved studies of the transparency of Earth's atmosphere from directly below the spaceship to the horizon, as well as biological studies, for the spaceship carried a package of frog ova, insects, plants, seeds (evonymus seeds), pea plants, algae, bacteria, and cancerous cells. The radiation levels experienced by cosmonauts were measured: 35–40 millirads for Bykovsky, 25 millirads for Tereshkova.

Two important scientific outcomes came from Vostok 6's Valentina Tereshkova. She was equipped with a photometer made by Oleg Vaisberg to measure the upper atmosphere's emissions and she made one set of observations of the layers of the stratosphere, including the mixing of aerosol layers between 14 and 19 km. The aerosol layers were mapped over the South Atlantic at an oblique angle looking towards Africa, the Vostok 6 cosmonaut finding a first thin aerosol layer at 11.5 km and a second, thicker dust layer at 19.5 km (accuracy ± 1 km). This was an early characterization of aerosol layers, identifying therein sulfurous ammonia. Second, strange things happened when cucumber and wheat seeds carried in the biological package were germinated on return to Earth. Compared to the control sample, these seeds developed spindly, scraggy roots, stems, and flowers. The significance of this only became apparent when cosmonauts later tried to grow such plants in orbit, for they became frustratingly difficult to grow properly [6].

Autumn 1963 marked an important institutional development. Biological science was emerging as an ever more important aspect of the manned spaceflight program. Until then, space medicine had been an integral part of the Air Force's Institute of Aviation and Space Medicine. The idea of a separate body for space medicine was first suggested by Sergei Korolev and Mstislav Keldysh in May 1959. The Ministry of Health also saw an opportunity to develop an institute under its own control and the outcome was the Institute for Bio Medical Problems (IBMP), established on 28th October 1963 under the Ministry of Health, headed for its first two years by the famous radiobiologist Andrei Lebedinsky, a student of the great Ivan Pavlov [7].

The importance of space biology was underlined by the first multi-man flight. Voskhod used the same cabin as Vostok, the ejector seat being replaced by seating for three cosmonauts without spacesuits. Since the cabin needed only one pilot, it was possible to fly a wider range of people on board, so accompanying Voskhod pilot Vladimir Komarov was spacecraft designer Konstantin Feoktistov and, most importantly from the scientific point of view, doctor Boris Yegorov, who carried out the first ever medical examinations in space using a machine called *Polinom*. Boris Yegorov was actually writing a thesis at the time on balance in the inner ear and his inclusion on the mission was an indication as to just how seriously Titov's earlier experience had been viewed. Boris Yegorov took blood samples to test calcium levels, assess muscle strength (using a dynamograph), judge muscle reactions, test for the factors that caused space sickness (e.g. rapid head movements), and he was also on the lookout for small blood vessels bursting under the effects off weightlessness. In the event, both Feoktistov and Yegorov suffered from minor space sickness, the first to do so since Titov.

While Yegorov focused on medicine, his companion Konstantin Feoktistov spent time on observations of atmosphere and auroræ, photographing some spectacular eruptions over Australia. He made the first visual observations of mesospheric clouds 80 km high (Figure 6.5). Like the Vostok missions, there was a biological package on board: bacteria, fruit flies, cancer cells, frog sperm and ova, plants, winter wheat seeds, algae, and pine seeds. This time, spores were placed in protective glass vessels, but they still presented hard-to-explain mutations after only a day in orbit. As for radiation, the dosage was 30 millirads each.

The following mission, Voskhod 2, tested the ability of cosmonauts to work in open space, when Alexei Leonov made the first ever spacewalk (his pulse rate

Konstantin Feoktistov, who took some of
the early, detailed observations from orbit

Figure 6.5. Konstantin Feoktistov's
observations.

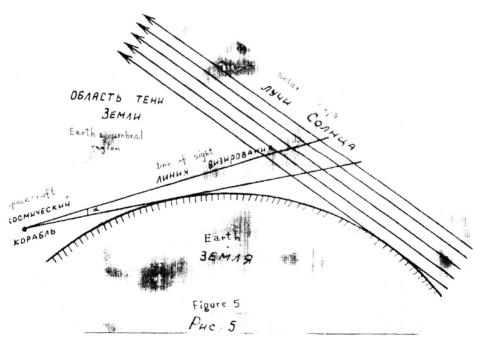

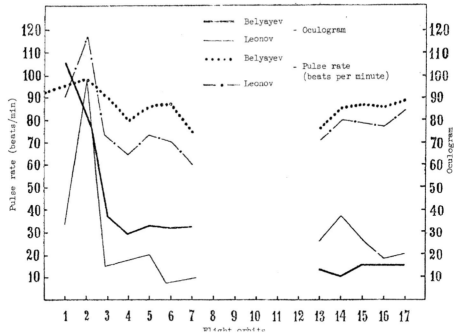

Figure 6.6. Pulse rates of the Voskhod 2 crew; the gap is the period when the spacecraft is flying away from the Soviet Union and transmissions cannot be received.

peaking at 168 when he struggled to get back in). An identical biological package was carried, except that this time, some samples were placed in a pouch in Leonov's suit when he spacewalked (in the event, there were no significant differences to them compared to the inside samples). With the first eight missions over, the biological outcomes were assessed: in orbit, bacteria grew by between 1.5 and 4 times the normal rate; chromosomes were re-arranged due to high G forces and there was a higher rate of cell division due to zero gravity. There was no effect on the germination of carrots, tomatoes, pine, beans, or wheat seeds, but there were some hereditary effects on lysogenic bacteria. The much higher orbit, closer to the radiation belts, meant that each cosmonaut received a dosage of 70 millirads [8]. There was a continuing emphasis on monitoring the cosmonauts' health, especially their pulse rates, which generally seemed to settle down after the early excitement of reaching orbit (Figure 6.6). The outcomes are summarized in Table 6.3.

Table 6.3. Scientific outcomes, early manned missions (Vostok and Voskhod).

Generally, no significant ill effects from missions of 4–5 days
Cosmonauts made a quick recovery on return
Radiation levels within reasonable limits
Space sickness an issue on one mission (Vostok 2), minor on another (Voskhod)
Some changes to biological samples

Overall, the prospects for ever longer manned spaceflight were good. Voskhod 3 was intended to mark a radical step forwards in space biology, making a high-altitude duration mission of three weeks by cosmonauts Georgi Shonin and Boris Volynov in May 1966. To pave the way for such a mission, it had been decided as far back as October 1964 that a 15–30-day flight by animals in advance was desirable, to anticipate both the effects of weightlessness and the radiation belts through which the cosmonauts would travel as well as to thoroughly test out the life support system over such a long period. This eventually took place in February–March 1966.

HITTING THE FIRST WALLS OF SPACE BIOLOGY: COSMOS 110

Dogs Veterok ("Breezy") and Ugolyok ("little black piece of coal") duly flew the Voskhod cabin for a 25-day mission, whose orbit took it out to from 186 to 904 km. This brought their cabin into the radiation belts, but at an inclination of 52°, the belts were not as severe as at more northerly latitudes. They were clad in new lightweight spacesuits. A range of sensors was attached to each to measure pulse, respiration, blood pressure, and the electrical activity of the heart and brain. The dogs were fed intravenously a 650-g daily mixture of meat, potatoes, flour, vitamins, and water. One of the two dogs was given anti-radiation drugs. Back on Earth, two control dogs underwent a similar mission on the ground. A full biological package was devised by the new Institute of Bio Medical Problems (IBMP), for also on board Cosmos 110 were bacteria, yeast, blood samples, and algae as well as onion, garlic, lettuce, radish, Chinese cabbage, carrots, chlorella, and bacteria. The plant package was a significant new development, for never before had there been such a large consignment of garden plants. Some samples and tissues were placed in steel and aluminum containers to protect them from radiation (and were compared to unprotected samples).

For the first few days, the dogs appeared to be uneasy, rocked their heads a little, and searched for a more comfortable position, calming down on the eighth day. Radiation levels for the dogs were low during the lower parts of the orbit (10–30 millirad) but rose rapidly as the spacecraft climbed to 1,000 km. The flight went relatively normally for nearly three weeks, but due to a decline in air quality in the cabin, a rise in cabin temperature to 30°C, and the detection of some arrhythmia in the dogs' hearts, Veterok and Ugolyok were brought down after 22 days, three days early, on 15th March after 330 orbits. The cabin landed near the river Volga, but fog prevented recovery until a day afterwards. Blood samples were taken immediately.

The initial press reports on the dogs' condition were predictably reassuring: nothing indicated any biological show-stoppers, so preparations went ahead for the launch of Voskhod 3. For a variety of technical and political reasons, the mission was eventually cancelled and the long-duration test originally planned did not take place until four years later, as Soyuz 9. In the event, the results of the Cosmos 110 mission indicate that the cancellation of Voskhod 3 may have been just as well for the two cosmonauts concerned. IBMP issued its medical report on the flight that May and announced that "new problems in space medicine" had arisen in long-term

Spacesuit for the Cosmos 110 dogs

flight. What the institute had to say was worrying (Table 6.4). The dogs were dehydrated, had lost weight (29%), their circulation had weakened, their muscles atrophied (15%), they lost coordination, and they took over a month to recover. Especially notable was the way in which their bone calcium had leaked away into their blood and urine, weakening their bones. They had suffered "cardiovascular deconditioning". Oleg Gazenko took the view that whereas the effects of space travel over a week or so might be minimal, three weeks was quite another matter because something different happened. He did speculate that their restricted space may have been a negative contributory factor. Not a lot of attention was paid to these reports in the West at the time.

Some of these echoed medical findings from the two longest American manned spaceflights to that point: Gemini 5 (8 days) and Gemini 7 (14) – but Cosmos 110 had flown more than a full week beyond this and the outcomes were much more severe, indicating that there was might be a threshold above which the effects of

Table 6.4. Cosmos 110 22-day mission: Veterok and Ugolyok, results.

The dogs were badly dehydrated and had lost 30% of their weight
Circulation was weakened and their muscles had atrophied
Some movements on landing were poorly coordinated
Full motor recovery took 8–10 days
Significant loss of bone calcium, excessive quantities in their blood and urine

Table 6.5. The early manned and biological missions.

Vostok	12 Apr 1961	Yuri Gagarin
Vostok 2	6–7 Aug 1961	Gherman Titov
Vostok 3	11–15 Aug 1962	Andrian Nikolayev
Vostok 4	12–15 Aug 1962	Pavel Popovich
Vostok 5	14–19 Jun 1963	Valeri Bykovsky
Vostok 6	16–19 Jun 1963	Valentina Tereshkova
Voskhod	12–13 Oct 1964	Vladimir Komarov, Konstantin Feoktistov, Boris Yegorov
Voskhod 2	18–19 Mar 1965	Pavel Belyayev, Alexei Leonov
Cosmos 110	22 Feb–16 Mar 1966	Dogs Uterok and Ugolyok

space travel became more significant. Happily, there were no negative long-term health effects on the two dogs. Ugolyok had five puppies afterwards, all being perfectly normal, but Veterok did not breed again.

The biological results were published later. The *transcendentia* plant, for example, grew taller than the control plant, was badly bent over on landing, but then straightened out. Its chromosomes had been rearranged while in orbit. Seeds had been moisturized on entry to orbit and most germinated sooner than the controls, but haricot beans germinated poorly, 10% less than on Earth. Those that did grow developed sooner, faster, and flowered twice more than controls. Lettuce grew 15% higher, 14% wider, 22% leafier, and had 50% more yield. Garlic seeds did not sprout until after landing, but then 24 came out, compared to nine controls. Chinese cabbage had 8% greater mass than controls. Cosmos 110 repeated the Vostok 2 experiment of testing changes to yeast cultures in oleic acid. Typhoid antigens and immune antidotes from an ox were carried in sealed glass ampoules, but they were unchanged after 22 days. Compared to ground controls, onion bulbs germinated sooner, with 18% longer roots and shoots, and had greater chromosome aberrations.

Overall, the biological package led to two sets of conclusions. First, long-duration missions had definite but variable effects on plants, some growing longer and bigger than had they stayed on Earth, but whereas some germinated better, others did less well. Second, radiation was an important variable. Small doses of radiation stimulated growth, principally through generating ascorbic acid, but larger doses of radiation (more than 500 millirads) were dangerous and had negative effects [9]. Table 6.5 summarizes the early biological manned or related missions.

SPACE BIOLOGY'S LUNAR JOURNEY: THE FLIGHT OF THE TORTOISES

The Soviet Union intended to beat the United States both to the first mission around the Moon and also to the first landing on its surface. The Russians came quite close to beating the Americans around the Moon, but trouble with the large rocket, the N-1, led them to fall behind in the race to land on the Moon. Just as biological missions had preceded the first cosmonauts into Earth orbit, so, too, biological tests were carried out in advance of cosmonauts flying out to the Moon. These were done on the Zond missions flown over 1968–1970, Zond being the name used for the L-1 cabin, a stripped-down version of the later Soyuz spaceship – one that would enable cosmonauts to fly around the far side of the Moon, without orbiting it, and then return to Earth. Four Zonds flew the six-day around-the-Moon profile and biological specimens were flown to test for radiation and any other ill effects that might result from such a journey and the high-speed re-entry into the Earth's atmosphere, barreling in at 11 km/sec.

To test the biological effects of flight to and from the Moon, a 150-kg experimental

The first animals to fly around the Moon – the turtles of Zond 5

package was fitted in the cabin. The principal biological cargo comprised two steppe tortoises, each weighing between 34 and 40 g (they were often reported as terrapins or turtles). Two were flown on Zond 5, along with hundreds of flies and fruit fly eggs, meal worms, seeds (wheat, tomatoes, pine, carrot, pea, barley), algae, and bacteria. Flies were good indicators of the effects of radiation, for severe radiation will show up quickly in genetic mutation.

Unlike the choice of dogs ten years earlier, the reasoning behind the choice of cargo was never fully explained, so we must therefore speculate. Tortoises could survive for some time with only limited water supplies – a consideration in a six-day mission and they could manage without food for long periods (which saved supplying a food system) while their hard skins improved the likelihood of their surviving a difficult re-entry. They were tested quite severely, for it was a full 39 days between the loading of the Zond 5 turtles and hatch opening back in Moscow.

Zond 5 became the first spacecraft to fly around the Moon and return to the Earth, being recovered in the Indian Ocean. Externally, instruments were carried for the study of radiation, cosmic rays, the solar atmosphere, and stars. A primary purpose was to measure the radiation threat to cosmonauts venturing out of Earth orbit, levels being carefully measured but found to be non-hazardous (Figure 6.7). Zond 6 carried a radiation detector, micrometeorite detector, and a biological package, presumably similar to its predecessor. During the final phase of Zond 6's

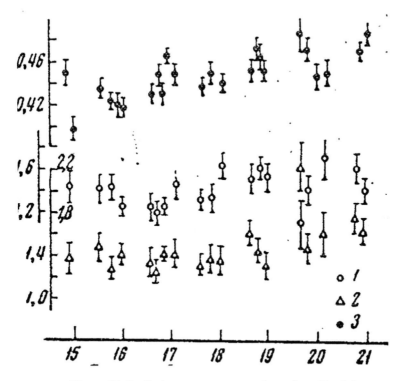

Figure 6.7. Radiation meter count readouts from Zond 5.

return to Earth, a gasket blew, Zond depressurized, and then the parachute prematurely released the cabin, which fell to Earth at some speed. The seeds were recovered from the wreckage but the animals must have perished.

Four turtles were carried on Zond 7, which made a textbook mission only three weeks after the American Apollo 11 lunar landfall, returning smoothly to the summer landing ground of Kazakhstan. On board were bacteria, algae, and seeds (mustard, tomatoes, carrot, pine, peas, wheat, and barley). The equally successful Zond 8 carried tortoises, flies, onions, wheat, barley, and microbes, flying a different return route over the northern hemisphere, which led to a precise splashdown in the Indian Ocean. Zond 8 carried a package similar to the manned Soyuz 9 mission earlier that year, including chlorella cultures (there was a slight decrease in survival rate compared to the control group). Zond 8 also carried a 300-cm^2 foil solar wind detector on the outside of its cabin.

The series over, the collected biological results from the Zond series were published ('Results of Biological Investigations Undertaken on the Zonds 5, 6, 7 and 8', *Kosmicheskiye Issledovaniya* in July–August 1971). The tortoises were examined for changes to their heart, vital organs, and blood. They were hungry and thirsty after their return and had lost 10% of their weight. Their muscles had atrophied and their organs deteriorated, but this was attributed more to prolonged absence of food

Table 6.6. Zond 5–8 missions and instrumentation.

Zond 5	15 Sep 1968	Tortoises, wine flies, meal worms, bacteria, plants, seeds Instruments for radiation, cosmic rays, solar atmosphere, stars
Zond 6	10 Nov 1968	Radiation, micrometeorite detectors, biological package
Zond 7	8 Aug 1969	Four turtles; bacteria, algae and seeds (mustard, tomatoes, carrot, pine, peas, wheat, and barley)
Zond 8	20 Oct 1970	Tortoises, flies, onions, wheat, barley samples, microbes Solar wind detector

than to the duration of the flight. There were some mutations in the seeds as a result of radiation but very much in line with experiments from low Earth orbit. Wheat and barley, when planted on Earth afterwards, grew better than the ground controls, apparently stimulated by space radiation. Onions were stimulated by their journey around the Moon, but there were chromosomal aberrations in barley and pine. Overall, radiation dosages seemed to be well within acceptable limits, not posing a danger to cosmonauts and not significantly different from conditions in Earth orbit. The average dose level of the missions was 3.5 millirad, most of which was picked up during transit through the radiation belts. The foil collection was found to contain solar wind, mainly helium. The mapping outcomes of the Zond missions, though, may ultimately prove to be a greater contribution to science (Chapter 3) [10]. The missions are listed in Table 6.6.

SOYUZ AS A PLATFORM FOR SPACE SCIENCE

Having competed unsuccessfully against the Americans in the Moon race, from 1969, the Soviet Union re-orientated its space program around space stations. The new manned spacecraft, Soyuz, which had replaced Vostok and Voskhod, would instead be used to bring crews up to orbiting space stations, the first of which would be ready in April 1971.

In the meantime, the Soyuz could also be used as a science platform in its own right. Its potential had become apparent during the first successful Soyuz mission, Soyuz 3, in October 1968. The single pilot, Georgi Beregovoi, had sufficient time and opportunity during his four-day mission to test the capability of the Soyuz as an observation platform, using a camera to take photographs of planets, stars, and the horizon, as well as features on the Earth's surface. He returned to the aerosol layer measurements made on Vostok 6, finding a layer at 35 km [11]. The use of spacecraft for systematic observations of the Earth's atmosphere was the achievement of two people, Kirill Kondratyev and, later, Yan Ziman. Kirill Kondratyev became head of the Department of Atmospheric Physics in Leningrad in 1958, where he began his life's work of using space observations for meteorology, atmospheric physics, and, later, ecology, writing over 100 books and still active in the next century.

On the Soyuz 4 and 5 mission in January 1969, the main goal was the first docking

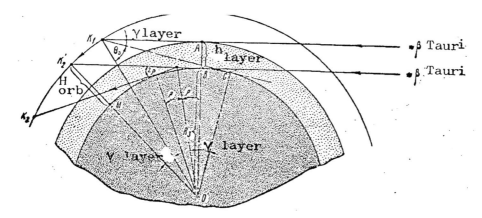

Figure 6.8. Measurement of airglow from Soyuz 5.

between two manned spacecraft. Two cosmonauts spacewalked from one to the other. Here, they collected an aluminum plate on the outside of the spacecraft to detect tritium and helium and to test the impact of the space environment on metals. Back inside, the scientific objectives involved geophysical studies of auroræ at 450 km with a new instrument, the RSS-1 spectograph, and astronomical observations of comet tails.

The RSS-1 instrument, a device developed by the Department of Atmospheric Physics at the University of Leningrad, was first flown on Soyuz 5 and enabled profiles to be made of the atmosphere up to 240 km. This was actually the beginning of attempts to use manned spacecraft to make significant observations of the state of the atmosphere, its layers, and the way in which it glowed after dark. Illustrations were published of the glowing ashy layer and the colors of the twilight auroræ (Figure 6.8). Soyuz 5 observations determined the existence of a nighttime layer of hydroxyl radiation 85–90 km up. The biological packages flown on the earlier manned missions were continued on Soyuz 5. Barley seeds showed aberrations, sprouting faster and generating longer roots than Earth samples. Those kept in an ethylene preservative had even more mutations [12].

The scientific work of Soyuz 4 and 5 was extended when three Soyuz spacecraft flew in Earth orbit together – what was called the *troika* mission (Russian for a "threesome", the missions being Soyuz 6, 7, and 8). This was an improvised mission while the new space station was in development, so the Soviet Union made a virtue of its science objectives: *They've brought space down to Earth!* trumpeted *Soviet Weekly* afterwards. An improved version of the RSS was used, the RSS-2, which the Soyuz 7 cosmonauts Vladislav Volkov and Viktor Gorbatko used on 13th October to carry out spectrophotometry of twilight auroræ. With Earth observation instruments, they made a profile of the region from the Caspian to the Aral Sea, the beginning of Russian space mapping from orbit [13].

Highlight of the *troika* mission was a welding experiment, whose purpose was announced as soon as Soyuz 6 reached orbit and echoed a soldering experiment

Soyuz 7 crew, with Vladislav Volkov operating the RSS instrument

carried out on the balloon *Komsomol* in 1939. The concept behind the experiment was that in a vacuum, metal parts could be joined together and, without air molecules to separate them, form a perfect seal. This would pave the way later for orbital construction of large structures. It was certainly an unusual experiment, although its actual application lay some stage in the future. Shortly before landing, the Soyuz 6 welding experiment was carried out using a 50-kg rig called *Vulkan*, built by the Boris Paton Institute of Electrical Welding in Kiev, Ukraine, and probably the most advanced of its kind in the world. Boris Paton (b. 1918) was the son of electric welding pioneer Yevhen Paton, builder of the first breathing welded bridge in Kiev.

The orbital compartment at the front was sealed off and depressurized to a vacuum, both cosmonauts observing operations on television from the safety of their descent module. Three welding methods were tested: consumable electrode plasma, low-pressure compressed arc, and electron beam welding on steel, titanium, and aluminum. Once the automatic welding apparatus had concluded its work, the module was re-pressurized; Valeri Kubasov retrieved the samples and brought them back into the descent module for the return to Earth a few hours later. A series of operating problems meant that results fell short of expectations, but the lessons learned were applied to future equipment used on orbital stations [14].

WEIGHTLESSNESS: A WARNING

With the prospect of the orbital station soon being ready, it was important to extend the biological knowledge of long-duration flight. This was the function of Soyuz 9, in effect carrying out the duration mission originally intended for Voskhod 3 four years earlier and for which Cos-

The Vulkan welding unit – the start of materials processing experiments in orbit

mos 110 was the precursor. Soyuz 9 was crewed by Vostok 3 veteran Andrian Nikolayev with newcomer Vitally Sevastianov. Their mission was to extend the then Soviet duration record of five days beyond the American record of 14 days and reach a target 18 days. Whilst the primary purpose of the mission was space biology, they were given no fewer than 50 experiments to keep them busy.

Mission doctors placed considerable importance on the men exercising each day to keep in shape and they were expected to use a chest expander and treadmill up to 2 hr a day, divided into two hour-long sessions. They wore the first version of a new lightweight suit called the *Penguin* (in Russian *Pingvin*), with elastic straps designed to put pressure on lower-body muscles. Considerable attention was given to psychological support and arrangements were made for the men to speak to their families on video link, follow World Cup soccer, vote in the general election, and even play chess with ground control.

The flight got off to a good start. Pulse rates were high for the first six orbits, probably reflecting the excitement of the new mission, gradually dropping to normal by day six. They experienced no space sickness and had a big program of work to do, mainly Earth observations. Using the new RSS, they made observations of the Earth's surface and atmosphere. Vitally Sevastianov specialized in following luminous clouds in the high atmosphere (around 80 km), finding that they comprised three distinct layers when observed against evening twilight. They were originally thought to exist only over polar regions, but he identified some over Arabia and later over wider areas, even though they could not be seen from the ground. A small biological package was carried with insects and bacteria. Their own radiation dosage was measured (for the record, it was 397 millirads for Sevastianov and 316 for Nikolayev). In an exotic experiment, purified aluminum plates were left on the outside to absorb hydrogen and helium particles from deep space.

Indications of difficulties began to emerge on day 12, when doctors noted the men becoming more and more fatigued. The exercising was not popular, because of the limited space, the accumulation of sweat (neither wind nor gravity to take it away), the lack of washing facilities afterwards, and the pressure of the experimental schedule. Telemetry indicated that carbon dioxide levels were so low that the men must not be exercising in line with instructions. They got so tired that ground control had to use a siren to wake them in the morning. Medical tests administered by the cosmonauts on one another showed that they had difficulty distinguishing color and contrast – a good indicator of fatigue. Pulse rates began to rise after day 14. After 285 orbits, mission control in Yevpatoria in the Crimea called it a day and ordered them home.

The return to Earth went normally enough, their cabin landing in a dusty ploughed field. They were flown off to Moscow and placed in two weeks' quarantine in the Soviet lunar receiving laboratory, the first and only time it was used, designed to accept returning cosmonauts from the Moon and place them in a safe environment where they would not spread Moon bugs to the Earthly population. Officially, though, since there was no Moon race, it was announced instead that they were in a special medical unit to protect their immune systems from Earth bugs. Coming out of quarantine, they gave a press conference from behind a glass panel, announcing that all was well, and press reports declared that the worries of doctors and biologists were unfounded.

Paralleling Cosmos 110, a different story emerged when the medical reports became available some months later. That autumn, the cosmonauts presented the

Recovery of the Soyuz 9 crew; the cosmonauts were so weak that they had to be lifted out, but the mission was an important learning experience

outcome of their mission to an international space conference in Koblenz, Germany. There, it was learned that they were so weak that they had to be pulled out of their cabin and were unable to walk. The head of cosmonaut squad, General Kamanin, feared for their health if they got on a plane straightaway and delayed their return to Moscow by a day. Once they reached Moscow, he boarded the plane to greet them but was shocked by their condition: pale, puffed up, apathetic, with no shine in their eyes, emaciated, sick. The welcoming ceremony was cancelled. Nikolayev could barely keep his balance on the airport apron. During the quarantine, doctors observed the men to be pale, slow to respond to enquiries, tired rapidly, had poor concentration, could not sustain conversation, and would not even lighten up with jokes. It took them three days to walk properly again and their heart rates were elevated for several days. For their part, they felt "extremely heavy", as if they were in a centrifuge. They had lost 78% of the muscle tone in their legs, 12.5% of their heart surface, 20% of their heart volume, and a 50% loss of blood pumped – all the preconditions for a heart attack. The cosmonauts were able for a post-flight reception on 3rd July, but their medical parameters did not return to normal for a full month.

Responsible for reporting the medical outcomes of the flight was aviation medicine specialist Oleg Gazenko, who, in 1969, had moved to the Institute for Bio Medical Problems (IBMP). He became its third director in 1969, taking over from the blood physiologist Vasily Parin (1965–1968). Gazenko remained the leading expert on weightlessness until he retired in 1988 (he died on 17th November 2007). His post-flight medical report held nothing back. For the first three days after their return, he narrated how the cosmonauts "looked tired, could not remain erect without effort and had feelings of heaviness", with pulse rates up to 120 at a time. Nikolayev lost 2.7 kg in weight, Sevastianov 4 kg, with their bone densities down between 4 and 9%. Their blood plasma was down 2 and 6%, respectively. Their urine showed an excessive excretion of potassium, sodium, sulfur, nitrogen, and phosphorous.

These outcomes made the flight a success, rather than a failure [15]. From this experience, it was learned that periods of weightlessness of much more than a week were potentially dangerous, unless counter-measures were taken (and enforced). For the upcoming orbital station, doctors insisted that a broader range of similar exercise devices be installed. The station would, hopefully, offer much more room than the cramped conditions of Soyuz 9, but the lesson was that good medical outcomes could not be taken for granted. The Americans also sat up and took notice, applying at least some of the lessons to their own upcoming space station mission, Skylab.

From the biological package, the main point of interest was chromosomal aberrations in seeds, but they varied widely, from 1% in *Crepis*, the least, to 11% in onions. Algae continued to grow in their container, with only a small decline in survival rates. The Soyuz early science missions are summarized in Table 6.7 and the results in Table 6.8.

Table 6.7. Early Soyuz missions science.

Soyuz 3	26 Oct 1968	Georgi Beregovoi
Soyuz 4	14 Jan 1969	Vladimir Shatalov
Soyuz 5	15 Jan 1969	Boris Volynov, Alexei Yeliseyev, Yevgeni Khrunov
Soyuz 6	11 Oct 1969	Georgi Shonin, Valeri Kubsov
Soyuz 7	12 Oct 1969	Anatoli Filipchenko, Vladislav Volkov, Viktor Gorbtako
Soyuz 8	13 Oct 1969	Vladimir Shatalov, Alexei Yeliseyev
Soyuz 9	1 Jun 1970	Andrian Nikolayev, Vitally Sevastianov

Table 6.8. Early Soyuz science outcomes.

Weightlessness has severe, health-endangering effects unless counter-measures are taken
Welding can be done in space
Potential to map land, seas, atmosphere from orbit
Luminous clouds found not just in Arctic, but Arabia and equatorial latitudes

SALYUT ORBITING STATION

The first space station made possible a radical extension in space science. Salyut, the first of the space stations, was designed and constructed in a relatively short period over 1969–1971, the imperative being to built a credible alternative to the American manned landings on the Moon. Despite its hasty construction, a coherent and extensive set of scientific instruments was installed. Within the shell of Salyut, there was sufficient space and weight allowance to outfit the station with about 1,500 kg of scientific equipment. In July 1970, the chief designer of the Soviet space program, Vasili Mishin, invited scientific institutes to make suggestions for the type of instruments that could be included on the forthcoming orbital station. The same month, he had a discussion with the director the Institute of Space Research, Georgi Petrov: one of the instruments discussed was a 15-m radio telescope that could be fitted to the station. There were proposals at the time for the building of a family of astrophysical observatories: *Protsion* (already approved in 1964), *Drakon*, and *Orion*.

In reality, there were two types of stations, although all were given the same series name (Salyut). The main line of development was that initiated by the decision to build space stations in 1969: Salyut, Salyut 4, Salyut 6, and Salyut 7. But three stations, though similar in weight, were actually military stations in a program that long preceded the 1969 decision. These military stations had the designator Almaz and were built by a different design bureau, that of Vladimir Chelomei. The Almaz stations were Salyut 2 (which failed soon after launch and was not occupied), Salyut 3, and Salyut 5. The Almaz stations focused on military observations, had lower orbits, were flown by military crews, and sent down recoverable film capsules. In the event, some of the experiments developed for the civilian Salyut stations were also used on the military Almaz stations – something that helped to conceal their core

Table 6.9. Salyut space stations, with launch and de-orbit dates.

Salyut	19 Apr 1971	15 Oct 1971
Salyut 2	3 Apr 1973	28 Apr 1973
Salyut 3	25 Jun 1974	24 Jan 1975
Salyut 4	26 Dec 1974	3 Feb 1977
Salyut 5	22 Jun 1975	8 Aug 1977
Salyut 6	29 Sep 1977	29 Jul 1982
Salyut 7	19 Apr 1982	2 Feb 1991

Details of mission crews for Salyut and Mir are given in the Annexe.

purpose. Accordingly, both series are treated together here and reviewed in chronological order.

The hasty construction of the first Salyut must have given little time for reflection on scientific objectives for the program as a whole. To rectify this, Mishin had a second round of discussions and met the Academy of Sciences in February 1971 to discuss scientific instrumentation for longer-operating stations of the second generation. Here, it was agreed that the space station program should make a major contribution to astronomical research. Specifically, they envisaged that future stations would combine a mixture of optical telescopes (1–3 m in diameter) and radio telescopes (30–50 m in diameter). The Salyut series is summarized in Table 6.9.

One area in which there was significant preparation was in space biology and gardens. Gavril Tikhov's role in the development of astrobiology has already been noted, but now was the opportunity to put much of the legwork undertaken since then to the test and it is worth saying a little more about developments in the Institute of Biophysics, set up in Krasnoyarsk in the early 1960s by Leonid Kirensky, Ivan Yerskov, and V.P. Dadykhin. It functioned as the Siberian wing of the Institute of Bio Medical Problems (IBMP). Dadykhin wrote the theory of space gardens in 1968 (*Growing Plants in Space*). Taking inspiration from Konstantin Tsiolkovsky and Vladimir Vernadsky, they first tried to develop algae as a food source for cosmonauts. Algae, or *Chlorella vulgaris*, had many attractions. It had no dead or inert matter, multiplied fast (cells divided every few hours), achieved a high conversion rate from photosynthesis (10%), tolerated movement, cleaned the air, removed carbon dioxide (90%, based on 8 m^2 a person), could absorb human waste, had all essential food elements, could recover from a systems failure of up to two days, and could defend itself against bacteria.

As a food source, though, it was a disaster, being not only bland, but virtually inedible. But could one combine algae as an air cleaner with other food sources to sustain life on orbital stations? Enter Iosif Gitelson (b. 1928), the father of Soviet space biophysics. A graduate from Moscow State University in 1951, he went to work in Krasnoyarsk the following year in a variety of institutes there (agriculture, photobiology, and biophysics). Later, he became Director of Biophysics in 1985, and in 1991, he opened the International Centre for Closed Ecological Systems there. Back in 1964, he led up the first attempt to establish a closed-cycle system, Bios 1. The experiment started timidly enough, with Gitelson leading by example to try out

Iosif Gitelson, father of Soviet space biophysics

Bios 1 for 12 hr himself, followed by experiments of ever longer duration, up to 30 days. The principal foods grown were cucumber, tomato, radish, kale, dill, Welsh onion, turnips, beet, and chaifa. The last, chaifa, was a little known sedge (*Cyprus esculentus*), which could grow rapidly in hydroponic farms. Vegetables (or higher plants) had the advantage over algae that they were edible, but produced inedible waste and were time-intensive to cultivate.

The experiments were extended to 90 days on Bios 2, with one tester at a time. 1972 saw the completion of the Bios 3 facility, with a volume of 315 m^3 and a sowing area of 63 m^2, designed to develop a closed ecosystem that could work for five months. The ground area measured 9 × 15 m, being 2.5 m tall, and was divided into areas for higher planets, algae/chlorella, and living quarters, sufficient for a crew of three. Plants grew under 20 6-kW xenon lamps, the facility using 400 kW, supplied by the local hydro station. Most of the cleansing of air and water was done naturally, but a thermo catalytic filter was used to complete the process of purifying the air and an ion-exchange to filter the water.

The biosphere in Krasnoyarsk

Table 6.10. Bios ground experiment durations.

Bios 1		
1964	Iosif Gitelson	12 hr
	Yuri Gurevich	24 hr
1965	G. Dralyuk	5 days
	V. Pushkova	12 days
	M. Basanova	14 days
	G. Tereshkova	30 days
Bios 2		
	G. Mazurkina	45 days
		90 days
Bios 3		
1972	N. Petrov, M. Shilenko, N. Bugraev	180 days
1977	M. Shilenko, N. Bugrev, G. Asinyarov	180 days
1983	N. Alekseev, L. Mozogovi, N. Bugraev	180 days

The first mission began on 24th December 1972 and lasted six months to 22nd June 1973. Although the crew of three stocked the larder with carry-in food at the start (mainly meat), each experiment raised the level of grown food. Wheat was the most cultivated crop (33.6 m^2), being supplemented by beet, carrot, dill, turnip, kale, radish, onion, cucumber, and sorrel. The first experiments indicated that one person needed 13 m^3 for oxygen and 30 m^2 for food for a sustainable system. Within ten years, Bios had achieved a 100% regeneration of the air and water and 93% of food. No cases of poisoning, sickness, or ill-health took place.

In the United States, NASA had begun closed-system studies in the early 1960s, but lost interest. When research resumed, there was amazement at the progress made in Krasnoyarsk. Although the USSR had reported the Bios experiments, Krasnoyarsk was a closed city (many nuclear research facilities were nearby) and it was not well known. The leading American expert, Frank Salisbury, professor of plant physiology at Utah State University, described the experiments as the most advanced in the world, leaving everyone else as beginners [16]. Table 6.10 provides details of the Bios experiments. Bios formed the basis of the space gardens on the Salyut orbital stations.

SALYUT SCIENCE

Salyut was occupied by only one crew and for a duration of 23 days. The mission came to a tragic end, the cabin accidentally de-pressurizing during the descent to Earth with the loss of the crew. Despite that, the Salyut design proved to be successful, a range of well documented experiments was carried out, and the experimental results either transmitted to Earth or recovered in the Soyuz cabin. The Soyuz 11 crew extended the duration record set the previous year by almost a week,

Table 6.11. Salyut science.

Medical	
Polinom 2	Test for pulse, blood pressure, blood
Amak	Blood analysis
Plotnost	Calcium density in bones
Atmosphere	
RSS-2	Hand-held spectrograph to determine polarization of light in atmosphere
Astronomy	
Orion 1	Ultraviolet astrophysical telescope
Anna 3	Gamma-ray telescope spectrometer
FEK 7	Cosmic ray detector
Biology	
Oasis 1	Greenhouse
Others	Effects of cosmic rays and weightlessness

from just under 18 days to almost 24 days. Table 6.11 describes the scientific equipment installed.

The design of Salyut, a series of tapering cylinders, was ideally suited to the installation in the middle of the floor of a large observational instrument for Earth observations. The most important space-pointing instrument on Salyut was the *Orion* telescope, designed by the Armenian Academy of Sciences, located outside the transfer compartment, the first human-operated telescope in orbit and the only one of the trio of astrophysical telescopes discussed earlier to fly. Cassettes of film could be changed using a mechanical arm and airlock. *Orion* had two mirrors, one of 28 cm, the other of 5 cm, presumably to sight it, operating in the ultraviolet spectrum in the 2,000–3,000-Å band. *Anna 3* was a gamma-ray telescope designed to register energy up to 100 MeV, with a $1°$ pointing accuracy to the Sun for 20-hr periods of observations. The cosmonauts took spectograms of β Centaurus and α Lyrae, selected because of their hot temperatures and the spectograms identified over 60 types of ionized metals: iron, nickel, chromium, titanium, and vanadium. Staying with astronomy, the FEK 7 cosmic ray detector, modeled on one flown on Zond lunar missions, tried to find elusive transuranic dirac monopole particles (it didn't). The cosmonauts also used new cameras, in reality modernized airborne cameras called AFA BA 40, to map the surface of the Earth precisely referenced to the Sun and the stars, developed by Yan Ziman and Yuri Bykov. The radio mass spectrometer was used during the equator crossing on 16th June 1971 to measure the ion composition of the F2 region of the ionosphere and it found a drastic anomaly where ions dropped off right on the night equator.

Just as *Orion* marked a step forward in human astronomy from orbit, so did Salyut begin the program of space botany, the experiments being led by Galina Nechitailo, subsequently to become the world leader in space botany. The first space greenhouse, *Oasis*, was carried on Salyut, with flax, leek, onion, and Chinese cabbage, designed for small size and fast growth. *Oasis*, with an area of 400 cm^2, was lit by three fluorescent lamps, installed in several layers of nutritious soil, and supplied with water by a manually operated pump, being photographed every 10 min during the mission. Although shoots soon appeared, the plants did not grow normally, being slower, feebler, and less well rooted than Earthly plants, even when the water supply was increased. Although the *Oasis* plants were brought back to Earth by the Soyuz 11 crew, they were lost during the depressurization of the cabin. In space biology, seeds, micro-organisms, and plants were brought on board to test the effect on them of cosmic rays. Tadpoles were hatched out (this took five days), frogs being considered important because their system of balance is comparable to humans. Fruit flies were carried in a container to test for changes in weightlessness. Unicellular algae were carried to see whether they could, in future space stations, absorb carbon dioxide and produce oxygen.

Medical experiments were a priority, especially granted the previous experience. *Polinom 2* was a medical device to test 22 key medical parameters, such as pulse, pressure, and blood volume. The blood analyzer was used to take blood samples so as to test for cholesterol, sugars, and other substances. *Plotnost* ("density") was an instrument to measure calcium content in bones, determining that it fell by between 15 and 20% during the mission. To combat weightlessness, the cosmonauts exercised 2 hr a day, using a 10-km/hr treadmill, got into a lower-body pressure suit, and wore *Penguin* suits to exercise muscles. The death of the crew members prompted immediate speculation that they had somehow become the victims of weightlessness. Using the outcomes from *Polinom 2*, the full medical record of the mission was recovered and published, showing how the crew had adapted to weightlessness and were actually in a reasonable condition to go through re-entry. It must have been a tormenting experience for the mission experimenters to go through the logs of the dead cosmonauts to read of the outcomes of their experiments. Viktor Patsayev, for example, conducted a lengthy experiment for Konstantin Gringauz that showed how the orbital station, in the course of its orbital path, would collect secondary electrons that it would then discharge as

Lower-body gravity suit, first tested on Salyut

it crossed from light into shade, proving his theory of high-frequency resonance discharge [17].

SOLO SOYUZ SCIENCE

In addition to the scientific work on the orbital stations, scientific activities were also carried out on four solo Soyuz missions during the early to mid 1970s. The first was Soyuz 12, a two-day mission in which the two cosmonauts, Vasili Lazarev and Oleg Makarov, re-qualified the Soyuz spacecraft. Although a short mission, it was a busy one.

Soyuz 12 was the first Soviet spacecraft to carry a multi-band imager, the technique having been tested out by Illyushin 14 flying laboratories from 1970. The nine-band imager compiled elevation and vegetation maps of the Caspian Sea and Nile delta, showing plants, mud, pollution, and suspended material. This was so promising that 1973 saw the formation of the Optico-Physical Department of IKI, headed up by the pioneer of space-based observations, Yan Ziman, who held the post until 1988 and was awarded the state prize for his work on these missions. The crew studied nighttime radiation from the atmosphere, finding an ash-gray thread-like glowing crown about 100 km high, later attributed to a layer of aerosols. Soyuz 12 mapped properly a layer of mesospheric clouds at 81 km (they could move 5 km up or down), first identified by Soyuz 9. An experiment was carried out into the growth of lysogenic bacteria, finding it to be little different from that on Earth [18].

The much longer, eight-day Soyuz 13 was equipped with a substantial body of scientific instruments. The front of the spacecraft was adapted to carry the second *Orion* telescope, *Orion 2*, designed by Grigor Gurzadyan and built by the Armenian Academy of Sciences. The telescope had a pointing accuracy of 5 arc seconds, was swiveled by 13 motors, and could be focused onto stars of magnitude down to 12th magnitude – a mixture of cool and hot stars. Soyuz 13 could lock onto a star for up to 20 min and hold it steady while long-duration exposures were taken. It had an airlock system for the retrieval of cassette tapes. During the mission, spectographs were taken of α Lyrae, Capella, and β Centaurus.

Likewise, Soyuz 13 carried the second space garden, *Oasis 2*. It was somewhat different from the first *Oasis* system on Salyut, being a bacterial waste regenerator comprising two interconnected cylinders that began to produce protein on the second day of the mission. Another part of *Oasis* tested the ability of chlorella algae to absorb carbon dioxide and produce oxygen.

Even though it was a short flight, medical experiments were continued, such as *Levkoi*. Because the heart no longer had to pump blood around the body against the force of gravity, it tended to pool in the upper parts of the body, especially the head, often giving cosmonauts a puffy appearance and contributing to space sickness, so *Levkoi* was a new system to measure the nature and level of such increased blood flow. Finally, the RSS-2M hand-held spectograph was used to measure dust particles and air pollution in the atmosphere, notably haze over India. The RSS actually came to be a more important instrument than imagined, for the scale of pollution it

Table 6.12. Soyuz 13 science experiments.

Orion 2	Astrophysical telescope
Oasis 2	Biological closed systems
Levkoi	Distribution of blood in the brain
RSS-2M	Dust, air pollution in the atmosphere
KSS-2	Water vapor in Earth's atmosphere

Table 6.13. Solo Soyuz science missions.

Soyuz 13	18 Dec 1973	Pytor Klimuk, Valentin Lebedev
Soyuz 16	2 Dec 1974	Anatoli Filipchenko, Nikolai Rukhavishnikov
Soyuz 22	15 Sep 1976	Valeri Bykovsky, Vladimir Aksenov

All on Soyuz rocket from Baikonour

detected began to make scientists aware of the dangers of climatic change. The nine-band imager was successfully flown a second time. The mission equipment is summarized in Table 6.12.

A year later, Soyuz 16 was a six-day engineering dress rehearsal for the Apollo Soyuz spaceflight due to take place the following summer. Advantage was taken of the flight to carry a small biological payload: *Arabidopsis* and a thermos flask with guppy fish. Guppy fish, or *Danio*, native to South America, were one of a small number of fish who give birth to live young and it was hoped that the five on board would do so in the course of the short mission. One did so, accustomed to weightlessness quite quickly, but became disorientated on return to Earth.

Although Soyuz 22, two years later, was primarily an Earth resources mapping mission for the USSR and GDR, the crew also carried some other experiments on their eight-day mission. These were *Biogravistat*, a biological experiment to measure the effects of weightlessness on higher plant shoots and the development of small fish fry, the KSS hand-held spectograph to measure water vapor and pollutants in the Earth's atmosphere, and an experiment to measure the frequency of cosmic ray flashes. This arose from the experience of America's Apollo 14 astronauts, who, returning from the Moon, had noticed cosmic ray flashes by night. On Soyuz 22, the cosmonauts did notice a high level of flashes, interestingly over the South Atlantic Magnetic Anomaly. The biological experiments included fish eggs, duckweed, and maize seedlings. One garden was attached to the wall of the spacecraft, another on springs, so that it could absorb the effects of the spacecraft's motion and get a truer representation of zero gravity, with a third, control box of seedlings on the ground [19]. The solo Soyuz missions are summarized in Table 6.13.

SALYUT 3 SCIENCE

Salyut 3 was the first of two Almaz military stations to fly, the principal aim being the photographing of targets of military importance using a large camera system

Table 6.14. Salyut 3 experiments.

Medical	
Amak 3	Monitoring of blood composition
Impuls	Vestibular system in weightlessness
Levkoi 3	Blood system and inter-cranial pressure
Polinom 2M	Heart performance and circulatory system
Rezeda 5	Inhalation/exhalation and capacity of lungs
Atmosphere	
RSS-2	Hand-held observation instrument
Polyaroid	Polarization of sunlight

called *Agat*. A scientific program was also carried out, which the Soviet Union was naturally keen to talk up, in order to conceal the military objectives of the station [20]. Salyut 3 was occupied only once, for 15 days, by Pavel Popovich and Yuri Artyukin, but it was the first successful Salyut occupation and recovery. Several new experiments and items of equipment were carried, such as *Rezeda 5*, a portable instrument to measure lung inhalation and exhalation, and *Priboi*, a water purification system (this eventually became operational on Salyut 6, providing initially about a liter of water a day). The RSS-2 was used to determine the level of chemical pollutants in the atmosphere and could make out the degree to which ozone at 90 km had deteriorated as a result of aerosols. This may have been the earliest use of orbital stations to monitor the process of climate change. Mission experiments are summarized in Table 6.14.

SALYUT 4 SCIENCE

Salyut 4 marked a significant step forwards in the volume of space science undertaken. The station had an additional, top solar array, able to provide more electrical power. Salyut 4 was occupied by two crews, the first for 29 days and the second for 63 days. Additional value was obtained from the fact that the first mission was in winter (January–February) whereas the second was in summer (May–July).

Salyut 4 carried a number of telescopes and spectrometers: the OST solar telescope, the X-ray telescope *Filin*, and the ITS-K infrared telescope. The principal scientific instrument in the base of Salyut 4 was a large 25-cm mirror conical solar telescope, OST-1, with a spectrometer to measure the Sun's rays. The telescope was built by the Crimean Astrophysical Observatory and the pointing system by the University of Leningrad. The cosmonauts orientated the station's base towards both the Sun above and the ground below. Turning towards the Sun, Salyut 4's solar observations began on 2nd February 1975 and the cosmonauts were able to follow plasma emissions and loops coming out of the Sun. OST could take in the entire Sun in its field and then focus on particular areas of interest such as floccolæ. The speed of the solar wind was estimated at 100 km/sec. Six hundred spectograms were taken of the Sun and of particular regions of interest, such as sun spots and solar flares (two were followed).

Salyut 4's telescope

Turning towards the Earth, OST was able to survey the aerosols in the Earth's atmosphere, including the ozone layer in the 20–60-km range, so Salyut 4 built up base line data on the nature of Earth's atmosphere. Alexei Gubarev carefully aligned the station with the Sun on one side and the Earth on the other so that the absorption of its rays could be precisely measured in the different layers of the atmosphere below. He was able to track the saturation of water vapor in the lower atmosphere (below 15–20 km) to its disappearance at 70–80 km, where the atmosphere is dry. Perhaps the most outstanding finding to emerge from this was confirmation that the ozone layer appeared to be weakening. The cosmonauts had a system called *Zentis*, which they used to re-spray the surface of the mirror when it became dirty on the outside.

Turning towards the stars, the X-ray telescope *Filin* had four X-ray detectors and two star sensors. They first observed X-ray sources Scorpio x-1 (the brightest in the sky, with unusual fluxes, explosions, and emissions), Hercules x-1, A 062000 (discovered by Britain's Ariel 5), and Cygnus x-1, noting their behavior and periodicity (Cygnus x-1 was probably the most reliable black hole candidate of stellar mass). They then moved on to X-ray sources Circinus x-1, Nova A0620-00, Rigel, nebula NGC-6720, and sources in Perseus, Taurus, Virgo, Carina, the Crab, and Lyra. *Filin* was devised by the Sternberg Institute and was combined with a 20-cm X-ray telescope, RT-4, sensitive to the range 44–60 Å, built by the Lebedev Institute. This ascertained that X-ray stars changed their intensity very rapidly, within a space of a few minutes. Soviet astronomers classified a group of stars that changed their brightness 30 to 40 times a second, called barsters.

The 30-cm ITS-K infrared telescope was a nitrogen-cooled infrared telescope developed by the Lebedev Institute of Physics under M.N. Markov and cooled at $-223\,^{\circ}\text{C}$ using a system developed by the Institute of Low Temperatures in Kharkov, Ukraine. It was aimed both towards Earth and towards heavenly objects. Turning it towards the skies, ITS-K was pointed towards the planet Saturn, the Larger Magellanic Cloud, the star Canopus, and nebulae in Cassiopeia. In doing so, ITS-K identified considerable quantities of interstellar gas and dust clouds. X-ray measurements were made of the blue supergiant β Orion (Rigel), cepheids W Virgo and RR Lyra, and galaxy NGC 6720. When ITS-K was used Earthward, it tried to measure infrared rays given off by the atmosphere so as to learn more about heat-exchange processes over land and oceans and about the level of water vapor. Temperatures in the upper atmosphere were measured at between 400 and 1,700°C. In the course of a 1,600-km pass on orbit 514 on 27th January, the ITS was used to measure the volumetric density of nitric oxide in the atmosphere (it was $1.5 \times 10^{-6} erg/cm^3/sec$). There were two mass analyzers that made scans of the ionosphere, each scan lasting 25 sec, and scientists were able to contrast the ionosphere during both storm conditions (orbit 2643) and quiet conditions (orbit 2702). There was a dramatic increase in ionospheric activity by night.

Two experiments concerned the environment around the station itself. First, the MMK-1 micro-meteorite detector attempted to measure the level of micro-impacts on the station itself: it had a surface area of $4 m^2$ and recorded impacts over the duration of the mission. Second, the *Spektr* experiment tried to measure the gases and plasma through which the station itself flew in Earth orbit. *Emissiya* was a wide-aperture photometer and interferometer to study the various layers of the upper atmosphere, especially at an altitude of 250–270 km.

The botanical work begun by *Oasis* on Salyut and Soyuz 13 continued. Salyut 4 carried the third *Oasis* botany experiment, this time using pre-planted seeds in a cartridge and an automatic watering system, the plants this time being peas and onions. On the first mission on Salyut 4, peas were sown, but only four out of 30 developed to maturity. On the second mission, peas and onions were sown: although most germinated successfully in about four days, all died within three weeks. The experiments were unsuccessful insofar as no plant made it through the growing cycle, but it was a definite albeit frustrating learning experience. Something appeared to be happening to plants about halfway through their growing cycle. The cosmonauts speculated that the lighting was too strong and that the plants became scorched. The cosmonauts were more successful with onions, which grew to 20 cm, albeit more slowly than on Earth and to a smaller size. Cosmonauts Klimuk and Sevastianov ate the onions in July 1975, the first time space-grown food was ever eaten in orbit – a milestone. Other biological experiments observed the behavior of different life forms in weightlessness: chlorella algae (*Biotherm 1*), microbes (MB), bacteria (FKT), fruit flies and beetle larvae (*Biotherm 2M*), tadpoles (*Biotherm 3*), and hamster cells (*Biotherm 4*). Continuing the theme of self-sufficiency, Salyut 4 carried the SRV-K system designed to recover water from the station's own atmosphere. Cosmonauts require about 2.5 liters a day to drink and the *Priboi* system was able to provide them with a proportion of this (between 1 and 2 liters).

A footnote to the Zond missions is that turtles flew once more, on Soyuz 20 in 1975 on a 92-day-long unmanned mission moored to the Salyut 4 orbital station, along with cacti, bulbs, gladioli, vegetables, corn, and green beans – 20 types of plants altogether. This was the longest time such samples had been exposed to the space environment and the recovered samples were studied to test whether there were any genetic changes. The mission appears to have been a comparator to Cosmos 782 (see Chapter 7), flying a similar payload but for a much longer period.

The countermeasures against weightlessness introduced on earlier missions were fully used, such as exercise suits (*Atlet*, replacing the *Penguin*), a lower-body negative pressure suit (*Chibis*), and anti-gravity suit (*Anti-8*). In an effort to help the cosmonauts come back in a better condition than their predecessors, their exercise level was stepped up in the last week of their mission and they increased their intake of water and calcium. *Chibis* was a suit that cosmonauts put on around their waist, the suit creating a pressure seal, the air pressure being reduced 30%, the idea being to draw blood down to the lower part of the body. Countermeasures seem to have helped. The red blood cells of all the cosmonauts were down on landing but their overall health was good. The second crew faced the additional hazard of damp: the walls of the station became wet and fungus began to grow on them, so that had to be wiped clean, but it was a warning of the importance of maintaining microbial control on future stations. For future stations, the decision was made that at least two portholes should admit ultraviolet light so as to kill microbes and fungus [21]. Salyut 4 experiments are summarized in Table 6.15.

Table 6.15. Salyut 4 science.

Astronomy	
OST-1	Orbital solar telescope
ITS-K	Infrared telescope spectrometer
Filin	X-ray spectrometer
RT-4	X-ray telescope
Atmosphere	
SSP-2	Solar spectrometer package
Emissiya	Neutral particles in Earth's atmosphere
Spektr	Gas and plasma of station's environment
KSS-2	Water vapor and pollution in Earth's atmosphere
Micrometeorites	
MMK-1	Micrometeorite detector
Medical	
Levkoi	Blood flow in the brain
Biology and liquids	
Biotherm 1	Algae
MB	Microbe breeder
FKT	Changes in bacteria
Biotherm 2M	Fruit flies
Biotherm 3	Aquarium
Biotherm 4	Hamster
Oasis 1M (3)	Greenhouse
Priboi	Water regeneration
Freon	Potential use of freon gas as a refrigerant

SALYUT 5 SCIENCE: MATERIALS PROCESSING PLANT ON ORBIT

Salyut 5, the second and last of the military stations, was likewise occupied twice, for 49 and 15 days, respectively, the main science experiments being carried out on the first and longer of the two missions. The main breakthrough for space science was that Salyut 5 marked the introduction of materials processing experiments in orbit. This was a frontier area of space science, designed to take advantage of the way in which materials might develop differently in weightlessness, not only to improve knowledge of physical processes, but with an eye on their potential for industrial applications. Salyut 5's package was called *Fizika*:

- *Sfera*, developed by the Institute of Metallurgy, to attempt to develop a perfect sphere in weightlessness, using metal made out of lead, bisimuth, zinc, and cadmium;
- *Reakstiya* was a development of the welding experiments first performed on Soyuz 6 and also devised by the Paton institute in Kiev;
- *Krystal*, to develop crystals in a solution;
- *Potok*, the Russian word for "flow", and *Diffusia* ("diffusion"), to follow what happened to liquids in weightlessness.

The materials science experiments on this and future missions varied considerably in the amount of crew intervention, duration, and the application of heat and currents. Some experiments were quite short, some ran for days (72 hr for *Diffusia*), while others ran for up to several months at a time. With *Sfera*, it proved very difficult during the re-smelting of the metal to obtain perfect spheres, for they tended to turn out elliptical, possibly due to microgravitational disturbances on board. *Reakstiya* was more successful: hairline cracks on two stainless-steel 15-mm tubes were soldered in a vacuum to test the hardness and consistency of the vacuum weld. Soyuz 21 returned two samples of soldered joints from *Reakstiya*, much stronger than Earth-based welding. *Krystal* was operated from 7th July to 1st August, then 9th August to 11th February automatically, and then under human control until 23rd February and it managed to develop crystals in solution up to 1.3 mm across. *Fizika* became the basis of no fewer than 500 experiments over the next 15 years and Liya Regel of the International Centre for Gravity Materials Science and Applications came to point out that their achievements included the first laser from a crystal growth, the first semi-conductor heterostructure, the first superionic crystal, the first crystals of CdTe and its alloys (HgCdTe, ZnCdTe, MnCdTe), the first zeolite crystals (several times larger than Earth), the first protein crystals (the first ones being insulin, glucogen, tymozine, antitrypsinn, interferon), the first making of gypsum by rapidly mixing hydroxapatite and calcium sulfate, the first polymer latex spheres, and the first chromium disilicide glass. New furnaces were built later, like Splav and Zona.

Salyut 5 resumed the fish experiments begun on Soyuz 16, the cosmonauts watching how well the two guppy fish adapted to weightlessness, accompanied by a camera that imaged them every 10 min. The aquarium had an air bubble in it: while on Earth, it would of course form on the top; in weightlessness, it formed in the

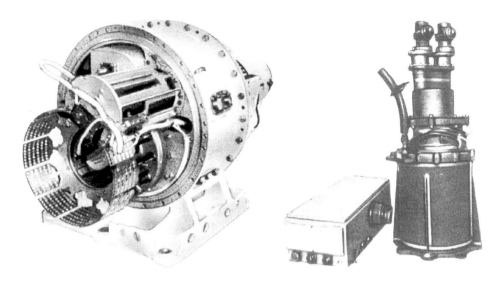

Splav furnace   Zona furnace

middle of the tank, which the fish would regard as "the surface" and poke their heads into from time to time for air. The second expedition crew brought up two tortoises (experiment *Terrarium*) and some zebrafish.

Experiments with fruit flies were of sufficiently long duration to notice significant changes (experiment *Kultivator*). Many died and no fewer than six different forms of mutations were observed, suggesting that weightlessness or radiation or a combination of the two had a significant effect. *Bioblok* was a biological experiment with three different containers and nutrient solutions to promote growth. Although the mushrooms germinated after 17 days, they did not grow normally: instead of their normal head, the top of the mushroom was formless.

The ITS infrared telescope flown on Salyut 4 was re-flown in a new version: the ITS-5. The ITS was turned towards the Sun, where it made a map of the Sun's temperature: whereas the surface temperature was in the order of 10,000 K, the ionized plasma chromosphere above it had a 10-fold higher temperature of 1,000,000 K. Turning the ITS towards the constellation Orion, the chemical signature there indicated new stars in formation. Finally, directing ITS-5 towards Earth, over 1,000 spectograms were obtained of oceans and land areas, mapping the distribution of water vapor, carbons, and ozone in the Earth's atmosphere. Stabilized towards the Sun, it made profiles of water vapor and ozone concentrations at altitudes of 20–70 km in the 6.3–9.6-mm band over 660 km ground track. This was done on 1st, 4th, 5th, and 22nd August during the Soyuz 21 occupation and 16th February 1977 on the Soyuz 24 occupation. A prime objective was to study the level of atmospheric pollution, the level of carbon monoxide, energy balance, and the state of the ozone layer, and in January 1978, the USSR published the first global air pollution base line map, which not only marked ozone against altitude, but also found sulfuric acid and ammonium sulfate mixed in the ozone.

A full range of medical measurements was carried out: *Impuls* (sense of balance), *Levkoi* (blood pressure in the brain), *Tonus* (muscle strength), *Rezeda* (lung capacity), *Amak* (blood composition), *Polinom 2M* (heart function), and *Palm 2M* (reaction time), which turned out to be slower than on Earth. A number of simple but effective changes were made to the exercise régime. First, belts were attached to the running track, pulling cosmonauts towards the track and thereby making running harder, and a bicycle was added, the task commissioned to the Likhachev car factory in Moscow.

Despite these improvements, the first expedition came to a premature end. Several weeks into the mission, medical controllers noticed that the cosmonauts' blood pressure was up and that each had lost 1.5 kg in weight – a function of an excessively demanding schedule. In effect, ground controllers had unlearned some of the lessons of Soyuz 9 and given the cosmonauts so demanding a schedule that the exercise régime was neglected. A halt was called for a two-day medical examination and to get things back on an even keel. Despite this, there was a rapid decline in the condition of the research engineer Vitally Zholobov, who became so disorientated that the mission had to be abandoned, with a rapid nighttime return to Earth. Both cosmonauts were so weak that they could not walk again for three days and Volynov had lost 7 kg. The second mission was successful, but limited to only two weeks.

One of the most unusual Salyut 5 experiments was conducted when it was unmanned, on 5th November and 7th December 1976, on orbits 2,180 and 2,702. The infrared spectrometer was turned towards the full Moon to scan the equatorial zone, specifically taking in the Luna 16, 20, and 24 landing sites. The main outcome was a radical revision upwards of the level of water on the Moon, from 0.03 to 0.1%. Although the "discovery" of water on the Moon is attributed to American Moon probes at the end of the century, Salyut 5 pointed towards such a conclusion long before [22]. Salyut 5's experiments are listed in Table 6.16.

Table 6.16. Salyut 5 experiments.

Medical	
Amak 3	Monitoring of blood composition
Impuls 2	Vestibular system in weightlessness
Levkoi 3T	Blood pressure
Polinom 2M	Heart performance, circulatory system, body temperature
Rezeda 5	Inhalation/exhalation and capacity of lungs
Palma 3M	Reaction times
Tonus	Muscle tone
Plotnost	Bone density
Materials processing	
Fizika/Diffusia	Mixing toluene and dibenzyl to make alloys
Fizika/Potok	Surface tension of metals
Fizika/Sfera	Mixing and cooling bisimuth lead alloys with cadmium and tin
Krystall	Monocrystals
Reaktsiya	Soldering of steel with magnesium nickel solder

Biology	
Aquarium	Development of guppy fish, *Danio rerio* fish, and their eggs
Bioblok	Plant growth in mushrooms, *Crepis* plants
Kultivator	Chromosomes in fruit flies
Terrarium	Study of tortoises in weightlessness

Atmosphere/astronomy	
ITS-5	Infrared telescope to study the atmosphere, ozone, Sun, stars, galaxy
RSS-2M	Aerosols in the atmosphere

SALYUT 6 SCIENCE

Salyut 6 marked a major advance. It was the first of the long-duration stations and its two docking ports made it possible for visiting missions to come to the station and for unmanned spacecraft, called Progress, to bring up fresh supplies of water, air, food, fuel, and experiments. Mission durations were systematically pushed back ever further into the medical unknown: 96 days, 139 days, 175 days, and then 185 days. The experiments on Salyut 6 may be divided into those undertaken by the resident crews and those undertaken by visiting missions. Most visiting missions took place as part of the Intercosmos program, flying up a cosmonaut from one of the Intercosmos countries, normally with a small suite of instruments devised by the country concerned.

Granted the attention give to the extension of mission duration, medical experiments featured prominently. To monitor the effects of weightlessness, the cosmonauts used the *Polinom 2M* and *Beta* electrocardiograph to measure heart function and the *Rheograph* to measure blood flow. As part of preventative measures, lower body negative pressure suits were again used, as were the *Chibis* muscle-loading suits. During the first mission, by Yuri Romanenko and Georgi Grechko, the following medical events were noted:

- both suffered space sickness for the first week; this comprised nausea, headache, appetite loss, a feeling of weakness, puffy face, increased blood pressure in the head;
- most of these symptoms disappeared within a week, but sleeplessness was an on-going problem;
- red blood cell production halted, the number of red blood cells declined, but then leveled off;
- white blood cell reduction also fell, making the cosmonauts vulnerable to infection;
- blood pooled in their head, heart, and upper body, drawn from the legs and arms, with the consequence that fingers and hands lost some sensitivity and feet became cold;
- on their return to Earth after 96 days, they were weak, finding it difficult to walk and even accomplish simple tasks with their arms (e.g. drink tea); most of these problems lasted only about four days; after a week, blood counts returned to normal.

Yuri Romanenko running on Salyut 6; typically, cosmonauts ran or exercised in other ways for up to 2 hr a day

The next crew, Vladimir Kovanyonok and Alexander Ivanchenkov, reported rushes of blood to the head when they entered orbit, but they quickly settled down. On their return, though, they felt "dizzy and fatigued", with minor movements requiring all their strength. Their mission, substantially longer at 139 days, saw the cosmonauts lose not only weight (about 5 kg each), but also size, up to 4 cm in hip size. They had reduced leg size, and a loss of fluid and muscular tissue. It took the two men three to four days to recover normal movement (what doctors called the "acute re-adaptation phase"). Several counter-measures were adopted to strengthen the cosmonauts in advance of returning to Earth, such as medical supplements (tonics) and drinking large quantities of water.

On the next mission, in which Vladimir Lyakhov and Valeri Ryumin went to 175 days, the cosmonauts increased the amount of exercising by 15%, walking or running 8.5 km a day on the running track, the result being an improved condition on their return to Earth, Ryumin even putting on weight. On the final long-duration mission, 185 days, by Leonid Popov and Valeri Ryumin, both cosmonauts gained weight and even height, both becoming 3 cm taller in the absence of gravity – a gain that Earth soon reduced. An important medical moment was that because the lifecycle of red blood cells is 120 days, these two cosmonauts were the first to generate their own red blood cells while in orbit – a first, and they were smaller than Earthly red blood cells. These missions saw real improvements in the reaction of crews to weightlessness, with the cosmonauts developing a better balance of work, exercise, and leisure. During these lengthy missions, a considerable amount of time was spent on loading and unloading cargoes from Progress freighters, which improved upper body muscles. On the 185-day mission, a loss of mineral heel bone of as high as 8.3% was recorded. Oleg Gazenko was able to characterize the medical experience of long-duration missions into four parts: the initial shock to the human system of entry into orbit and weightlessness; a period of stabilization as blood redistributed around the human body during prolonged flight; the shock of re-entry; and post-landing adjustment.

The space botany program developed on Salyut and Salyut 4 resumed. This comprised:

- *Oasis*, for peas and wheat, with its own light system and a means for directing root growth downward;
- small hydroponic cultivator greenhouse *Fiton* developed by the Ukrainian Institute of Molecular Biology for *Arabidopsis*, a wild herb with a 40-day cycle in which plants grew in a nutritional solution; *Fiton* had a single, strong lamp, nutrient solution and filter for onions, cucumbers, tomatoes, garlic, fungi, wheat, peas, and carrots;
- *Biogravistat*, an artificial centrifuge, designed to test whether gravity was a complicating factor in plant growth, using cucumber and lettuce;
- *Vazon* ("vase" in English), a cultivator bed for onions and tulips.

This time, the botanical experiments had their own closed environmental system, designed to keep out harmful chemicals that might be found in the station's air ventilation system, with the lights on for 14 hr a day. The results, though, were still disappointing, plants making a good start but then withering. Tulips started well, but then lost their petals. The plant seeds were sterile, though, intriguingly, the tulips, when returned to Earth, quickly recovered and grew again normally. Growth was short: onions to 10 cm but would not flower, tulips to 52 cm, a tree only to 10 cm, and wheat just grew very slowly. On following missions, other plant varieties were tried, such as tomatoes, cotton, garlic, radish, dill, strawberries, apples, and parsley, with equally disappointing results. On Earth, the botanists could not decide whether the problem was gravity, radiation, the space station environment, or the garden environment. The cosmonauts were disappointed with the results, especially considering the loving care they gave to the plants, for minding them was a

Growing wheat in orbit; the lengthy Salyut 6 missions provided the first opportunity for prolonged experiments in space gardens

highlight of the mission. Later in the mission, the Azolla plant was also grown – a fast-growing Vietnamese floating plant used to supply nitrogen in paddy fields. The most successful was the hardy herb *Arabidopsis*, which successfully bloomed.

But it was all very difficult. Attempts to stimulate growth by covering the roots in black plastic had little effect. Another problem was that in the absence of gravity, plants lacked the means to dispose of their own wastes, so, in effect, they poisoned themselves. The Soyuz 35 crew brought up, in the *Malachite* cultivator, mature orchids along with fresh seeds. The mature orchids wilted at once, while the new seeds grew, flowered, and then wilted without producing seed. Interestingly, though, when brought back to Earth, they, too, recovered. In desperation, the botanists tried applying a magnetic field around *Biogravistat*, called *Magnetobiostat*, with better results. *Biotherm 2M* hatched out a consignment of fruit flies, with 68 sent down to Earth for analysis. This fly was favored because of its short growing cycle (40 days) and simple DNA (only about 20 genes). Lettuce was another plant grown on board, with samples returned on the *Bioblok 3M* container on Soyuz 34. Although the level of daily radiation exposure was low, the effect on seeds accumulated over six months, to the point at which there was four times the normal level of aberrant cells and mutations. *Biogravistat* provided some clues as to what was happening. This was a small centrifuge, ferried up by Progress 5, similar to one tested on unmanned Cosmos missions (e.g. Cosmos 782: see Chapter 7). In the centrifuge, seeds grew better, in the direction of gravity, whereas the weightless roots went out in all directions. The centrifuge was used for barley and mushrooms and once the cosmonauts found the correct rate to set the speed of the centrifuge, it produced relatively normal plants. The cosmonauts used it to try to hatch out quail, fruit flies, tadpoles, and micro-organisms.

Cytos was a Soviet–French experiment in the *Biotherm* series with micro-organisms that were examined under electron microscope on their return, showing some quite clear differences between those that had been in orbit and grown either faster or slower than control samples on Earth, although the reasons for the difference were not evident. Cells were kept cool, at 8°C, during ascent and descent to inhibit growth, but encouraged to grow by 25°C temperatures in weightlessness. Seventy-six cultures were brought back from the *Cytos* experiments and these had grown 54% faster. Plant cells changed color and shape, and got smaller.

The materials processing experiments initiated on Salyut 5 were continued, with an electric smelting kit called *Splav*, the Russian word for "alloy", which was mounted on the outside of the station, but with airlocks on the inside so that the cosmonauts could retrieve samples, which were heated to 1,100°C. This time, the station was put in free fall for when *Splav* was used, so as to prevent micro-gravitational disturbances. The first monocrystals made in *Splav* had an encouragingly uniform structure, but with minor deformities caused by the station's movements. *Splav* was later used to produce "foam steel", which is metal made of silumin, titanium, and silicon nitride – extremely strong, but as light as wood. Some experimental outcomes were puzzling: lead chloride was crystallized in the Czechoslovakian experiment *Morava*, but the drops did not turn out in the elongated blobs expected, instead in a cylindrical screw shape. The other furnace, *Krystall*, which was delivered by Progress 7, exposed samples to temperatures of up to 1,200°C.

The welding experiments begun on Soyuz 6 were resumed, using a new welding device called *Isparitel* ferried up there in summer 1979. An electron beam was used to apply film coatings on metallic and non-metallic surfaces by thermal evaporation and condensation, the experiment being conducted by rotating the equipment outside the station through a hatchway. Two hundred specimens were produced and retrieved by the cosmonauts on the stations. *Vaporizer* was an external electron beam welder that was used to test whether it would be possible to re-spray the surface of telescopes automatically and thereby preserve their condition.

The main instrument on Salyut 6 was the 150-cm helium-cooled BST-1M telescope, set in the floor of the station, which could be trained on the Earth, its atmosphere, and on heavenly objects. Cooled at –269°C, it was one of the largest telescopes of its kind. This was used to study the distribution of the polar lights,

Extended observations were made of the atmosphere from Salyut 6; here is the narrow band of the Earth's atmosphere, with stars just visible above

silvery clouds in the high atmosphere, the transport of dust, and, looking outward, to image stars α and β Centauri. The BST 1M was used by cosmonauts Yuri Romanenko and Georgi Grechko on 20th February 1978 to make a profile of the brightness of sub-millimeter radiation as the station flew northwards from the equator. Looking at the atmosphere, the final crew, Kovalyonok and Savinyikh, spotted a new form of lightning, in the high atmosphere, later called a "sprite" by the Americans: instead of discharging conventionally downwards, it formed tendrils in a wide column that then formed a circular flash that rippled outwards.

Salyut also made the first Soviet space-borne ocean microwave soundings to determine the height of waves, performed in conjunction with eight ships in the Pacific, part of the program developed by the pioneer of space-based ocean surveying, V.S. Etkin (1931–1995). *Biosfera* was a new project to establish base line levels of air pollution, using the Spetktr-15 instrument, which measured molecules and aerosols up to 100-km altitude.

Salyut 6 carried not only optical telescopes, but also the first radio telescope operated from a space station. Progress 7 ferried up the 350-kg KRT-10, made of fine mesh wire, and a 1.5-m-diameter antenna. This telescope was, in effect, the outcome of the discussion between Mishin and the Academy of Sciences and IKI over 1970–1971. The telescope antenna was set in place at Salyut's aft docking port and pulled open as Progress backed away from the station, unfurling to its full 10-m diameter. Because the docking port was needed for subsequent spacecraft arrivals, the telescope could be used for only a short period, from 24th July to 9th August 1979. The KRT-10 was used in conjunction with a 70-m radio telescope in the Crimea and other large tracking dishes in the deep-space network so as to create a very long base line for the pair's observations. Objects studied during the two-week period included α Cassiopeia and pulsar 0329 + 054 as it scanned along the plane of the Milky Way.

The *Yelena F* gamma-ray and radiation detector was used to measure gamma rays and radiation in the near-Earth environment: both were low over the equator, higher in higher latitudes, and by far the strongest over the South Atlantic Magnetic Anomaly. It weighed 22 kg, was designed by the Engineering and Physics Institute in Moscow, and had a sensitivity in the 30–500-MeV range. As for radiation on the station itself, the station had 12 detectors, which found that the average daily radiation exposure to the cosmonauts was in the order of 15–30 mrads/day. An unusual discovery emerged from the *Astro* experiment, developed by Estonian Rikho Nymmik of Intercosmos 6 fame. Solid-state dielectric track detectors were installed in the airlock and Nymmik found the existence in Earth orbit in Salyut 6's path of hitherto unknown high-energy heavy particles (10^{25}–10^{30} eV).

Spacewalking was resumed on Salyut 6 and the station saw the introduction of external experiments – instruments placed on and retrieved from the space station's hull. The first such routine space walk was undertaken by Vladimir Kovalyonok and Alexander Ivanchenkov in July 1978. This involved bringing in a container of biological samples that had been exposed for 300 days on the hull of Salyut (*Medusa*) and checking on the 6-cm^2 MMK-1M meteorite detectors for impacts. *Medusa* was a cassette with titanium, steel, glass, paints, ceramics, rubbers, and sealants, with a radiation detector, and it typically also contained small ampoules with amino acids, biopolymers, and monocultures. *Medusa* samples returned to Earth found that exposure to sunlight did cause the beginnings of biological growth, suggesting the importance of light in the creation of life, even in a vacuum. The MMK-1M plate was returned to Earth, where it was found to have 200 tiny crater impacts, mainly caused by paint flakes, which began to emerge as an ever growing source of orbital

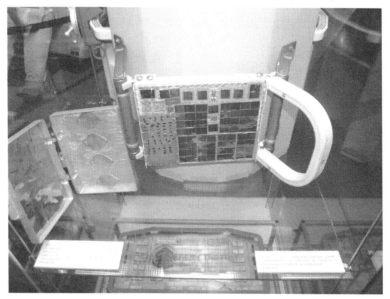

Salyut 6 marked the resumption of spacewalking on a substantial scale; these were fixtures used to attach samples for collection, replacement, and retrieval

pollution (from around this time, the practice developed of not painting the upper stages of rockets) [23].

The Intercosmos program experiments generally comprised a small suite of experimental projects developed by national scientific institutes: they were limited in extent, granted that the visiting missions were generally in the order of only a week. The most ground-breaking was probably the *Interferon* experiment conducted during the Hungarian space mission (Soyuz 36) in 1980, for this tested whether the growth of interferon anti-cancer bacterial cultures could be accelerated in orbit. The most futuristic was one delivered by the Mongolian mission (Soyuz 39). Holograms are well known in science fiction and the spaceship in *Star Trek* even had a recreational holodeck. The purpose of the hologram test was more mundane: to transmit three-dimensional images of materials processing experiments, where there was a real value in getting a three-dimensional picture of changes in substances. Here, on 27th March 1981, the first ever holograms were transmitted from Salyut 6 to the ground using the KGA-1, a LG-78 helium–neon gas laser. Hitherto, holographic machines had been heavy, complex, and required considerable supervision, so it was a challenge to build a small, light one that could be used by people not experts in holographic technology. The KGA-1, devised by the AF Ioffe Institute of Physics and Technology in Leningrad, weighed only 5 kg and was a box shape, 450 × 210 × 120 mm, and used only 60 W of power, the same as a domestic light bulb. On that day, cosmonauts Vladimir Kovalyonok and Viktor Savinyikh used the hologram in connection with the materials processing experiment, sending to the ground holograms of a sodium chloride crystal dissolving in water and then images of the window of the space station to assess impacts and microscopic imperfections. Three-dimensional images were beamed to the ground for periods varying from fractions of a second to less than a minute, the results being proclaimed as "satisfactory". Ultimately, this had considerable potential, especially in the area of medicine [24]. Salyut 6's experiments are summarized in Table 6.17.

Table 6.17. Salyut 6 science.

Biology	
Oasis	Plants
Fiton	Garden
Cytos	Effects of space environment on micro-organisms
Biotherm 2M	Fruit flies; *Arabidopsis* and *Crepis* plants and seeds
Malakhit	Cultivator
Biogravistat	Centrifuge
Medicine	
Beta 3	Cardiograms and electrograms
Polinom 2M	Heart and circulatory system
Reograph	Blood flow through heart and cardiovascular system
Amak	Blood and urine sampling
Astronomy	
ST1M	Telescope

KRT-10	Radio telescope
Yelena F	Gamma rays and radiation in near-Earth environment
Materials processing	
Splav	External smelting of materials
Krystall 3	Internal furnace
Vaporizer	External electron beam welding apparatus
Pion	Fluid physics
KGA-1	Holograms
Atmospheric science	
Spektr 15K	Density and temperature profiles of the atmosphere
RSS-2M	Hand-held spectograph

SALYUT 7 SCIENCE

Salyut 7 followed a similar design. Medical science was again a prominent objective, with the duration of missions again extended: 211 days (Anatoli Berezovoi, Valentin Lebedev) and then 239 days (Leonid Kizim, Vladimir Solovyov, and cardiologist Oleg Atkov, the second Russian doctor in orbit after Boris Yegorov). The *Polinom 2M* medical monitoring unit was replaced by a more advanced system called *Aelita*, which tested each cosmonaut for pulse, circulation, blood flow, and circulation, the accumulated data then transmitted directly down to doctors on the ground. An ultrasound scanner, *echograph*, was used for the first time to make a complete picture of a cosmonaut's heart system and blood circulation.

Both missions were tests of ever more elaborate counter-measures. The first crew came down in a swirling blizzard: retrieving the cosmonauts took some time and the immediate post-flight medical examinations were quite delayed. By contrast, Soyuz T-10 came down in calm autumn weather and we have a better record: the medical teams described the cosmonauts as "like people who had just been through a major surgical operation" and they were not permitted to walk for several days. Oleg Atkov wrote up the results of this mission, finding that one crew member lost 7 kg of weight. Calf circumference fell by between 14 and 21%. There was an enlargement of the liver, kidney, pancreas, renal cavity, and jugular vein. Weightlessness caused blood to pool and urine to dump, resulting in a lower level of blood flow.

Valentin Lebedev's mission had a postscript 25 years later when he gave an interview to *Pravda* alleging that he was losing his sight as a result of damage to his eyes from his 211-day mission (independently, it was known from Cosmos 782 and 936 that radiation had damaged the eye retinas of rats). His sight began to deteriorate many years later and he developed a cataract because, he said, of prolonged exposure of his eyes to radiation and direct sunlight. Many years earlier, Oleg Gazenko had warned that the only effect of space travel that was not reversible and where permanent damage might be done was the effect of protons, heavy ions, and other high-energy galactic radiation particles on nerve cells in the eyes.

The main scientific change was that a battery of X-ray telescopes replaced the

BST-1M as the principal instrument. This was the XT-4M built by the Lebedev Institute, to scan in the 2–30-eV range using proportional gas counters covering 3,000 cm^2. It was accompanied by the XS-02M X-ray spectrometer built by the Sternberg Institute. In the rear tunnel was the RS-17 X-ray telescope to scan the 2,000–800,000-eV range. Forty observation sessions were conducted, examining such objects as Scorpio x-1, Cygnus x-3, and a rare flare star in NGC-4151. PIRAMIG (Photography Infra red Atmosphere Interplanetary Medium Galaxy), a French experiment, took 600 images to map the Magellanic clouds.

The botanical package was called *Svetblok M*. To address the problems that had arisen on Salyut 6, ventilation and garden design were improved, with lighting levels radically increased to 24 hr a day when station electricity permitted. There was a water pump and fans to make the air flow and air the roots like a hair drier. This time, there was something of a breakthrough with *Fiton 3* when *Arabidopsis* flowered (three flowers), podded, and then released mature seeds for the first time – 200 eventually, half of which turned out to be viable. Peas grew extremely long, reaching 20 cm and some even 30 cm, sprouting up to six branches and many leaves. Then they encountered problems, the roots growing out of the soil and mold forming on the leaves. Attempts were made to grow oranges (the *Rost* experiment). In the second *malakhit* experiment, tomatoes, coriander, radishes, and cucumber were grown, but only the cucumber with any success.

But what role might radiation play in this? The *Bioblok* experiment identified the especially damaging effect of HZE super-heavy charged particles in the course of

The *Malakhit* experiment on Salyut 7

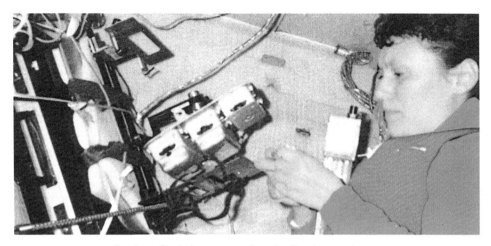

Svetlana Savitskaya operating the *Tavria* experiment

three experiments with lettuce, run for 40, 201, and 457 days. HZE made penetrating impacts deep into tissue and, indeed, on the biology missions, HZEs made severe but localized damage on lettuce tissue, producing aberrant cells. Despite this, the growth of the lettuce was unaffected and the destroyed tissue was eventually replaced.

The *Medusa* experiment was again carried, cosmonaut Valentin Lebedev bringing the samples back inside to check the effects of cosmic rays on biopolymers and amino acids, and the effects of micro-meteorites on the MMK detectors. In addition to the normal external experiments, cosmonauts Vladimir Dzhanibekov and Viktor Savinyikh fitted a plate to detect cometary material in advance of the approach near Earth of comets Giacobini–Zinner and Halley.

The materials processing kit comprised new installations, *Magma F* and *Korund*, and involved the development of pure crystals and the mixing of such materials as cadmium selenide and iridium antimonide. The *Tavria* experiment was an attempt, in weightlessness, to separate different biological compounds through electrical methods (electrophoresis), which could be the basis of much purer vaccine production. *Tavria* made eight capsules of anti-flu virus, equal to a year's supply for the Leningrad Pasteur Research Institute of Epidemiology, Microbiology & Hygiene and also animal antibiotics. *Tavria* was used to process interferon, anti-infection preparations, and other pharmaceutical products, apparently very efficiently. The crew of Soyuz T-9 (Lyakhov and Alexandrov) made transmissions of heat and mass transfers of liquids in the *Tavria* electrophoresis unit.

A new welding device, *Isparitel M*, was flown on Salyut 7, using a changeable head, a greater variety of metals (e.g. silver and tin), and new techniques (e.g. bracing). In addition, the Paton Institute developed a hand-held electron beam tool called VHT, in effect an electron beam gun for cutting, welding, brazing, and heating with a concentrated heat source with up to 10 keV. This was ferried up to the station in early 1984 and used during a spacewalk on 25th July 1984 by Svetlana Savitskaya. She attached it safely to the side of the station and used it for cutting steel and titanium 0.5 mm thick, applying silver coatings and then welding and brazing materials 1 mm thick. The VHT was then brought back inside and the samples returned to Earth. It was used again two years later on 30th May 1986 by cosmonauts Leonid Kizim and Vladimir Solovyov, who welded together 10 girder structures and tubular booms, producing high-quality joints. Other materials were placed on the outside of the station, such as polymers, glass fibers, and plastics for up to 1,501 days (all lost moisture, some cracked due to temperature changes, but others hardened in the vacuum).

The hologram experiments begun on Salyut 6 were continued with a new hologram machine, the KGA-2. The crews of Soyuz T-5 (Berezovoi and Lebedev) and T-7 (Popov, Serebrov, and Savitskaya) used the KGA-2 to transmit holograms of the separation of substances during the preparation of medicines, notably albumin over a period of 40 min.

Two other attempts to develop biological space science were considered during the 1980s. The first was called the "old age plan", the idea being to send an ageing cosmonaut into space to see how a much older man would withstand weightlessness, the candidate being Voskhod engineer Konstantin Feoktistov, 58 years old in 1984. The second, much more substantial idea was to develop a specialized space module for medical science, called Medilab, to be sent up to Salyut 7's successor. The first idea was dropped when Feoktistov became ill but was, ironically, taken up independently by the Americans, who put astronaut hero John Glenn into space in 1998. Although the design of Medilab became quite advanced, the money problems that began to envelop the Soviet space program in the 1980s and 1990s meant that it never got as far as construction.

Georgi Grecho returned to the observations of the atmosphere that he began on Salyut 4. In the course of a northern-latitude set of autumn observations (September 1985), he determined the main ozone and aerosol layer to be at 20–30 km, identifying eight sublayers, levels, and structures varying according to latitude. The layer was lying 0.5 km lower than during his Salyut 6 observations in 1978.

Towards the end of the Salyut 7 mission, a large module docked with Salyut 7, the Cosmos 1686: this included a large experimental block called Pion K. In the event, the Pion K instruments got only limited use. Cosmos 1686 arrived at Salyut 7 on 2nd October 1985, but the station had to be abandoned less than two months later, when station commander Vladimir Vasyutin became unwell, the second case of a station evacuation due to cosmonaut illness. The Pion K system was used a second time the following year during the 50-day visit by Leonid Kizim and Vladimir Solovyov when they flew there from the Mir space station. Some of the Pion instruments were able to operate without human operation and continued to return data for another three years. Designed by S.P. Ryumin and Yuri Mineev, the instruments were designed to detect electrons in the 0.3–2-MeV range [25]. The instruments used in Salyut 7 science are listed in Tables 6.18 (Salyut 7) and 6.19 (Pion K).

Table 6.18. Salyut 7 science.

Botany	
Svetblok M	Space garden
Oasis	Garden
Fiton	Garden
Cytos 2 and 3	Incubator for bacteria cells
Malakhit	Incubator for plants and vegetables
Materials	
Tavria	Electrophoresis
Magma F	Semi-conductors
Korund	Rotating crystal furnace
Pion M	Fluid science
KGA-2	Holograms
Medical	
Aelita	Testing system replacing *Polinom 2M*
Echograph	Ultrasound scanner
Astronomy	
RT-4M	X-ray telescope
RS-02M	X-ray spectrometer
Ryabina	Cosmic radiation spectrometer
External	
Medusa	Effects of cosmic rays on bacteria
MMK	Micro-meteorites
Electrotopograph	Effects of cosmic rays on metals and alloys
Comet	Detect cometary debris
Atmosphere	
MKS-M	Atmospheric spectrometer
Epho	Measurement of aerosols in the atmosphere
Yelena F	Gamma telescope
Spektr-15M	Spectrometer
SKR-2M	X-ray spectrometer
Mariya	Spectrometer for magnetic fields predicting Earthquakes
Astra 2	Measurement of gases at orbital altitude

Table 6.19. Pion K science (Cosmos 1686).

Astronomy	*Materials processing*
MRSF-1K mass radio spectrometer	*Korund*
Ozon radiometer	*Kristallizator*
Faza spectrometer	*Magma F*
Sevan cosmic ray detector	
Canopus for gamma rays	
Ega for neutrons	
ITS-7, to study the Sun and other stars in the infrared band	

MIR SCIENCE

Mir became the most famous of all the Russian space stations, operating in orbit for an amazing 15 years, three times its design life, from 20th February 1986 to 23rd March 2001. The volume of science carried out on Mir was enormous. Some old Salyut 7 experiments were also used: the first crew to Mir, Leonid Kizim and Vladimir Solovyov, traveled across space in the Soyuz T-15 spacecraft to Salyut 7 in May 1986, retrieved 400 kg of old equipment, and brought it back to Mir at the end of the following month. They ferried back an Earth mapping camera, the Pion M materials science experiment, the *echograph* health-monitoring unit, the PCN low-light camera, and the EFU robot electrophoresis pharmaceutical purification equipment.

Although the Mir core module was similar in size and shape to the preceding Salyut stations, in reality, the Russians followed a different approach to the installation of scientific equipment. The decision was taken that the core module, the original Mir, would be a crew module only, uncluttered by scientific instrumentation. Scientific work would be undertaken on board several large modules docked to Mir:

- Kvant astrophysical module, arrived April 1987;
- Kvant 2, arrived December 1989;
- Krystall, arrived June 1990;
- Spektr, arrived May 1995;
- Priroda, an Earth resources remote sensing module, arrived April 1996.

Spektr and Priroda were furnished with American equipment for a joint flight program with the United States for 1995–1998. Most of the Russian scientific equipment was located in Kvant, Kvant 2, and Krystall. For convenience, Kvant's astrophysical work is reviewed in conjunction with the series of Soviet observatories in the 1980s (Chapter 7).

Mir's science program was impeded by a number of problems. First, there was always less electrical power than the scientists would have liked in order to run the experiments. Power built up to 25 kW in 1997, but this fell to 15 kW when the Spektr module (with its large solar panels) was put out of action by a collision. Running Mir's systems had the first call on electrical power, scientific experiments being a lower priority. Second, Mir had only 10-min communications with the ground every orbit – and that was when Mir flew over Russia – which made round-the-clock, ground-monitored experiments impossible. Third, except during the later period of shuttle visits, Mir had very little capacity to send experiments or their results back down to Earth, for Soyuz had room for only 50 kg of downward returning cargo.

Mir continued the same broad categories of space science as those developed during the Salyut missions. Several hundred experiments were flown, including many brought on board by visiting missions and only the most significant can be reviewed here, starting with space medicine. Typical resident missions were set for six months, which settled down as the ideal mission length from a medical and psychological point of view. These long-duration missions had been preceded by a year-long

Table 6.20. Mir – extension of duration records.

Soyuz TM-2	Yuri Romanenko	326 days
Soyuz TM-4	Vladimir Titov, Musa Manarov	366 days
Soyuz TM-18	Valeri Poliakov	438 days

mission on the ground in a mockup interplanetary spacecraft as far back as 1967–1968 by Dr Gherman Manovtsev, technician Boris Ulybyshev, and biologist Andrei Bozhko. These experiments had a strong focus on the psychological aspects of spaceflight and how to build and maintain teamwork in a confined and stressful environment. One such experiment, albeit a shorter one, was conducted near Brno in Czechoslovakia in 1988. Here, sociologist Jaroslav Sykora organized an experiment with 13 men and one woman 30 m underground in an old mine in Tisnov, Moravia, called *Stola 88* (*stola* means "mine" or "tunnel"). With this background and that of the earlier Salyut missions, Mir provided an opportunity for endurance records to be pushed back again, this time in three stages, as noted in Table 6.20.

Vladimir Titov and Musa Manarov's accomplishment broke the psychologically important barrier of one year in orbit, while Dr Valeri Poliakov's duration flight was consciously set at the amount of time it would take to fly to Mars, spend a month there, and return. Having a doctor on board was not only a test for Valeri Poliakov himself, but enabled there to be a high level of medical supervision of his companions: the German *Reflotron* instrument permitted him to carry out live blood analysis in orbit, the machine giving an immediate readout of blood composition, rather than sending samples down to the ground for later study. The original Soviet medical equipment was now joined by a range of new apparatuses from countries whose cosmonauts visited Mir: APT (fluids, Britain), *Electrocardiograph* (cardiovascular activity, France), *Monimir* (coordination of reflexes, Austria), and Son-K (sleep studies, Bulgaria).

Psychological aspects of spaceflight were given considerable attention. Long-term (six-month) crews faced a number of challenges, such as isolation, insomnia, and confinement, which could be reflected in poor relationships on board. These were closely monitored from the ground and cosmonauts trained in

Dr Valeri Poliakov

Dr Jaroslav Sykora

advance in the building of personal relationships. With the resumption of social contacts on their return to Earth, their psychological condition improved quickly in line with physical health.

The core of the countermeasures was the exercise program – 2 hr a day, be that on the running track or the bicycle, and increasing the exercise level as the return to Earth approached and increasing the wearing of the special suits to 12 hr a day. In addition, returning crews were expected to take calcium and vitamin supplements. When Vladimir Titov and Musa Manarov came back to Earth in December 1988, both were in reasonable condition: it seemed that the deterioration of conditions had flattened out and that energetic exercise and the increased consumption of fluids did much to ward off the worst effects of re-entry after long missions.

But the absence of gravity was not easily conquered. When Alexander Volkov returned from Mir, he had lost 50% of his red blood cells and his coordination was poor for the first three days. But his medical parameters were back to normal in three months and he declared himself ready to fly again. Yuri Romenenko tired a lot when he passed the 300-day mark in his mission and had lost 15% of his leg muscle on his return. Valeri Poliakov found, after his 438-day flight, that he lost a lot of his body liquid – 30%, or 17 liters. This was a function of both the blood system discharging water through urine and also exercise, "but you just have to go on drinking". Poliakov recorded how in the course of 2 hr of exercise, he lost 1.7 liters of water. One of the unsolved problems was how to deal with calcium loss and bone mass loss: Valeri Poliakov lost 20–25% of his calcium. In no case did the level appear to level off after time, though the effects varied a lot from one Mir participant to another, one American astronaut losing 10% of leg bone mass (Norman Thagard), another almost none (Shannon Lucid).

Writing up the Mir experience, Anatoli Grigoriev, the director of IBMP (1988–2008), noted that:

- 50% suffered from space sickness;
- bone loss was in the order of 7.8%, ±1.7%;
- there was a reduction in back muscles of about 4.4% and their density of 0.4%;
- there were some, but moderate, changes in blood;
- there was a depression in the immune system, but it recovered after 30 days.

The medical results from Mir were so extensive that a two-volume edition of *Space Biology and Medicine of IBMP* was devoted to them: *The Orbital Station Mir*.

Granted the length of human occupation on the station, much more attention was paid to the accumulation of microbes on board. An instrument was devised to measure them so as to anticipate any harmful effects: the AK-1 microbe sampling unit. An inventory on Mir once logged 234 species of bacteria and 126 species of fungi! Some were potential pathogens and could damage both humans and electrical equipment. According to Valeri Poliakov, living in artificial air for lengthy periods was unhealthy, like living in a polluted city, and one of the most important foods to have on board is yogurt, to kill unhealthy space station bacteria that can get into your system.

Radiation levels were watched closely, especially around the solar maxima of 1990 and 2000. Radiation levels on the cosmonauts were measured by a battery of instruments called *Akkord*. This found that the normal radiation dose in the course of a year would be 10–15 rads, compared to the 5 rads a year maximum level set for workers in the atomic industry. The monitoring system did find that some parts of the station were safer than others when it came to protection from radiation; the well protected Kvant module offered the best protection. On several occasions, in September and October 1989 and November 1998, the crew was advised to move there for protection. A cosmonaut receives an average of 35–40 millirads a day, but up to 100 millirads at solar maximum. Valeri Poliakov received a total of 8 rads on his first Mir flight (eight months) and 14 rads on his second (14 months). Alexander Alexandrov, the Bulgarian cosmonaut, carried a 6.5-kg radiation detector called *Liulin* and found that radiation levels doubled over the South Atlantic Magnetic Anomaly, falling back over northern latitudes, mention of which, Mir remapped the South Atlantic Magnetic Anomaly in 1990, grading the different radiation intensities, maxima and minima. Electron densities were measured according to height (Figure 6.9). Not only that, but a global map was assembled of the many anomalies of our planet. The R-16 instrument on Mir meant that there was a consistent system for measuring radiation in Earth orbit and by the end of its mission, there was a full dosage chart for 1991–2001.

Materials science experiments were resumed, most being located in the Krystall module. These experiments were called *Pion M, Korund 1M, Gallar, Krater V, Kristillizator, Zona*, and *Optizon*. The last, *Optizon*, had a furnace capable of reaching 2,100°C. Overall, they were less successful than hoped, due to insufficient electrical power (*Krater*, for example, needed 2 kW), malfunctioning equipment, microgravity interference, and insufficient training for the operators in a highly specialized discipline. Similar protein crystal growth experiments were flown by American scientists this time and their conclusion was that whilst crystals grown in orbit were 24% better than those on Earth, the space environment was probably overrated as a production site. The station had a pilot electrophoresis production plant called *Svetlana*, weighing 800 kg, which was used to make interferon, anti-flu drugs, and human and animal antibiotics, while another production plant, *Ainur*, was able to produce 100 kg a year of vaccines. Some crystals were much bigger, five times so, than those on Earth. The first *Korund*, for example, was designed to make up to

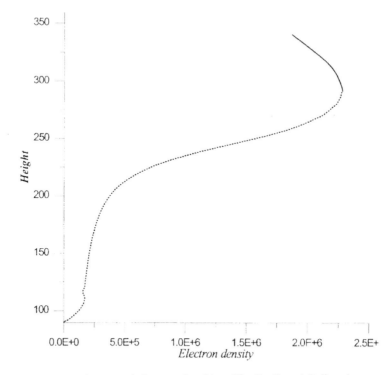

Figure 6.9. Mir map of electron densities. (Credit: Sergei Pulinets)

18 kg of materials without human supervision. A new welding unit was used: *Yantar*. The value of the materials processing experiments is difficult to assess. The ambition of the Mir experiments was to move from the experimental towards commercial production, but the collapse of the Soviet economy meant that companies on the ground could not have been in a worse position to exploit the results. The experiments dating back to Salyut 5 achieved many firsts, but taking full advantage of them required more electrical power, cosmonaut engineers dedicated to their operation, faster return of samples, and an engaged user community.

A number of external experiments were run to test the effects of the open space environment on materials. This involved the placing, on the outside of the station, of metals (e.g. copper, silver, tungsten, aluminum), electronic equipment, adhesives, optics, lubricants, and paints, the most elaborate system for doing this being the Ukrainian-built *Elektrotopograph* experiment. Spacewalks to collect samples on the hull were a regular feature of Mir missions. *Trek* was a 1-m^2 panel left outside in May 1991 to collect cosmic rays, the panels being exposed for two two-year periods to collect galactic cosmic rays before retrieval. *Kromza* comprised foil arrays designed to detect interstellar gas, especially helium, and retrieved three times for later ground analysis. During a spacewalk in September 1994, Vasili Tsibliev and Alexander Serebrov found a 10-cm hole in one of Mir's solar panels and 65 small impacts on the hull – a warning of growing débris levels in low Earth orbit. On the

Euromir 95 mission, a dust detector was installed on the hull on 20th January 1995 and retrieved in February 1996: this suggested that Mir did intersect a dust cloud and there was one substantial hit.

The promising botanical experiments carried out on Salyut 7 were not resumed for some time. Granted their promise, this was strange, but Bulgarian botanists prevailed on Mir mission planners that they should be allowed to take the lead on the botanical experiments scheduled for Mir, although their equipment was much older. Their experiment was called *Svet*, modeled on the original *Oasis*. They sponsored a conference on the topic in 1990 to preview their planned experiments on Mir. *Svet 1* was flown up on the Krystall module and managed to grow radishes in 23 days and cabbage in 54 days, but wheat failed after 40 days. *Svet 2* was flown up on the shuttle in 1995 and was managed by American astronauts Shannon Lucid and John Blaha over 1996–1997.

Everything went wrong for the first crop: lamps failed, temperatures soared to 37°C, the plants became disorientated, and no heads grew. On the second attempt, although 260 wheat seeds were released from 40 plants; they proved to be sterile. Searching around for an explanation, botanists suspected that they had suffered ethylene poisoning, either from Mir's coolant system or else because there was no system to remove the ethylene generated naturally by plants themselves. The use of new ventilators enabled the next American astronaut, Michael Foale, to grow two mustard seeds through a full cycle. The botanists were far from home and dry, because only a minority of plants survived and often in a weakened form. Botanical experiments resumed once the Americans had left Mir, the lead passing back to Russia this time. A wheat crop was sown on 30th November 1998, ears appearing on 15th January 1999 and coming ripe on 1st March 1999: a red-letter day in space botany, providing a harvest of 510 g – the first time a plant had been grown through a full cycle. Wheat often produced more ears than on Earth: eight rather than three. They also grew cabbage, broccoli, and red mustard, typically taking two to four weeks, but their leaves were a different color from on Earth. The *Svet* radish crop was a success – although there were some abnormalities and the plants were smaller and a third the weight than on Earth. The ultimate dream of Bios inventor Iosif Gitelson was to place a biological cultivator on the *outside* of the spacecraft, but, as he found out at the time, "engineers were too conservative about biological technology".

Biological experiments on Mir focused on frogs and quail. Japanese journalist Toyohiro Akiyama brought up, on behalf of the Japanese Institute of Space and Astronomical Science, six Japanese tree frogs. This type of frog was specially selected because it had suction caps on its legs, thereby enabling it to better cope with weightlessness. The main reaction he noticed though was that the tree frogs appeared to suffer from space sickness, for they appeared to attempt to vomit on several occasions.

Most of the first batch of 48 quail eggs failed to develop beyond the first five days of the normal 17 days of incubation, but the first of eight quail chicks was born on 22nd March 1990 – a landmark day in space biology. Although the quail chicks blinked and reacted perfectly normally, they were very frail and unable to feed themselves. Film showed the birds to be highly disorientated by weightlessness,

Peas gone to seed in the greenhouse (Credit: NASA)

thrashing their wings as they tried to fly. No amount of trying to stabilize themselves using suction to the floor did any good and all the first batch died. In the mid 1990s, NASA astronauts continued the quail experiments, but there was a high abnormality rate at birth (13%). On the last scheduled resident Mir crew in 1999, no fewer than 60 quail eggs were ferried up on board Soyuz TM-29. Ten hatched out, but seven died from cold during the descent and only three returned to the laboratory on Earth alive. It was difficult to avoid the conclusion that growth was gravity-dependent. This was confirmed when quail were brought into space at various stages of development. Those brought into space immediately on fertilization all died, while those incubated on the ground developed normally.

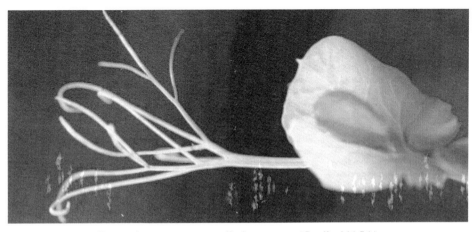

Some plants grew unusually long stems (Credit: NASA)

Water droplets attached to plants (Credit: NASA)

Finally, in the continuing effort to make Mir self-sufficient, a new step forwards was taken in the recovery of water. In addition to extracting water from the air (the SRV-K2M), a system was devised to extract water from urine (the SPK-VM), while the *Elektron* system took oxygen from water and the *Vozdukh* system scrubbed carbon dioxide from the air, meaning that Mir was substantially nearer to a closed-cycle system [26]. Mir science experiments are summarized in Table 6.21.

Table 6.21. Science on Mir.

Materials science	
Biocryst	Crystal growth
Ainur	Crystal growth (vaccines, medicine, insulin)
Pion M	Semi-conductor production
Korund 1M	Pilot commercial production
Gallar	Furnace
Krater V	Small furnace
Kristallizator	Furnace
Zona 2	Manufacturing oven
Zona 3	Manufacturing oven
Svetlana	Pilot electrophoresis production plant (interferon, antibiotics, anti-influenza)
Optizon	Furnace for silicon
External	
Elektrotopograph	Exposure of materials
MMK	Micrometeorite detector
Trek	Panel to detect super heavy cosmic rays
Kromza	Foil to collect interstellar gases
Astronomy	
Granat	Gamma rays, fast and slow neutrons
Ryabina	Cosmic radiation spectrometer
SPIN 6000	Portable X- and gamma-ray spectrometer
Sprut	External spectrometer to measure fluxes in elementary particles
ITS-7D	Infrared spectrometer
MKS-M2	Spectrometer to detect background radiation

Buket	High-resolution soft gamma and X-ray telescope
Marina	Cosmic ray detector
Glazar 2	Telescope

Medical

AK-1	Detection of microbes
Aelita	Multi-channel recording unit
APT	Behavior of fluids in the body
Echocardiograph	Cardiovascular measurement
Son-K	Sleep studies
Reflotron	Blood composition
Akkord	Radiation monitoring

Atmosphere

Alfa E	Ionosphere and magnetosphere
Astra	Sampling of space environment
EFO-1	Dust and aerosols
Spektr 256	Spectrometer
ITS-7D	Spectrometer
ARIZ	X-ray spectrometer
Balkan 1	Cloud lidar
Faza	Earth's atmosphere
Taurus	X-ray spectrometer
Grif	Gamma-ray detector
MIRAS	Mir Infrared Atmospheric Spectrometer
Ozon M	Ozone and aerosol depletion

Botany and biology

Fiton	Greenhouse for onions, radish, cedars
Inkubator 2	Hatching of birds (e.g. quail)
Gravistat	Greenhouse for wheat, flax, *Arabidopsis*
Svet	Greenhouse for radishes and lettuce

INTERNATIONAL SPACE STATION (ISS) SCIENCE

Whilst the Mir space station comprised a base block (Mir) and five modules (Kvant, Kvant 2, Krystall, Spektr, and Priroda), the International Space Station comprised, at its beginning, two Russian modules (Zarya and Zvezda) and two American ones (Unity and Destiny). In the course of time, two airlocks were added (Pirs and Quest), a node (Tranquility), vast solar panels and two large trusses, specialized European and Japanese modules (Columbus and Kibo), and two small Russian multi-purpose modules (Poisk and Rassvet). It became the largest object in orbit, the most visible object in the night sky and the biggest international scientific project ever. Rassvet was especially important, for it was kitted out with 12 workstations, a glovebox, two incubators, a vibration-free platform, racks, and shelves.

The original idea was that while construction was in progress, the station would have a crew of three who would contribute mainly to station operation, maintenance, and construction. Once construction was complete, the crew would be increased to six and a real scientific program would begin. In reality, construction took much longer than anticipated – no less than ten years, 1998–2008 – and the station did not reach a six-person crew until May 2009. Positively, though, this distinction between construction (when no science would be undertaken) and operations (when it would be the dominant activity) was not adhered to and mission managers were able to commence a scientific program from the very beginning.

It was always intended that the Russian scientific research program be conducted from specialized modules, but because of financial shortages, these were long delayed. Neither of the original Russian modules was designed for scientific research, Zarya being the control block for the station and Zvezda the crew quarters. A further complication reducing the volume of Russian science on the station was that the Russians, ever short of money, in effect gave over their modules to the Americans for their scientific experiments. Not until 2010 did the Russian segment have a substantial science platform of its own, with the arrival of the modules Poisk and Rassvet. As was the case with Mir, one of the big problems was the difficulty in getting experimental results back to Earth, for Soyuz had only 50 kg of cargo space for its return missions. Russian electronic downlink capacity was also poor.

Despite a slow start, an impressive Russian experimental program was developed for the International Space Station (ISS), with items ferried up bit by bit by Soyuz manned spacecraft, Progress freighters, and the American shuttle. The ISS started with 12 Russian scientific experiments on expedition 1, such as materials science (*Plazmenny Kristall*), the examination of human body fluids (*Sprut*), and radiation studies (*Prognoz, Bradoz*). A significant volume of work was under way within four years, and by 2010, there were 64 experiments in operation, with 38 already completed and 32 in preparation, many in collaboration with other countries (mainly European). Scientific papers arising had made a slow start, but as many as 127 were published in 2007 and they had totaled 609 by 2010. The dominant fields of activity were geophysics, microgravity, and technological, with the most promising early results in the area of medicines such as vaccines (AIDS and Hepatitis B), insulins, anti-cancer drugs (interleukin), and immunomodulators.

No fewer than four Russian–American boards supervised medical aspects of the flights. There was a daily medical check and report, with in-depth checks using medical equipment at intervals (e.g. blood, electro-cardiogram tests). Spacewalks, because of their strenuous nature, required an especially high level of pre-walk and post-walk monitoring and rest was often ordered after them to recover from fatigue. Radiation was checked constantly: levels of radiation ranged from 1,865 millirads for the first crew to 3,850 on the 12th. The medical outcomes for the first 20 ISS missions were as follows:

- The health of the crews on their return was satisfactory and there were no significant clinical events at all. Spacewalking was, medically speaking, the most demanding part of the mission. There was quite a range of individual responses to both weightlessness and mission duration.

- There was a gradual fall in blood pressure and respiration rates during flight, both rising sharply after landing. There was a risk of heart impairment after six months. Cosmonauts could hold their breath much longer, though, typically 90 sec as against 60 sec beforehand.
- Sleeping was a problem, all reporting a moderate decline in sleep activity (–20%), some a severe decline (–50%), causing worries of error and poor performance.
- The main irritant to cosmonauts during missions was, unexpectedly and mundanely, noise. This was because of the ventilators and other instrumentation on the station, but it was so loud that most cosmonauts used ear plugs. Noise was 4–16 dB above reasonable limits and was a challenge to future space station design.
- On landing after long missions, cosmonauts typically suffered from fatigue, poor balance, bad posture, awkward motor movements, and loss of bone mineral content.
- A two-month period of convalescence in a health resort was normally sufficient for cosmonauts to recover. The one problem that persisted was low bone mineral density, which took much longer.

Because of the level microbiological infection identified on earlier missions (e.g. Salyut 4 and Mir), there was a higher level of environmental monitoring, using air and chromatograph sampling, which found 70 microbial species, but levels were within limits. One of the surprises was that the main source of the bacteria was not bugs that had hatched out and been living in the modules for years. Instead, most came from arriving cargo ships and visiting crews who brought with them fresh bugs and contaminants.

Russia resumed space botany experiments from expedition 2 with Yuri Usachov. These included fruit fly genetic studies (*Polygen*) and more greenhouses (*Lada* and *Rasteniya 2*), with peas, tomatoes, radish, and lettuce. Four cycles of peas were harvested and DNA abnormalities detected. The clues to the successful growing of plants in space appeared to be the provision of sufficient lighting, the use of suitably grained soil, removing ethylene poisons, and aerating the soil sufficiently so as not to rot the roots. Radiation experiments from earlier missions were continued with a full-size human torso on board called *Matroshka*, a dummy full of sensors to measure the absorption of radiation by the human body, both inside and outside the station.

One of the most remarkable outcomes came from the *Biorisk* experiment, which involved placing bacteria and fungi on the outside of the station for several months at a time to test their reaction to the space environment (the first experiments had been done on Mir with *Bacillus subtilis* and the spore *Aspergillus versicolor*). Samples were then brought back on board by spacewalking cosmonauts and returned to Earth. Although damaged by a year and a half in raw space, they still functioned. Now, Japanese scientists recommended the Institute for Bio Medical Problems (IBMP) to try out the African bloodworm mosquito, one used to hibernating during periods of extreme or lengthy drought. In summer 2007, cosmonauts Fyodor

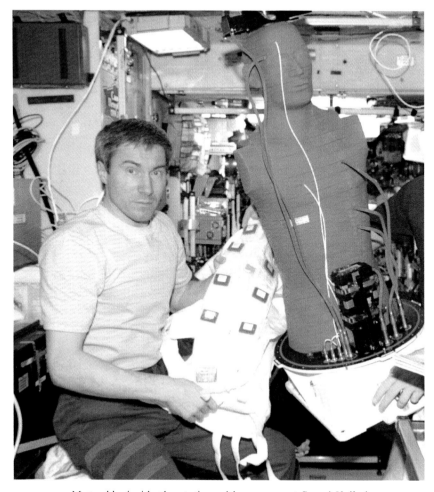

Matroshka inside the station with cosmonaut Sergei Krikalev

Yurchkin and Oleg Kotov placed a *Biorisk* cylinder on the outside of the ISS with African bloodworm mosquitoes, barley seeds, crustaceans, and bacteria. A year later, the cylinder was brought back in by Sergei Volkov and Oleg Kononenko and delivered to Moscow for analysis. Amazingly, the mosquitoes had survived. They were able to convert their moisture into tricallosa sugar, which crystallized and cooled to –210°C, the temperature of liquid nitrogen, and thereby survive the vacuum, radiation, and heat-and-cold cycles of a year in orbit. Back on Earth, the mosquitoes were crawling and flying again. According to IBMP's Vladimir Sychev, the survival of insects in such extreme conditions gave fresh credibility to the theory of panspermia, the idea originally promoted by Vladimir Vernadsky that life was spread throughout the universe by comets and meteorites carrying life.

Another frontier area developed on ISS was the exotic science of "dusty plasma". Plasma is the fourth state of matter (after solid, liquid, and gas) and the study of

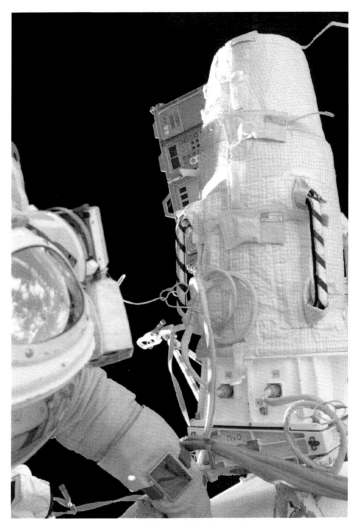

Matroshka outside the station

plasma and plasma crystals had important applications in the area of electricity, energy transmission, solar energy cells, computer superconductors, the removal of dust from industrial units, food, medicines, and clothing. Experiments on Earth are limited by gravity, while zero gravity enables three-dimensional studies of the behavior of dusty plasma. Earlier experiments had led to the development of a sterilizing cold plasma torch used to sterilize antibiotic-resistant wounds in surgery. In the Poisk module, cosmonaut Oleg Kotov installed the German-built PK-3 + plasma vacuum chamber. Experiments were typically made in 90-min runs, the results videoed and returned to Earth on computer disk by Soyuz cabin [27].

Progress freighter spacecraft were also used for science. Once they had delivered their freight to the space station, they were normally filled with rubbish and de-

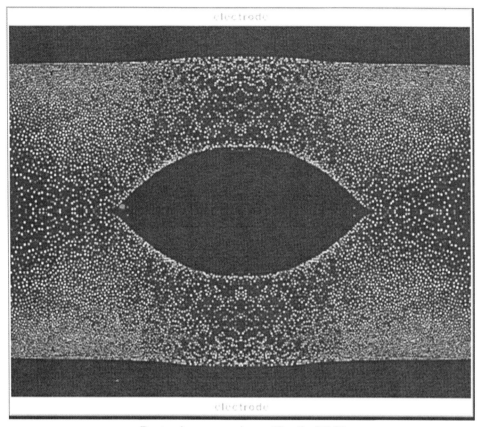

Dusty plasma experiment (Credit: DLR)

orbited immediately over the Southern Ocean. Some, though, embarked on a period of independent flight before de-orbiting. Progress M1-7 (March 2001) deployed a 22-kg satellite, *Kolibri* (or "humming bird"), to monitor solar plasma. Progress M-65, for example, was fitted with laser beams to test the reflectivity of the atmosphere in its extended mission in November–December 2008. Progress M-04M continued the experiment (*Reflection*) in a two-month independent flight while Progress M-06M made studies of the effects of engine emissions on the ionosphere (*Progress Radar*) in 2010. Table 6.22 summarizes the Russian science experiments on the ISS.

ORBITING SPACE STATIONS: WHAT WAS LEARNED?

Biological space science played an important role in paving the way for manned flight – both for the first manned flights (the Korabl Sputniks) and for long-duration missions (Cosmos 110). Once manned flights got under way, the principal focus of Russian space science was on the human reaction to weightlessness and the space environment – an entirely justified reaction judging by the discovery of space

Table 6.22. Selection of Russian science experiments, International Space Station.

Area	Name and type
Human life sciences and medical	Heart, muscle performance – electrocardiogram, dynamometer
	Analysis of blood, urine, and air
	Torso to test radiation dosage of human body (*Matroshka*)
	Coordination, motor and eye movements (*Pilot M*)
	Fluids in the human body (*Sprut*)
	Dental health in orbit (*Paradont*)
	Cardiac activity (*Kardio*) and pulmonary respiration
	Effectiveness of medicines (*Farma*)
	Bone marrow (MSK)
	Vaccines (*Vaccine, Antigen, Interleukin, Mimitek*)
Plants, cells and biology	Greenhouses: *Lada, Rasteniye, Svet*
	Fruit fly genetic studies (*Poligen*)
	Bio-reactors (genetics, proteins (*Rekomb K, Konjugatsiya*)
	Biotechnological (*Mutatia, Aseptic, Biomulsia*), plasmids (*Plasmida*)
	Plant growth hormones (*Micefit*)
Space environment	Effect of meteorites on the station (*Meteoroid*)
	Magnetic levels outside station (*Iskazhenie*)
	Microbes inside the station, outside (*Biorisk*)
	Neutron flux (BTN)
	Methods for dispersing oil pollution (*Bioecology*)
	Gamma rays and high-energy charges particles (*Vsplesk*)
	Hypergolic residues, contamination, propellant residue (*Kronka*)
Cosmic rays and radiation	Effects of rays on outside and inside (*Platan*)
	Radiation levels (*Prognoz*)
	Radiation and biology (*Bradoz*)
Geophysics, atmosphere, remote sensing	Lightning (*Molniya*)
	Real time weather, environmental, disaster monitoring (*Uragan*)
	Oceanography and bioproductivity (*Diatomia*)
	Airglow, atomic oxygen (*Fialka*)
	Carbon dioxide, methane, greenhouse gases (*Rusalka*)
	Glows in the upper atmosphere (*Reaksiya*)
	Magnetic field and microgravity (*Kolumb*)
Materials science	Plasma crystals: *Plazmenniy krystall, Glycoproteid, Mimitek Glycoproteid, Mimitek* (PK series)
	Super-high-temperature welding (*Synthesis*)
	Porous materials (*Membrane*)
	External coatings (*Control*), glues (*Restoration*), corrosion (*Quartz*), antifriction devices (*Tribocosmos*), strengths (*Perspektiva*), electrojets (EDK)
	Biology, medicine, pharmacology (*Kristillizator*)
	Melting (*Reper Kalibri*)
	Heat-mass exchange (*Gel*) and liquids (*Krit*)

sickness, first by dogs (Belka) and then by humans (Titov), followed by the severe effects of weightlessness unless effective counter-measures are taken (Cosmos 110, Soyuz 9). The priority given to understanding weightlessness and the space environment was evident when a medical doctor, Boris Yegorov, was flown into orbit at one of the first available opportunities. Counter-measures were a dominant thread of space station operations, a series of instruments and experiments being put in place, to the extent that the world's greatest repository of knowledge on long-duration flights probably lies in the Institute for Bio Medical Problems. Even though humans have now proven their ability to adapt to long-duration missions and even though the current six-month tours of the International Space Station present few problems, the question of adaptation to weightlessness is far from closed and many outstanding issues remain.

The construction of the first orbiting station in 1971 meant that it was possible to begin a substantial program of space science on orbit, tended by cosmonauts on board. Although Salyut's program may have been conceived in some haste, its basic outline of experiments set the pattern for the subsequent stations: medical research, botany, astronomy, and atmospheric science, with materials science added from Salyut 5 and the retrieval of external samples through spacewalking beginning on Salyut 6. The *Oasis* and related experiments on Salyut, Mir, and the ISS showed that scientists had much to learn about plants, animals, and microbes in the orbital environment and it took over 20 years to make critical breakthroughs. Space-based telescopic studies added considerably to our knowledge. The materials science experiments, while proving commercially disappointing, may ultimately prove to have been ahead of their time and lead to positive future outcomes. Welding experiments paved the way for the construction of large structures on orbit in the distant future. In the area of atmospheric science, Salyut marked the beginning of an awareness of the global problem of climate change, atmospheric pollution, and aerosol depletion. Finally, there was an important range of more exotic experiments, ranging from panspermia to holograms.

In summary, the following has been learned from the experience of the manned precursor missions, manned spaceflight, and orbiting stations:

- weightlessness and human biology:
    the discovery of space sickness;
    dogs a good predictor of the effects of manned spaceflight on humans;
    turtles verified the safety of manned lunar flight;
    radiation levels in orbit tolerable for humans (10–15 rad a year);
    need for radiation shelters;
    people can fly 14 months in Earth orbit;
    negative effects from longer flights (e.g. loss of calcium, muscle atrophy);
    long-duration flights require counter-measures, principally exercise;
    also helpful are muscle-loading suits, water, tonics, vitamins;
    recovery in two phases: acute and general;
    considerable benefits from having doctor on board;
    danger of damage to eyesight from radiation, direct sunlight;

- the atmosphere and the Earth's environment:
  - three magnetic anomalies;
  - mapping of lower layer of magnetic field;
  - atmospheric density in low Earth orbit;
  - luminous clouds over Arabia;
  - growing levels of pollutants in the atmosphere;
  - first global pollution map, with constituent elements;
  - discovery of lightning sprites;
  - decrease of ionospheric activity by night;
  - need to pay serious attention to débris levels in orbit, especially paint flakes;
- biology and botany:
  - difficult to grow plants: many grow fast, unusual shapes, but then fail;
  - gravity and light are key elements in plant, animal growth;
  - birds can hatch in orbit, but find it difficult to adapt to zero gravity (0 G);
  - small doses of radiation stimulate plant growth but large rates damage;
  - after 20 years' experiments, plants can be grown full cycle in orbit (wheat, peas);
  - microbacteria will grow on board: use windows, controls, eat yogurt;
  - it is possible to regenerate water in orbit and approach sustainability;
  - it is possible to grow and eat food in orbit;
- materials science:
  - first materials processing experiments in orbit;
  - welding can pave the way for the construction of large orbital structures;
  - experimental production runs of vaccines, antibiotics, interferon;
  - holograms tested, transmitted to the ground;
- astronomy and astrophysics:
  - speed of the solar wind;
  - solar behavior – solar loops, sunspots;
  - rapid changes in X-ray stars;
  - new type of X-ray stars: barsters;
  - temperatures of the Sun and coronosphere.

REFERENCES

[1] Burchett, Wilfred; Purdy, Anthony: *Cosmonaut Yuri Gagarin – first man in space*. Panther, London, 1961; Boris Chertok: *Rockets and People, Vol. I–III*. Series editor Asif Siddiqi. NASA, Washington, DC, 2006–2009; Antipov, V.V., *et al.*: Some results of the medical and biological investigations in the second and third satellites, unreferenced paper.

[2] Central Intelligence Agency (CIA): Scientific intelligence report, long range Soviet scientific capabilities 1962–70, Monograph I, *Geophysical Sciences*. Washington, DC, 1961; Monograph X, *Space Biology and Astrobiology*. Author, Washington, DC, 1959; Central Intelligence Agency (CIA): Scientific intelligence report, *Soviet Bioastronautics Research Program*. Washington, DC,

1962. For the story of Nikolai Vavilov, see Pringle, Peter: *The Murder of Nikolai Vavilov – the story of Stalin's persecution of one the 20th century's greatest scientists.* JR Books, London, 2009. His brother was Sergei Vavilov, later President of the Academy of Sciences.

[3] Yefremov, A.I.: Investigation of solar X-rays and lyman alpha radiation on 19–20th August 1960, in Priester, Wolfgang, ed.: *Space Research*, Vol. III. COSPAR, Paris, 1962; Krassovsky, Valerian: Certain problems of upper atmosphere physics and near Earth space, in Skuridin, G.A., *et al.*, eds: *Space Physics*, papers from conference held in Moscow, 10–16 June 1965; Novosti, Moscow: *Ten Years of Space Exploration.* Author, Moscow, 1967; Vernov, Sergei, *et al.*: Structure of the Earth's radiation belts at 320km, in Muller, P., ed.: *Space Research*, Vol. IV. COSPAR, Paris, 1963; Panasyuk, Mikhail: Radiation reflections, in Zakutnyaya, Olga, ed., *Space, the First Step.* IKI, Moscow, 2007; Vernov, Sergei, *et al.*: *Detection of the Inner Radiation Belt at 320km Altitude in the Region of the South Atlantic Magnetic Anomaly.* NASA, TTF 8,127.

[4] *Soviet Space Achievements.* Novosti, Moscow, 1965; Vernov, Sergei, *et al.*: Discovery and investigation of the Brazil anomaly by spaceships and the Cosmos series of satellites, in King-Hele, D.G.; Muller, P.; Righini, G., eds: *Space Research*, Vol. V. COSPAR, Paris, 1965.

[5] *Soviet Space Achievements.* Novosti, Moscow, 1965; Gordon, L.K., *et al.*: Effects of spaceflight conditions in Vostok 3 on the seeds of higher plants. *Cosmic Research*, Vol. 1, 1963; Lebedev, V.N., *et al.*: Doses of cosmic rays on biological packs on Vostok 3 and 4. *Cosmic Research*, Vol. 1, 1963; Delone, N.N., *et al.*: Effects of spaceflight on transcendentia palusdrosa microspores on Vostok 3 and 4. *Cosmic Research*, Vol. 1, 1963; Glembtsky, Yuri, *et al.*: Effects of spaceflight factors on the frequency of sex-linked recessive lethal mutations. *Cosmic Research*, Vol. 1, 1963.

[6] Vaisberg, Oleg: Sputnik 1 and something else, in Zakutnyaya, Olga, ed., *Space, the First Step.* IKI, Moscow, 2007; Rosenberg, G.V., Tereshkova, Valentina: Stratosphere aerosol-based measurements on spacecraft, in Skuridin, G.A., *et al.*, eds: *Space Physics*, papers from conference held in Moscow, 10–16 June 1965; Gordon, L.K., *et al.*: Effects of spaceflight physiological processes associated with the germination of seeds of some higher plants. *Cosmic Research*, Vol. 3, 1965.

[7] Siddiqi, Asif: *The Challenge to Apollo – the Soviet Union and the space race, 1945–74.* NASA, Washington, DC, 2000.

[8] Novosti, Moscow: *Ten Years of Space Exploration.* Author, Moscow, 1967; Delone, N.L., *et al.*: Effects of factors of cosmic flight in Voskhod on microspore transcendentia paludosa. *Cosmic Research*, Vol. 4, 1966; Zhukov-Zerzhnikov, N.N., *et al.*: Biological investigations on Voskhod and Voskhod 2 spaceships. *Cosmic Research*, Vol. 4, 1966; Kondratyev, Kirill: *Observation of Mesospheric Clouds from Space.* NASA, TTF 13,893; Gazenko, Oleg; Gyurdzhian, A.A.: *Physiological Effects of Gravitation.* NASA, TTF 376, 1965, and *Results of the Medical Investigations on the Spacecraft Voskhod and Voskhod 2.* NASA, TTF 9,539, 1965.

[9] Illyin, E.A.: Biological satellites and their contribution to space biology and medicine, in Napolitano, L.G., ed.: *Proceedings of the XXVI International Astronautical Conference*. Lisbon, 1975; Gregoriev, Yuri; Kovalev, Yevgeni: *Physical and Radiological Investigations of Artificial Earth Satellites*. NASA, TTF, 724; Atakov, Yuri, *et al.*: Results of experimental studies of dosimetry and shielding in the Cosmos 110 satellite. *Cosmic Research*, Vol. 6, 1968; Delone, N.L., *et al.*: Stimulation of growth of onion after spaceflight of bulbs on spacecraft Cosmos 110. *Cosmic Research*, Vol. 5, 1967; Atakov, Yuri, *et al.*: Results of experimental studies in dosimetry and shielding in Cosmos 110. *Cosmic Research*, Vol. 7, 1969.

[10] Boltenkov, B.S.: Measurements of isotopic composition of particle fluxes on Soyuz, Zond 8 and Luna 16, in Bowhill, S.A.; Jaffe, L.D.; Rycroft, M.J., eds: *Space Research*. COSPAR, Paris, 1971; Delone, N.L.: Effects of the conditions of spaceflight in Zond 5 on seeds, onions and transcendentia. *Cosmic Research*, Vol. 9 (1), 1970; Gazenko, Oleg, *et al.*: *Results of Biological Studies Performed on the Zond 5,6,7 Stations*. NASA, TTF 13,372.

[11] Rosenberg, G.V.; Sandomirsky, A.B.: Altitude variation of the scattering coefficient from Soyuz 3 measurements of aerosol stratification, in Kondratyev, Kirill; Mycroft, M.J.; Sagan, Carl: *Space Research*. COSPAR, Paris, 1970.

[12] Kondratyev, Kirill, *et al.*: Some results of geophysical observations from Soyuz manned flights, in Bowhill, S.A.; Jaffe, L.D.; Rycroft, M.J. eds: *Space Research*. COSPAR, Paris, 1971; Garna, K.P.; Romanova, N.I.: Influence of space flight factors on Soyuz 5 on barley seeds. *Cosmic Research*, Vol. 9 (4–6), 1971; Kondratyev, Kirill: *The Use of Airglow Layer Effect for Autonomous Navigation and Orientation of Manned Spacecraft*. NASA, TTF 14,110.

[13] Kondratyev, Kirill, *et al.*: Spectrometry of the Earth from manned spacecraft Soyuz 7 and 9, in *Astronautical Research*, papers from the XXI International Astronautical Federation conference, Konstanz, Germany, 1970.

[14] Paton, Boris, *et al.*: *Welding in Space and Related Technologies*. Cambridge International Science Publishing.

[15] Hendrickx, Bart: The Kamanin diaries, 1969–1971. *Journal of the British Interplanetary Society*, Vol. 55, No. 9–10, September–October 2002; Shayler, Dave: Flight of the falcons – the 18 day space marathon of Soyuz 9. *Journal of the British Interplanetary Society*, Vol. 54, No. 12, January–February 2001; Shtern, M.I.: *Investigations of the Upper Atmosphere and Outer Space Conducted in 1970 in the USSR*. Report to 14th meeting of COSPAR. NASA, TTF 666, 1971.

[16] Gitelson, Iosif, *et al.*: Long-term experiments on man's stay in biological life support systems, in McElroy, R.D.; Tibbitts, T.W.; Volk, T., eds: *Life Sciences and Space Research, Advances in Space Research*, XXIII, Vol. 9, No. 8, 1989. Dadykhin, V.P.: *Growing Plants in Space*. NASA, TTF 704, 1968, originally published as *Kosmichekoye rasteniyeevodsto*. Znaniye, Moscow, 1968. See also Gitelson, Iosif: Problems of creating closed biological life support systems, in Napolitano, L.G., ed.: *Proceedings of the XXVII International Astronautical Congress*, Anaheim, 1976, and DLR: *Study on the Survivability and Adaptation*

of Humans to Long-Duration Interplanetary and Planetary Environments. DLR and others, Germany, 2001.

[17] Ziman, Yan: How the optical physical department overcame the consequences of perestroika, in Zakutnyaya, Olga, ed., *Space, the First Step.* IKI, Moscow, 2007; Casado, Javier: Agriculture in space. *Spaceflight,* Vol. 48, No. 5, May 2006; Gazenko, Oleg; Burstedka, K.H.: *Chelovek v Kosmoce.* Nauka, Moscow, 1974; Gringauz, Konstantin; Patsayev, Viktor, *et al.: Equipment for Investigation of Secondary Electron High Frequency Discharge at Salyut Orbital Station – conduct of experiment and its results.* NASA, TTF 15,772.

[18] Busnikov, A.A., *et al.:* Optical characteristics of the mesopause and lower thermosphere on the nightside of Earth, in Rycroft, M.J., ed.: *Space Research.* COSPAR, Paris, 1974; Kravtsova, V.I.: Multiband spectral imagery – a contribution to study ocean dynamics. Paper presented in Anaheim, 1976; Gazenko, Oleg, *et al.: Biological Research in Space – some conclusions and prospects.* NASA, TTF 125,961, 1974.

[19] Burgess, Colin; Hall, Rex: *The First Cosmonaut Team – their lives, legacy and historical impact.* Praxis, Chichester, with Springer, Berlin, Heidelberg and New York, 2009.

[20] Siddiqi, Asif: The Almaz space station complex – a history, 1964–1992, Part I. *Journal of the British Interplanetary Society,* Vol. 54, No. 11–12, November–December 2001; Clark, Phillip S.: Classes of Soviet/Russian reconnaissance satellites. Paper presented to British Interplanetary Society, 2 June 2001.

[21] Bruns, A.V.: Ultraviolet spectra of solar floccolae and prominences from Salyut 4, in Mycroft, M.J.: *Space Research,* Vol. XVI. COSPAR, Paris, 1975; Novosti: *Soviet Space Studies.* Novosti, Moscow, 1983; Beigman, I.L., *et al.:* Upper limits on soft X-ray fluxes from some celestial objects observed by Salyut 4, in Baity, W.A.; Peterson, L.E., eds: *X-Ray Astronomy, Advances in Space Exploration,* Vol. 3. Pergamon, 1979; Loewsky, A.S., *et al.:* Mass spectrometer measurements of the F2 region neutral ion composition from Salyut 4, in Rycroft, M.J., ed.: *Space Research.* COSPAR, Paris, 1976; Markov, M.N., *et al.:* Infrared spectrum of nitric oxide obtained in the upper atmosphere at middle latitude by the orbital space station Salyut 4. *Cosmic Research,* Vol. 15, 1977; Babichenko, S.I.: Cosmic X-ray radiation from on board the Salyut 4 orbital station. *Cosmic Research,* Vol. 15, 1977.

[22] Avduevsky, V.S., *et al.:* Technological experiments on board Salyut 5. Paper presented in Prague, 1977; Markov, M.N., *et al.:* Water vapour and ozone in the mesosphere observed, in Rycroft, M.J., ed.: *Space Research.* COSPAR, Paris, 1978; Siddiqi, Asif: The Almaz space station complex – a history, 1964–1992, Part II. *Journal of the British Interplanetary Society,* Vol. 55, No. 1–2, January–February 2002; Marov, Mikhail: Infrared reflection spectra of Moon and lunar soil, in Rycroft, M.J.: *Space Research.* COSPAR, Paris, 1979; Regal, Lyia: *Research Experiences on Material Science in Space.* International Centre for Gravity Materials Science and Applications, Clarkson University, Potsdam (NY), 1993; Markov, M.N., *et al.:* Spectrum and variability of solar radiation in the range 3–13μm. *Cosmic Research,* Vol. 18, 1980; Avduyevsky,

V.S.: Technical experiment: Diffusion on space station Salyut 5. *Cosmic Research*, Vol. 18, 1980.

[23] Powell, Joel W.: Soviet space science. *Journal of the British Interplanetary Society*, Vol. 36, No. 10, October 1983; Nevzgodina, L.V., *et al.*: Effects of prolonged exposure to spaceflight factors for 175 days on lettuce seeds, Atakov, Yuri: Results of cosmic ray dose field measurements on Salyut 6, Pavel, H., *et al.*: Spaceflight effects on paramecium tetraurelia flown on Salyut 6 in Cytos I and Cytos M, and Kordyum, E.L., *et al.*: Optical and electron microscope studies of Funaria hygrometrica protonema after cultivation for 96 days, all in Holmqvist, W.R., *et al.*, eds: *Advances in Space Research*, Vol. 1, No. 14, 1981; Paton, Boris, *et al.*: *Welding in Space and Related Technologies*. Cambridge International Science Publishing; Bakun, V.N.: Some results of measurements of atmospheric submillimetre radiation from Salyut 6, in Rycroft, M.J., ed.: *Space Research*. COSPAR, Paris, 1979. Medical results are reported in Gazenko, Oleg, *et al.*: Major medical results of the Salyut 6 Soyuz 185 day mission, in Napolitano, L.G., ed.: *Proceedings of the XXXII International Astronautical Conference*, Rome, 1981; Barta, C., *et al.*: Experiment Morava on Salyut 6, in Napolitano, L.G., ed.: *Proceedings of the XXIX International Astronautical Congress*, Dubrovnik, 1978.

[24] Tuchkevich, V.M.; Semenov, Yuri P.; Gurevich, S.P.: Holography is conquering space. *Zemla i Vselenaya*, 3/84.

[25] Pravda: Soviet cosmonauts burnt their eyes in space for USSR's glory. *Pravda*, 17 December, 2008; Paton, Boris, *et al.*: *Welding in Space and Related Technologies*. Cambridge International Science Publishing; Miller, A.T.; Nevzgodina, L.V.: Biological effects of galactic radiation HZE particles in experiments on orbital station Salyut 7, in Oser, H., *et al.*, eds: Life Sciences and Space Research, Vol. XXIII, 5, *Advances in Space Research*, Vol. 9, No. 11, 1989; Atkov, Oleg: Some medical aspects of eight months spaceflight, in Young, R.S., *et al.*, eds: *Life Sciences and Space Research – gravitational biology*, Advances in Space Research, Vol. 12, No. 1, 1992; Grechko, Georgi: Ozone and aerosol fine structure space experiment for the observation of the fine structure of ozone and aerosol distribution in the atmosphere, in Forbes, J.M., *et al.*, eds: *The Earth's Middle and Upper Atmosphere, Advances in Space Research*, Vol. 12, No. 10, October 1992; Siddiqi, Asif: The Almaz space station complex – a history, 1964–1992, Part II. *Journal of the British Interplanetary Society*, Vol. 55, No. 1–2, January–February 2002; *Echograph* is reviewed in Kaplan, D., *et al.*: The echography Doppler experiment on Salyut 7, in Napolitano, L.G., ed.: *Proceedings of the XXXIII International Astronautical Conference*, Paris, 1982; Startsev, Oleg; Nikishin, Eugene: *Structure and Properties of Polymeric Composite Materials during 1,501 Days Outer Space Exposure at Salyut 7 Orbital Station*. Salyut Design Office, Moscow, and Altai State University, Altai, undated. For the issue of nerve damage, see Gazenko, Oleg, *et al.*: *Biological Research in Space – some conclusions and prospects*. NASA, TTF 125,961, 1974.

[26] Salmon, Andy: Science on board the Mir space station, 1986–1994. *Journal of the British Interplanetary Society*, Vol. 50, No. 8, August 1997; Salmon, Andy:

Mir – workshop and laboratory, in Hall, Rex, ed.: *The History of Mir, 1986–2000*. British Interplanetary Society, London, 2000; Salmon, Andy: Research in orbit, in Hall, Rex, ed.: *The International Space Station – from imagination to reality*. British Interplanetary Society, London, 2002; Grigoriev, Anatoli; Potapov, A.N.: Advances and perspectives of space biology and medicine, in Zakutnyaya, Olga; Odintsova, D., eds: *Fifty Years of Space Research*. Institute for Space Research, Moscow, 2009; Poliakov, Valeri: Long term spaceflight – personal impressions, in Young, R.S., *et al.*: *Life Sciences and Space Research: Gravitational Biology, Advances in Space Research*, Vol. 12, No. 1, 1992; Grigoriev, Anatoli: Summing up cosmonaut participation in long-term space-flight, in Young, R.S., *et al.*, eds: *Life Sciences and Space Research: Gravitational Biology, Advances in Space Research*, Vol. 12, No. 1, 1992; Shrine, N.R.G.: *Euromir 95 – first results from dustwatch P detectors of the European space exposure facility*, in McDonald, J.A.M., *et al.*, eds: *Hypervelocity Impacts in Space, Advances in Space Research*, Vol. 20, No. 8, 1997; Hansson. Anders: Wheat on Mir. Paper presented to British Interplanetary Society, 6 June 1998; Salisbury, Frank, *et al.*: Plant growth during greenhouse II experiment on Mir, in Hornech, G., *et al.*, eds: *Science and Life Sciences, Advances in Space Research*, Vol. 31, No. 1, 2003; Earth's first babies born on Mir. *Soviet Weekly*, 16 August 1990; Grigoriev, Anatoli, *et al.*: Main results of medical support to the crews of the International Space Station. Paper presented to International Astronautical Federation, Valencia, Spain, 2006; Henry, M.K., *et al.*: Launch conditions might affect the formation of blood vessels in the quail chorioallotonic membrane. *Folia Veterinaria*, No. 42, 1998; for the *Stola 88* experiment, see Sykora, Jaroslav, *et al.*: Results of the laboratory simulation of long-lasting spaceflights in the Czech Republic. Paper presented to 61st International Astronautical Congress, Prague, September 2010.

[27] Kidger, Neville: The expedition crews – life on orbit, in Hall, Rex, ed.: *The International Space Station – from imagination to reality*, Vol. 2. British Interplanetary Society, London, 2005; Grigoriev, Anatoli, *et al.*: Main results of medical support to the crews of the International Space Station. Paper presented to International Astronautical Federation, Valencia, Spain, 2006; Fortov, Vladimir: Dust plasma crystals and liquids on the Earth and in space, Zakutnyaya, Olga; Odintsova, D., eds: *Fifty Years of Space Research*. Institute for Space Research, Moscow, 2009; Peslyak, Alexander: *Mosquito Survives in Outer Space*. Published on space.com, 20 February 2009; 25th series of German–Russian plasma physics experiments. *Space Daily*, 4 February 2010; *Research in Space*, DLR (German space agency), 2010; Funtova, I.I., *et al.*: Autonomic function testing on board ISS for crew health monitoring with Puls and pneumocard – results, limitations and next steps; Baranov, V.M., *et al.*: Changes in the sensitivity of the central respiration mechanism during space flight; Baevsky, Roman, *et al.*: How do cosmonauts sleep in microgravity?; The process of adaptation of the cardiovascular system to the conditions of weightlessness; Fortov, Vladimir; Morfill, Gregor: Dusty plasma compressibility from analysis of externally driven dust-acoustic shock wave propagation;

Sorokin, Igor; Markov, Alexander: Scientific potential of Russian mini research module Rassvets; Elkin, Konstantin, *et al.*: Effectiveness of long term program of scientific and applied experiments on the ISS Russian segment, all papers presented to 61st International Astronautical Congress, Prague, September 2010.

7

Later Soviet space science: the observatories

Chapter 2 showed how key space science programs evolved in the late 1950s and early 1960s: object D, MS, Elektron, the DS and DS-U satellites, Proton (with descendants *Efir* and *Energiya*), *Ionosfernaya*, and, later, Prognoz and Intercosmos. We know that a number of key projects approved in 1964 never flew and it is apparent that space science became quite dispersed across the Cosmos program, flying as instruments on military missions or as *nauka* modules. During the later Soviet period, from the late 1970s, throughout the 1980s, and up to the collapse of the Soviet Union, a more visible scientific program reasserted itself and continued into the Russian period of 1992 onwards. The program passed from pioneering projects to programs focused on astrophysics and space physics. Four dedicated astronomical observatories were flown (Astron, Kvant, Granat, and Gamma). These were the high-visibility, prestige projects of later Soviet space science. The main volume of scientific work in the later period was focused on the Intercosmos program, two spin-off projects, *Ionozond* and *Oval*, and climaxed in the APEX and *Aktivny* projects. The retrenchment of the 1990s led to severe cuts in the science program, but some projects survived, notably Prognoz M (Interball) and the Koronas solar observatories. Biological science became a program in its own right (Bion), as did materials science (Foton), both becoming international cooperative ventures.

ASTRONOMICAL SCIENCE: ASTRON, KVANT, GRANAT, AND GAMMA

The 1980s saw the Soviet Union bring to fruition four space observatory projects: Astron, Granat, Gamma, and Kvant. The precise point at which these observatories originated is not known, but may have been prompted by American plans in the 1970s to develop a series of "great observatories", of which the flagship was the Hubble telescope, work on which began in the 1970s. Although their origins lay in the 1970s, the existence of a program of Soviet observatories was not unveiled until the 26th general assembly of the international committee on space research, COSPAR, in Toulouse, France, in 1986. The focus of these observatories, announced the Soviet delegation, would be X-rays and gamma rays. Indeed, Granat was described at the time as "Russia's answer to Hubble", with Gamma as Russia's answer to the American Gamma Ray Observatory, called Compton, and Kvant as equivalent to AXAF, called Chandra. The implementation of these observatory projects was intended to bring Soviet deep-space observing science to the level of the United States and European countries. Following this, the Spektr project, through which the USSR would study the entire range of electromagnetic radiation, would move decisively ahead [1]. As far back as 1959, the CIA had noticed the all-round quality of Soviet astronomical science, but its weakness was in stellar physics, due partly to a lack of southern-hemisphere observatories, partly to the legacy of political repression. Physics and astrophysics had developed in Germany, Britain, and Denmark, spurred on by annual international conferences of the great experts. Leningrad physicists had built connections with the international community from 1928; they had only a few years to learn before political repression began. Lev Landau brought astrophysics to the USSR, but he was broken by a brutal KGB prison in 1938–1939, while George Gamov fled to the West. At the annual meeting of the Academy of Sciences in 1960, Academician V.A. Ambartsumyan, possibly the country's leading astronomer, complained that "we are still not working at the level necessary" and that "we lag by quite a bit in the development of one of the most important areas of astronomy – the study of the galaxy and the extra-galactic world" [2].

It appears that the idea of a great observatories program was approved in around 1970. A fifth observatory was approved in February 1972: the KRAS 3, a 3.1-m radio telescope to orbit out to 12,000 km to search for quasars, masers, and protoplanetary disks. Chief designer Vasili Mishin took personal responsibility for the project, but when he was replaced as chief designer in May 1974, the project was re-invented as a joint Soviet–American sequel to the Apollo–Soyuz Test Project, but it, too, ran into the sand and eventually was lost [3].

GREAT OBSERVATORIES: ASTRON

By the time of the 1986 announcement, one of the observatories was already in orbit. This was Astron, originally intended to be manned and using an adaptation of the OST telescope taken from Salyut. Experience on the space stations suggested that manned platforms were too much affected by on-board movement and an unmanned platform would be more stable. Construction of Astron took three years. There had only been two ultraviolet observatories before – the American *Copernicus* and International Ultraviolet Explorer – but Astron was much the largest.

Astron went up on 23rd March 1983, using a Proton rocket from Baikonour. Grandiosely announced as "the world's first orbiting astronomical space station", it used the same design as the Venera 15 and 16 radar observatories then on their way to the planet Venus but with bigger solar panels due to the energy demands of the instruments and weighed 3.5 tonnes. Its high orbit of apogee 200,000 km brought it outside the Earth's radiation belt and its shadow for 90% of the time. Its high-apogee orbit was above the USSR for 20 out of 24 hr.

Astron

The cover was dropped off five days after arrival in orbit and, after calibration, on 29th March, Astron began its program of observations. Astron was designed for one year, but the mission was not terminated until the last communication session on 23rd March 1991. About 200 communication sessions were held every year. Astron was orientated by means of a solar sensor, always referenced to the Sun and a navigation star, normally Canopus, but Vega, Sirius, Betelgeuse, and Arcturus were used, as well as planets Jupiter, Saturn, and Mars. Astron's observing sessions were done live, 3–4 hr at a time, much like an Earth telescope, normally observing one source per session. Astron could swivel across the whole celestial sky in 12 min in the course of which as many as 70,000 measurements were taken. There was no on-orbit data storage, but magnetic tapes of all the observations were kept. Astron was given a long list of targets for study: Taurus, Leo, Hercules x-1, the Crab, Orion, distant galaxies, pulsars, and quasars. The telescope was designed to measure the temperature and chemical composition of stars and obtain their spectrograms. In its first year, Astron observed 70 stars, 22 quasars and galaxies, and 22 galactic background fields. The observation program focused on stars with high metallic elements, because their chemistry may hold clues to the origins of the universe.

Astron carried as a main instrument a 400-kg, 80-cm main mirror ultraviolet telescope, *Spika*, 3.5 m long, built by the Crimean Astrophysical Laboratory and equipped with a French spectrometer called UFT, intended to observe stars in the 1,200–3,000-Å range with a pointing accuracy of 2 sec and a resolution of 2.7 m/sec.

Principal investigator was Professor Andrei Severny of the Crimean Astrophysical Observatory. The second instrument was Vladimir Kurt's X-ray spectrometer, SKR-02M, made with the help of the Sternberg Institute, which aimed to study the transition of red giants to neutron stars. Here, the principal investigator was Alexander Boyarchuk of the Crimean Astrophysical Laboratory. It had an area of 1,780 cm^2 with two detectors taking data every 2.28 milliseconds, accompanied by a small field recognition camera. The telescope had three slits: 40 μm for bright stars, 0.4 μm for faint stars and extragalactic objects, and 3 mm for nebulae and the galactic background.

The testing out of the instruments took place a week after entry into orbit, with the telescope being pointed at an empty part of the sky without X-ray sources. It was then calibrated against the X-ray source in the Crab nebula, a stable and well known object. The telescope was then focused on Cygnus x-1, Cygnus x-3, Ophiuchus x-2, and then extragalactic sources 3C273 quasar, NGC 4486 elliptical galaxy in Virgo, NGC 7552 galaxy in Crane, the supernova in M83 spiral galaxy in Serpens, and object SS433. On 13th April, Astron detected a splash from a fast burster, the level of pulses in the 2–25-keV range shooting up three-fold for 30 sec and then tailing off. Dramatically, on 30th June 1983, the Hercules x-1 source suddenly stopped transmitting. This was a visible variable star combined with an X-ray neutron star, which had been very stable for ten years. Subsequent analysis suggested that the neutron star's X-ray radiation had converted into a gas disk around the star, called the accretion disk. On 28th April 1983, Astron found its first X-ray burst on MXB1728-34 +. On 9th July 1983, Astron swung around the center of the galaxy to make a comparative study of X-rays from seven sources: Scorpius x-1 (the most intense), Norma x-1 (the highest bursts), 4U 1728-24, 4U 1730-22, 4U 1728-34, Eagle x-1, and Scorpius x-6. On 16th August 1983, fast burster MXB 1728-34 was eclipsed by the Moon, which enabled the most precise measurement to be made of the source, finding that its radiation, while appearing to be in fast pulses, was actually constant. One of the main objects of attention, observed over six months, was fast burster MXB 1733-35, near our galaxy's center. This was a close binary system of a neutron star and red dwarf and Astron followed the way in which helium and hydrogen flowed from the dwarf to the neutron star as nuclear fuel. When it over-accumulated, it flashed in the form of nuclear explosions.

Of particular interest were symbiotic stars, which are far distant, gigantic binary systems, one being a cool red giant (3,000 K), 50 solar radii across, with high levels of titanium dioxide and strong periodic flows of energy to the secondary hot component, which was measured as being several hundredths smaller than the solar radius. Whenever radiation transferred from one to the other, energy levels would flare up four times.

Astron made ultraviolet observations of several galaxies, such as NGC 4689 and 3664, the purpose being to map out their different components of hot stars, cold stars, and dust clouds, finding them much more varied in their composition than visual observations ever suggested (and which Hubble confirmed many years later).

One of the main interests of the mission was Gamma Ray Bursts (GRBs), known since the 1970s, with about 100 a year being the normal frequency, each burst

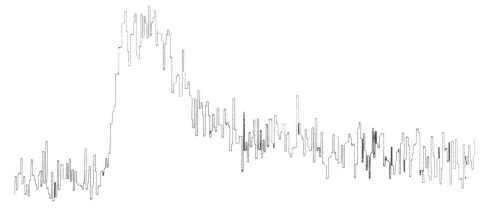

Figure 7.1. Astron detection of gamma burst.

normally lasting only a few tens of seconds but with phenomenal energy levels (one is illustrated here in Figure 7.1). Astron's computer was given data on about 600 X-ray and ultraviolet sources to survey, generally for a period of up to 3 hr each, but with the proviso that if a gamma burst were detected elsewhere, it should at once swivel to the new target. This happened when on 23rd–24th February 1987, the biggest supernova for 1,000 years burst (1987a) and Astron focused on it right away – possibly the first orbital telescope to do so. Astron did so in two ranges, ultra-violet and X-ray, simultaneously, so as measure the product of nuclear synthesis.

This supernova was to become the center of attention for astronomers the world over for several years. Supernovae held an important place in the heart of Soviet astrophysicists. Iosif Shklovsky had written a world-leading explanation of the phenomenon of supernovae in 1953. With Valerian Krassovsky, he had later outlined how radiation from supernovae could have a profound effect on the evolution of the universe and the species living therein, both stimulating life and causing genetic mutation through their rays, and they speculated that carboniferous vegetation on Earth was the outcome of supernova radiation. Sadly, Shklovsky was no longer alive to study 1987a, for he had died shortly beforehand in a botched minor operation to relieve an embolism in his leg.

In April 1986, Astron swiveled its telescope to observe the visiting comet Halley, to measure the level of ammonia venting from the comet (for the record, between 0.44 and 0.94% of the out-gassing). Its spectral data led to the conclusion that the comet lost 400 million tonnes during an approach to the Sun. Astron was also turned back to Earth to study Earth's ozone layer and the contribution of rocket launches to ozone formation.

Compared some successor missions in the great observatories series, the scientific papers from the Astron mission were limited. The reason is not known, for there is no indication that there were problems with the spacecraft or its instruments. A short documentary was released by TsentreNauchFilm, *Astron*, focusing mainly on its construction. Astron's successor was Granat, originally to have been called Astron 2. The highlights of Astron's six years were as follows:

- observations in late 1983 of a symbiotic binary star in Andromeda in which the energy from one star was sucked across to the other, building up to an explosion;
- detection in 1984 of an X-ray pulsar in Hercules that had been silent for some time, a pulsar in Taurus, and a rapid burster in Sagittarius;
- characterization of a number of gamma burst sources, ranging from rapid bursters to flickering ones (e.g. Hercules x-1); one neutron star was found to be giving off bursts of radiation 10 sec apart but only twice a year in such a regular way as to originally give rise to the suspicion that they came from an extraterrestrial civilization;
- recording of the disappearance of an X-ray source in Hercules and the cessation of signals of a neutron star;
- observation of the discharge of large volumes of matter from stars, especially huge volumes of uranium in 73 Draconis and tungsten in κ Cancer;
- determination that the hot sources of dwarf novas are 30 times smaller than cold sources and are as small as a thousandth of an Earth's radius;
- all-sky X-ray survey, leading to a catalog of X-ray sources [4].

GREAT OBSERVATORIES: KVANT

Next launched was Kvant, four years later. This was a unique experiment – an observatory linked to an orbital station, Mir (Mir's other science operations were reviewed in Chapter 6). Kvant was originally designed in 1979 as an experimental module for attachment to the Salyut 7 space station, the decision being taken in 1984 to fly it to the following space station, which later became known as Mir. Kvant was a specially built 12-tonne module, which used a 10-tonne tug to bring it up to the Mir space station, where it eventually arrived on 9th April 1987. Once delivered, the tug undocked three days later, eventually burning up over the Pacific in 1988. Kvant itself had an important engineering function, for its six gyrodines, spun at 10,000 rpm, were used to orientate not only its own telescopes with extreme precision, but the whole Mir complex. It also served as a docking station for Progress spacecraft.

Kvant was 5.3 m long, 4.35 m wide, and with a pressurized working area of 40 m^3. It had five instruments, the first four in a suite sometimes called the Röntgen observatory, with a high level of international collaboration:

- coded masked imaging spectrometer (COMIS or TTM from Russian *Teleskop s Tenevoi Maskoi*) (University of Birmingham, Britain, and Space Research Laboratory in Utrecht, Netherlands), 2.5 m long, 40 cm in diameter, 540-cm^2 detector for 2–30-keV X-rays with a precision of 1 arc second;
- X-ray scintillation spectrometer (High-Energy X-ray Experiment (HEXE)), developed by the Max Planck Institute of Extraterrestrial Physics, Munich, with the University of Tübingen, 15–20 keV (Germany), with four identical detectors with 200-cm^2 area each;

- gas scintillation proportional counter Sirene 2 (European Space Agency, in Russian GSPS from *Gasovyi Stsintillyacionnyi Proportsionalnyi Spectrometer*) for the 2–100-keV range with high-energy resolution; geometrical area of detector was 300 cm^2;
- Pulsar X-1 complex, which included Pulsar X-1-B X-ray telescope (another name *Spektr*), 300 kg, with four detectors in a broad band 20–800 keV with geometrical area of each detector 314 cm^2 and a 50-kg telescope, Pulsar X-1-V (*Ira*) for gamma-ray bursts (USSR);
- 40-cm ultraviolet telescope *Glazar* (Byurakan, Armenia/Switzerland), capable of 8-min exposures down to 17th magnitude, intended for search for galaxies and quasars in the UV range, as well as for studies of UV fluxes from known sources;
- additional experiments in magnetospheric physics and biology.

Thus, Kvant's full range was from 2 to 800 keV, the broadest ever launched up to that point, and made it the most powerful astrophysical platform in orbit for a decade. The instruments were first turned on 9th June 1987. Operation of Kvant was subordinate to the requirements of Mir as a manned orbiting space station: Kvant was typically used when the cosmonauts were asleep. There was an extended period of operations of Kvant from April to September 1989 when the station was temporarily unmanned, probably the most valuable part of the mission. Kvant was de-activated in September 1989 in preparation for the arrival of new new modules, but reactivated in October 1990, the instruments showing no deterioration in the meantime. The TTM detector broke down in October 1987, but it was replaced by cosmonauts Vladimir Titov and Musa Manarov during spacewalks on 30th June and 20th October 1988 [5]. Some observation periods were quite long: for example, the German experiment took observations every 25 min, and worked until 1997.

Kvant's arrival in April 1987 was fortuitous, for it coincided with the discovery of the supernova by Astron and astronomers the world over. The experimenters persuaded the Energiya design bureau, which operated Mir, that priority should be given to observing such a rare supernova and this was agreed, though this had implications for the way Mir operated, how it was orientated, and the work time of the cosmonauts. The supernova rose in intensity ten times up to 10th August 1987, after which Kvant followed the decline in luminosity of the supernova, analyzed its chemistry, tried to find a black hole in the collapsing star, and saw the way in which a cloud of cold molecular gas expanded from the shell to collide with the interstellar wind. Kvant detected hard X-rays coming from the supernova, up to 100 keV, identifying cobalt in its composition. According to mission scientist Rashid Sunayev, the detection of hard X-rays was a major discovery and enabled scientists to remodel how a star died and was transformed into a neutron star of a black hole. No low-energy X-ray waves were picked up from the supernova, suggesting that it didn't emit any. Here, the main instruments used were HEXE and Pulsar X-1, as the energetic range of TTM was too soft.

The Röntgen telescope made an average of four observations a day over 1987–1990. Its main focus was supernova 1987a (370 sessions), followed by Cygnus x-1

(110), the Crab object (98), and Hercules x-1 (72). The Crab object had been discovered by the Japanese spacecraft Ginga and was one of the most intense X-ray sources in the sky: Kvant found that it had a tail of radiation behind it, a pulsed signal up to 500 keV, and measured its period at precisely 33.375035 sec. Kvant was able to watch Hercules x-1 begin to spin up and its observations of Cygnus x-1 suggested that it was a black hole. The *Glazar* telescope made observations of the Markarian galaxies, named after their founder, the Soviet astronomer B.E. Markarian. Part of the Röntgen program were studies of black hole candidates, which are to a large extent X-ray novae (objects that appear suddenly, reaching high brightness, and then decreasing in brightness until they disappear from the sky for 30–50 years). It was also used for studies of the galactic center region, which hosts a multitude of sources of a different nature. Besides studies of the known objects, TTM also discovered 11 new sources in our Galaxy, named KS (meaning *Kvant* source).

Although the formal end of the Kvant mission was in March 2001, when the whole Mir complex was de-orbited, the principal observations were from the early years, especially 1989, when it was unmanned and could be pointed from the ground at will. The highlights were:

- portrait of supernova 1987a and its hard X-rays;
- long-term observations of the Larger Magellanic Cloud (LMC) and its bright sources;
- profile over January–February 1989 of Cir X-1, originally thought to be a black hole, now considered to be a low-mass X-ray binary pulsar;
- portrait of X-ray objects such as Vulpecula X Nova and others in Perseus, Vela, and Centaurus;
- observations of X-ray pulsars, such as GX1 + 4, which spun down; periods of nine pulsars were measured;
- discovery of two new X-ray transients: KS 1732-273 and KS 1741-293, the latter making two bursts, on 20th and 22nd August 1989;
- cartography of the galactic plane; Kvant found 30 X-ray sources and 24 X-ray bursts as well as two new X-ray transients; one was only $1°$ from GX1-4 and was KS1731-260, which peaked in August 1987;
- detection of three hard X-ray transients with tails: GRS 1716-24, GRS 1009 45, and GROJO422 + 32;
- observations of bright transient sources, such as the spectra of GS 2000 + 25 and GS 2023 + 338;
- discovery of X-ray burster KS 1731-260 in August 1989, which emitted for 15 days; X-ray bursters were low-mass binary systems of compact weakly magnetized neutron stars whose surface is supplied with helium-rich matter of its companion star [6].

GRANAT AND THE GREAT ANNIHILATOR

Granat was, like Astron, based on an interplanetary probe, in this case the Mars 1971S orbiter, with the mid-course engine and propulsion unit removed and a set of telescopes installed in the place of the Martian experiments. In addition, some left-over instruments were scavenged from the 1984 VEGA missions to Venus. At one stage, it was to be called Astron 2, but it acquired its own identity, Granat, Russian for the mineral garnet. Granat was a substantial probe weighing 4.4 tonnes of which 2.3 tonnes were equipment. It was large: 6.5 m tall and 8.5 m across its solar panels. Granat's orbit, like Astron's, took it out to 200,000 km, where its telescope could work unaffected by Earth's radiation belts. Mission lifetime was set at eight months.

The aim of the mission, as its name suggested, was to study hard X-rays and gamma rays, neutron stars, black holes, traces of supernovae, the center of the universe, and the interstellar medium. The telescope was designed to have an accuracy of 5 arc minutes for up to 4 hr at a time and able to function autonomously from ground control. Five reference stars (e.g. Crab) were programmed into its computer so that it could orientate itself. Granat's equipment was made in the USSR, Bulgaria, France, and Denmark. The main energy range was 2 keV– 100 MeV, but with the Sigma telescope operating 40–200 keV. The main instruments are shown in Table 7.1.

Due to a technical malfunction, ART S and Tournesol instruments could not work properly and were not used for astrophysical purposes.

The Proton rocket put Granat into a four-day orbit of 1,800–202,000 km, similar to that of Prognoz, starting at 51.9°. The central computer system failed at orbital insertion, but apart from being deadweight, mission control in Yevpatoria was able to manage remarkably well without it. Granat's orbit went through extreme perturbations, up to 2° a month, which led to its orbit being modified to 74.5° by 1991, raising its perigee to 8,300 km and reducing the apogee to 195,000 km. By

Table 7.1. Granat astrophysical instruments.

Sigma telescope (France), 3.5 m, 1.2 m diameter, 1 tonne, 30 keV–2 MeV, with coded aperture

ART P (Astronomie Röntgen Telescope) coded mask X-ray telescope, 3–60-keV range, which included four identical modules built in IKI's design bureau in Frunze (now Bishkek, Kyrgyzstan)

ART S (Astronomie Röntgen Telescope) X-ray proportional counter spectrometer, 3–100 keV

Konus B gamma-ray burst detector, 20 keV–8 MeV, made at the Ioffe Institute in Leningrad, with seven detectors in the 100 keV–120-MeV range

Phoebus (Payload for High Energy BUrst Spectroscopy) gamma-ray bursts detector, 100 keV– 120 MeV (France)

WATCH all-sky monitor intended for detection of transient sources 5–150 keV (Denmark)

Tournesol (or Russian *Podsolnukh*, which is "sunflower"), 2–25 keV (USSR, Bulgaria), a set of instruments to study gamma-ray bursts' afterglow

Charged particle monitor KS 18M

Granat

September 1994, its orbit measured 59,025–144,450 km, 86.7°. Granat operated until 27th November 1999, with pointed observations for the first five years and survey mode from September 1994. Total observation time of the galactic center was 5 million sec (1,388 days).

Following three months of calibration, observations began on 11th March 1990 and data were transmitted throughout for eight years, either in real time at 3,073 bytes/sec or dumped from its 150-MB memory at 65,546 bytes/sec. Three communication sessions were held every three days. On a typical orbit, Granat offered 56-hr observing time above the top of the radiation belt (60,000 km).

Granat was badly affected by the collapse of the Soviet Union, for its mission control was the 70-m dish in Yevpatoria, Ukraine, which required new funding to continue its work. Thankfully, the French space agency stepped in to provide fresh finance. Three years later, pointing fuel was so low that on 30th September 1994, it was put into a gentle spin with the solar panels facing the Sun, slowly sweeping across the whole sky every six months. A year later, using the small amount of remaining fuel, Granat was refocused on the center of the Milky Way for three weeks to make long-duration exposures of the center of our own galaxy. This was done again in spring 1997, but despite the great haul of information coming back, at this stage, the Russian Space Agency pulled the remaining funding (1997 marked the worst period of financial shortages). France again came to the rescue and three more three-week studies of the galactic center were made, in September 1997 and March and September 1998. By this stage, the nitrogen fuel tanks were almost dry, so the spacecraft was put into a stationary position, waiting for celestial objects to glide into position. The following year, Granat had only one transmitter left and it was prone to overheat after 20 min. Whenever this happened, it had to be switched off to cool, so downloading data took a long time. Granat eventually gave out on 27th November 1999. Despite some Western reports that the mission disappointed or even failed, it had a significant scientific output and led to more than 400 scientific papers [7].

For chronology of the scientific mission, Granat began its 10-year observational career by pointing towards the galactic center 25,000 light years away. Granat studied the area carefully throughout spring and autumn 1990, when it came within view. At its center was the powerful radio source Sagitarius A, where a black hole was expected to be found – but it wasn't, although there was a radio source 300 light years away from the center. Over 1990–1991, Granat studied Seyfert galaxies, radio galaxies, quasars, and clusters, such as M87, Virgo, 3C273, Centaurus A, Seyfert NGC4151, and spectra of their systems. During 1990, Granat followed Kvant by focusing on the 1987 supernova in the Large Magellanic Cloud. Also during the early stage, the *Konus B* was switched on, giving early results: from 11th December 1989 to 20th February 1990, it detected 60 solar flares and 19 gamma-ray bursts and studied the spectra of black hole candidates. The detailed study of the galactic center led to the conclusion that the bulk of the X-ray and gamma emission from this region came from compact sources in the vicinity of the galactic center. The Granat maps of the region remained the most detailed for the whole 1990s.

The most dramatic single event of the mission took place on 13th October 1990 when for 15 hr, there was a huge eruption of 600 keV and a massive annihilation of positrons in galactic microquasar IE1740.7-2942. Later, a mysterious object, with a double-sided radio jet and a molecular cloud, was found in this direction. Now, its brightness dropped ten times. Granat observed that what appeared to be taking place was electron–positron annihilation, so it was called "the great annihilator". A similar process was later observed around X-ray Nova Muscae, which dropped more than ten times. Granat observed another electron–positron annihilation event over 20th–21st January 1992. Granat made 800-hr live observations of the great annihilator over 54 sessions and determined it to be a double-sided radio jet structure.

Granat then focused on its main mission: to study black hole candidates. First, old black hole candidates were studied, such as Cygnus x-1 black hole candidate (ten observations from March 1990 to March 1992), GX 339-4 (August–September 1991), and GRS 1124-684, a low-energy massive binary with variable radio emissions. From 1991, new candidates were found, the first of eight. X-ray Nova Persei GRO JO 422 + 32 was one of the brightest, with energies up to 1,300 keV. In the case of another, GX339-4, it was found that plasma was flowing from the black hole to its accretion disk. As a result of Granat, it was possible to remodel black

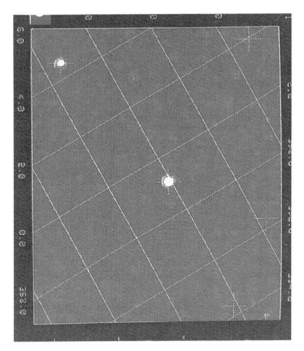

Granat image of galactic center

holes around what was called the "accretion disk" model. Some black hole candidates were found to fluctuate (GRS0834-430) while others, like Cygnus x-1 and Hercules x-3 were stable. Granat observed so-called Quasi-Periodical Oscillations (QPO) of several black hole candidates, such as X-ray novae Muscae and Persei, as well as Cygnus x-1, which were previously believed to be only neutron star markers. As for now, such QPOs are observed on many black hole candidates.

In August 1992, Grant found the first micro quasar in our galaxy: GRS 1915 + 105 (GRS for GRanat Source). In 1994, the WATCH instrument observed two X-ray flares. The information was spread, and further radio observations led to the discovery of two clouds traveling from the central object seemingly faster than the speed of light (only two other examples have since been discovered). Later, this source was identified with the variable source of radio and infrared emission.

Observations of the galactic center were the focus of the later stages of the mission. Granat made a long and deep 2,000-hr survey of the galactic center, the longest ever at that time, where there were known to be 60 X-ray sources. It detected six low-mass X-ray binaries, including an X-ray pulsar, two X-ray bursters, and four new sources. The galactic center was dominated by three transient sources and by three persistent, strongly variable sources: Terzan 2, KS 1731-260, and GX 354-0. Altogether, Granat made five studies of the galactic center, examining 3,000 objects in two years, enabling star maps to be tidied up. Its observations largely took place when it was higher than 70,000 km. The principal observations were of:

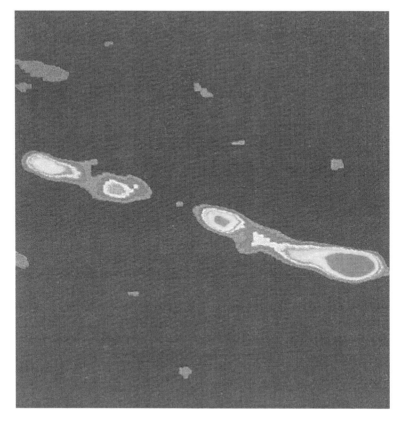

Granat image of radio source in the center of our galaxy

- GRS 1716-249, black hole candidate;
- IE 1740.7-2942, the great annihilator;
- GRS 1758-258, hitherto unknown X-ray burster, black hole candidate;
- GX 1 + 4, X-ray pulsar;
- GX 354-0, X-ray burster;
- Terzan II, X-ray burster;
- SLX 1735-269, unknown;
- A 1742-294, X-ray burster;
- GRS 1743-290, unknown;
- Terzan 1, X-ray burster.

Some of these had remarkable features: GX 1 + 4, in the center of the galaxy, emitted 2-min pulses. GRS 1758-258 dominated its region of the sky with its output of 30 keV. It looked like Cygnus x-1 and was a good candidate for the black hole club. Granat observed it for 300 hr and saw it decline two magnitudes in early 1991.

An important outcome was that Granat was able to make a comparative table of the periods of 16 X-ray pulsars, starting with Hercules x-1. It was found that the pulses of the pulsars ranged from 0.7 sec (SMC x-1) to 677 sec (GX301-2) and that

Table 7.2. Periods of X-ray pulsars studied by Granat.

Hercules	1.23 sec
SMC-1	0.709 sec
Centaurus X-3	4.8 sec
OAS 1657-415	37.848 sec
4U 0115+63	3.61 sec
GX1+4	114.6 sec
GX301-2	677 sec

they could spin up and then spin down. Granat watched Hercules x-1 decline over a 35-day cycle. It was able to measure the periods of X-ray pulsars as shown in Table 7.2.

Bursts, flares, and other radio sources were an important focus of Granat. Phoebus recorded 208 gamma-ray bursts and 47 were located to an accuracy of 0.5°. At one stage, Konus detected 19 bursts and 60 solar flares in 27 days and there were 25 gamma burst events in the first year, from 18th December 1989 to 5th June 1990. Durations ranged from 0.3 to 600 sec, energies from 2 to 15 MeV. From this arose the suggestion that there was a new class of "short, hard bursters". Some of them were of very short duration, only 2–10 ms. GRB200390, for example, had a weak pulse, then a strong burst for 35 milliseconds, then a main, powerful burst 400 milliseconds later. Some were characterized: for example, a burst in Nova Muscae had a high content of lithium. A range of other sources were studied and these are just a few:

- galactic binary source GRB 260190, which had a 30-sec burst on 17th January and a 4-min burst on 18th January 1990;
- GX 339.4 made a three-month outburst in 1991 up to 400 keV and left scientists concluding that it was a black hole or neutron star;
- SLX 1732-304 in the globular cluster Terzan 1, one of 150 galactic globular clusters that Granat measured and timed an X-ray burst on 8th September 1990 with energies of 3–30 keV;
- soft gamma-ray transients, GX354-0 and Nova Persei; GX354-0 had never been seen before, was a persistent source of X-rays below 20 keV, while Nova Persei, first discovered by the American Compton space observatory in 1992, was the brightest soft X-ray source in the sky;
- transient sources: burster and neutron star KS 1731-260, discovered originally by Kvant and Tra x-1, originally found by Britain's Ariel 5 in 1974.

Overall, in eight years, Granat discovered about 20 X-ray sources previously unknown, which were black hole and neutron star candidates. Although the purpose of Granat was to look deep into the universe, it also observed the closest stellar object at hand: our own Sun. The Phoebus detector was used to analyze and make intensity maps of solar flares, such as the powerful eruption of the Sun on 18–19th March 1990. There was a flare of 10 MeV on 11th June 1990 over a 26-sec period and the instrument registered 35 hits over 50 MeV. A flare on 15th May 1990 was much quieter, with nothing above 1.3 MeV. Phoebus recorded how tonnes of deuterium were synthesized by the Sun when it flared and this was carried away in the solar wind. Highlights of Granat's observations were:

- the great annihilator;
- development of the accretion disk model;
- finding many more black hole candidates, confirming existing ones as black holes;
- discovery of flaring X-ray pulsars;
- detection of X-ray bursts from neutron stars;
- the measurement of rotation periods;
- finding new radio sources, including stronger radio sources than ever before;
- the first long-term observations of the neutron star 4U 1700-37, first discovered by the Uhuru satellite in 1976, finding it to be a strongly binary system but with the rate of rotation slowing down, the star concerned being quite variable;
- finding 20 new X-ray sources including seven X-ray novas; one was the strongest source in the universe to date;
- finding 11 Low Mass X-ray Binaries (LMXBs) in the 35–100-keV range;
- studies of three active galactic nuclei in the 40–1,300-keV energy range: the Centaurus A radio galaxy (which brightened three times), Seyfert NGC4151, which was found to have a hard X-ray source, and quasar 3C273, where it found a second X-ray source close by GRS 1227, hinting at a new population of extragalactic objects.

This clearly marked it as the most productive astrophysical mission of Soviet and Russian space science [8].

THE LAST GREAT OBSERVATORY: GAMMA

Gamma, the last Soviet-period observatory, was based on the Progress cargo craft used to ferry equipment to the Salyut, Mir, and ISS space stations and is the smallest of these four great observatories. As far back as 1965, Gamma had originated as a free-flying module that could be docked to a large orbiting space complex, part of a two-project series, the other being an infrared telescope called *Aelita*. The intention was that in between periods of free flight, Gamma would dock with the orbital complex, where cosmonauts would retrieve film and carry out servicing tasks. The mission was redefined in 1972 when academician physicist Vitaly Ginzburg persuaded the Academy's praesidium that the Soviet Union must develop an observatory to study gamma rays in the 50–500-MeV range, determine their coordinates, and examine gamma-ray radiation coming from the Sun, the galaxy, and extragalactic objects. It was agreed that its capabilities must be better than the American SAS-2 and the European COS-B and launch was set for 1982 (e.g. angular resolution of Gamma was to have been three to four times better than that of COS-B).

France joined the project in 1974, Poland later (the Warsaw Polytechnical Institute), and Gamma was officially approved by the government on 17th February 1976. The project went through further evolutions whereby it might be docked periodically to the Mir space station, or alternately be a free-flier and serviced by a manned Soyuz crew, which would dock on a port on the front and retrieve and replace data cassettes. Detailed descriptions of the project are available from 1976, so the project had an extraordinarily long gestation [9].

In the end, it was simplest to fly Gamma as an independent mission. Gamma was 7.67 m long, 2.72 m in diameter, 11.75 m across its solar arrays, and weighed 7,320 kg. After orbital insertion, it climbed to an orbit of 416–434 km, following which there was a six-week period of systems check and calibration. Gamma carried:

- gamma-ray telescope *Gamma 1*, 50–5,000 MeV for highly energetic particles (France), with Polish star tracker, *Telezvezda*;
- X-ray telescope, *Pulsar X 2*, 2–25 keV;
- low-energy telescope (called *Disk-M*), 20–5 keV.

Its focus was gamma radiation in the 50–50,000-MeV range, both within and without our galaxy. The French telescope filled the main body of the spacecraft, while the other instruments were located on the front. The mission suffered two setbacks. First, an electrical failure robbed the spark chambers of power, so the gamma-ray telescope had to manage without them, which meant that its power of resolution was reduced by 10°. Second, Disk M broke down. Both meant changes to the flight program.

In its first six months, Gamma spent 320 hr observing the pulsar x-1 in the constellation Vela (it was also used as a calibration source), the remains of a supernova and the most persistently bright high-energy object in the sky. Over 1990–1991, Gamma studied the Heminga pulsar in the Crab nebula for 141 hr. From February to April 1992, Gamma made 15 scans of the galactic center totaling 30 hr. Three other gamma-ray stars were studied: Cygnus x-3, Hercules x-1, and Nova Muscae source. Gamma found that the Crab pulsar rotated every 33 milliseconds, Vela every 89 milliseconds.

Although designed to study distant stars, Gamma also made measurements of our own star, the Sun, focusing 100 hr of observation time on gamma-ray radiation from the Sun. One of the high points of the mission was on 26th March 1991 when Gamma happened to be obser-

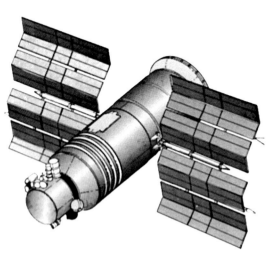

Gamma

ving the Sun at the time of a 10-sec solar flare. The gamma rays reached 300 MeV and Gamma was able to measure the intensity and energy of the flare and the disintegration of neutral pions. This flare, a sudden short impulse of accelerated electrons, was followed by another on 15th May 1991, a much longer 2-hr extended flare of neutral pions with gamma emissions. Emissions reached 2,000 MeV. Finally, although Gamma was a skyward-looking mission, it confirmed data from earlier missions that magnetic waves could be detected in advance of Earthquakes (see below). In the 300–500-km altitude range, its instruments noted charged particle events in the geomagnetic field tube some 2–5 hr before Earthquakes [10]. End-of-mission dates are given variously as 1st July 1991 and 1st January 1992. Table 7.3 summarizes the great observatories and Table 7.4 their periods of operation.

Table 7.3. The great observatories.

Astron	23 Mar 1983	1,951–201,120 km, 51°, 97 hr 57 min
Kvant	31 Mar 1987	170–299 km, 51.6°, 89 min (initial orbit before docking)
Granat	30 Nov 1989	1,957–201,693 km, 52°, 98 hr 23 min (initial orbit)
Gamma	11 Jul 1990	417–436 km, 51.6°, 93 min

Table 7.4. Periods of operation of the great observatories.

1983	1987	1989	1990	1992	1999	2001
Astron	————————————					
Kvant		——————————————————				
Granat			——————————————			
Gamma			————			

INTEGRAL: THE GREAT ATTRACTOR

A successor mission to Granat was proposed. This faltered for lack of money, but the Russian space agency made an arrangement with the European Space Agency for 25% observing time on its Integral observatory in exchange for launch on a Proton rocket. Integral was launched 17th October 2002. It carried a gamma-ray telescope, gamma-ray spectrometer, X-ray monitor, and optical monitor, and had an expected observing life of ten years. Data were made publicly accessible in the Russian Science Data Centre, which also established a classroom with monitors so as to teach young researchers.

Integral provided important results, starting with a catalog of hard X-ray sources, of which 215 were in our galaxy and 136 outside, most on the inside being black holes and those outside being active galactic nuclei (49 are still uncategorized). It found a new class of hard X-ray object (IGRJ 16318-4848), mapped the distribution of X-ray binaries in our galaxy, recorded the variations of X-ray pulsars, discovered ultra-weak gamma-ray bursts, and observed a fast transient supergiant. It discovered

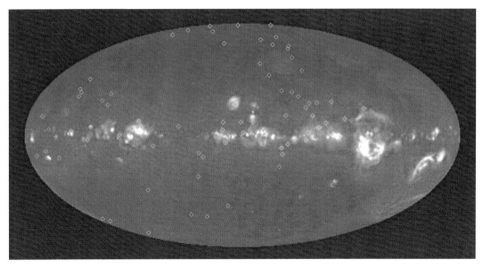

Integral sky map (Credit: ESA)

a population of hitherto obscured X-ray sources and reconstructed the past history of the supermassive black hole irradiated by Sgr B2 in 1700.

Within the Milky Way, there was an "X-ray Milky Way" called the Galactic Ridge, which broadly matched the visible light of the Milky Way as seen from Earth by amateur astronomers. The components were different, for they included weak X-ray sources and accreting white dwarf binaries. Integral was able to provide an X-ray map of the Galactic Ridge, while its studies of accreting white dwarf binaries enabled them to be modeled as plasma falling on the star's surface at 1,000 km/sec and heating to temperatures over 1 million K. Examining positron atoms at the 511-keV line, Integral found that every second, about 10^{43} positrons were annihilated in the central region of the galaxy. Integral also enabled more precise measurement of the amount of X-ray radiation reflected back from the Earth into space. More Integral results are:

- our local universe (200 parsec across) is not very uniform and has many voids, where galaxies tend to form;
- most galaxies have at their center a supermassive black hole, averaging 10^6–10^9 solar masses, which, at some stage, becomes an active galactic nucleus;
- our local universe has three points of concentration: the Virgo cluster of galaxies, the Perseus–Pisces supercluster, and the agglomeration around Hydra Cantaurus, called "the great attractor", possibly the mission's main finding.

Intended as a follow-on to the observatories project was a star-mapping satellite, Lomonosov, named after the great Russian astronomer. Again, this had a parallel Western project, Hipparcos, a European star-mapping satellite launched in 1989. Lomonosov would have used a Venera-type spacecraft like Astron and Granat, gone into a high, 200,000-km orbit, and would have mapped over 400,000 stars of magnitude 10 or brighter. Formally announced in 1990, it was unable to attract either domestic or international funding: although the project reached an advanced state of planning in the Sternberg Institute, it is not known whether any hardware was actually built. When the Hipparcos catalog was published in 1997, Lomonosov became redundant and was abandoned.

INTERCOSMOS: INTRODUCING THE SECOND PHASE

Chapter 2 outlined the first phase of the Intercosmos program, using the DS satellite and its derivatives. Intercosmos 15 marked the introduction of a new design: a standard bus with the capacity for orientation and station keeping in orbit, to replace the DS series. This was the second satellite design from Mikhail Yangel's OKB-586 design bureau in Dnepropetrovsk, in October 1966 renamed the Yuzhnoye ("southern") design bureau.

Work on a successor to the DS-U series began in Yuzhnoye in 1971, which came up with three prospective designs of multi-purpose spacecraft, or KAM in Russian, KAM 1, 2, and 3. In the event, KAM 1 was selected, called the Automatic Universal Orbital Station (AUOS, in Russian *Avtomatitecheskaya Universalnaya Orbitalnaya Stantsiya*). Approval was given by the government on 26th June 1972 and the design lasted 40 years. The chief designer was V.I. Dranovsky. The AUOS was a substantial advance and able to carry a much more extensive suite of instruments, up to 600 kg, including small subsatellites. Typically, the AUOS was 2.3 m tall and 4 m in diameter (including solar panels), with a weight of between 420 kg (Intercosmos 15) and 900 kg (later missions), a lifetime of six months, and the ability to hold and download vastly greater volumes of data. The four solar panels were each 64 cm long, providing up to 50 W of power. Data transmission systems were much improved and new receiving stations were set up in Hungary, the GDR, Poland, and Czechoslovakia (until then, data could be received only in the USSR). Two variants were introduced: Earth-pointing (AUOS-Z, Z for *Zemlya* or Earth) and Sun-pointing (AUOS-S, S for *Solntse*, Sun). Although the AUOS series was developed primarily for the Intercosmos program, including Aureole, it was also used for two key Soviet science missions called *Oval* (Cosmos 900, AUOS-Z-R-O) and *Ionozond* (Cosmos 1809, AUOS-Z-I-E). The satellites comprising the AUOS series are described in Table 7.5.

THE *ELLIPSE* AND *MAGIK* MISSIONS

Intercosmos 17 was called the *Ellipse* mission and designed to study the magneto-sphere, cosmic rays, energetic and neutral charged particles, and micrometeorites. Principal investigators were Naum Grigorov, A.A. Gusev, and A.F. Titenkov. One of its objectives, set by Naum Grigorov, was to detect high-energy electrons through the transition radiation method. *Ellipse* was designed to fly through magnetic storms, its orbit timed 7 min apart from Cosmos 900 (see below). Together, they flew through the magnetic storm of 27th–28th October 1977, noting changes in the

Table 7.5. AUOS series.

AUOS design	
AUOS Z	*AUOS SM*
Z-T-IK Intercosmos 15	Koronas *Fyzika*
Z-R-O Cosmos 900 *Oval*	Koronas I
Z-E-IK Intercosmos 17 *Ellipse*	
Z-M-IK Intercosmos 18 *Magik*	
Z-I-E Cosmos 1809, Intercosmos 19 *Ionozond*	
Z-R-P-IK Intercosmos 20, 21 *Priroda*	
Z-AV-IK Intercosmos 24 *Aktivny*	
Intercosmos 25 APEX	
Z MA IK Aureole 3	

ionosphere and the ring current. *Ellipse* followed many subsequent storms, noting how they were at their most intense during the night and early morning, with proton flows up 10 times and electrons 100-fold. *Ellipse* found five distinct latitudinal zones of plasma inhomogeneities – features that appeared only during such disturbances. It mapped, measured, and compared proton fluxes up to 500 MeV between the eastern and western regions below the radiation belts. It found that high-energy solar cosmic rays penetrated not only at polar latitudes, but down to 60° during magnetically quiet times. Intercosmos 17 also flew a high-resolution spectroscopic telescope designed by nuclear physicists and when turned to the Sun, picked up high levels of helium 3 coming from the Sun – quite a surprise at the time. It flew experiments on the radiation environment in space to find ways to protect cosmonauts from harmful radiation – an electrical analyzer to register low energy protons and electrons, two dosimeters, and analyzer to measure the spectrum of particles penetrating through different tissue-equivalent materials [11].

Intercosmos 18 was called the *Magik* mission (for Magnetospheric Intercosmos) and carried equipment from Hungary, Poland, Romania, Sweden, and the GDR to study.

- the geomagnetic field;
- electrostatic fields of magnetospheric origin;
- electron and ion densities and temperatures;
- ion and neutral composition of the upper atmosphere;
- low-energy particle fluxes and their angular distributions (100 eV–50 keV);
- magnetic components of very-long-frequency electric waves, 100 Hz–16 kHz.

Magik project scientist was I.A. Julin (IZMIRAN) and technical manager E.M. Vasilyev. *Magik* was most important because it marked the introduction of subsatellites. On 14th November, three weeks after in-orbit testing was complete, Intercosmos 18 released a Czechoslovakian subsatellite, Magion 1, a 15-kg rectangular box prism measuring 300 × 300 × 160 mm. Magion satellites were developed by Professor Pavel Triska of the Institute of Aeronomy of the

Intercosmos 17 *Ellipse* preparations

Czechoslovakian Academy of Sciences and his colleague, J. Vojta (technical manager) of its Institute of Atmospheric Physics. Magion 1 was given a kick of 20 cm/sec and began to draw away from Intercosmos 18 at a rate of 60 km/day, drifting away to measure atmospheric drag. Magion 1 transmitted for three years, until it fell out of orbit on 10th September 1981. Live transmissions went directly to the tracking station in Panska Ves in Czechoslovakia (50°N, 14°E). Magion 1 had six instruments (Table 7.6).

Table 7.6. Magion 1 instruments.

Instrument to measure electric and magnetic fields, 0.05–16-kHz broadband
Instrument to measure narrowband channels, 0.45, 0.8, 1.95, 4.65, and 15 kHz
16-channel frequency analyzer (0.185–15 kHz, 60 dB)
Resonance exciter 0.8–8 kHz
Electric field detector 0.01–80 Hz
Tubes to measure electrons above 30 keV

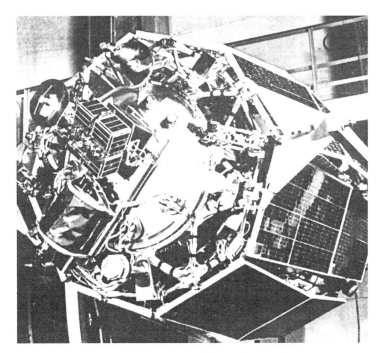

Intercosmos 18 *Magik* with Magion on front

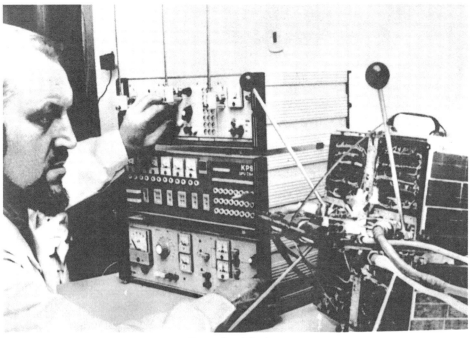

Preparing Magion

Magion

Small satellites became a favored method of space exploration in the 1990s, their most famous developer being Surrey Satellite Technologies in Britain. They were able to take advantage of advances in miniaturization and computerization and spare space on launching rockets, and demonstrated considerable versatility. The achievements of the Magion series, which date to 1978, have often been overlooked. They were developed by the most technologically advanced country of the Eastern Block – Czechoslovakia – and the release of Magion was a landmark event in Czech space exploration. The project was a part of more extensive program of magnetospheric research, which included rocket, satellite, balloon, and ground-based experiments. The idea of launching two spacecraft in one set is a powerful tool to study magnetospheric phenomena, as it helps to tell apart spatial and temporal variations in space plasma parameters. Later, the Interball and European *Cluster* missions used the same idea.

THE *IONOZOND* MISSIONS

The next mission, Intercosmos 19, was called *Ionozond*. *Ionozond* was a two-satellite program led by IZMIRAN to model the F layer of the ionosphere, with one mission in the Intercosmos program (Intercosmos 19, part of the program mentioned earlier) and one in the Cosmos program (Cosmos 1809) and the two are treated together here. A formidable team was assembled. Project head was S.I. Avdyushin, director of the Russian Institute of Applied Geophysics, the scientific leader was Yuri Galperin from the 2MS series and the Aureole program, the principal investigator was Yuri V.

Table 7.7. *Ionozond* instruments.

IS-338 digital topside sounder (plasma density above F peak) 0.3–15.95 MHz
NAM-4 mass spectrometer for ion densities and composition
SF-3M photo-electron spectrometer for fluxes in range 10–15 eV
ANCh-2ME to measure low-frequency waves and electromagnetic fields 70–20 kHz
DEP-2 instrument to measure electric fields
AVCh-2K device to measure high-frequency fields
IZ-2 impedance probe to measure electron density and plasma concentrations
KM-9 to measure electron temperature 600–1300 K and ion concentrations
EMO-1 optical spectrometer: atmospheric glows between 20°S and 30°N, 51, 64, and 39 Å
P4: electron concentration in atmosphere (Bulgaria)
AVC-2 plasma analyzer
IRS-1 radio spectrometer (Poland)

Mineev, and it also included Vadim Istomin (1929–2000), whose spectrometers had flown on Mars and Venus probes.

Intercosmos 19, led by Y.V. Mineev, was designed to study the upper ionosphere by radio probing in pulses, with other devices to record low-frequency radiation in the Earth's magnetosphere, as well as measure electron concentrations and temperature and energetic particles. It concentrated on electrons with energies of 0.3–2 MeV at low altitudes. It did not carry a subsatellite, but operated in conjunction with Intercosmos 18 to get a cross-section of Earth's magnetosphere, comparing spring and autumn. It had the largest on-board memory to date, able to store 17 hr or 10 orbits of sounding data at a time and, in the course of time, built one of the world's biggest repositories of topside sounding data. The instruments are detailed in Table 7.7.

Both spacecraft were essentially identical, the only difference being the tape-recording system. Intercosmos 19 used instruments from Bulgaria, Czechoslovakia, Hungary, and Poland, as well as the USSR. Launched on 27th February 1979, it worked for three years, until 8th April 1982, during a period of high solar activity. Cosmos 1809, launched on 18th December 1986, likewise carried instruments from these countries but functioned twice as long, until 23rd May 1993. Each carried a topside sounder that gave electron density profiles of the magnetosphere. This was the principal instrument, the first version of which had flown on the Cosmos 381 *Ionosfernaya Stantsiya* mission in 1970. The sounder, called the IS-338, transmitted in pulses of 0.133 milliseconds across 338 different frequencies over the range 0.3–15.95 MHz in 25-kHz steps, the answers pinging back to two 50-m perpendicular dipole antennae, sounding cycles lasting from between 5 and 64 sec. The SF-3M electron spectrometer was used to detect electrons in energy ranges of 10 eV–15 keV, with sampling periods at 16 and 32-sec intervals, with data storage of up to 2 hr. Its dipole was orientated along the magnetic lines of Earth.

The data outcomes of both missions were enormous, leading to hundreds of scientific papers that did not reach exhaustion until 2003. In 2000, one of the scientists from the program, Sergei Pulinets, began to put the data together from the

Intercosmos 19 *Ionozond* rollout

Polish scientists involved in Intercosmos 19 at rollout

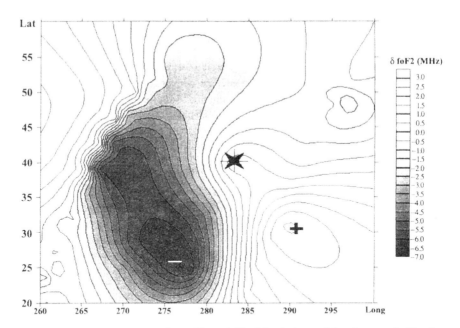

Figure 7.2. The nuclear plume from Three Mile Island, imaged by *Ionozond*. (Credit: Sergei Pulinets)

project, making it internet-accessible at IZMIRAN. Recognizing the world importance of the archive, NASA contributed a grant to finance the project. Some of the old *Ionozond* tapes had spoiled, but 90% were retrieved: between them, they provided a 3-year and 7-year set of profiles of the ionosphere. Intercosmos 19 made the first global picture of the distribution of ions in the upper atmosphere. A comprehensive map was made of the spatial and temporal features of the Main Ionospheric Trough (MIT) based on 62 passes with samples taken every 8 sec. The spacecraft compiled colored maps of the ionosphere, with red and yellow showing the spikes of plasma activity against the deep blues and blacks of space. The *Ionozond* made profiles of magnetic storms: Intercosmos 19 flew through the electromagnetic storm of 22nd March 1979 and saw how the Earth's magnetic field expanded and contracted and how the electrons were diffused. Cosmos 1809's passes were timed with the launching of plasma beam injections on sounding rockets from the research ship *Professor Zubov* in the zone of the South Atlantic magnetic anomaly. One of the most remarkable images of the mission showed the nuclear plume over the Three Mile Island power station, which experienced the most dangerous nuclear reactor failure before Chernobyl (Figure 7.2).

The main discoveries of the *Ionozond* missions were:

- intense flares of suprathermal electrons over 100 eV;
- low-frequency emissions in the ionosphere at 0.1–2 MHz whenever rising energetic electrons exceeded those falling down;
- Discrete Plasma Emissions (DPEs); these were of unknown origin and had

never been detected by ground stations; between March 1979 and May 1981, Intercosmos 19 found 60 DPEs in the 10–25-KHz range, mainly over Europe between 500 and 1,000 km, principally between 3 am and 9 am during periods of enhanced geomagnetic activity and sometimes associated with the formation of a strong ring current;

- two humps of ions in the F region, called the equatorial anomaly;
- enhancement of plasma around the satellite when its transmitter was in resonance with the frequency of ambient plasma, ions, and electrons accelerating straight after the sounding of a radio pulse;
- location of the boundary of the mid-latitude electron trough, the main feature of the nightside sub-auroral F region;
- a cliff of concentrated electrons at 50–60°S at 240–300° longitude before and after the south polar summer, but which vanishes in December;
- rising levels of man-made electrical emissions over Europe and China. These had first been detected by Britain's Ariel 3 satellite many years earlier; and the 15-m dipole was used to identify those parts of the planet with the highest levels of artificial electrical output;
- detection of intense flares of superthermal electrons of 1,000 eV;
- a "longitudinal fine structure" in nighttime lower latitudes connected to the solar wind, electrical fields, and atmospheric heating.

There was a fresh understanding of the magnetic fields and lines around the Earth (Figure 7.3). Intercosmos 19 had a sequel, for, in 1998, a replica of its topside sounder was flown up to the manned space station Mir and used to compile an ionospheric map from the much lower altitude of 350 km [12].

Perhaps the most remarkable outcome of the *Ionozond* missions was the discovery that Earthquakes were preceded by electrical waves several hours before they struck, opening the door to the possibility of predicting Earthquakes and thereby saving countless lives. Our planet has between 100 and 200 Earthquakes a year. The notion that there were electrical precursors to Earthquakes is not new, for auroral lights had been seen above Rome before its Earthquake in 373 BC. Intercosmos 19 found anomalous increases in low-frequency emissions, "noise bursts" in the 0.1–0.15- kHz range, just before Earthquakes – the first spacecraft to do so. First, the F layer of the ionosphere became compressed over the epicenter, followed by 15-min gravity waves, inhomogeneities, heating, and plasma turbulence. Looking over the Intercosmos 19 data as a whole, there was a 90% correlation between low-frequency electrical waves and Earthquakes within three days (an example is given of the Kermadec Earthquakes in Figure 7.3). Later, Cosmos 1809 found intense electromagnetic radiation below 450 Hz over the aftershocks of the Spitak Earthquake in Armenia. As a result, similar instruments were used on subsequent missions to test the theory. Aureole 3 detected noise increases in 100-keV protons and electrons over 17.9°S and 179.3°W just 3 hr before a 5.4 Earthquake. What were called "noise belts" were generated 3–6 hr before and after Earthquakes while flying over the epicenter. Following the mission, a model was put forward whereby geomagnetic field lines were generated from the epicenter to create ionospheric

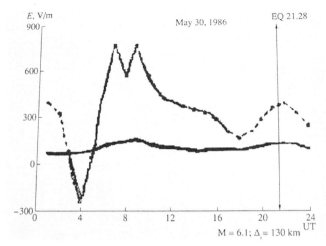

Figure 7.3. Electrical activity before the Kermadec Earthquake detected by Intercosmos 19. (Credit: Sergei Pulinets)

bubbles of rare plasma regions up to 900 km with an intensity of 0.5 nT in the 8-Hz range. While overflying seismically active regions of the Earth's plates (e.g. Malaya, central America), Intercosmos 22 *Bulgaria 1300* found emissions before 12 Earthquakes up to magnitude 6.9. Electric disturbances were noted as much as 89 hr before quakes. Intercosmos 24 noted an increase in light ions over seismic zones. These findings were challenged in the West as being unreliable and the Russians admitted they found it a difficult phenomenon to explain, searching for explanations as the release of radon gas in advance of an Earthquake. Pioneer of this research was Sergei Pulinets, who, over the next 20 years, published numerous papers, making a convincing case linking subtle changes in the Earth's electrical activity to subsequent Earthquakes in such diverse locations as Italy, Alaska, Australia, New Guinea, and New Zealand [13].

THE *PRIRODA* AND *BULGARIA 1300* MISSIONS

Intercosmos 20 and Intercosmos 21 marked a change towards applications, for they were focused on the Earth, carrying equipment to measure the temperature of the sea to within 1°C, using two telescopes developed earlier by Cosmos 1076 and 1151. Called *Priroda* ("nature"), they used the Tsyklon rocket for the first time – a more powerful rocket able to put payloads up to 3.5 tonnes into orbit. Intercosmos 20 was able to interrogate and pick up transmissions of stored data from up to 16 buoys bobbing in the Earth's oceans and worked in cooperation with the GDR research vessel *Alexander Humboldt*. Instruments were carried to detect "upwelling" radiation coming out of the atmosphere and to profile aerosols in the atmosphere. Participating countries were Czechoslovakia, the GDR, Hungary, and Romania.

Intercosmos 22 *Bulgaria 1300* was a special mission part-led by M.V. Teltsov to

Table 7.8. Intercosmos 22 *Bulgaria 1300* instruments.

Ion drift meter and retarding potential analyzer
Spherical electrostatic ion trap
Cylindrical Langmuir probe
Double spherical electron temperature probes
Low-energy electron photo electrostatic analyzer array
Ion energy-mass composition analyzer
Wavelength scanning ultraviolet photometer
Proton solid-state telescope
Visible airglow photometers
Spherical vector electric field probes
Triaxial magnetometer

mark the 1,300th anniversary of the foundation of Bulgaria, when the Bulgars arrived and settled in what we now know as Bulgaria. This used not the AUOS, but the Meteor design of weather satellites, which had a weight of 1,500 kg and carried 12 experiments built by the Bulgarian School of Space Research to investigate the Earth's magnetic field, especially around the equator, but the mission also examined weather anomalies and changes in Earth's radiation balance (Table 7.8). The solar panel supplied 2 kW of electric power and there were two tape recorders, each with 60-MB storage.

Intercosmos 22 appears to have been a substantial breakthrough for Bulgaria, along with Romania, the least scientifically developed of the socialist block countries of eastern and central Europe. It appears to have galvanized Bulgarian scientists, leading not only to a volume of papers and analysis of this mission, but a range of subsequent Bulgarian experiments on orbital stations (Chapter 6). The main achievements and discoveries of Intercosmos 22 *Bulgaria 1300* were:

- mapping of SARs, or Stable Auroral Red arcs; these are the "northern lights" but much further south than normal, first observed in southern France in 1956 and mid-latitude, about 300–700 km high, apparently caused when energetic particles meet the cold plasma circular current of the plasmasphere; also recorded by the American Dynamics Explorer 2, Intercosmos 22 made the first observations of SARs on 21st August 1981 on orbit 203 and 18 more by 24th May 1982; it measured their intensity (6,300 Å) and found that they had a complex and irregular nature;
- profile of the auroral oval during two night passes on orbits 203 and 231 on the 21st and 23rd August 1981; it recorded how the oval was heated by particle precipitation, which it measured at between 2 and 8×10^{-7} wm^{-3}. There was an abrupt decrease in mid and high-latitude electron and ion densities as it entered the auroral oval;
- there was a stable system of longitudinal currents in the auroral oval, divided into two sheets; currents flowed inward at the poles and outward at the equator in the morning, the other way around in the evening;
- measurement of perturbations, disturbances, and anomalies in the magnetic

Intercosmos 22 *Bulgaria 1300*

field around the South Pole; during the geomagnetic storm of 25th August–19th September 1981, it found a new magnetic anomaly over Antarctica, south of the Brazilian anomaly; a table was compiled of densities in the F layer [14].

THE *AKTIVNY* AND APEX MISSIONS

The last two missions used identical 450-kg AUOS spacecraft and were run as related projects, Intercosmos 24 (*Aktivny*) and Intercosmos 25 (APEX). They marked the end of the Intercosmos program and, in another way, the end of *Soviet* space science, for Intercosmos 25 was launched a mere two weeks before the Soviet Union went out of existence. For the period 1980 onwards, Intercosmos had been led by its second director, Academician Vladimir Kotelnikov, and the organization lapsed with the end of the Soviet Union.

The objective of *Aktivny* was to study low-frequency radio waves in the Earth's magnetosphere and radiation belts. The contributing countries were Hungary, Poland, Romania, Czechoslovakia, Bulgaria, and the GDR, the mission being directed by IZMIRAN. Project scientist was B.I. Shevchenko and technical manager E.M. Vasilyev. Data were received by not only the Intercosmos countries, but also the United States, Brazil, Canada, Finland, Japan, and New Zealand.

Aktivny looked wiry: it had a 14-m boom for vertical stabilization, a loop antenna for deployment in the orbital plane, and other instruments (Table 7.9). There were two batteries, one for the high-power VLF transmitter, which used a considerable amount of electricity, and the other battery for the rest of the instruments, all fed by the solar panels. It took several hours to charge up the VLF for its transmission. A substantial volume of information was expected, so it was equipped with a 10-MB memory. It was launched on 28th September 1989.

One of the main purposes of the mission, as the title "active" suggests, was to affect near-Earth plasma by radio emissions or by releasing gases to test their

Table 7.9. Intercosmos 24 *Aktivny* instruments.

ONCh-G low-frequency generator with large loop aerial (9.6 kHz, 5 kW)
ONCh-2 low-frequency analyzer
NVK-ONCh low-frequency wave complex
PG plasma generator
SPE 1 electron and proton spectrometer
KM 6 cold plasma parameters detector
PRS 2 plasma radio spectrometer
NAM 5 mass spectrometer to analyze comet matter in the Earth's vicinity
ZL A Langmuir probe to detect plasma electron parameters
PIVI instrument for plasma emission, exciting, and measuring
ANAPURNA electron pitch-angle analyzer
DME soft electron detector, 0.01–10 keV
KSANI anomalous ionization detector

Intercosmos 24 *Aktivny*; note Magion 2 mounted on top.

propagation in the ionosphere, using a subsatellite as a measuring instrument. The theory behind the release of such gases was that by following the behavior of artificial clouds of known composition and proportions, it would be possible to learn much more about natural events, in effect, a miniaturized and much less dangerous version of *Starfish* (Chapter 2). For this, the Czechoslovakian Magion 2 subsatellite was carried (Table 7.10). There had been extensive experiments in the sounding rocket program from the 1970s and the first gas release in orbit had taken place from two Meteor spacecraft in 1977–1979, generating small electromagnetic fields near the spacecraft. Now, in the KSANI (Xenon ANomalous Ionization) experiment, performed under a cooperative program with the United States in early 1990, *Aktivny* released xenon gas over Alaska on 21st February, 2nd and 3rd March, and over Utah on the 12th, 14th, and 16th March, where ground observers duly observed its effects. There were seven xenon releases over daytime high latitudes.

Aktivny experienced a number of operational problems. First, there was a delay in deploying the ONCh G loop aerial and it failed on 30th October 1989. Second, there were problems with the subsatellite. At 65 kg, it was four times heavier than its predecessor, a new design of a polyhedron with folder sensors and seven instruments. It was equipped with a 0.2-N ion electric pulsar propulsion system for maneuvers over a distance of 100 m–100 km. Although it was successfully separated, there were thruster problems. Three months after separation, they were 400 km apart instead of 10 km as planned. Signals were received for six years, up to October 1995.

Table 7.10. Magion 2 instruments.

SGR magnetometer for magnetic field fluctuations
KEM 1 wave complex measurer
KM12 instrument to measure temperature, density of cold plasma
ZL A C Langmuir probe
PRC 2C radio spectrometer
DANI C, MPC, DOK 1S, SEA to measure particles

Table 7.11. APEX instruments.

UEM 2 microsecond electron gun accelerator (200 sessions)
UPM xenon plasma gun (250 sessions)
PEAS 12-direction electron and ion spectrum analyzer
DANI low-energy electron and ion energy analyzer
KM-10 cold plasma analyzer
NAM 5 radio frequency ion mass spectrometer
DEP 2E electric fields detector very-long-frequency electric field experiment
DEP 2R electromagnetic fields analyzer
MNCh 2 search coil magnetometer
VCh VK high-frequency wave spectrometer
FS photometer and UF-3K photometer comprising optical complex
SGR 5 magnetometer
NAM 5 mass spectrometer
NVK ONCh low-frequency wave complex

Intercosmos 24 provided some confirmation of the validity of Earthquakes research begun with Intercosmos 19. On 24th October 1990, the USSR carried out an underground nuclear test at its traditional site on Novaya Zemlya Island. Within minutes, *Aktivny*, flying 900 km overhead, had recorded electromagnetic disturbances due to the magnetohydrodynamic excitation of the ionosphere's E layer by the acoustic wave.

Intercosmos 25, APEX stood for Active Plasma EXperiment and it followed *Aktivny* in the release of gas plasma. Its instruments were similar or modernized versions. The overall mission objective was to better understand the ionosphere, atmosphere, auroral phenomena and the polar lights, and the interaction of solar wind, magnetosphere, and ionosphere (its instruments are listed in Table 7.11). It also worked with ground-based radio and geophysics observatories. The project aimed to study the influence of electron and plasma beams upon the Earth's ionosphere and magnetosphere. APEX was equipped with two electron guns at the back, one being aimed at the Earth and the other in the direction of the spacecraft. The modulated plasma accelerator was able to inject continuous current plasma beams (200–300 eV). The electron accelerator was able to inject beams of 10 eV, firing for durations of 2 microseconds for 23 sec, at 40 kHz. These experiments were

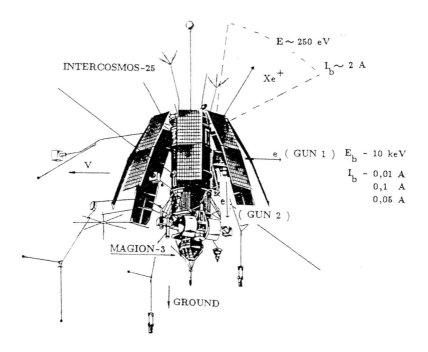

Intercosmos 25 APEX

carried out at all latitudes, the gun being fired 200 times at altitudes between 500 and 3,000 km, the intention being that they be measured both by the main satellite and its subsatellite, Magion 3. APEX project scientist was Viktor Oraevsky and technical manager V.S. Dokukin, both of IZMIRAN. Electron guns had been tested extensively in sounding rocket programs going back to 1973 (see below).

Magion 3 was released from Intercosmos 25 ten days into the mission on 28th December 1991. Its instruments were designed to study aurorae, radiation, electromagnetic waves, charged particles, and plasma: they are listed in Table 7.12. Almost identical to Magion 2, it was 52 cm across, 52 kg in weight, deployed a 1.7-m aerial, and used a small pulsar engine to drift away to 2,000 km, until 9th January, when a small thruster was fired to bring them back together again to 100 m for 30th January. These maneuvers were repeated in May and June, with separation distances of several hundred kilometers and then returning to 400 m. The main craft injected plasma into the tail of APEX's flight path so as to be measured by Magion. Two hundred such injections of electrons, plasma, and xenon took place, lasting 4–20 min each and all were detected by Magion. The Magion 3 battery charging system suddenly began to decline on 20th August 1992 and it fell silent on 9th September.

The mission was regarded as successful and the preliminary findings were discussed at an APEX workshop in Prague in April 1992. The main discoveries of the mission were:

Table 7.12. Magion 3 instruments.

SGR fluxgate magnetometer
KEM 1 wave experiment
KM12 plasma analyzer
ZL A S Langmuir probe
PRS 2 S radio spectrometer
DANI S particles analyzer
MPS particles analyzer
FDS photometer

- double current sheet bands 500 km apart in a double auroral oval that formed during the period of recovery from magnetic storms (Magion 2);
- maps of plasma noises that accompanied ion fluxes, finding high levels over Europe (APEX);
- in the electron gun experiments, the gun created low-frequency plasma turbulence and wideband noise; the beams were successful in exciting ions, disturbing plasma, and causing reactions from whistler waves (APEX);
- the finding of new non-linear electromagnetic structures: collisionless shock waves, plasma density jumps, and slanting ionospheric throws (APEX);
- approaching the North Pole at apogee, ions rained down, plasma noise rose, and horizontal currents formed (theta structure) (APEX);
- profile of the magnetic storm of 8th June 1992, following temperatures, densities, the precipitation of high-energy electrons and ions (30–150 keV), complex currents, and the recovery of the magnetic field (Magion 3);
- it was proved experimentally that kilometer radiation generation is produced by ascending electron flows; the modulated electron beam can be used for nonlocal determination of electron density and magnetic field value [15].

The research carried out by the last two Intercosmos marked the culmination of work begun by Konstantin Gringauz in space research dating to the 1950s, for he died on 10th June 1993. A memorial conference was organized in his honor on his 90th anniversary in June 2008.

THE *OVAL* MISSION

There was one AUOS mission that, unlike *Ionozond*, did not have a comparator in the Intercosmos program. This was Cosmos 900, whose specific task was to investigate, as the name suggests, "the oval", the crown of daylight aurorae that forms over the North Pole. Although not part of the Intercosmos program, Cosmos 900 also carried equipment from Czechoslovakia and the GDR to study the polar lights and the magnetosphere.

Cosmos 900 was one of the high points of scientific investigations of the northern

Olga Kohrosheva

lights – phenomena that held a strong attraction for Russian scientists. A number of earlier missions had studied the northern lights, notably Cosmos 261. It is worth saying a little more about the northern lights. The idea that the Earth was magnetized was known as early as 1600 (Gilbert's seminal work, *De Megnete*) and research into how particles were magnetized as they fell to Earth began with Kristian Birkeland in Norway three centuries later.

The general focus of Cosmos 900 was the northern lights and, in particular, its oval, hence the mission name. Although the oval had been known for some time, the person who characterized it in the form in which it is known today was Olga Khorosheva (1929–2008), a colleague of Yuri Galperin and Alexander Lebedinsky during the International Geophysical Year in Loparskaya, Murmansk. Olga Khorosheva had made a detailed study of the images taken by Alexander Lebedinsky during the IGY. She was the first to suggest that the solar particles fell down to Earth, not in a spiral, as everyone thought, but formed into an oval over the North Pole. Her view was the northern lights were "stretched" out from their natural fall to Earth by the magnetic fields and, in particular, by the outer radiation belt with which it was co-located. She presented her opinion in a scientific paper in *Geomagnetism and Aeronomy* in 1962. She was comparatively junior and her radical hypothesis attracted little attention at the time.

Eventually, a special space mission was organized to test her theory with Konstantin Gringauz as main project scientist and Boris Tverskoy of the Institute for Nuclear Physics as project scientist. Also part of the team were eminent magnetospheric scientists Yevgeni Gorchakov (1932–2003) and Elmar Sosnovets (1935–2004). Cosmos 900 was launched in 30th March 1977 into a 460–523-km orbit of 83°, which brought it well within the polar regions. Table 7.13 lists its instruments.

Cosmos 900 confirmed that Olga Khorosheva's suppositions were correct. It was the first satellite to compile a full map of the auroral oval. It was a uniquely productive mission, justifying her original work. She lived to the age of 79 and died on 6th January 2008. Other highlights of the mission were:

Table 7.13. Cosmos 900 instruments.

Flat retarding potential analyzer (V. Afonin)
High-frequency electron temperature probe (V. Afonin)
Spherical ion trap with floating potential (Gennadiy Gdalevich)
Cylindrical electrostatic probe (Gennadiy Gdalevich)
Differential energy spectrometer (Elmar Sosnovets)
Differential low-energy spectrometer (M. Telstov)
Panoramic electrostatic spectrometer (N. Shutte)
Relativistic proton and electron Cerenkov counter (Yevgeni Gorchakov)
Auroral photometer (Vladimir Tulupov)

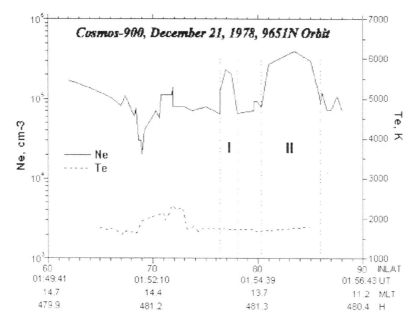

Figure 7.4. Cosmos 900 passing through storm. (Credit: Viktor Khalipov)

- location of the boundary of the Main Ionospheric Trough (MIT), finding it really to comprise two troughs;
- 14 magnetic storms, where Cosmos 900 calculated the temperature of the electrons (Figure 7.4); when it passed through the northern aurorae in April 1977, it found energies fluctuating by 1.2–1.5 keV around trapped electrical fields;
- the great magnetic storm of 1st December 1977, when Cosmos 900 followed the precipitation into the polar trough and ring current;
- the finding by the Cerenkov counter of a narrow belt of electrons of 15 MeV in the outer radiation belt during storms that are then accelerated and injected into the inner part of the magnetosphere;

Figure 7.5. Cosmos 900 plasma blobs. (Credit: Viktor Khalipov)

- the finding and characterization of ionospheric plasma blobs up to 650 km across with densities 5.7 times higher than the normal electron environment, generated in winter daytime (Figure 7.5);
- studies of the magnetosphere during quiet conditions; even still, its instruments detected protons and electrons in the 100 eV–20-keV range precipitating into the ionosphere over high and low latitudes;
- a study of the 12-hr magnetic storm of up to 1,000 γ in September 1977, where it found that proton precipitation was strongest by day and drifted west in the auroral regions; by contrast, electron precipitation was strongest by night and drifted east; it took a period of 4–5 hr for protons over 1 MeV to penetrate the polar caps;
- during three powerful solar flares that month, the sequence of events in the ionosphere could be chronicled: there was a sudden ionospheric disturbance, the electrical field moved to southerly latitudes, small inhomogeneities developed, and the ring current decayed;
- finding, in the course of two aurorae crossings on 20th and 22nd September 1977, a high-intensity zone between 100 and 500 km, with peaks in between of varying intensities;
- from studies of the ionosphere at low and equatorial latitudes at 500 km, it was found that the equatorial magnetic anomaly frequently disappeared during daytime and at night filled with plasma bubbles [16].

AUREOLE 3

Aureole 3 was much the most successful of the Aureole series (for Aureole 1 and 2, see Chapter 2) in the ARCADE project. Aureole 3, 1,000 kg in weight and launched on the Tsyklon rocket, carried 12 experiments (of which seven were French, four Russian, and one joint), the aim being to focus on the magnetosphere–ionosphere at high latitudes. The mission was led by Yuri Galperin, who regarded it as one of the high points of his career.

Table 7.14. Aureole 3 experiments.

Three soft particle spectrometers
Particle spectrometer
Ion energetic spectrometer
Ion mass spectrometer
Isoprobe
Electric field probe
Magnetic field probe
Magnetometer
Auroral photometer
Energetic particle detector

Aureole 3 was a pressurized cylinder, 2.7 m tall, 1.6 m in diameter, with eight solar panels generating an average of 50 W and a maximum of 250 W. Aureole 3 benefitted from advances in French instrumentation and computers. Several improvements were made in the structure and service systems of the satellite, such as solar panels, which could distort the results of the measurements due to their electromagnetic properties. Yuri Galperin insisted on preparing special new electromagnetically clean solar panels. As a result, their electromagnetic disturbances were decreased 1,000-fold, and their durability improved (similar panels were later used on the Intercosmos 22 *Bulgaria 1300*). It carried proton and electron spectrometers (*Kukushka, Pestchanka, Spectro*), energetic particle detector (*Fon*), thermal mass spectrometer (*Dyction*), inferometric probe (*Isoprobe*), electric and magnetic field probes (ONCh-TBF), magnetometer (TRAK), and photometer (*Altair*) (Table 7.14). Readings were taken on both a pre-ordered basis and on command, with direct readout to French stations and recorded dumps over Soviet territory, with common scheduling of experiments. The satellite operated from September 1981 to summer 1986. Following *Ionozond*, it focused on the effect of Earth's electric activity on the space environment, identifying the main sources as transmitters and industrial plants. Operations were carefully planned so that Aureole 3 data could be coordinated with other satellite observations, ground-based measurements, and radio transmitters. In autumn 1981, several industrial explosions were carried out near Alma-Ata (Kazakhstan): several laboratories used this opportunity for the MASSA experiment to study magnetosphere–atmosphere links during a seismically active event. The effect of the large-scale acoustic wave on the upper atmosphere and ionosphere was studied. The time of the explosion was coordinated with Aureole 3's orbit. As a result, electrostatic VLF and ELF noises as well as an intense MHD wave were recorded in the corresponding flux tube. These results were later confirmed by similar experiments [17]. Galperin's laboratory convened the many international conferences at which the mission outcomes were shared. Although such an approach might be considered routine now, they were not in the early 1980s.

Table 7.15 is a summary of the second phase of the Intercosmos program.

Table 7.15. Intercosmos – second phase (AUOS).

AUOS design			
Intercosmos 15	19 Jun 1976	Plesetsk	484–518 km, 74°, 94.6 min
Cosmos 900 *Oval*	30 Mar 1977	Plesetsk	460–523 km, 83°, 94.4 min
Intercosmos 17 *Ellipse*	24 Sep 1977	Plesetsk	519–468 km, 83°, 94.4 min
Intercosmos 18/Magion 1	24 Oct 1978	Plesetsk	407–768 km, 83°, 96.4 min
Intercosmos 19 *Ionozond*	27 Feb 1979	Plesetsk	502–996 km, 74°, 99.8 min
Cosmos 3M launcher			
Intercosmos 20/*Priroda*	1 Nov 1979	Plesetsk	502–996 km, 74°, 99.8 min
Intercosmos 21/*Priroda*	6 Feb 1981	Plesetsk	475–520 km, 74°, 94.5 min
Aureole 3	21 Sep 1981	Plesetsk	380–1,920 km, 82.6°, 110 min
Intercosmos 24/Magion 2	28 Sep 1989	Plesetsk	511–2,479 km, 82.6°, 116 min
Intercosmos 25/Magion 3	18 Dec 1991	Plesetsk	437–3,073 km, 82.5°, 121 min
Tsyklon launcher			
Meteor design			
Intercosmos 22 *Bulgaria 1300*	7 Aug 1981	Plesetsk	825–906 km, 81.2°, 101.9 min
Vostok M launcher			
Intercosmos 23 was the Prognoz 10/Intershock mission (see Chapter 2)			

THE SEARCH FOR FIREBALLS: PROGNOZ M (INTERBALL)

Chapter 2 looked at the Prognoz program. This was followed, during the Russian period, by the Prognoz M program, which acquired the separate designator of Interball, aimed to study the tail end of Earth's magnetosphere. These missions built on the Intercosmos missions just described and were the first substantial science missions of the post-Soviet, Russian period.

The mission was supposed to follow straight after Prognoz 10, but the space program suffered from a severe retrenchment during the early period of the Russian Federation, although spending on space research had begun to decline from around 1988, several years earlier. Scientific programs suffered badly; Interball managed, precariously, to hold out, but only just, for project funding was suspended several times. There were many times when the money stopped and it seemed that the mission might never happen. When the data collection system was not delivered, the system had to be built at short notice in-house (in the event, engineers built a system with ten times more capacity than Prognoz).

The mission was defined at a series of conferences in France, Warsaw in the early 1980s, and, crucially, in Plovdiv in 1982. The scientists returned to the idea, first discussed at the start of the Prognoz program, of the Troika, three satellites flying simultaneously. The three-satellite mission (*Troika*) shrunk to two, of which one would be in a long elliptical orbit in the Earth's magnetic tail, on the opposite side to the Sun, and a second, the auroral spacecraft, in a shorter orbit over Earth's pole. Extra value would be obtained because each would carry a small subsatellite. The four satellites, operating simultaneously, would provide a comprehensive picture of

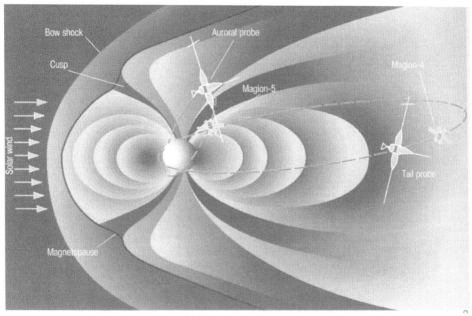

Interball final mission concept

solar-electrical activity in the ionosphere. The concept was to have two pairs of
satellites to make *in situ* measurements of the Earth's electrical and magnetic fields

Leading Czech scientist, Pavel Triska

so as to better understand cause-and-
effect changes, plasma structures in the
tail, and what happens in storms.
Ideally, such data should be coordinated with other satellites, such as the
American Geotail, WIND, and Polar.
The subsatellite mission was demanding, this time separating out to distances of 10,000 km. As mission
scientist, Ingrid Sandahl, explained,
one of the major problems was where
and how plasma and energy from the
solar wind entered the magnetosphere.
In spite of many years of study, the
problem was still largely unsolved,
simply because it was so complex.
Above all, the large-scale picture was
still missing [18]. A specific task was to
search for hot fireballs in the magnetospheric plasma and the explosive heating of plasma that led to auroral bursts
and magnetic storms. The search for

Table 7.16. Prognoz 11 (M 1) (Interball 1, tail probe) instruments.

SKA 1: ion distribution measurements, 50–5,500 eV
Electron: three-dimensional measurements of electron components of plasma
Promics-3: mass spectrometer, ion composition
VDP: ion and electron fluxes
AMEI 2: energy spectra and heavier ions
AP 3: cold plasma spectrometer
Corall: three-dimensional ion distributions
Alpha 3: thermal plasma ion flux
OPERA: electric field fluctuations
MIF-M/PRAM: magnetic field fluctuations
IMAP 2: magnetometer
IFPE: proton and electron fluxes
AKR X: kilometer radiation analyzer
ADS: nine-channel spectra analyzer
FGM-1: magnetic fields
SKA 2: low-charged electric particles
DOK 2X: energy spectra and distributions
RF 15 1: solar X-ray bursts and time profiles
SOSNA: radiation levels
RKI-2: ionizing and ultraviolet solar radiation

hot, magnetic fireballs suggested the mission title, which came from "international" and "ball".

Prognoz M was, like its antecedents, built by the Lavochkin design bureau in Moscow, with the Czech Republic as the main international collaborator. No fewer than 20 countries contributed to the project, the principal ones being Canada, Sweden, and France, as well as the European Space Agency as a formal partner. Each mother craft weighed over a tonne (1,270 and 1,400 kg, respectively). The director of IKI, Albert Galeev, put in charge of the mission plasma physicist Lev Zelenyi. On the Czech side, the leader was Pavel Triska.

The suite of instruments (Table 7.16 and 7.17) was impressive, Interball 1 carrying nine plasma instruments, three energetic particle detectors, and an individual solar X-ray detector, magnetometer, wave analyzer, radio detector, ion composition spectrometer (Sweden), and particle detector (France). Interball 2 had seven plasma instruments and a single ion emitter, magnetometer, wave analyzer, radio detector, energetic particle detector, camera (Canada), and the same Swedish and French instruments. Each Interball could transmit 32 kps in real time and had two memories, one of 30 MB, the other of 120 MB. Whereas the Prognoz series normally operated for half a year, it was hoped that Prognoz M would operate for up to five years.

After a 10-year gap following Prognoz 10, the first Prognoz M satellite, Interball 1, the magnetospheric tail was launched from Plesetsk on 2nd August 1995 into a four-day orbit 193,000 km out. The tail probe had a high out-of-ecliptic orbit so as to reach the high-altitude cusp and subsolar magnetopause on the dayside of Earth

Table 7.17. Prognoz 12 (M 2) (Interball 2, auroral probe) instruments.

SKA 3: electrons and ions
ION: electrons and ions
PROMICS 3: ions, thermal plasma
Hyperboloid: ions
KM-7: cold ions
Alpha 3: ion trap to study thermal plasma ion fluxes
IMAP 3: magnetometer
IESP-2M: electric fields and ULF waves
Polrad: auroral radiation
Memo: electromagnetic waves
NVK-ONCh: VLF electromagnetic waves, energetic fields
DOK 2: electrons, ion beams
UVAI: auroral imager
UFSIPS: auroral oxygen emissions
RON: control over spacecraft potential
RD 1M: dosimetry
ANOD: comparison between different variants of solar batteries

Table 7.18. Magion 4 instruments.

KEM-3: wave complex (Czech Rep, Bulgaria)
SAS: spectrum analyzer (Poland)
ULF: wave-form analyzer (Czech Rep, Hungary)
SG-R8: three-component magnetometer (Romania)
DOK-S: energetic electron and proton spectrometer
MPS, SPS: plasma electron and proton spectrometer (Czech Rep, Russia)
RF: X-ray spectrometer (Czech Rep)
VDP-S: plasma flow detector (Russia, Czech Rep)
LSP: Langmuir probe (Russia, Czech Rep)

and the neutral sheet in the nightside tail. Nine hours and 30 min after orbital insertion, the small 58.7-kg Czech subsatellite Magion 4 separated to begin its separate mission. Magion 4 was a boxed small satellite with X-shaped solar panels carrying a magnetometer, photometer, and search coils with a 32-MB computer and transmitter (details of its instruments are given in Table 7.18). The mission got off to a bad start because one of its booms failed to unfurl, but this did not seem to affect the scientific return and the computer was able to adjust the orientation of the spacecraft to compensate.

Interball 1 continued to transmit until it burnt up in the atmosphere on 16th October 2000, having held 525 communications sessions with Earth. Magion 4 maneuvered and drifted distances of between several hundred and up to 4,800 km during the mission. The intention was that Magion and Interball would cross the bow shock 200 km apart in spring and summer and the magnetotail 2,000 km apart in autumn and winter, with the spacecraft closing and separating for these seasonal

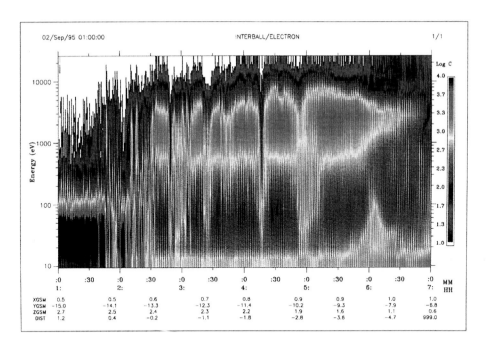

Interball data printout from the *Elektron* instrument, 1995

maneuvers. For the period August 1995–August 1996, Magion crossed the equator up to 40 min ahead of Interball, coming back for rendezvous on 20th July 1996 and flying in close formation until October and then moving up to 120 min behind (January 1997), reuniting in May 1997. Magion 4 worked until September 1997 and decayed on 15th October 2000. The maneuvers were achieved with a remarkably small engine, with a thrust of 0.1 N and a specific impulse of 60–70 sec.

From mid September 1995 to March 1996, the spacecraft went through the plasma sheet every four days and through the bow shock together 130 times. Between them, Interball 1 and Magion 4 made possible precise measurements of the velocity of wind in the magnetopause. Magion 4 carried very long and extremely long-frequency antennae to study wave phenomena in the inner magnetosphere between the equator and 50°N at altitudes of 4,000–12,000 km. Most of the noises it picked up were in the frequency band up to 22 kHz and comprised whistlers, hiss, chorus, and ground transmitters.

The mission was intended follow magnetic storms and improve the forecasting of their effects. Beforehand, on 13th March 1989, Quebec in Canada had suffered a complete short-out because of a magnetic burst. Interball did not have long to wait, for Earth was buffeted by a large magnetic cloud following a coronal mass ejection from the Sun on 6th January 1997, its effects lasting until 11th January. On the early morning of the 11th, Interball was in the perfect position when the trailing edge of the cloud hit the Earth's magnetosphere with ion densities 20 times higher than the average of the solar wind, 170 ions/cm^3, arriving at 300 km/sec.

The second, auroral spacecraft, Interball 2, was supposed to be launched three months after the first. Funding was so tight that the Russian Space Agency proposed to cancel the second mission as unnecessary, but the Institute for Space Research persisted and Interball 2 was launched a year later into an elliptical 5-hr orbit out to 19,202 km but at a much higher inclination. The auroral probe was set to cross the auroral oval to observe the acceleration of particles and the flow of electric current to connect the magnetospheric tail with the conducting ionosphere. Like Granat, its orbit was continually perturbed and had to be re-corrected five times. Nitrogen gas ran out in September 1998 and it was expected that the satellite would lose control. Remarkably, even in a passive mode, it was able to pick up enough solar power to continue to operate until 30th January 1999, when, during its 451st communications session, it was apparent that the battery had discharged and ground control put it into stand-by (*konservatsiya*) mode. There was no reply during the next communications session on 1st February, so it was presumably dead. One can only imagine the surprise when Interball 2 came back on air on 25th January 2000, almost a year later! Apparently, the probe was tumbling in such a way as to trap 30 sec of solar power every minute, sufficient to regenerate the electrical systems. House-keeping data showed that the probe was in good condition and had neither frozen nor over-heated.

The story of Magion 5's adventure was even more dramatic. Interball 2 likewise released its 68.5-kg Czech subsatellite, Magion 5, along with an even smaller 33-kg Argentine microsat, which was taking a piggyback ride on the launch. Initially, no signals could be picked up from Magion 5. First, it was thought that it had failed to separate, but weak signals showed that it was flying separately. There was a full inquiry into the Magion failure, which, in the spirit of *glasnost*, was posted by IKI on its website. The post-mortem verdict: the main spacecraft cut its power supplies to the subsatellite before separation in the normal way but failed to issue the command for Magion to immediately activate its own solar power supplies, so its batteries were quickly exhausted.

Czech ground control at Panska Ves tried to rescue the situation for several days, but to no avail. Their colleagues in the Institute for Atmospheric Physics were not yet ready to grieve and despite the ridicule of their colleagues, for three times a week for the next three years, they sent a signal to Magion 5 to command it to switch on its telemetry. They nearly fell over when, 28 months later on 6th May 1998, Magion 5 responded to a routine command and resumed normal transmission as if nothing had happened. By 17th May, scientific data were being returned. Magion 5 was then 15 min 20 sec or 3,000 km distant from Interball 2.

Not all the scientific instruments had deployed and final deployment was not achieved until August 1999. The little engine was used twice a week to correct the attitude until the gas ran out in 2001. Long after Interball 1, 2, and Magion 4 had concluded, Magion 5 was still on the air. Not only that, but the camera designed to photograph the northern lights transmitted superb images to Earth as it flew over the Arabian peninsula and the Volga basin while its radio detectors sent back multicolor maps of the ionosphere. Separation and closing maneuvers do not appear to have been conducted and Magion 5 essentially followed Interball 2 at a distance of 15 min

Our current model of the Earth's magnetosphere owes much to data from satellites like Interball 1 and 2 (Credit: ESA)

20 sec (3,000 km apart). By 2001, some 41 GB of telemetry had been received in 1,500 hr of operation. It is no surprise that the small Magion satellites became the basis of small satellite development, such as Kolibri (launched from ISS in 2002), Kompass, and Tatyana (see below). Interball was a part of the vast program coordinated by the Inter-Agency Consultative Group (IACG) for space science. It included a number of spacecraft distributed in the different regions of magnetosphere (outside as well) between the L1 and L2 Sun–Earth libration points for solar–terrestrial studies. Significant results were obtained together with other spacecraft from the program (WIND, Polar, SOHO, and others).

INTERBALL: THE MAGNETOSPHERE REMODELED

The results of the two probes, when published, provided detailed charts and information on aurorae, electrons, cold plasma, and electromagnetic fields. The main presentations were made at a conference, *Interball and Beyond*, held in Sofia in February 2002. They were then put on a web data archive by IKI, with over 1,000 scientific papers arising from the mission, with an online archive (250 GB), broken down year by year for project participants and others interested. This was the first time that the results of a Russian scientific mission had been put on the internet and involved the shipping of all the magnetic tapes from the receiving station in Yevpatoria to Moscow for collection. Interball led to a fundamental reconsideration of the nature of the magnetosphere, with models of a smooth structure giving way to new concepts of energy bursts, fluctuations, and filamentary structures. These were the headline findings from the Interball project:

- over 1995–2000, there were 101 magnetic storms, of which 42 were magnetic clouds from coronal mass ejections, the rest from other solar disturbances, mainly Co-rotating Interacting Regions (CIRs); magnetic clouds were the main cause of storms during the rise of the solar cycle, CIRs during its decline;
- magnetic storms were most severe in March/May and September/November, marking the tilt of the Earth's magnetic dipole with the Sun–Earth line;
- magnetic clouds caused the compression and deformation of Earth's magnetosphere, oscillations in the magnetic tail, disturbances in the plasma sheet, and injections of ions and electrons into the polar cap;
- the solar wind was significantly modified when it reached Earth's magnetic field; when ions reached the bow shock, they divided into two: a bell-like hot core and a flat tail, the shock creating a population of suprathermal ions that moved along the magnetic field line downstream towards Earth;
- the inner magnetosphere had a population of cold ions (> 1 keV) surrounded by two spherical layers of more energetic ions (2–15 keV);
- even a quiet solar wind was quite inhomogeneous, with many sub-structures, boundaries, and features;
- the foreshock area was turbulent;
- the low-latitude magnetopause was less stable than at high latitudes;
- high-energy electrons leaked in through the inner boundary layer;
- there were depleted low-density, high-temperature notches in the plasma-sphere during magnetic storms;
- during a storm, nighttime ion temperatures went down;
- dusk ion fluxes were higher than at dawn;
- the electric current of the tail's plasma sheet was full of filaments and this may also be the case in the magnetosphere of the large planets like Jupiter;
- during periods of instability in the magnetopause, plasma clouds and bursts slowed as they moved deeper into the magnetosphere; there, they became less dense, warmed, and turned from fluid shapes into beams;
- there were narrow-band fluxes just inside the high-latitude magnetopause;
- there was a "chorus" of emissions in the magnetosphere from the equator; in April 1997, at high latitudes, minute-long bursts of electrostatic broadband ultra-long frequency were detected for the first time;
- new features of the solar wind were found: middle-scale structures several million kilometers across, sharp pressure pulses, and large plasma variations.

There were two entirely unexpected discoveries:

- 300 events of what were called Almost Monoenergetic Ions (AMIs), 1–20-min beams made of helium and downstream of the bow shock;

- Fine Dispersion Structures (FDS): these were energetic particles on the nightside of Earth in the magnetotail plasma sheet penetrating the inner magnetosphere to drift one to three full turns around the Earth, replenishing the energetic particle population before escaping; between them, these phenomena showed that there were permanent spontaneous disruptions and recoveries of the plasma sheet, even in quiet periods.

Specific measurements were made:

- the width of the magnetosphere as it spread out behind the Earth was measured at up to 500 radii, or 3 million km;
- the cusp was well defined between 4 and 10 radii;
- storms began as far as 15 radii out;
- the velocity of particles in the magnetopause was measured;
- the outer boundary layer was a deflected magnetic field with high electromagnetic turbulence where energetic particles descend and grow in warm plasma sites; the lower boundary layer experienced wave turbulence of 1–5 min;
- the boundary between the magnetopause and the magnetosphere was in constant movement, both locally and generally, with variations of up to 5,000 km;
- proton temperatures ranged from 4,000 to 6,000 K (Interball) to 8,500 K in the depths of the plasmasphere (Magion 5), but at lower geomagnetic latitudes dropped to 2,000 K;
- the effects of radiation on the 12 32-W solar cells were measured; Magion 4 was within the radiation belts for only 4% of its orbit, but even still, solar proton flares were damaging; Magion 5 spent 40% of its time within the radiation belts, but the amount of damage was related to the density of particles at any given time;
- the first measurements were made of low-density magnetotail lobes.

Another interesting result came from coordinated experiments with ground-based facility EISCAT in Norway and the Interball auroral probe on the other side, aimed at magnetospheric–ionospheric coupling. The Interball spacecraft was located at local magnetic midnight at ~8,000-km altitude, higher than the ionosphere, while EISCAT was "heating" the ionosphere. Immediately after the switch-on, variations of the local magnetic field and a burst of 0.1–6-keV electrons were registered over the background of the almost empty flux tube. That was, so far as is known, the first direct observation of the artificially induced field-aligned current similar to that formed during substorms. It is still unclear whether this current will develop to a full natural current circuit or not, depending on the particular state of the magnetosphere.

Individual discoveries were:

- Interball 1 and 2 found beamlets – bursty plasma events of 60 sec with cold electrons of 200–300 eV;
- Interball 1 found patches of dense plasma, or plasmaspheric plumes, after

periods of moderate magnetic activity; by contrast, Magion 4 found areas of high-temperature depleted regions that tended to form around midnight and last two to three days, also called plasmaspheric notches;

- Magion 4 found a large region of low-temperature dense plasma near the magnetopause and regions of stagnant plasma;
- Magion 4 found and measured magnetopause sheets 100 × 500 km and 50 × 200 km;
- Interball 2 found small-scale bursts of electrostatic waves at 2–3.5 radii during magnetic disturbances;
- Magion 5 found noise above the lower hybrid resonance frequencies;
- Magion 4 found electrostatic broadband ultra-long-frequency waves with frequencies below ion cyclotron frequency in high latitudes.

The French instruments achieved significant results. *Elektron* found a sheet of electrons of 100 eV extending 5 radii on the flanks of the magnetosphere between the solar plasma and the internal magnetosphere. *Hyberboloid* found a pulsing of the magnetosphere every 2 min as low-energy cold plasma was transported along the lines of the auroral forces. The experiment *Ion* followed hydrogen and helium ions up to 20 keV as they streamed in from 60 radii out to fall as aurorae at a latitude of 72.4°, especially at around 6 am each morning in autumn 1996. The experiment *IESP* measured the components of ultra-low-frequency waves (0–30 Hz), finding them to be a mixture of electrostatic and electromagnetic [19].

This mission concluded the Prognoz and Magion series. A number of successor projects were sketched, *Vulkan* and *Diagnostika*, but were unable to progress because of the shortage of funding. Europe's Cluster mission, though, was modeled on Interball. On a personal note, Interball marked the conclusion of the career of Yuri Galperin, since the mid 1950s one of the pioneers of Russian space science. Returning from a conference in Japan, he died suddenly on 28th December 2001. The missions are summarized in Table 7.19, followed by the Magion missions (Table 7.20).

Table 7.19. Prognoz M program/Interball.

Interball 1/Magion 4	3 Aug 1995	797–193,000 km, 62.8°, 91 hr 40 min
Interball 2/Magion 5	29 Aug 1996	770–19,200 km, 63°, 5 hr 47 min

Both on Tsyklon from Plesetsk

Table 7.20. Summary of Magion missions.

1	24 Oct 1978	15 kg	406–768 km	Intercosmos 18 *Magik*
2	28 Sep 1989	52 kg	500–2,500 km	Intercosmos 24 *Aktivny*
3	18 Dec 1991	52 kg	438–3,070 km	Intercosmos 25 APEX
4	3 Aug 1995	59 kg	1,000–198,000 km	Interball 1
5	11 Jul 1996	64 kg	1,000–20,000 km	Interball 2

ATMOSPHERIC SCIENCE FROM SOUNDING ROCKETS: VERTIKAL

The *Aktivny*, APEX, and Interball missions built on a series of experiments carried out over the previous 20 years by sounding rockets. Here, we look at the progress of sounding rockets during the later period of Soviet space science.

High-altitude flights by sounding rockets were glamorous in the 1950s, but lost their shine when it became possible to put small satellites into orbit in considerable numbers in the 1960s. Despite that, they continued to play a valuable part in space research, being used to good effect by several countries, such as Britain (the *Skylark* rocket). Such sounding rockets are still valued for their ability to make *in situ* measurements at those altitudes that are too high for balloons but too low for satellites.

The same was true in Russia. The Vertikal program was approved by government decree in 1964 and introduced in 1967. "Vertikal" was the name of a program, a new rocket, and the scientific payload, inevitably causing some confusion. "Vertikal" was an integral part of the Intercosmos program, mainly involving the USSR, Bulgaria, Hungary, the GDR, Poland, and Czechoslovakia. Just to confuse things, the first launch was not actually part of the Vertikal program at all. This was on 12th October 1967 and was called *Verikalnyi kosmicheskyi zond* ("Vertical space probe"), put up on the Cosmos 3 launcher. The hermetically sealed container was made by OKB-10 (see Cosmos 381, Chapter 2). Its objective was to study radiation doses under different protective materials. The second and third launches were Korolev's old R-5A used in the earlier dog flights. Thereafter, the launcher used in the program was the new Vertikal, also called Vertikal Mir, built in the Polyot design bureau in Omsk, based on the Yangel's R-14U, with a length of 21 m. The missions were a trade-off between altitude and payload; it could reach various altitudes: 4,400 km with a small payload; 1,500 km with 860-kg payload; or 500 km with the largest payload, 1,300 kg. It was essentially an R-14 rocket without a second stage. Mission time was up to 52 min. The R-14 Vertikal had a purpose-built payload, which looked like a slimmed-down version of the DS sputnik, but was built by the OKB-10 design bureau. There were two modifications of the probes: atmospheric (VZA) and astrophysical (VZAF), the latter in turn having two variants: recoverable (VZAF-S) and non-recoverable (VZAF-N). The typical mission profile was for a rapid acceleration, the cabin to be fired free at 100 km, so as to prevent contamination of findings. The lid would open on the scientific payload as it continued to soar to altitude, and close again at the appropriate point in the descent. A parachute would open at 6 km and the container would land 20 km from where it started. The scientific container could be reused. One of the functions of the missions was to test out instruments for orbital flight: spectrometers for observing the Sun flown on the first two Vertikals were later installed on Intercosmos 4 [20].

These missions were well publicized in the Soviet press. In the case of Vertikal 8, images were published of the rocket being towed to the launch site before its ascent to 505 km. Later, three members of the recovery team posed with the parachute beside the bucket-shaped landing container on the flat Volga steppeland. There were 11 missions, the 1983 launch bringing the series to an end. Four additional missions were made, starting on 13th October 1966, called *Yantar*, the purpose being to test ion engines

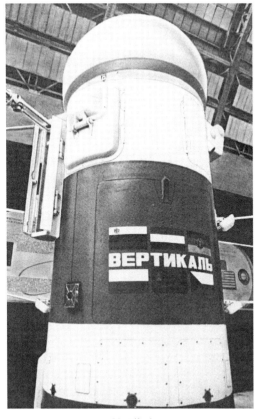

Vertikal

Vertikal container

Vertikal 8 launch crew

and the reaction of the ionosphere to its operation, as well as to scoop up atmospheric nitrogen. Details are provided in Tables 7.21 (launches) and 7.22 (mission focus).

Table 7.21. Vertikal launches.

Vertikal	12 Oct 1967	Cosmos 3	4,400 km, 52 min
Vertikal 2	28 Nov 1970	R-5V	463 km
Vertikal 3	3 Sep 1975	R-5V	502 km
Vertikal 4	14 Oct 1976	Vertikal	1,512 km
Vertikal 5	20 Aug 1977	Vertikal	500 km
Vertikal 6	25 Oct 1977	Vertikal	1,500 km
Vertikal 7	3 Nov 1978	Vertikal	1,500 km
Vertikal 8	28 Sep 1979	Vertikal	505 km
Vertikal 9	28 Aug 1981	Vertikal	505 km
Vertikal 10	28 Dec 1981	Vertikal	1,510 km
Vertikal 11	20 Oct 1983	Vertikal	500 km

All Kapustin Yar. There was also a Vertikal launch to 1,514 km on 18th September 1981 under the program Gruziya Spurt (see below).

Table 7.22. Focus of Vertikal: missions.

Vertikalnyi kosmicheskyi zond*	Earth's upper atmosphere, ionosphere, radiation doses
Vertikal 1	UV and X-ray solar radiation, ionosphere, metcorites (Bulgaria/Hungary/GDR/Poland/USSR/Czechoslovakia)
Vertikal 2	UV and X-ray solar radiation, ionosphere, meteorites (Bulgaria/Hungary/GDR/Poland/USSR/Czechoslovakia)
Vertikal 3	Interaction between Earth's atmosphere and solar corpus cular and wave radiation (Bulgaria/GDR/USSR/Czechoslovakia)
Vertikal 4	Interaction between Earth's atmosphere and solar corpuscular and wave radiation (Bulgaria/GDR/USSR/Czechoslovakia)
Vertikal 5	Short-waves solar corona radiation, meteorites (Poland/USSR/Czechoslovakia)
Vertikal 6	Upper atmosphere and ionosphere, interaction between solar short-wave radiation and the Earth's atmosphere (Hungary/Bulgaria/USSR/Czechoslovakia)
Vertikal 7	Upper atmosphere and ionosphere, interaction between solar short-wave radiation and the Earth's atmosphere (Hungary/Bulgaria/USSR/Czechoslovakia/Romania)
Vertikal 8	Solar corona short-wave radiation (USSR/Poland/Czechoslovakia)
Vertikal 9	Solar ultraviolet rays (Poland/Czechoslovakia/USSR)
Vertikal 10	Upper atmosphere and ionosphere (nighttime launch), parameters of solar short-wave radiation absorption (Bulgaria/Hungary/Poland/Romania/USSR/Czechoslovakia)
Vertikal 11	Solar short-wave radiation (Poland/Czechoslovakia/USSR)

* Not formally in program

Results were published from a number of missions. Vertikal 2 was lucky in that its rapid ascent coincided with a solar flare and it was able to get an X-ray image of the event. Some other results were as follows:

- oxygen densities in the atmosphere varied according to seasons, being higher in October; H_e++ ions were found at the high point of the trajectory (Vertikal);
- the temperature of electrons rose from 1,600 K at 200 km to 4,000 K at 500-km altitude (Vertikal 3);
- there was a gradient change in the concentration of ions at 650 km, with numerous irregularities from 800 to 1,100 km (Vertikal 6);
- discovery of a small source of powerful X-ray radiation at the edge of the Sun's disk (1970) [21].

SMALLER SOUNDING ROCKET PROGRAMS

Vertikal was the main sounding rocket program, but the Soviet Union also used much smaller sounding rockets to improve our knowledge of the atmosphere and ionosphere. In Chapter 1, we noted the introduction of meteorological rockets in the 1950s, the MR-1 series and the MR-05. The latter was replaced by the MR-08, which was fired 540 times between 1959 and 1965. Next was the MR-12, approved by government Decree 240-90 of 23rd February 1960 and developed by OKB-9 of Fedor Petrov (1902–1978). The MR-12 was able to bring 50 kg of scientific instruments to 180-km altitude and operated under the guidance of the Applied Geophysical Institute (1965–1968) and then the Experimental Meteorological Institute (1968–1999). First flights were from Kapustin Yar and Heiss Island in 1966 and several variants were later developed: the M-175, the M-250, the MR-20, and the M-100. The MR-12 was launched from Heiss to release sodium clouds into the polar thermosphere – an experiment developed by French scientists, the first of 30 such experiments over 1967–1972.

The scale of the meteorological rocket program was astonishing: from 1962 to 1997, two years after production closed, over 2,200 rockets were launched, 220 in the peak year. Not only were land locations used in the Soviet Union (Kapustin Yar, Heiss Island), but so too were launch centers abroad, notably Thumba in India, Kerguelen Island in the Indian Ocean (French territory) as well as international locations (Molodezhnaya base, Antarctica). Over 1969–1972, 120 firings were made from Molodezhnaya alone, up to five a month. Sounding rocket campaigns, using various titles, continued throughout the next decades (see Table 7.23). The program appears to have ended with project *Flacus*, the final missions being on 30th January and 6th February 1997 from Kapustin Yar. After a gap of ten years, weather rocket launches resumed on 24th September 2007 when the ship *Mikhail Somov* brought ten M-100B rockets to the Krenkel meteorological station on Heiss Island.

The rockets were small enough to be launched from ships and the Hydro-meteorological service had a number available for this purpose: the *Professor Vize*,

Professor Zubov, and *Academician Chirchov*. These rockets permitted dedicated scientific campaigns of rockets to be launched, gathering data under a variety of conditions. These included launches in cooperation with France off the Kourou Launch Centre (*Professor Zubov*, 1971) and with the United States off the Wallops Island launch center, Virginia, June 1978 (*Professor Vize*), as well as project Jaspic, to study corpuscular sources in the night ionosphere [22].

Apart from measuring atmospheric conditions *in situ*, the sounding rockets were also used to inject gas clouds into the atmosphere and to use electron guns to test their impact on the lower ionosphere, the antecedents of *Aktivny* and APEX (see above). The concept was called *Feyerwerk* ("firework"), in which an electron gun fired a flux of energetic particles and cold plasma into the upper hemisphere. The idea was developed by new IKI director Roald Sagdeev with IZMIRAN and the Paton Electric Welding Institute in Kiev. This used the same type of electron beam equipment used on Soyuz 6 (see Chapter 6), with electron injectors with a power supply of up to 4.5 kW and a voltage of 10 kV. The purpose was to excite the ionosphere luminously at an altitude of 75–162 km, make artificial aurorae, and observe it to model natural aurorae. In the first test, *Zarnitsa 1*, using a MR-12 sounding rocket flying up to 161.5 km on 29th May 1973, beams were injected in three phases and viewed in the 300-km path downrange of Kapustin Yar. The experiment was repeated in *Zarnitsa 2* in June and again on 11th September 1975 and in successor programs, *Splokh 1* and *2* (see Table 7.24). In the course of eight launches, the concept was validated and the ascending rockets could be seen surrounded by an electrical glow that generated electron precipitation.

Feyerwerk 1 became a preparation for a geographically most exotic series, called ARAKS, led by Yuri Galperin from the *Ionozond* program. ARAKS stood for Artificial Radiation and Aurora between Kerguelen and the Soviet Union – though it was also the name of a river between Turkey and Armenia where the first planning meeting for the series had taken place in 1970. The idea was to match Soviet sounding rocket experiments from 70°N, near the Arctic, with equivalent latitudes in the southern hemisphere, so that these two locations were linked with one and the same magnetic flux tube. France conveniently owned an old whaling station in the southern Indian Ocean, Kerguelen Island, so an agreement was signed between the two countries in October 1969. There Galperin and his colleagues encamped for several months to fire two Eridan rockets (26th January and 15th February 1975), surrounded by colonies of penguins. Soviet tracking cameras, AFU-75s, electron guns, and other instruments to measure particles and waves were installed there. The rockets fired from there used electron guns to inject electron beams and plasma blobs into the ionosphere and magnetosphere. Bursts were emitted between 160 and 185 km and used an electro cyclotron to generate whistlers. The electrons and plasma clouds duly followed the force line of the magnetosphere in a high arc and descended over Sogra, Archangelsk, but the brightness of artificial aurora was no higher than a 7th-magnitude star. ARAKS thus, for the first time, succeeded in using artificial low-frequency emission to stimulate electron precipitation from the radiation belts. Radio waves generated by the beams were also studied.

Feyerwerk and ARAKS were followed by *Ariel* (1977–1979), *Aelita* (1978–1979),

ARAKS launch site in Kerguelen

ARAKS rocket in preparation

The ARAKS site was a penguin colony

and *Gruziya Spurt* (Vertikal launch, 18th September 1981). In the case of *Ariel 3*, the electron gun fired with an energy of 400 joules, 1,800 in the case of *Ariel 4*. For *Gruziya Spurt*, so-named because of the participation of the Georgian Academy of Sciences, a much more powerful Vertikal rocket was used because it offered not only high altitude (1,500 against 400 km), but also the opportunity to fire a beam as powerful as 6 keV. These experiments received a hostile reception in the scientific community and the press abroad, although such experiments were conducted worldwide at the time (one was flown on the third shuttle mission). A version of the electron gun was carried on the Phobos mission (see Chapter 5) to ionize the soil of the Moon and measure its composition.

After a four-year gap, experiments then moved offshore so as to conduct them at a wider range of latitudes and longitudes. An MR-20 meteorological rocket was used from the North Atlantic (18°N, 30°W) in 1985 in a Soviet–Polish experiment using an ion gun to inject lithium ions into the ionosphere. The following year, with COMI P, two nighttime launches took place from Kapustin Yar with a cesium plasma jet injected into the bottomside middle-latitude ionosphere.

Then, the Soviet Union returned to one of the geomagnetic points of interest of the original Sputniks, namely the South Atlantic magnetic anomaly, where the inner boundary of the radiation belt dips very low over the anomaly. Project COMBI SAMA (South Atlantic Magnetic Anomaly) saw the launching of sounding rockets to 235 km. The documentation does not say from where the rockets were launched, but hint that it may have been the research ship *Professor Zubov*. The electron gun was fired continuously from the 120-sec point at 165 km in the ascent right up to 235 km, registering a deep modulation of the natural energetic particle flux. These were the experiments timed with the overhead passes of *Ionozond* (Cosmos 1809, see above).

The experiments continued with three MR-20 launches from the ship *Professor Vise* at 69°N in the Chukotskoe Sea in September 1988 and from the *Professor Zubov* in the Norwegian Sea in September 1989 (exact dates are not available). An MR-12 launch in November 1988 succeeded in modeling an artificial plasma bubble. The last launch in the Soviet period was *Contrast 3* on a MR-20 rocket from the *Professor Zubov* at 18°N, 53.2°W in August 1991, which injected a plasma cloud of pure barium vapor into the low-latitude ionosphere at an altitude of 224 km. The neutral barium was quickly ionized by solar radiation, extended to 300 km in length and 40 km across up to 290-km altitude, and then drifted along the magnetic field, with the shipboard scientists observing the cloud through the dusk, night, and following dawn. This launch came at a time of high drama in the Soviet Union, for it was the week in which the Communist government collapsed and may have been the last launch of this strange series. By this stage, active experiments were now being conducted over prolonged periods from orbit by Intercosmos 24 and 25 (see above) [23]. The meteorological and sounding rocket campaigns are summarized in Tables 7.23 (campaigns) and 7.24 (experiments).

Table 7.23. Meteorological rocket campaigns.

Sun atmosphere	1969, 1971, 1973, 1976 (22 launches)
Tropical dawn	1971
Polar morning	1972, 1974
Corpuscular energy sources	1976–1979
Stereotop	1 December 1978
Ipocamp 1 (high atmosphere)	March 1974
Ipocamp 2	March 1977
Ipocamp 3	March 1979
Ipocamp 4	1981

Table 7.24. Electron guns, beams, and plasma blob experiments (*Feyerwerk*).

Zarnitsa 1	30 May 1973 (cesium)
	29 Jun 1973
ARAKS 1	16 Jan 1975 (Kerguelen)
ARAKS 2	15 Feb 1975 (Kerguelen)
Zarnitsa 2	11 Sep 1975 (caesium)
Spolokh 1	4 Sep 1975 (barium)
Spolokh 2	29 Jun 1978 (barium)
Ariel 1	29 Oct 1977
Ariel 2	30 Oct 1977
Aelita 1	6 Oct 1978 (Kapustin Yar) (lithium electron accelerator)

Ariel 3	30 Nov 1978 (Kapustin Yar)
Aelita 1	1 Dec 1978 (Kapustin Yar)
Aelita 2	29 Oct 1979 (Kapustin Yar)
Ariel 4	18 Nov 1979 (Kapustin Yar)
Gruziya Spurt	18 Sep 1981 (Vertikal, Kapustin Yar)
Plasma	18 Mar 1985 (North Atlantic)
COMBI P	17 Sep 1986 (Kapustin Yar)
COMBI P	31 Oct 1986 (Kapustin Yar)
COMBI SAMA	5 Aug 1987 (South Atlantic)
COMBI SAMA	6 Aug 1987 (South Atlantic)
CONTRAST	1 Sep 1988 (Chukots Sea, *Professor Vize*)
COMBI	1 Nov 1988 (Kapustin Yar)
CONTRAST	2 Sep 1989 (Norwegian Sea, *Professor Zubov*)
CONTRAST 3	21 Aug 1991 (North Atlantic, *Professor Zubov*)

SOLAR SCIENCE: KORONAS

Science programs suffered badly during the period of retrenchment of the space program in the 1990s (see Chapter 8). Apart from Interball, the only other science program to be fulfilled was the Koronas series of solar observatories, awarded to NPO Yuzhnoye in Dnepropetrovsk. Historically, the satellite dated back to the government decision on the future of the space program on 3rd August 1964, but the three missions took 45 years to bring to fruition, which must be something of a record.

Koronas stood, in Russian, for Comprehensive Orbital Near Earth Observations of the Active Sun and its aim was to study the Sun's internal structure (helioseismology), to investigate solar activity, predict the impact and arrival time of solar disturbances in the Earth's atmosphere, and to discover the reasons for periodic discharges and eruptions from the Sun. Its specific aims were to learn more about solar eruptions – why they took place, their structure, characteristics, cycles, and solar plasma emitted. Koronas was a three-part series: Koronas I (1994), Koronas F (F for *Fizika*), and Koronas Foton, and also involved scientists from abroad. The program was managed by IZMIRAN with the participation of the Lebedev Institute, Moscow Engineering Physical Institute (MEPhi, or MIFI in Russian), the Institute for Space Research, and several others, including foreign organizations. The first two Koronas were built according to the OKB-586's AUOS design (its precise designator was AUOS SM K1) and had a pointing accuracy towards the Sun of 10 arc minutes. It was broadly similar to the SOHO European–American solar observatory. Polar orbits enabled periods of uninterrupted observations for 20 days at a time.

The first Koronas solar observatory, Koronas I, was launched on 2nd March 1994. The weight was 2,160 kg, lifetime one year, with data returned to IZMIRAN's center in Troitsk. NASA reported that Koronas I suffered an orientation failure after a number of months and this may explain the small number of papers published

Table 7.25. Koronas I instruments.

DIFOS solar optical photometer (Viktor Oraevsky)

Combined solar X-ray telescope and optical coronagraph *Terek*, 5-25, 170–180, 3–4, 4,000–
6,000 Å (I. Zhitnik)

DIOGENESS X-ray photometer–spectrometer
   Solar and ionospheric radio spectrometer (Zbigniew Klos, Sergei Pulinets, Valeri Fomichev)
   High-resolution spectrometer (Janusz Sylwester)]

RES solar X-ray spectrometer in 190–205, 8.41–8.43, 1.85–1.87 Å (I. Zhitnik)

VUSS Vacuum ultraviolet solar spectrometer (Tamara Kazachevskaya, Pavel Svidsky)

IRIS X-ray burst spectrometer (G. Kocharov)

SUFR Solar ultraviolet radiometer (Tamara Kazachevskaya, Anatoli Nusinov)

GELIKON X-ray and gamma-ray spectrometer (E. Mazets)

SORS solar radio spectrometer (0.03–20 MHz)

SKL cosmic rays spectrometer, comprising three instruments: MKL cosmic rays monitor, SKI
   3 space radiation spectrometer, SONG (from Russian *SOlar Neutrons and Gamma
   radiation*) particle monitor for gamma rays (0.1–100 MeV); neutrons (less than 30 MeV),
   protons (up to 1 MeV), and electrons (0.05 MeV or more) (Sergei Kuznetsov)

IMAP magnetometer

AVS amplitude and time spectrum analyzer

from the mission. Koronas I fell out of orbit on 4th March 2001. Its instruments are detailed in Table 7.25, with principal investigators where known.

Terek, based on an instrument that had flown on the Phobos missions to Mars, was one of the most important, for it was to study fine structures in the solar atmosphere, determine hot plasma in active regions, and study coronal holes. The solar X-ray spectrometer provided black-and-white images of the Sun. The high-resolution spectrometer explored energy sources and sinks in solar flares. The beginning of its flight coincided with the solar minimum, so that only at the end of the flight was the period of solar activity increasing. Despite this, it gained some interesting results. On 14th April 1994, the instruments registered a particle flux increase, although there were no solar flares. Later, on 17th April, an uncommon magnetic storm occurred. It was linked to the arrival of solar plasma, associated with a coronal transient, which had accelerated particles near the Sun and then, near the Earth, initiated the beginning of the magnetic storm. This event was also observed by IMP 8, *Ulysses*, and SOHO. The SKI 3 instrument registered oxygen nuclei in the Earth's radiation belts, which were interpreted as geomagnetically trapped anomalous cosmic ray particles. This result confirmed earlier results from the Cosmos spacecraft (see Chapter 2).

Although Koronas I was a solar satellite, it was also turned Earthward. Its first task was to make a global map of radio noise in the 0.1–15-MHz range and its second to detect high-frequency pulses shot through the ionosphere (done by the Sura facility in Nizhny Novgorod on 14th June 1992). The map of Earth's radio noise followed the earlier work of Intercosmos 19, Cosmos 1809, and Intercosmos 25 APEX. The highest levels of radio noise were found over Europe and south-east Asia (from the Chinese coastline to Singapore), caused by transmitters, power stations, electricity lines, and heavy industry, and they were already beginning to have a distinct effect on the ionosphere. Koronas I enabled the compilation of a map of charged particles under the radiation belts from 50°N to 50°S.

Although the successor was delayed for the better part of a decade and at one stage looked unlikely to fly, Koronas F (*Fizika*) was eventually launched on a Tsyklon 3 from Plesetsk on 31st July 2001 (AUOS-SM-KF). It weighed 2,260 kg. Its aims were to spend a year studying solar activity (including active regions, flares, ejections, and the seismology of the solar interior) and the interaction between the solar wind and the Earth's magnetosphere. Its orbits were carefully chosen, so that Koronas F could observe the Sun continuously for the periods of 20 days. Transmissions were sent back every 36 hr to the IZMIRAN institute ground control. Koronas F indeed appears to have carried a similar range of experiments to Koronas I and made good any shortfalls of data from the earlier mission (Table 7.26). The main task of Koronas F was helioseismological studies (glimpsing the Sun's interior by looking at its oscillations), investigation of dynamic processes (flares, plasma eruptions, etc.), and solar cosmic ray studies.

The mission got off to a shaky start with a power out after only three weeks, but from which it quickly recovered. The military passed control over to IZMIRAN in May 2002 and the first raw data were published that autumn on its own dedicated website. This was the beginning of what turned out to be a successful mission. Koronas F coincided with solar maximum (cycle 23) and its subsequent decline. Thirty flares were recorded in the first three years and it was there for one of the greatest periods of solar eruptions over October–November 2005. By December 2001, Koronas F had returned 400,000 images of the Sun and 100 MB of primary information, enabling three-dimensional images of the Sun to be constructed. Many maps were made of the Sun's disk and on it, sunspots, and microbursts. Eventually, a million X-ray spectra were compiled with 500,000 high-resolution X-ray images. Eventually, it fell back to Earth in the Indian Ocean off Kerguelen on 7th December 2005.

Table 7.26. Koronas F instruments.

DIFOS multi channel solar photometer (Viktor Oraevsky)
SPIRIT complex for Sun imaging in X-rays and UV, comprising:
  SRT solar X-ray telescope (I.I. Sobelman, I.A. Zhitnik)
  RES X-ray spectroheliograph (I.I. Sobelman, I.A. Zhitnik)
SUFR ultraviolet radiometer for full disk emissions (Tamara Kazachevskaya)
VUSS ultraviolet spectrophotometer (Anatoli Nusinov)
DIOGENESS X-ray spectrometer photometer for active regions and flares (Janusz Sylwester)
RESIK X-ray spectrometer (Janusz Sylwester)
IRIS X-ray spectrometer for microflares and their precursors (G. Kocharev)
GELIKON X- and gamma-ray spectrometer (E. Mazets)
SKL solar cosmic ray spectrometer, comprising MKL, SKI 3, and SONG instruments (Sergei Kuznetsov)
SPR X-ray spectropolarimeter (Vladislav Pankov, Yuri Kotov)
SPR N X-ray polarimeter (I.I. Sobelman, I.P. Tindo, S.I. Svertilov)
AVS amplitude and time spectrum analyzer (Y.D. Kotov)
RPS 1 X-ray spectrometer, solar flares, and their precursors (V.M. Pankov, Y.D. Kotov)

A conference on the first results was held in the IZMIRAN headquarters in Troitsk in February 2005. Seventy-two scientific papers were published, mostly by Russian authors, but some with American collaborators, covering the elements in coronal structures, the energetics of flares, their temperatures, duration, and energies, matching them against the solar minimum and maximum. Koronas F was also lucky, as the decrease in solar activity was accompanied by very energetic solar events. By 2006, 275 papers had been published. Koronas F was the first solar observatory to make soft X-ray spectra in the range 3.2–6.1 Å. Sunspots were counted daily and matched against the levels of radiation reaching Earth. Rare elements were found in solar flares: potassium, argon, chlorine, and sulfur. These were some of the highlights of the mission:

- 12th November 2001: the finding of a new class of fast dynamic plasma features with temperatures over 20 million K, caused when hot plasma filled magnetic elements in the form of clouds, spiders, loops, waves, and arcs;
- 30th September 2002: the camera recorded a blob of mass ejected out of the corona;
- five extraordinary solar events when the Sun emitted energetic gamma rays and neutrons up to 14 GeV, matched against data from spacecraft in Earth's magnetosphere, at one stage causing a temporary vanishing of the outer radiation belt of electrons above 1.5 MeV.

Turning to Earth, Koronas compiled a map of Earth's nighttime atmospheric glow, which was found to vary according to season and solar activity. Koronas determined the levels of molecular nitrogen and atomic oxygen in Earth's atmosphere up to 500 km and how they varied according to solar activity. These were the main discoveries of the mission:

- Coronal Mass Ejections (CMEs) of speeds of hundreds of kilometers a second in the course of which the solar magnetic field would open, close, and re-open;
- diffuse hot plasma clouds from 0.3 to 0.4 radii out, spider-like arches of gases with a lifetime of a few hours and pulsing flares;
- temperatures in the most dynamic regions of hot plasma were up to 10 million K;
- disturbances in the solar corona out to 3 solar radii, with up to 170 different chemical elements in the corona, such as helium, potassium, and chlorine;
- solar flares were preceded by what was called "pre-flare heating", which could give warning of impending flares; some of these flares were the most intensive ever observed and time profiles were made of X-rays within solar flares, finding a build-up period of 1 min, tailing off over the following five;
- solar flares traveled ten times further out than was realized before;
- only 20% of solar flares caused storms in the Earth's magnetic field [24].

The aim of Koronas *Foton* was to study processes of energy accumulation and its transformation to the accelerated particles' energy during solar flares, mechanisms of particles' acceleration, their propagation and interaction in the Sun's atmosphere, the impact of solar energy on the Earth's magnetosphere and ionosphere, and high-energy particles, the intention being to provide uninterrupted coverage for as long as 25 days at a time (after which it would enter temporary eclipse). The long saga of delays that afflicted its predecessor was replayed. Provision of the AUOS bus took such a long time that, eventually, the Russian Space Agency gave up and the project ended up with the Scientific Research Institute of Electrical Mechanics (VNIIEM), developer of the Meteor weather satellite. As a result, Koronas *Foton* used a Meteor satellite – quite an old design. It required some modification to achieve the intended level of pointing accuracy and an orbital height accurate to 1 km. Koronas *Foton* had much improved downlink relays, 1 MB a second, with a capacity of 8 GB of data a day.

Weighing 1,900 kg, the Meteor had a 540-kg payload of 11 instruments, using 400 W of power per orbit (Table 7.27). Research supervisor was Yuri Kotov of the Astrophysics Institute at Moscow Engineering–Physics Institute, the organization in charge of scientific instruments of the project. In 2005, the Indian Space Research Organization joined the project, supplying a 55-kg low-energy gamma-ray telescope to detect solar and galactic radiation. Specific mission objectives were to investigate energy accumulation in solar flares, fast particles in the solar atmosphere, the acceleration of particles from the Sun, and the chemical composition of solar emissions.

Launch was originally set for 2004, but it kept slipping. Final clearance for launch was not given until a review on 9th October 2008 and the spacecraft was shipped to Plesetsk two months later. On 30th January 2009, Koronas *Foton* rode the very last Tsyklon 3 rocket out of the cosmodrome, achieving the intended orbit. There were some initial difficulties in powering, controlling, and pointing the satellite, but by March, *Foton* had sent back spectacular red and green color images of a blazing Sun and the outer regions of the solar corona against the deep-blue black of space. In one respect, the delay was fortuitous, for it was launched not only at solar minimum, but during an unusually prolonged and quiet minimum. The first pictures were stunning, showing the burning rim of the Sun, wispy ejections, and the blue outer region of the corona merging into the black of the universe.

Sadly, the mission encountered difficulty. In July, one of the three power sources went down (due either to a battery, circuit, or circuit failure). The other two were insufficient to provide adequate power, so instruments had to be temporarily switched off, especially when in the Earth's shadow. The underpowered spacecraft became severely stressed and on 1st December, the instruments were shut down while efforts were made to resolve the problem. The episode raised questions as to whether the move to Meteor system had been the right decision. The project was finally closed on 30th June 2010. During its time, the TESIS telescope/spectrometer obtained around 300,000 images of the Sun, part of which are stored in open access. The series is summarized in Table 7.28.

Table 7.27. Koronas Foton instruments.

Natalya 2M high-energy emission spectrometer
Penguin M hard X-ray emission polarimeter
Konus RF X- and gamma-ray spectrometer
BRM fast X-ray monitor
FOKA multichannel ultra-violet monitor
TESIS extreme ultra-violet telescope/spectrometer
Elektron M Peska charged particles analyzer
SM8M magnetometer
RT 2 low-energy gamma-ray telescope (India)
STEP F electron and proton detector (Ukraine)
SOKOL solar photometer

Table 7.28. Koronas series summary.

Koronas I	2 Mar 1994	487–528-km, 85.2°, 94.7 min
Koronas *Fyzika*	31 Jul 2001	486–530 km, 82.5°, 95 min
Koronas *Foton*	30 Jan 2009	533–560 km, 82.5°, 96 min
Tsyklon 3, Plestesk		

SMALL SPECIALIZED SATELLITES: PION AND KOMPASS

A number of small, specialized science satellites flew at the end of the Soviet period and subsequently. These were Pion and KOMPASS. First, Pion, a specialized program of upper-atmosphere studies, was introduced with the Pion series of small satellites in 1989. These were subsatellites, 45-kg spheres, 33 cm in diameter, released in pairs from Earth resources satellites to carry out about two months of independent flight to test for air density. They were built by students at the Korolev Aviation Institute in Samara and made of glass and magnesium aluminum alloys. Each Pion lasted about a month before falling out of orbit (Table 7.29).

Second, following the original *Ionozond* discoveries of Earthquakes from orbit, specialized instruments and missions were flown to test for Earthquakes. Experimental equipment to test for correlations between electrical activity and changes in Earth's tectonic plates were first tried on board the Salyut 7 orbital station and then a Meteor 3 weather satellite, which, between them, detected 36

Table 7.29. Pion series.

9 Jun 1989	Pion 1, 2	256–268 km, 82.3°, 89.8 min
7 Aug 1989	Pion 3, 4	254–272 km, 82.6°, 89.9 min
2 Sep 1992	Pion 5, 6*	222–233 km, 82.6°, 89.1 min

Date is date of in-orbit release, not launch
* Also called Pion Hermes 1, 2.

unexplained increases in particle detection, of which 34 were followed by
Earthquakes. Five similar instances were noticed by Intercosmos 22 *Bulgaria 1300*.
An Earthquake detector, *Mariya*, shaped like a box on a step ladder, was then flown
on Mir orbital station on its Krystall module, brought up by Progress 33 and
operated for 3,000 hr from January 1988 to October 2000. This found that there
were changes in the electromagnetic field up to five days before Earthquakes, with a
10-fold increase in field levels 150 min beforehand. *Mariya* took 200 measurements
of intensity fluctuations that did match small Earthquake events (less than
magnitude 4). Mir's module Kvant also had an experiment called *Arfa* to measure
correlations between charged particles and seismic activity. An Earthquake detector
called *Arina* was carried on the Resurs DK Earth resources mission, launched on
15th June 2006 [25].

To develop this field further, the Russian space agency ran a competition for a
small satellite for Earthquake prediction – a program called Vulkan, with the
mission specifications set down by IZMIRAN. There were two rival bids –
KOMPASS, from the Makeev design bureau in Miass, and Predvestnik, from KB
Arsenal design bureau in St Petersburg. The competition was won by the
KOMPASS team and two satellites duly launched. KOMPASS weighed 80 kg, of
which 20 kg was experimentation. KOMPASS stood for, in translation, Complex
Orbital Magneto Plasma Autonomous Small Satellite. The first KOMPASS had a
monitor to search for low-frequency waves that might be Earthquake predictors as
well as an NVK detector to make radio topography of the equatorial regions of layer
F2 of the ionosphere and observe the boundary of the plasmasphere. It was launched
as a piggyback on the Meteor 3M1 mission in 2001, but it failed soon after entering
orbit.

By contrast, KOMPASS 2 was the only payload for its rocket, using as a launcher
the Shtil submarine ballistic rocket from the submerged submarine *Ekaterinberg*
from underneath the Barents Sea. KOMPASS 2 carried an Earthquake detector built
by the Nuclear Physics Institute of Moscow State University, with an electrical field
analyzer and radio wave detectors to study electromagnetic radiation, orbital
plasmas in the 50 kHz–17.9-MHz band and the flow of particles. Although it
achieved a correct orbit (399–494 km, 78.9°), things went wrong, the satellite went
out of control, and was declared lost after three days. Despite that, amateurs still
picked up some short signals the following month. Amazingly, because so few stories
like this have happy outcomes, control was recovered six months later in early
December. The Total Electron Content Detector of IZMIRAN made high-precision
electron concentration distribution measurements through the vertical structure of
the ionosphere [26].

The objectives of another microsatellite, *Tatyana*, were primarily educational.
Tatyana was built by students of Moscow Lomonosov University to mark its 250th
anniversary and weighed only 23 kg. Box-shaped with a long boom, it was launched
on 20th January 2005. Its scientific payload was intended for studies of various
processes in the near-Earth space and its upper atmosphere (electron and proton
fluxes in different regions of the magnetosphere), background Earth atmosphere's
ultraviolet glow, including phenomena due to micrometeorites, anthropogenic

KOMPASS

factors, Transient Luminous Effects (TLEs) in the Earth's electrical environment, and lightning. As a fortunate coincidence, on 20th January 2005, a powerful solar flare occurred, so that *Tatyana* could measure solar cosmic ray fluxes. It is also reported to have contributed to the study of lightning in the Earth's atmosphere, especially the phenomenon discovered by NASA's RHESSI astrophysical satellite that terrestrial lightning included many powerful gamma radiation flashes. *Tatyana* assembled an ultraviolet map of the equator marking out its levels of electrical activity [27]. A second was launched in 2009 and the missions are summarized in Table 7.30.

Table 7.30. Specialized small satellites, 2000–2009.

KOMPASS 1	10 Dec 2001	985–1,015 km, 99.6°, 105 min
Tatyana	20 Jan 2005	912–967 km, 83°, 103.8 min
KOMPASS 2	26 May 2006	402–620 km, 79°, 93.5 min
Tatyana 2	17 Oct 2009	815–821 km, 98.8°, 101.3 min

KOMPASS 1 on Zenit 2 from Baikonour; KOMPASS 2 on Shtil from Barents Sea; Tatyana on Cosmos 3M from Plesetsk; Tatyana 2 on Soyuz 2.1B from Baikonour

BIOLOGICAL SCIENCE: BION

With the imperative of flying biological missions as a precursor to manned spaceflight passed, the Soviet Union was now in a position to develop a biological space science program that set its own self-contained objectives and test an ever broader range of animals and plants. To do so, biologists were able to use the 4-tonne cabin of the type used originally for the Korabl Sputnik missions in 1960–1961. It was reliable and mass-produced for a variety of civilian and military missions. As noted in Chapter 2, early biological experiments were carried out as supplementary experiments in the Cosmos program, when Cosmos 92, 94, 109, and 368 carried seeds of radish, beans, tomatoes, cabbage, carrot, yeast, and chlorella, with Cosmos 368 also carrying bacteria and cell cultures. A few biological changes had been noticed, but they were of short duration.

The idea of a series of biological missions was formally adopted by the government in January 1970 [28]. The Bion project was to carry quite a series of cargoes into orbit over the following three decades. Plants, seeds, and seedlings were carried on all missions. Over a hundred rats were carried on the first nine flights. Insects were carried on all but Bion 6. Rarer varieties flown were turtles (Bion 1), fish (Bions 6 and 8), amphibians (Bions 7–11), worms (Bion 8), and eggs (Bions 2 and 8). In the part of the program that attracted most international attention, monkeys were flown from Bion 6 onwards. Three (Bions 3–5) carried centrifuges. These missions are now reviewed.

The first dedicated space biology mission was Cosmos 605, involving participation from the Intercosmos countries. The cargo included 25 white rats, tortoises, insects, beetles (which hatched out successfully), fungi, and bacterial cells. Cells became a new focus of space biology, with scientists eager to learn how microgravity and radiation affect their appearance, adhesion, skeleton, and resistance to toxins. Cosmos 605 threw a 1.4×10^7-V electromagnetic field around the spacecraft in an attempt to provide protection from radiation, deflecting the flow of charged particles, apparently with some success. A mobile laboratory was built to retrieve and examine the cargo immediately after landing. The first results were:

- although the rats were physically active in orbit, afterwards, there were structural changes in their skeletal muscles, limbs, and long tubular bones, which weakened; their central nervous systems were depressed and they showed the equivalent of fatigue in humans; body temperatures fell; recovery took about 25 days; adrenalin flow increased during spaceflight;
- animals ate less and grew more slowly than the control group on the ground; weightlessness appeared to slow down the metabolic processes;
- animals took three to four weeks to recover to the same weight as control animals;
- there was a decrease in bone strength generally and in bone marrow formation;
- fruit flies bred to a second generation entirely normally; flour beetles developed with 100% viability, but frog spawn did not develop, probably because growth was gravity-dependent;

- fungi were quite affected, growing an odd shape, their stems being much thinner and their heads larger [29].

Oleg Gazenko wrote that one of the major findings was that for some life forms, gravity (whether planetary or artificial) was necessary for embryonic life to develop, but for others it was not.

The second dedicated biology mission, Cosmos 690, carried insects, yeast, seeds, plants, and bacterial spores. In an experiment to test what radiation levels animals could endure, the albino rats received doses of high rates of cesium 137 radiation, with up to 880 millirad, close to the limits of lethality (1,200–1,300), and duly developed blotches and hemorrhages on their lungs afterwards, but they did recover. Weightlessness appeared to increase vulnerability to radiation and made recovery longer [30].

This biological program was then extended to Western countries, notably the United States. The roots of such cooperation went back to 1962, with bilateral agreements made between NASA and the USSR Academy of Sciences. This included compilation of a treatise called *Foundations of Space Biology and Medicine*. Although this was never stated publicly at the time, cooperation in space medicine was considered to be the area least likely to cause political problems or security difficulties between the two countries. As a result, there was structured information exchange between the two sides during the 1960s at a time when other forms of cooperation were limited. A joint American–Russian working group on space biology and medicine began meeting in 1971 and was formalized under the Nixon–Kosygin agreement on space cooperation signed in Moscow in May 1972, at the same time as the Apollo–Soyuz Test Project was agreed. Two years later, the Russians offered to fly American experiments on future Cosmos biology missions at no cost. The astonished NASA scientists, who lacked any equivalent program of their own, instantly accepted before anyone on the government side had the chance to say no ("better to ask forgiveness than to ask permission"). There was a prolonged hiatus between the Apollo program and the start of the space shuttle (1975–1981), during which time the Americans had no access to manned or biological missions. Remarkably, the series survived the freezing of cooperation in other fields that occurred during the Carter Administration in 1978–1979 and during the subsequent Reagan presidency (1981–1989).

This new program provided the only international opportunity for space scientists and biologists to have regular access to dedicated medium-duration orbital space biology missions during the Cold War period. Over time, the collaborative program acquired the name "Bion", with earlier missions retrospectively renamed Bion. In the event, there were two international phases with Bion: the Americans were the principal partners for Bions 3–7 and 11 and the Europeans for Bions 8–10. The Bion program had a number of stated objectives:

- generally, to understand the adaptation to weightlessness of a wide range of life forms (e.g. animals, fish, insects);
- specifically, to address particular problems affecting humans in weightlessness, such as space sickness;

- later, to examine the effects of radiation on life forms.

NASA duly joined the next Cosmos biology mission, Cosmos 782, which flew 14 experiments of 20 species of animals and plants, including 25 rats, 500 minnow eggs and tortoises (half in a centrifuge, half in weightlessness), fruit flies (to study ageing), carrots, and cancerous cells. The most important innovation was the use of a centrifuge for the rats provided by a laboratory in Bratislava, Czechoslovakia. Their experience was compared to an identical group in a non-centrifuge section. In the event, the centrifuge could be seen to slow changes to the cardiovascular system, respiration, metabolism, and blood coagulation. Cosmos 782 was launched from Plesetsk and recovered in a Siberian snowstorm 19 days later. Because of the importance of assessing the condition of the returned animals immediately, a field laboratory was set up beside the returned cabin comprising two large heated tents. Cosmos 782 also carried two solid-state spectrometers, one on the inside and one on the outside, to measure the energy in cosmic ray particle impacts. A French–Soviet–Romanian experiment looked at the effects of galactic radiation on seeds and single-cell organisms [31].

The first jointly planned biological mission was Cosmos 936 in 1975. Experiments were proposed by experimenters in both countries and subjected to peer review and criticism from their colleagues. At the other end of the process, missions were reviewed in joint symposia at which results were shared and compared, thus bringing

Bion cabin, with flags of international participants

Biopan experiment (Credit: ESA)

cooperation to a structured standard, international, professional level. The centrifuge again found that artificial gravity did reduce the effect of weightlessness on muscle and bone.

On Cosmos 1129, the USSR installed a *nauka* container (see Chapter 2) on the side, which could be opened to the space environment (and closed in time for re-entry) (this system was subsequently developed as *Biopan* and flown on Foton missions by the European Space Agency). Biological specimens (lettuce and shrimp eggs) were exposed directly to the space environment to test their reactions. Cosmos 1129 also saw the first, though unsuccessful, attempt to breed mammals in space. A centrifuge was carried to test whether insects would hatch normally under different levels of gravity: 0.3, 0.6, and 1 G (they did). The *Oasis 3* garden was flown, with its own light and the plants photographed every 10 min. Tomato and maize grew normally, fungi more slowly than on the ground, while of the *Arabidopsis* plant, only 112 plants were fertile and 47 sterile (compared to 142 fertile and four sterile in the ground controls). Slovakian rats were again flown: when recovered, they had reduced bone volume, reduced bone formation, and increased fat in their bone marrow. For scientists, though, these results always posed a dilemma: how much of this change was due to the real effects of weightlessness, how much due to mission stress and handling? Cosmos 1129 carried plastic detectors to measure the impact of cosmic rays so as to assess the danger they posed to orbiting cosmonauts, as well as fluoride and silicon coatings to test how much they might protect humans [32].

Starting with Bion 6 (Cosmos 1514), the Russians moved some missions to an almost polar orbit (82°) to test exposure to the much harsher cosmic radiation of northern latitudes (the Cosmos 110 dog mission had orbited at the more benign 52°). The results were "quite bad", indicating that even in short periods, cosmic rays release free radicals that damage cells and the DNA system, induce cancer, and slow the rate at which the body repairs itself from damage. Later, Bion 9 (Cosmos 2044) coincided with a high-energy solar proton event of 9 GeV arising from an unusually intense solar flare at 27°S, 10°E, the spacecraft receiving a high dose of radiation at northern latitudes [33].

Bion 6 was also important because it marked the start of the monkey missions. This broke the long Russian tradition of hitherto using dogs as their primary research animal. Monkeys were closer to humans on the genetic scale and although more difficult to handle, held out the prospect of learning more about the effects of space travel on humans, especially the impact of radiation. The monkeys came from a zoo in Sukhumi, Georgia, but later moved to the Institute of Medical Primatology of the Russian Academy of Medical Sciences in Adler, on the Black Sea. The animals were expected to be not more than 5 kg in weight and 440 mm tall, so as to fit their cabin. The first monkey mission, though, was an unhappy one. Two days into the flight, the condition of monkey Bion deteriorated and the cabin was brought back two days later. Bion died soon thereafter, the victim of a strangulated bowel, apparently completely unconnected to the mission. More happily, Bion 6 included 18 pregnant white rats who subsequently produced normal litters.

The week-long Bion 7 (Cosmos 1667) attempted to test whether limbs re-grew if they suffered small amputations (newts were the unfortunate victims of this experiment). One thousand five hundred flies were carried and their hatching in weightlessness observed. There were guppies in an aquarium. Maize seeds and crocuses were carried in the garden on board. Here, there was quite a difference between the plant and animal outcomes. Whereas the plants and seeds changed little, the rats had not only atrophied on their return to Earth, but the first stages of osteoporosis had set in. This became a warning for cosmonauts. Over the years, it became apparent that some parts of the bone structure were more vulnerable to osteoporosis than others – the pelvis and lumber structure most, followed by legs, ribs, hands, with the skull least. Later, it was determined that the bone recovery in cosmonauts could take up to two years [34].

European Space Agency (ESA) countries were the principal partners on the Bions 8, 9, and 10 missions. Bion 8 (Cosmos 1887) resumed active shielding first tested on Cosmos 605, throwing a 305-keV electrical field around the spacecraft for more than a full orbit, 105 min – an experiment that appears to have been successful. Bion 8 famously attracted public attention when one of the two monkeys freed himself from his restraint and began tampering with the controls, sparking the classic headline of "Monkey to ground control: 'I've taken over!'". Worse was to come, for the capsule was misaligned at re-entry (hardly the fault of the monkey) and came down 3,000 km off course in Siberia in temperatures of –15°C. The monkeys were kept warm by villagers who found the cabin but the fish experiments were lost in the cold. Bion 9 carried a record 30 experiments addressing tissues, muscular skeletal growth, muscles, heart tissue, cells, metabolism, and radiation in rats and monkeys. The ESA experiments are summarized in Table 7.31.

Table 7.31. ESA Bion 8 and 9 experiments.

Cosmos 1887/Bion 8		
CARAUCOS 1	Insect eggs	Effects of gravity, cosmic rays on stick insects
DOSICOS 1	Bacteria spores	Radiation damage on seeds, spores
SEEDS 1	Plant seeds	Effects of heavy cosmic rays
Cosmos 2044/Bion 9		
CARAUCOS 2	Insect eggs	Effects of gravity, cosmic rays on stick insects
DOSICOS 2	Bacteria spores	Radiation damage on seeds, spores
FLIES 1	Fruit flies	Effects of radiation, 0 G on fly ageing
PROTODYN	Plants	Effects of 0 G on plants
SEEDS 2	Plant seeds	Effects of heavy cosmic rays

Planning for Bion 10 began in the old USSR, but was eventually launched from the new Russia, the mission being organized against a background of increasing uncertainty, even chaos. There were ten experiments on Bion 10, which brought together ESA on one side and Russia and Ukraine on the other, both under the auspices of the Institute for Bio Medical Problems (IBMP). ESA provided a 42-kg purpose-built biological container/incubator with temperature control and computer memory, *Biobox*, this first mission being dedicated to understanding bone weakening and the reaction to radiation of algae and seeds. The SEEDS and DOSICOS experiments were mounted on four 22-cm-wide circular *nauka* containers with openable lids installed at different locations on the recoverable module. Details of the Bion 10 experiments are provided in Table 7.32, including those that repeated previous Bion missions (notes).

The time of year meant that Bion 10 found itself in what is called an out-of-ecliptic orbit, which meant that it was continuously exposed to sunshine. The Russians must have hoped that this would not present a problem, but after day 9, the internal temperatures had reached 31°C and were still rising. As a result, Bion 10

Table 7.32. Bion 10 experiments.

Name	Purpose	Notes
DOSICOS 3	Measurement of radiation levels	Bion 8, 9
SEEDS 3	Effects of heavy cosmic particles .g. protons on seeds	Bion 8, 9
FLIES 2	Motor activity and lifespans of fruit flies	Bion 9
BONES	Effects of weightlessness on mouse bones	*Biobox*
MARROW	Effects of weightlessness on cells	*Biobox*
OBLAST	Loss of bone mass in weightlessness	*Biobox*
FIBRO	Effects of weightlessness on mouse cells	*Biobox*
ALGAE	Effects of weightlessness on algae cells	
CLOUD	Breeding of fruit flies in weightlessness	
WOLFFIA	Effects of radiation on duckweed	

was brought down two days early, some 800 km east of the originally planned landing site. Soon after the forest landing, journalists were introduced to the happy monkey pair of Krosh and Ivasha eating apples, but the salamanders were less fortunate and seven of the 15 had perished.

The results from Bion 10 were mixed. The airtight seals of three of the four KNA *nauka* containers had broken, making the results unusable, while the temperature rise had compromised some of the other experiments (e.g. ALGAE). Despite this, seven yielded significant results, 24 scientific papers from the mission were published later, and some experiments were re-flown on Foton missions. In the experiments on cultures and cells, all reacted in some way to spaceflight, but there were some ways in which they tended to react (e.g. cell shape) and others in which they did not (e.g. DNA). The FLIES experiment found that young flies were hyperactive in space, for some unknown reason, but died younger (the "high-activity-makes-life-shorter" theory). Heavy particles hit five duckplants in their buds and killed them, while the other duckplants showed stress from being in space, but eventually returned to normal. In the BONES experiment, it was found that the mineralization of the center of the bone was reduced by spaceflight and bone-forming activity was suppressed. The experiments in bone mass reduction probably broke the most ground, for they sparked a debate among scientists as to whether this was due to bones sensing and reacting to weightlessness (the original theory) or whether other factors were at work, such as changes in fluids in the body. This was an important issue that could help with understanding osteoporosis on Earth [35].

However, what brought this joint Russian–American program to a premature end was nothing to do with international politics, but an unexpected adversary: the American animal rights movement. Animal rights activists were never happy about the program, or about American participation in it. In autumn 1996, they barricaded themselves into the office of NASA administrator Dan Goldin in an effort to prevent future missions, objecting to the cruelty involved. They attacked the way in which sensors were implanted in the monkeys before the mission and removed afterwards, each exercise requiring an operation under anesthetic, and that samples were also taken of the monkeys' tissue.

Goldin permitted American participation in the next mission, Bion 11, to go ahead, but on condition that the monkeys be treated "humanely". Flying with them in December 1996 were newts, snails, flies, insects, and bacteria. Macaque monkeys Lapik and Multik went into orbit in December 1996, but their behavior was quite different from the start. Multik adapted quickly to weightlessness and got down to his tests quickly, but Lapik appeared to be quite space sick and barely moved his head or body the first day. Both then carried out their motor tests with their arms and hands, but neither was interested to carry out his pedal tasks. When they landed 130 km north of Kustanai after 14 days, both monkeys greeted their handlers, their doctors describing them as "bright, alert and responsive". They were flown to Orenburg and then the IBMP in Moscow. A full three days after landing, both were anesthetized in order to take biopsies and remove sensors. Lapik became sick and lethargic, recovering after a day, but Multik went into cardiac arrest and none of the efforts of the doctors was to any avail. No post-landing operation had ever been carried out so soon after touchdown, the previous norm being seven days. The post-mortem came to the conclusion that spaceflight may indeed weaken returning animals or humans and their response to drugs, including anesthetic drugs. Control experiments with other monkeys, following similar procedures but without space-flight, did not lead to these unfortunate results. Following this, NASA took the decision that should an astronaut require an operation after returning to Earth, that this be delayed as long as possible after landing. Regardless of the circumstances, this tragic end made the campaigning point for the activists. Several months later, in May 1997, NASA announced that it would not participate in flying monkeys again in the future and that was the end of the program. The outcomes of this mission and the Bion project were published in papers over 30 years and summarized in a special edition of the *Journal of Gravitational Physiology* in 2000 [36]:

- weightlessness induced significant changes in animals, but they were generally reversible;
- there was a reduction in skeletal, bone, and muscle mass;
- bones weaken and bone growth is retarded; muscles atrophy;
- unless exercise is taken, animals suffer from lost muscle strength and a reduced immune system; in particular, the heart is likely to degenerate;
- the reduced level of physical activity retards the release of hormones; cholesterol levels rise;
- injuries to skin and muscles heal more slowly in weightlessness;
- post-flight operations carried out shortly after landing are dangerous and should be delayed as long as possible; spaceflight may reduce the ability to cope with anesthetics;
- flies and larvae hatch less successfully in space than on the ground; the evolution of life appears to be gravity-dependent, at least to a point;
- the shape and metabolism of plant cells changed but the space environment did not cause significant mutations in genes, cell division disintegration, the disruption of hereditary information transfer, or non-reversible physiological changes;
- there were plenty of changes in monkeys and rats especially, but these changes were reversible and there was none that would limit the human presence in space;
- we still do not have enough information on the damage that can be done by radiation to the central nervous system, cardiovascular system, blood, immune system, or intestines;
- centrifuges were useful in the provision of countermeasures;
- space sickness is a problem and cosmonauts should not be expected to carry out high-performance actions early in flight;
- countermeasures are important even for short-duration missions (five days);
- electrical fields help in the reduction of radiation.

The Bion missions are summarized in Tables 7.33–7.35.

Table 7.33. Space biology: Bion missions summary.

Cosmos 690/Bion 2	23 Oct 1974	223–389 km, 62.8°, 90.4 min	21 days
Cosmos 782/Bion 3	21 Nov 1975	227–405 km, 62.8°, 90.5 min	20 days
Cosmos 936/Bion 4	3 Aug 1977	219–396 km, 62.8°, 90.6 min	19 days
Cosmos 1129/Bion 5	25 Sep 1979	218–377 km, 62.8°, 90.5 min	19 days
Cosmos 1514/Bion 6	14 Dec 1983	214–259 km, 82.3°, 89.3 min	5 days
Cosmos 1667/Bion 7	11 Jul 1985	211–270 km, 82.4°, 89.3 min	7 days
Cosmos 1887/Bion 8	29 Sep 1987	216–383 km, 62.8°, 90 min	13 days
Cosmos 2044/Bion 9	15 Sep 1989	205–264 km, 82.3°, 89.2 min	14 days
Cosmos 2229/Bion 10	29 Dec 1992	218–376 km, 62.8°, 90.5 min	13 days
Bion 11	24 Dec 1996	217–379 km, 62.8°, 90.5 min	15 days

Table 7.34. Space biology: Bion payloads.

Cosmos 605/Bion 1	Tortoises, 25 rats, insects, fruit flies, four beetles, mushrooms, fungus
Cosmos 690/Bion 2	Albino rats
Cosmos 782/Bion 3	25 rats, minnows, flies, carrots
Cosmos 936/Bion 4	Fruits flies, seeds, white rats, carrots
Cosmos 1129/Bion 5	38 rats, 60 quail eggs, carrots, insects, bird and mammal embryos
Cosmos 1514/Bion 6	*Abrek, Bion*; 10 rats
Cosmos 1667/Bion 7	*Verny, Gordy*; 10 rats, white mice, fish, insects, tritons, 10 newts
Cosmos 1887/Bion 8	*Yerosha, Dryoma*; insects, 10 rats, fish
Cosmos 2044/Bion 9	*Zhakonya, Zabriaka*; rats, tritons, fish, flies, beetles, ants, worms, stick insects
Cosmos 2229/Bion 10	*Krosh, Ivasha*; Spanish newts, beetles, larvae, frog eggs
Bion 11	*Lapik, Multik*; salamanders, flies, plants, seeds, algae

Monkey names in italics.

Table 7.35. Space biology: participating countries, apart from USSR/Russia.

Cosmos 605/Bion 1	(USSR only)
Cosmos 690/Bion 2	(USSR only)
Cosmos 782/Bion 3	USA, Czechoslovakia
Cosmos 936/Bion 4	USA, France, Poland, Romania, GDR, Bulgaria, Czechoslovakia
Cosmos 1129/Bion 5	Bulgaria, Hungary, GDR, Poland, Romania, France, Czechoslovakia
Cosmos 1514/Bion 6	USA
Cosmos 1667/Bion 7	France, USA
Cosmos 1887/Bion 8	USA, Poland, GDR, France, ESA, Romania, Hungary, Bulgaria, Czechoslovakia
Cosmos 2044/Bion 9	USA, France, Canada, ESA states, Norway
Cosmos 2229/Bion 10	USA, Canada, Czech Rep, Lithuania, Poland, Ukraine, Uzbekistan
Bion 11	USA, China

Later, the Russians announced plans to revive the Bion program. The TsSKB Progress design bureau in Samara detailed plans for a more sophisticated version, Bion M, able to orbit at 450 km for up to 60 days, equipped with solar panels to supply power to 800 kg of experimentation. Intended passengers were rats, mice, gerbils, amphibians, reptiles, crustaceans, mollusks, fish, insects, bacteria, plants, and cultures.

MATERIALS SCIENCE: FOTON

Foton was, like Bion, a part of the Cosmos program that was broadened into an international collaborative venture. Materials processing experiments first began on the Salyut orbiting stations (see Chapter 6). They continued on Mir and the ISS but were also flown on dedicated unmanned missions. As with Bion, the Soviet Union adapted the Korabl Sputnik cabin for use as a dedicated Earth-orbiting materials-processing satellite for 14–16-day missions. The purpose of the Foton missions was to enable experiments to be conducted in processed alloys and optical materials and in the testing of semi-conductors. There was no equivalent of the Foton program in the United States or Europe (though there was in China, with the FSW program). The first Foton mission was within the Cosmos program (Cosmos 1645 in 1985). Foton was formally introduced in its own name on 14th April 1988 as an international, collaborative program for the manufacture in orbit of semi-conductors and extra-pure materials. Foton was built by the Kozlov TsSKB Bureau in Samara, weighed up to 6.4 tonnes, and orbited at $62.8°$ at between 220 and 400 km from Plesetsk. Chemical batteries provided 400 W of power and its 2.3-m-diameter cabin had a payload of 700 kg in a volume of 4.5 m^3. Standard equipment included four experiments, listed in Table 7.36.

In order to protect equipment during the final stages of the descent to Earth, solid rockets fired under the three 27-m-diameter parachutes to cushion the final fall. A feature of the series was that it used already flown capsules. Foton was a modest earner for the Russian space program, as much as €20m a mission, enabling Western companies and agencies to have access to zero gravity in a way not otherwise possible. Although Foton concentrated on material science experiments, Foton also began to carry some European (mainly French) life sciences experiments from the Bion program.

Foton 9, on 14th June 1994, carried a French experiment, *Gezon*, to study the melting of materials and a European *Biopan* experiment to study the behavior of amino acids in space. Foton 10 was a joint European Space Agency/French CNES

Table 7.36. Standard Foton instruments.

Splav-2 furnace
Zona furnace (Zona 1, 4)
Kashtan electrophoresis unit
Konstanta furnace (for glass)

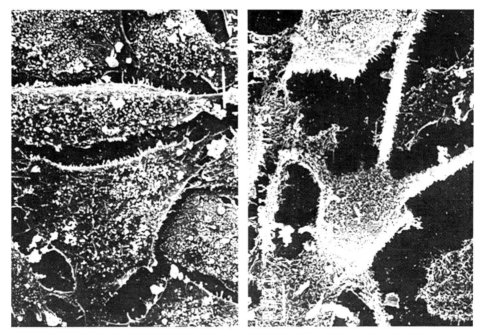

Cells affected by radiation (Credit: ESA)

mission that ended in disaster. Launched on a 15-day mission out of Plesetsk on 16th February 1995, the Foton carried semiconductors and the ESA *Biobox* with shrimps and urchins. The cabin returned to Earth perfectly on 3rd March near Orenburg in the Urals but when it was lifted out of the recovery area the next day, the helicopter flew into a blizzard. Gusts of wind seized the capsule, causing it to sway alarmingly and the Mil crew dropped the payload from an altitude of several hundred meters. The experiments were smashed.

The Foton 11 replacement mission took place on 9th October 1997, most of the experiments being re-flights from Foton 10 (this was part of the arrangement if something went wrong). Launched from Plesetsk, it carried German experiments devised by Kayser Threde and a small cabin called Mirka (standing for microgravity re-entry capsule). The Mirka capsule was a spherical 150-kg cabin, 1 m in diameter, carried on the front of the main cabin, separated after retrofire, with its own ablative material, parachutes, and beacons. Mirka carried three experiments – one on heat shield instrumentation, another on heat flow, and a third to test new types of ablative material. The main cabin carried crystal growth experiments, human cell biology experiments to learn more about cancer, flies (to test for ageing), and beetles (to see how their biological clocks were affected by zero gravity). Both the main cabin and the Mirka touched down safely on brown steppe grass near Omsk on 23rd October. Foton 12 flew in September 1999, carrying experiments from Russia, Germany, China, France, and Sweden. Foton 12 included a furnace, a machine for measuring microgravity, a biological container, and a device to measure electromagnetic emissions.

Details inside the Bion/Foton cabin

Following Foton 12, the cabin was improved and called the Foton M series, with an increased payload (660 kg), longer lifetime (14 days), higher orbit (400 km), higher-performance lithium batteries, and enhanced attitude control to improve the pureness of substances made in microgravity. The first M-1 mission began on 15th October 2002, carrying 650 kg of experiments from the European Space Agency and the Canadian Space Agency. In a rare failure of the Soyuz rocket, an engine exploded, the rocket crashed back to Earth, and the payload was lost in the fireball. A replacement mission was again organized: this was Foton M-2, launched from Baikonour on 31st May 2005, the first time a Foton had launched from Baikonour. This carried 385 kg of experiments in the areas of biology, fluid physics, material processing, meteorites, radiation dosimetry, and exobiology. Six experiments were Russian, five European, and there were seven others. The mission was controlled from the ESA payload operations center, based in Esrange, Kiruna, Sweden. It made a smooth landing 16 days later and ESA specialists were on hand to retrieve the cargo.

Foton M-2 carried no fewer than 44 ESA experiments. The most unusual was probably STONE and this experiment related to strange events in Antarctica several years earlier. There, in 1981, American investigators found meteorites, apparently from Mars, challenging previous views that meteorites could only come from the asteroid belt. The idea that they could have been shot into space and reach Earth as the result of a high-speed impact on another planet had been seen as outrageous – but it could now be seen as possible and, moreover, a big boost to panspermia theorists. By 2008, NASA had collected 17,000 meteorites from Antarctica, preserved in the cold, proportions of which came from the Moon and the asteroid belt, with 39 identified as Martian. In a dramatic development, one of them appeared to have fossilized worm-like life forms that had survived crashing through the Earth's atmosphere and had been preserved in the rock ever since.

Martian life? Many people were, naturally, skeptical. Now, rock samples resembling Martian meteorites were embedded with microbiological life to see if and how they could survive re-entry. Three types of rock were tried: igneous basalt (lost), sedimentary dolomite (burned up), and simulated Martian regolith (basalt, carbonate, and sulfate, which survived). The microbes, though, did not survive either the heat or the deceleration of re-entry – but that did not prove that life could never travel between the stars, just not quite like this. Panspermists pointed out that whereas most bacteria try to reach light, some might be able to burrow into rock to survive [37].

Another unusual experiment on Foton M-2 concerned lichens. Here, lichens (*Rhizocarpon geographicum*) were taken from 2,000 m up in the high Siera de Gredos mountains in Spain and installed on the *Biopan* experiment. Investigator Rosa de la Torre found that despite 15 days in space, the lichens made a recovery of more than 90% on their return to Earth, showing how lichens could live in vacuum, extreme temperatures, and solar and cosmic radiation, presumably giving renewed heart to panspermists.

Foton M-3 brought findings about how life may be spread in the universe (Credit: ESA)

Table 7.37. Foton materials science missions.

16 Apr 1985	Cosmos 1645*	215–390 km, 62.8°, 90.5 min
21 May 1986	Cosmos 1744*	219–373 km, 62.8°, 90.4 min
24 Apr 1987	Cosmos 1841*	218–381 km, 62.8°, 90.5 min
14 Apr 1988	Foton 1	217–376 km, 62.8°, 90.4 min
24 Apr 1989	Foton 2	218–383 km, 62.8°, 90.5 min
11 Apr 1990	Foton 3	217–376 km, 62.8°, 90.4 min
4 Oct 1991	Foton 4	215–396 km, 62.8°, 90.6 min
8 Oct 1992	Foton 8	220–360 km, 62.8°, 90.3 min
14 Jun 1994	Foton 9	220–363 km, 62.8°, 90.3 min
15 Feb 1995	Foton 10	220–369 km, 62.8°, 90.4 min
9 Oct 1997	Foton 11	218–375 km, 62.8°, 90.4 min
7 Sep 1999	Foton 12	217–384 km, 62.8°, 90.5 min
15 Oct 2002	Foton M-1	Fail
31 May 2005	Foton M-2	259–292 km, 63°, 90 min
14 Sep 2007	Foton M-3	258–280 km, 62.9°, 90 min

All on Soyuz rocket. All from Plesetsk, except Foton M-2 (Baikonour) * Retrospectively named Fotons 5, 6, and 7.

The Foton M missions carried more cargoes taking refuge from the Bion program. Foton M-3 carried cockroaches, snails, lizards, butterflies, gerbils, and fish, as well as an exotic experiment developed by young European scientists to lower a 5-kg cabin on a 30-km tether to test a new re-entry technique (this seems to have worked, but the cabin was not found to prove it). The two outstanding results from Foton M-3 concerned, first, cockroaches and, again, Martian meteorites. First, cockroaches were flown on board, the inspiration of students in Voronezh Medical Academy. They had enough time to conceive during their 12 days in orbit and subsequently gave birth to healthy baby cockroaches on their return. Second, Martian meteorites – all those found in Antarctica were basaltic in nature and black in color, but was that all? The wanderings across Mars of NASA's rovers *Spirit* and *Opportunity* confirmed that Mars had been a warmer, wetter, watery world before, with remnants of life more likely to be found in sedimentary rocks. Basaltic rocks had been flown through re-entry before, but, here, Foton M-3 attempted to fly two samples of sedimentary rocks with known bacteria within them through the 1,700°C heat of re-entry. Both survived, although severely ablated into a whitish color. The bacteria did not survive, but their pompeified remains could be discerned. The lesson was that Antarctic scientists should now look for not just black basaltic rocks, but – more difficult in the snowy landscape – white-colored sedimentary meteorites. The Foton missions are summarized in Table 7.37.

LATER SOVIET SPACE SCIENCE: WHAT WAS LEARNED?

During this final period, Soviet space science became focused on a series of priorities: astronomy and astrophysics; the Sun; biology; materials science; the magnetosphere and ionosphere; and materials processing. Although individual instruments and experiments continued to be flown within the Cosmos program, space science missions were more and more flown as identifiable, dedicated, and coherent projects. The results from these missions were considerable, especially *Ionozond, Oval*, individual Intercosmos missions, and Interball. Several late Soviet epoch missions provided significant data into the Russian period, the star performer being Granat.

The fall in funding for space science in the 1990s was such that no new programs could be undertaken. Those that were flown were leftovers designed during the Soviet period, notably Koronas (a project first approved in 1964) and Interball (which dated to the 1970s). Persistence by the scientists enabled these missions to fly in a difficult economic climate, rewarded by the results from both. The biological and materials processing programs Bion and Foton were sustained by Western funding, without which it is doubtful whether they would have continued. Ingenuity is also evident in the way in which access was obtained to the Integral mission. Not until well into the Russian period did prospects begin to improve, to which we turn in Chapter 8.

In summary, the following was learned from later Soviet space science:

- astronomy and astrophysics:
    characterization of X-ray pulsars, supernovae, rapid bursters, neutron stars;
    characterization of gamma radiation sources and stars;
    mapping of the galactic center;
    new black holes;
    annihilator sources;
- the Sun:
    characterization of solar flares, sunspots, solar storms, coronosphere;
    structure of solar atmosphere, active regions, coronal holes and sinks, plasma regions;
    helioseismology, new glimpses into the Sun's interior;
- Earth's magnetic environment and ionosphere:
    mapping, characterization of the auroral oval;
    charting of the ionosphere, main ionospheric trough, boundary of magnetosphere;
    electron density profile of magnetosphere;
    finding of plasma blobs, discrete plasma emissions;
    measurement of magnetic anomalies;
    finding of new electromagnetic structures, shock waves, and irregularities;
    testing of effects of artificial energy releases (ion, cesium, lithium, plasma blobs);
    impact of solar magnetic cloud;
    measurement of complex structures, components of solar wind;
    growing effects of human electrical activity;
    nature of lower-latitude aurorae;
- atmosphere:
    measurement of energy balance in atmosphere, release of radiation;

profiles of temperatures, densities;
electrcomagnetic waves a predictor of Earthquakes;
- biology:
  potentially serious radiation damage from high-altitude flight into radiation belts;
  plants, cells, DNA vulnerable to radiation, free radicals;
  electrical shielding can be an effective countermeasure;
  animals, humans suffer same deterioration from 0 G, especially to bones, heart;
  changes are reversible;
  slower metabolic processes in 0 G;
  some forms of life can breed successfully in space, such as cockroaches;
  artificial gravity an effective countermeasure for small animals;
  unusual effects from spaceflight, such as fly life, shape of fungi;
  post-flight surgical operations should be delayed as long as possible;
  lichens can survive re-entry;
  there may be sedimentary Martian meteorites on Earth, such as in Antarctica.

REFERENCES

[1] Johnson, Nicholas L.: *Soviet Year in Space, 1990.* Author, Teledyne Browne Engineering, Colorado Springs, Colorado, 1991; Lardier, Christian: L'avenir est aux raons x et gamma. *Avimag*, 926, 15 octobre 1986; Zeleny, Lev: *Is the Golden Age of Russian Space Science Still Ahead?* Novosti, Moscow, September 2006.

[2] Central Intelligence Agency (CIA): Scientific Intelligence Report, Soviet Space Research Program, Monograph XI, *Astronomical Aspects.* Washington, DC, 1959; Miller, Arthur I.: *Empire of the Stars – friendship, obsession and betrayal in the quest for black holes.* Little Brown, London, 2005; Academy of Sciences of the USSR: *Annual Meeting, 24–6 February 1960, minutes.* US Department of Commerce, Office of Technical Services, Washington, DC, 29 August 1960. For the story of the Leningrad physicists, see Segrè, Gino: *Faust in Copenhagen – a struggle for the soul of physics.* Jonathan Cape, London, 2007.

[3] Galeev, Albert; Tamkovich, G.M., eds: *35th Anniversary of the Institute of Space Research of the Russian Academy of Sciences.* Author, Moscow, 1999.

[4] White, N.W.; Filipov, L.G., eds: *Physics of Compact Objects, Advances in Space Research*, Vol. 8, No. 2–3, 1988; Sagdeev, Roald Z., ed.: The principal phases of space research in the USSR, in USSR Academy of Sciences, History of the USSR, New Research, 5, *Yuri Gagarin – to mark the 25th anniversary of the first manned spaceflight.* Social Sciences Editorial Board, Moscow, 1986; Kurt, Vladimir; Sheffer, E.K.: Astron X-ray experiment. *Earth & the Universe*, Vol. 2, No. 84; Boyarchuk, A.A.: Astron – window to ultraviolet space. *Earth & the Universe*, Vol. 5, No. 84; Zvereva, A.M., *et al.*: The abundance of ammonia in Comet Halley derived from ultraviolet spectrophotometry by Astron and

International Yltraviolet Explorer. *Astrophysical Journal*, Vol. 404, 10 February 1993, summarized in NASA Technical Reports.

[5]   For background to and details of Kvant, see Hendrickx, Bart: The origins and evolution of Mir and its modules. *Journal of the British Interplanetary Society*, Vol. 51, No. 6, June 1998; Evans, Ben: Delving into Mir's astrophysical attic. *Spaceflight*, Vol. 49, No. 5, May 2007.

[6]   Sunayev, R.A., *et al.*: Highlights from the Kvant mission and hard X-rays from supernova 1987a, in Bleeker, J.A.M.; Hemsen, W., eds: *X-ray and Gamma Ray Astronomy, Advances in Space Research*, Vol. 10, No. 2, 1990; Sunyaev, R.A.: Results of Mir Kvant 1987–9, in Bassani, L.; Palumbo, G.G.C.; Vedrene, G., eds: *Recent Results and Perspectives in Instrument Development in X- and Gamma-Ray Astronomy, Advances in Space Research*, Vol. 11, No. 8, 1991; Spiteri, George: Mir's first module – the inside story. *Spaceflight News*, March 1990; Sunayev, R.A.: The radiography of stars. *Soviet Science & Technology, Almanac*, 1990; Maisack, M.: Broadband X-ray observations of Cir X-1, in Drew, J.E., ed.: *New Developments in X-ray and Ultra Violet Astronomy, Advances in Space Research*, Vol. 16, No. 3, 1995; Barnet, D., *et al.*: Sigma/ Grant observations of hard X-ray emission from type I X-ray bursters, in Trümper, J.; Cesarsky, C.; Palundo, G.G.C.; Bignami, G.F., eds: *Space Astronomy, Advances in Space Research*, Vol. 13, No. 12, December 1993; Kaniorsky, A.S., *et al.*: Three hard ray transients GRS 1716-24, GRS 1009 45 and GROJO422 + 32 in broad band observations by Röntgen Mir, in Day, C., *et al.*, eds: *Black Holes, Future Missions to Primitive Bodies, Advances in Space Research*, Vol. 19, No. 1, 1997.

[7]   For the background to Granat, see Verigin, V.: The nine years of Granat. *Novosti Kosmonautiki*, No. 2, 193, 1999; Johnson, Nicholas L.: *Soviet Year in Space, 1990.* Author, Teledyne Browne Engineering, Colorado Springs, Colorado, 1991.

[8]   *Lettre de CNES*, 129, 131; Goldwurm, A.: Sigma observations of the low-mass X-ray binaries of the galactic bulge, in Gehrels, N., ed.: *Gamma Ray Astronomy, Advances in Space Research*, Vol. 15, No. 5, May 1995; Grebenev, S.A., *et al.*: Observations of hard galactic centre source IE 1740.7-2942 with Granat, in Gehrels, N., ed.: *Gamma Ray Astronomy, Advances in Space Research*, Vol. 15, No. 5, May 1995; Cordier, B., *et al.*: First soft image of galactic centre; Terekhov, O.V.: First results of the Phoebus Soviet French gamma burst experiments on Granat; Golenetski, S.A., *et al.*: Gamma ray bursts with Konus B on Granat; Gilfanov, M.R., *et al.*: First results of X-ray pulsar observations with ART-P telescope on Granat; Lund, Niels: First results from wide field X-ray monitor on Granat; Rogues, J., *et al.*: First observations with Sigma/ Granat; Natalucci, L.: X-ray observations of the Crab pulsar with Sigma, all in Bassani, L.; Palumbo, G.G.C.; Vedrene, G., eds: *Recent Results and Perspectives in Instrument Development in X- and Gamma-Ray Astronomy, Advances in Space Research*, Vol. 11, No. 8, 1991; Sazonov, S., *et al.*: Long term observations of 4U 1700-37 by the Granat all-sky monitor, in Drew, J.E., ed.: *New Developments in X-ray and Ultra Violet Astronomy, Advances in Space Research*,

Vol. 16, No. 3, 1995; Grevenev, S.A., *et al.*: Spectral states of galactic black hole candidates – results of observations with ART-P Granat, in Day, C., *et al.*, eds: *Black Holes, Future Missions to Primitive Bodies, Advances in Space Research*, Vol. 19, No. 1, 1997; Lutovinov, A., *et al.*: Timing of accreting neutron stars with ART P telescope on Granat, in Drew, J.E., ed.: *New Developments in X-ray and Ultra Violet Astronomy, Advances in Space Research*, Vol. 16, No. 3, 1995; Gilfanov M., *et al.*: Hard X-ray spectral properties and discovery of narrow annihilation line in spectrum of Nova Muscae; Laurent, P., *et al.*: Sigma observations of the soft gamma ray source GRS 1758-258; Claret, A., *et al.*: Sigma observations of new hard X-ray and soft gamma transients GX354-0 and Nova Persei; Paul, J., *et al.*: High angular resolution observations in soft gamma ray band with Sigma; Cordier, B., *et al.*: IE1740.7-2942 revisited; Bouchet, L., *et al.*: Sigma hard X-ray observations of GX 339.4, all in Trümper, J.; Cesarsky, C.; Palundo, G.G.C.; Bignami, G.F., eds: *Space Astronomy, Advances in Space Research*, Vol. 13, No. 12, 1993; Trotter, G., *et al.*: Radio and X-ray/gamma ray observations of two solar flares, in Pick, M.; Machado, M.E., eds: *Fundamental Problems in Solar Activity, Advances in Space Research*, Vol. 13, No. 9, September 1993; Polishchuk, G.M.: Perspective of Russian space activities for scientific research, in Zakutnyaya, Olga; Odintsova, D. eds: *Fifty Years of Space Research*. Institute for Space Research, Moscow, 2009; Pavlinsky, M., *et al.*: X-ray burster SLX 1732-304 in the globular cluster Terzan 1 – observations with the ART-P on Granat, in Drew, J.E., ed.: *New Developments in X-ray and Ultra Violet Astronomy, Advances in Space Research*, Vol. 16, No. 3, 1995. Finogrenov, A.: Granat/Sigma observations of X-ray Nova Persei, 1992, in Day, C., *et al.*, eds: *Black Holes, Future Missions to Primitive Bodies, Advances in Space Research*, Vol. 19, No. 1, 1997; Churazov, E.: LMXBs and black hole candidates in the galactic centre, in Day, C., *et al.*, eds: *Black Holes, Future Missions to Primitive Bodies, Advances in Space Research*, Vol. 19, No. 1, 1997.

[9] For the background and origins of the Gamma mission, see Galper, A.A.: Experiments in space in the field of gamma ray astronomy. *Earth & the Universe*, 1991, No. 4; Hendrickx, Bart: The origins and evolution of Mir and its modules. *Journal of the British Interplanetary Society*, Vol. 51, No. 6, June 1998.

[10] Laikov, N.G., *et al.*: Energy spectra of solar flare emission in range 0.03–2GeV registered by Gamma, in Pick, M.; Machado, M.E., eds: *Fundamental Problems in Solar Activity, Advances in Space Research*, Vol. 13, No. 9, September 1993; Akimov, V.V., *et al.*: Geminga pulsar observations with gamma telescope Gamma 1; Akimov, V.V., *et al.*: Time variations of high energy gamma emission of Vela pulsar observed by Gamma, in Trümper, J., *et al.*, eds: *Space Astronomy, Advances in Space Research*, Vol. 13, No. 12, 1993; Alexandrin, S.U.: High-energy charged particle bursts in the near-Earth space as Earthquake precursors. *Annales Geophysicae*, Vol. 21, 2003.

[11] Biryukov, A.S., *et al.*: Dynamics of Precipitating Particle Fluxes in Magnetic Storms, *Advances in Space Research*, Vol. VI, No. 1, *Physics of Planetary Atmospheres*. COSPAR, Paris, 1981; Horwitz, J.L., ed.: *Geospace Plasmas, Advances in Space Research*, Vol. 8, No. 18, 1988; Kuzentsov, S.N.: Energetic

charged particle fluxes beneath the radiation belts. Proceedings of the ICRC, 2001 (Intercosmos 17); Biryukav, A.S., *et al.*: Penetration boundary of solar cosmic rays into Earth's magnetosphere during magnetically quiet times. *Cosmic Research*, Vol. 21, 1983.

[12] Kidger, Neville: The Intercosmos program, 1967–80 – an overview. *Spaceflight*, Vol. 23, No. 6, June 1981; Kocheva, N.A.: Longitudinal variation of daytime equatorial ionosphere from Intercosmos 19, in Rawer, K.; Bradley, P.A., eds: *Ionospheric Informatics and Empirical Modelling, Advances in Space Research*, Vol. 10, No. 8, 1990; Rothkaehl, Hanna: HF noises as indicator of the ionospheric trough through location, in Sandahl, A.; Saunders, M.A., eds: *Auroral and Related Phenomena, Advances in Space Research*, Vol. 13, No. 4, April 1993; Deminova, G.F.: Fine structure of F2 longitudinal distribution in night time low latitude ionosphere from Intercosmos 19 data, in Bilitza, D., *et al.*, eds: *Description of Low Latitude and Equatorial Ionosphere in the Internal Reference Ionosphere, Advances in Space Research*, Vol. 31, No. 3, 2003; Horwitz, J.L., ed.: *Geospace Plasmas, Advances in Space Research*, Vol. 8, No. 18, 1988; Korobeinikov, V.G., *et al.*: *Registration of ELF Waves in Rocket Satellite Experiment with Plasma Injection, Advances in Space Research*, Vol. 12, No. 12, 1992; Oraevsky, Viktor; Triska, Pavel: Active Plasma Experiment – project APEX, *Advances in Space Research*, Vol. 13. No. 10, 1993; Galperin, Yuri, *et al.*: Detection of election acceleration in the ionospheric plasma under the influence of high-power radio radiation near the local plasma frequency aboard the space vehicle Intercosmos 19; Kishcha, P.V.; Kochenova, N.A.: Model for the height of the ionosphere maximum in the main ionospheric trough zone. *Geomagnetism and Aeronomy*, Vol. 35, No. 6, 1996; Pulinets, Sergei; Benson, R.F.: *Radio-Frequency Sounders in Space*. IZMIRAN, Troitsk; Karpachev, A.T.: Electron concentration distribution in the high latitude topside ionosphere of the southern hemisphere under nighttime summer conditions. *Geomagnetism & Aeronomy*, Vol. 35, No. 6, June 1995; Galperin, Yuri: Detection of electron acceleration in the ionospheric plasma under the influence of high power radiation with local plasma frequency on Intercosmos 19. *Cosmic Research*, Vol. 19, 1981.

[13] Parrot, M.: Use of satellites to detect seismic-electromagnetic effects, in Singh, R.P.; Furrer, R., eds: *Natural Hazards Monitoring and Assessment using Remote Sensing Techniques, Advances in Space Research*, Vol. 15, No. 11, 1995; Korepanov, V.: *Remote Sensing as a Tool of Seismic Hazards Monitoring*. Institute for Space Research, Lviv, 2000; Larkina, V.I.: *Low Frequency Radio Emission Hindrance as a Means of Diagnostic Processes in Near Earth Space*. IZMIRAN, Troitsk, undated; Gousheva, M., *et al.*: Quasi static electric fields phenomena in the ionosphere associated with pre and post Earthquake events. *Natural Hazards Earth System Sciences*, Vol. 8, 2008; Boskova, J., *et al.*: VLF emissions at frequencies in the plasmasphere as observed by low orbiting Intercosmos satellites, in Campbell, C.R.; Gringauz, Konstantin, eds: *Physics of Thermal Plasma in Magnetosphere, Advances in Space Research*, Vol. 6, No. 3, 1986; Pulinets, Sergei; Legenka, A.D.: Spatial-temporal characteristics of large-

scale disturbances of electron density observed in the ionospheric F region before strong Earthquakes. *Cosmic Research*, Vol. 41, No. 3, 2003; Pulinets, Sergei, *et al.*: *The Earthquake Prediction Possibility on the Basis of Topside Sounding Data*. IZMIRAN, Troitsk, 1991; Pulinets, Sergei, *et al.*: Radon and ionosphere monitoring as a means for strong Earthquake forecast. *Il Nuevo Cimento*, Vol. 22, No. 3–4, 1999; Pulinets, Sergei: Strong Earthquake Prediction Possibility with the Help of Topside Sounding from Satellites, *Advances in Space Research*, Vol. 21, No. 3, 1998; Pulinets, Sergei; Legenka, A.D.: Dynamics of near-equatorial ionosphere prior to strong Earthquakes. *Geomagneticsm & Aeronomy*, Vol. 42, No. 2, 2002; Pulinets, Sergei, *et al.*: *Ionospheric Variations before Strong Earthquakes Observed by Topside Sounders in the Solar Cycle Maximum*. Instituto de Geoficz, Mexico, 1996.

[14] Kutiev, A., *et al.*: Energy deposition in the polar ionosphere as determined by measurements on Intercosmos Bulgaria 1300, in Booth, C.A., *et al.*, eds: *Advances in Space Research*, Vol. 2, No. 10, 1982; Stoeva, P., *et al.*: Characteristics of SAR arcs registered by the Emo-5 filter and spatial scanning photometric system on board the IC Bulgaria 1300 satellite on 21st August 1981, in Sandahl, A.; Saunders, M.A., eds: *Auroral and Related Phenomena, Advances in Space Research*, Vol. 13, No. 4, April 1993; Kirov, B.: Main ionospheric trough studied from Intercosmos Bulgaria 1300, in Schmerling, E.R., *et al.*, eds: *Magnetospheric and Ionospheric Plasmas, Advances in Space Research*, Vol. 5, No. 4, 1986, 1986; Arshinkov, I.S., *et al.*: Magnetic field measurements in the ionospheric–magnetospheric region, in Reigber, C., *et al.*, eds: *Solid Earth Geophysics and Satellite Orbits, Advances in Space Research*, Vol. 6, No. 9, 1986; Gogoshev, M.M.: Observations of South Atlantic magnetic anomaly with Intercosmos Bulgaria 1300 during geomagnetic storm from 25th August to 19th September 1981, in Schmerling, E.R., *et al.*, eds: *Advances in Space Research*, Vol. 5, No. 4, 1986; Arshnikov, I.S., *et al.*: First results of magnetic field measurements on Intercosmos 1300 Bulgaria. *Cosmic Research*, Vol. 21, 1983.

[15] Grebnev, A.I., *et al.*: The Study of a Plasma Jet Injected by an On-Board Plasma Thruster, *Advances in Space Research*, 1981; Echim, M., *et al.*: Multiple current sheets in a double auroral oval observed from the Magion 2 and 3 satellites. *Annales Geophysicae*, Vol. 15, 1997; Echim, M., *et al.*: Early stage of storm recovery stage – a case study. Interball site, from 9 May 1998; Oraevsky, Viktor; Triska, Pavel: Active plasma experiment APEX, in Bernhardt, P.A.; Möhlman, D.; Ip, W.-H., eds: *Space Plasma Physics, Advances in Space Research*, Vol. 13, No. 10, October 1993; Sagdeev, Roald, *et al.*: Experiments with Injection of Powerful Plasma Jet into the Ionosphere, *Advances in Space Research*, 1981; Bankov, L., *et al.*: Critical Ionization Velocity Experiment XANI On Board the Intercosmos 24 Active Satellite, *Advances in Space Research*. Vol. 13, No. 10, 1993; Deminov, M.G., *et al.*: Dynamics of midlatitude ionospheric storms – a qualitative picture. *Geomagnetism & Aeronomy*, Vol. 35, No. 1, August 1995; Karpachev, A.T.; Demimova, G.F.; Pulinets, Sergei: Ionospheric changes in response to IMF variation. *Journal of Atmospheric & Terrestrial Physics*, Vol.

57, No. 12, 1995; Mikhailov, Yuri, *et al*.: VLF Effects in the outer ionosphere from the underground nuclear explosion on Novaya Zemlya Island on 24th October 1990, *Physics of the Chemistry of the Earth*, 25.

[16] Sosnovets, Elmar, *et al*.: Effect of magnetospheric processes on the ionosphere during the magnetic storm of 1st December 1977, based on Cosmos 900, in Rycroft, M.J., ed.: *Space Research*. COSPAR, Paris, 1979; Stepanov, A.E.; Khalipov, Viktor: Large scale blobs of ionospheric plasma – ground based and satellite measurement, in Zelenyi, Lev; Geller, M.A.; Allen, J.H., eds: Auroral phenomena and solar terrestrial relations. Proceedings from a conference held in memory of Yuri Galperin, 3–7 February 2003. IZMIRAN, Troitsk, with Scientific Committee for Terrestrial Physics; Shutte, N.M.: High and low latitude energy spectra of protons and electrons precipitating into the ionosphere observed from Cosmos 900, in Rycroft, M.J., ed.: *Space Research*. COSPAR, Paris, 1978; Gringauz, Konstantin: Observations of Aurorae in Far Ultraviolet from Cosmos 900, Schmidthe, G. and Champion, K.S.W., eds. *Advances in Space Research*, Vol. 1, No. 12, *Mesosphere & Thermosphere*, COSPAR, Paris, 1981; Logachev, Yuri I.: *40th Anniversary of the Space Age in the Research Institute of Nuclear Physics of Moscow University*. Moscow State University, Moscow, 2009; Bryunelli, B.E.: Magnetospheric electrical fields and the rotation of the Earth, in Vernov, Sergei; Kocharev, G.E., eds: Proceedings of the VIth winter school in space physics, Apatity, 18 March–1 April 1969, Part 1; Altyutseva, V.I., *et al*.: Variations in intensity and anistropy of precipitating particle fluxes above 30keV. *Cosmic Research*, Vol. 20, 1982; Gdalevich, Gennadiy, *et al*.: The ionosphere at low and equatorial levels at 600km during magnetospheric–ionospheric disturbances in 1977 – data from Cosmos 900. *Cosmic Research*, Vol. 20, 1982; Vlasova, N.A., *et al*.: Penetration of solar protons and α particles over 1MeV into polar caps. *Cosmic Research*, Vol. 19, 1981; Gdalevich, Gennadiy: Variations in charged particle concentrations in high latitude ionosphere during magneto–ionospheric disturbances in September 1977 according to Cosmos 900. *Cosmic Research*, Vol. 19, 1981; Kovtuyk, A.S., *et al*.: Structure and dynamics of auroral photons and electrons with energies of tens and hundred of keV from measurements on Cosmos 900. *Cosmic Research*, Vol. 19, 1981; Ivanova, T.A., *et al*.: Radiation measurements on Cosmos 900. *Cosmic Research*, Vol. 18, 1980; Kovtyukh, A.S., *et al*.: Radiation measurements aboard Cosmos 900. *Cosmic Research*, Vol. 18, 1980.

[17] Galeev, Albert; Tamkovich, G.M., eds, *35th Anniversary of the Institute of Space Research of the Russian Academy of Sciences*. Author, Moscow, 1999.

[18] Sandahl, Ingrid, *et al*.: Cusp boundary layer observations by Interball, in Russell, C.T., ed.: *Results of the IASTP Program, Advances in Space Research*, Vol. 20, No. 4–5, 1997.

[19] Eismont, Nathan, *et al*.: Flight dynamics operations in Interball project. Proceedings of the 12th International. Symposium on spaceflight dynamics, Darmstadt, Germany, 2–6 June 1997. ESOC, Darmstadt, ESA; Molodtsov, V.: On the flight of the spacecraft Interball 2. *Novosti Kosmonautiki*, No. 5, 208, 2000; Kremev, R., *et al*.: Project Interball. *Aviatsia i Kosmonavtika*, 1993, No. 8; Sibeck,

D.G.; Zastenker, Georgi, eds: *Plasma Processes in Near Earth Space – Interball and beyond, Advances in Space Research*, Vol. 31, No. 5, 2003; Russell, C.T., ed.: *Results of the IASTP Program, Advances in Space Research*, Vol. 20, No. 4–5, 1997; Kopik, A.: Research data from the Interball projects are still requested today. *Novosti Kosmonautiki*, No. 10, 285, 2006; Kotova, G.A., *et al.*: Study of notches in Earth's plasmasphere based on data from Magion 5. *Cosmic Research*, Vol. 46, No. 1; Bezrukikh, V.V., *et al.*: Thermal structure of dayside plasmasphere according to data of tail and auroral probes and Magion 5. *Cosmic Research*, Vol. 44, No. 5; Titova, E.E., *et al.*: Verification of the backward wave oscillator model of VLF chorus generation using data from Magion 5. *Annales Geophysicae*, Vol. 21, No. 5, 2003; Triska, Pavel, *et al.*: *Space Weather Effects on the Magion 4 and 5 Solar Cells*. Institute of Atmospheric Physics, Prague, 2005; Teodosiev, D., *et al.*: ULF wave measurements aboard Magion 4 narrow band wave events observed in the magnetopause regions. *Planetary & Space Science*, Vol. 53, No. 1–3, January 2005; Hristov, P., *et al.*: ULF Turbulence in Magnetospheric Boundary Layers during April 1997 as Measured by Magion 4, *Advances in Space Research*, Vol. 31, No. 5, 2003; *Lettre du CNES* 27, December 1996; Deminov, M.G., *et al.*: Dynamics of the mid-latitude ionospheric trough during a magnetic storm. *Geomagneticsm & Aeronomy*, Vol. 35, No. 6, 1996; Yermolaev, Yuri, *et al.*: The Earth's magnetosphere response to solar wind events according to Interball project data. *Cosmic Research*, Vol. 38, No. 6, 2000; Yermolaev, Yuri, *et al.*: Ion distribution near the Earth's bow shock. *Annales Geophysicae*, 15, 1997; Yermolaev, Yuri: Observations of multicomponent distribution function of ions on Interball tail probe satellite. *Cosmic Research*, Vol. 37, No. 6, 1999; Yermolaev, Yuri: Strong geomagnetic disturbances and their correlation with interplanetary phenomena during the operation of the INTERBALL project satellites. *Cosmic Research*, Vol. 39, No. 3, 2001; Zastenker, Georgi, *et al.*: Investigation of solar wind correlations and solar wind modifications near Earth by multi-spacecraft observations. *Final Scientific Report*, IKI/MIT; Zastenker, Georgi: New features of the solar wind observed by the Interball satellite, in Zelenyi, Lev; Geller, M.A.; Allen, J.H., eds: Auroral phenomena and solar terrestrial relations. Proceedings from a conference held in memory of Yuri Galperin, 3–7 February 2003. IZMIRAN, Troitsk, with Scientific Committee for Terrestrial Physics.

[20] Sheldon, Charles S., II: Vertikal sounding rocket program. Published on globalsecurity.org, 2009. Clark, Phillip S.: The Skean program. *Spaceflight*, Vol. 20, No. 8, August 1978; Pillet, Nicolas: Le program Vertikal. www.kosmonavtika.com (accessed 9 April 2007); Serafimov, K.: Dynamical behaviour of the daytime topside ionosphere from Vertikal 6, in Rycroft, M.J., ed.: *Space Research*. COSPAR, Paris, 1978; Zelenyi, Lev, *et al.*: Interball mission generates results on magnetospheric dynamics and magnetosphere-ionosphere interaction. *Eos*, Vol. 85, No. 17, 27 April 2004.

[21] Ershova, V.A.; Sivtseva, L.D.: H_e^{++} ions in Earth's atmosphere, in Kondratyev, Kirill; Mycroft, M.J.; Sagan, C., eds: *Space Research*. COSPAR, Paris, 1970; Jakimec, J.: Analysis of solar X-ray spectrum of 20th August 1971

from Vertikal 2, in Rycroft, M.J.; Reasenberg, R.D., eds: *Space Research.* COSPAR, Paris, 1973; Chapkunov, S.: Electron temperatures and density measured by Vertikal 3, in Rycroft, M.J., ed.: *Space Research.* COSPAR, Paris, 1976; Felkse, D., *et al.*: O_2 densities from solar lyman α absorption measurements by Intercosmos 4 and Vertikal 1, from Bowhill, S.A.; Jaffe, L.D.; Rycroft, M.J., eds: *Space Research.* COSPAR, Paris, 1971; Vakulov, P.V., *et al.*: Streams of nuclei of the middle group of energies of 3020MeV nucleon at altitudes of 350km during solar flares. *Geomagnetism & Aeronomy*, Vol. 28, No. 4, 1988; Morozov, A.I.; Shubin, A.P.: *Space Electrojet Engines.* NASA, TTF 16,542, 1975; Shtern, M.I.: *Investigations of the Upper Atmosphere and Outer Space Conducted in 1970 in the USSR.* Report to the 14th meeting of COSPAR. NASA, TTF 666, 1971.

[22] Lardier, Christian: Soviet meteorological rockets, a history 1946–1991. Presentation to the International Astronautical Congress, Glasgow, 2 October 2008; Cambou, F., *et al.*: Zarnitsa rocket experiment on electron injection, in Rycroft, M.J., ed.: *Space Research.* COSPAR, Paris, 1974; Paton, Boris, *et al.*: *Welding in Space and Related Technologies.* Cambridge International Science Publishing; Klos, Zbigniew, *et al.*: HF Emission Relation to the Li Ion Beam Injected into Ionosphere – Plasma rocket experiment, *Advances in Space Research*, Vol. 13, No. 10, 1993; Oraevsky, Viktor, *et al.*: Complex Plasma Injection Experiments for Investigation of Plasma Beam Interactions, *Advances in Space Research*, Vol. 10, No. 7, 1990; Oraevsky, Viktor, *et al.*: Modulation of the Background Flux of Energetic Particles by Artificial Injection, *Advances in Space Research*, Vol. 12, No. 12, 1992; Alexandrov, V.A., *et al.*: Structure of Plasma Blobs Injected into the Ionosphere from a Rocket, *Advances in Space Research*, Vol. 1, No. 2, 1981; Avdyushin, S.I., *et al.*: Interaction of Artificially Injected Plasma Flows with Ionospheric Plasma, *Advances in Space Research*, Vol. 12, No. 12, 1992; Portnyagin, Y.I., *et al.*: Experiments with SF_6 Injection into the Polar Ionosphere, *Advances in Space Research*, Vol. 12, No. 12, 1992; Oraevsky, Viktor, *et al.*: Modelling of Artificial Plasma Bubble in Ionosphere, *Advances in Space Research*, Vol. 12, No. 12, 1992; Sagdeev, Roald, *et al.*: Rocket Environment during Electron Beam Injection, *Advances in Space Research*, Vol. 1, No. 2, 1981; Cambou, F., *et al.*: ARAKS – controlled or puzzling experiment? *Nature*, 28 February 1978; Dokukin, V.S., *et al.*: Results of Zarnitsa 2 – a rocket experiment on artificial electron beam injection in the ionosphere, *Advances in Space Research*, 1981. For an account of such experiments worldwide at the time, see Active Experiments in Space – report of symposium in Alpbach, Austria, 24–8 May 1983. Paris, ESA, ESA SP 195.

[23] Rothkaehl, Hanna; Klos, Zbigniew: Broadband high frequency emission as indicator of global changes in the ionosphere, in Sibeck, D.G.; Zastenker, Georgi, eds: *Plasma Processes in Near Earth Space – Interball and beyond, Advances in Space Research*, Vol. 31, No. 5, 2003; Kuzentsov, S.N.: Energetic charged particle fluxes beneath the radiation belts. Proceedings of the ICRC, 2001; Rothkaehl, Hanna; Klos, Zbigniew: HF radio emissions as a tool of ionospheric plasma diagnostics. *Annali di Geofisica*, Vol. XXXIX, No. 4,

August 1996; Dzyubenko, N.I.; Ivchenko, V.N.: Effectiveness of transferring the energy of incoming electrons into optical atmospheric emissions according to observations from an artificial polar aurora. *Cosmic Research*, Vol. 16, 1978.

[24] COSPAR: *Solar Observatory Koronas F – three years of observation of solar activity*. COSPAR, Paris, 2004; Kuznetsov, V.D.: Space research of the Sun, in Zakutnyaya, Olga; Odintsova, D., eds: *Fifty Years of Space Research*. Institute for Space Research, Moscow, 2009; Zhitnik, I.: Results of XUV full Sun imaging spectroscopy for eruptive and transient events, from Dennis, B.R., ed.: *Energy Release and Particle Acceleration in Solar Atmospheric Flares and Related Phenomenon, Advances in Space Research*, Vol. 32, No. 12, 2003. For a description of the Koronas Foton mission, see Zak, Anatoli: *Koronas Foton*. www.russianspaceweb.com.

[25] Galpter, A.M.: High energy particle flux variations in Earthquake predictions, in Singh, R.P.; Furrer, R., eds: *Natural Hazards Monitoring and Assessment using Remote Sensing Techniques, Advances in Space Research*, Vol. 15, No. 11, 1995; Can cosmonauts give quake alert? *Soviet Weekly*, 13 September 1990.

[26] Afanasiev, Igor: Mini spacecraft KOMPASS. *Novosti Kosmonautiki*, No. 11, 226, 2001.

[27] Zelenyi, Lev: Fifty years to change our views of the world, in Zakutnyaya, Olga; Odintsova, D., eds: *Fifty Years of Space Research*. Institute for Space Research, Moscow, 2009; Thunderstorms and elementary particle acceleration. *Space Research Today*, No. 177, April 2010. COSPAR, Paris.

[28] Ilyin, Eugene A.: Historical overview of the Bion project. *Journal of Gravitational Physiology*, Vol. 7, No. 1, 2000.

[29] The real effects of weightlessness. *Soviet Weekly*, 18 May 1974; Space lab's 'spectacular success'. *Soviet Weekly*, 9 March 1974; Illyin, E.A.: Biological satellites and their contribution to space biology and medicine, in Napolitano, L.G., ed.: *Proceedings of the XXVI International Astronautical Conference*, Lisbon, 1975; Gazenko, Oleg: *The Biosatellite – results of the experiment*. NASA, TTF 15,863; Kovalev, E.E., *et al.*: Investigation of the basic characteristics of electrostatic shielding from cosmic rays on the artificial Earth satellite Cosmos 605. *Cosmic Research*, Vol. 14, 1976.

[30] Rats were guinea pigs in space safety flight. *Soviet Weekly*, 11 January 1975; Grigoriev, A.I.; Potapov, A.N.: Advances and perspectives of space biology and medicine, in Zakutnyaya, Olga; Odintsova, D., eds: *Fifty Years of Space Research*. Institute for Space Research, Moscow, 2009.

[31] Grigoriev, A.I.; Potapov, A.N.: Advances and perspectives of space biology and medicine, in Zakutnyaya, Olga; Odintsova, D., eds: *Fifty Years of Space Research*. Institute for Space Research, Moscow, 2009.

[32] Tairbekov, M.G., *et al.*: Biological unit on Cosmos 1129, in Holmqvist, W.R., *et al.*, eds: *Advances in Space Research*, Vol. 1, No. 14, 1981; Wronski, T.J., *et al.*: Skeletal alternations in rates in spaceflight, in Holmqvist, W.R., *et al.*, eds, *Advances in Space Research*, Vol. 1, No. 14, 1981.

[33] Hansson, Anders: Cosmic rays. Paper presented to British Interplanetary Society, 3 June 1989; Grigoriev, Anatoli; Potapov, A.N.: Advances and

perspectives of space biology and medicine, in Zakutnyaya, Olga; Odintsova, D., eds: *Fifty Years of Space Research*. Institute for Space Research, Moscow, 2009; Smart, D.F., *et al.*: Dose rate observed on 18–21 October 1989 and its modulation by geophysical effects, in Horneck, G., *et al.*, ed.: *Life Sciences and Space Research*, *Advances in Space Research*, Vol. 14, No. 10, October 1994. The date cited here may be in error, as the mission launched on 15th September.

[34] Gazenko, Oleg; Ilyin, E.A.: Investigations on board biosat Cosmos 1667, in Malacinski, G.M., *et al.*: *Life Sciences and Space Research*, *Advances in Space Research*, Vol. 6, No. 12, 1986; Gazenko, Oleg; Grigoriev, Anatoli; Potapov, A.N.: Advances and perspectives of space biology and medicine, in Zakutnyaya, Olga; Odintsova, D., eds: *Fifty Years of Space Research*. Institute for Space Research, Moscow, 2009.

[35] Demets, R.; Jansen, W.H.; Simeone, E.: *Biological Experiments on the Bion 10 Satellite*. European Space Agency, Paris, 2002.

[36] Ilyin, Eugene A., *et al.*: Bion 11 mission – primate experiments. *Journal of Gravitational Physiology*, Vol. 7, No. 1, 2000; Ilyin, Eugene A.: Historical overview of the Bion project. *Journal of Gravitational Physiology*, Vol. 7, No. 1, 2000; Borisov, Oleg: What can the Bion 9 satellite tell us? *Zenith*, January 1990, No. 35.

[37] Mullen, Leslie: STONE experiment on Foton. *Astrobiology Magazine*, 11 May 2007.

RUSSIAN-LANGUAGE REFERENCES

1. Курт В.Г. Точка бифуркации отечественной программы внеатмосферной астрономии – Троицкий вариант, №15, 28.10.2008.

2. Автоматические космические аппараты для фундаментальных и прикладных научных исследований/Под общ. ред. д-ра техн. наук, проф. Г.М. Полищука и д-ра техн. наук, проф. К.М. Пичхадзе – М.: Изд-во МАИ-ПРИНТ, 2010.

3. Полежаев П.Н., Полуэктов В.П. Космическая обсерватория «Гамма» – Земля и Вселенная, №3, 1991.

4. Гальпер А.М. Космический эксперимент в области гамма-астрономии – Земля и Вселенная, №4, 1991.

5. Лантратов К. Завершены исследования на КА «Интеркосмос-24»/Новости космонавтики, № 21, 8–21.10.1995.

6. Лантратов К. АУОСы продолжают работу/Новости космонавтики, № 21, 8–21.10.1995.

7. Лисов. И. Россия-Чехия-Аргентина. Запущены «Интербол-2», «Магион-5» и «Mu-Sat»/Новости космонавтики, №18, 26.08–08.09.1996.

8. Лисов. И. «Интерболы» и «Магионы» продолжают работу/Новости космонавтики, №21, 26.09–23.10.1998.

9. Лисов. И. Россия-Чехия. В полете ИСЗ «Интербол-1» и «Магион-4»/Новости космонавтики, №16-17, 30.07–26.08.1995.

10. Кузнецов В.Д. Спутник «Коронас-Ф» наблюдает Солнце вблизи максимума активности/Земля и Вселенная, 36, 2002.

11. Логачев Ю.И. Исследования космоса в НИИЯФ МГУ: Первые 50 лет космической эры/Ю.И. Логачев; под ред. проф. М.И. Панасюка – Москва, 2007.

12. Космонавтика: Энциклопедия/Гл.ред. В.П. Глушко; Редколлегия: В.П. Бармин, К.Д. Бушуев, В.С. Верещетин и др – М.: Сов. энциклопедия, 1985– 528 с.

8

Perspectives, past and future

In this final chapter, we review, first, future missions that are proposed and then make an historical overview of Soviet and Russian space science so as to analyze key organizations, institutions, systems, and patterns of development. We look at its key features, personalities, decisions, and the dissemination of outcomes in an international context.

DECLINE OF SCIENCE DURING THE TRANSITION

Russian space science suffered badly during the period of post-Communist transition. The space program as a whole suffered three waves of contraction (pre 1991, 1991–1996, 1997), lost three-quarters of its staff, and, in the late 1990s, there were predictions that it might cease to exist [1]. It was only saved because resourceful space managers attracted in Western investment and collaboration and turned the space program around from being the most state-dependent to the most commercial in the world in less than ten years.

Within the retrenchment, some parts fared worse than others, especially the weather satellite and science programs. Russian space science fell into a "deplorable condition" in the view of the director of IKI, Lev Zelenyi [2]. Missions were repeatedly delayed, reorganized, and re-scoped and few survived at all, Interball and Koronas being the lucky exceptions. The institutes on which the science program was based suffered badly. The Ioffe Physical Technical Institute in St Petersburg, called "the cradle of Soviet physics", saw its budget fall from 66 million rubles to 3.4 million rubles. Directors of institutes made a priority of holding on to their human resources – their scientific staff – but they did so at the cost of all other spending – laboratories, equipment, even journals – letting their buildings decay. The Institute for Space Research lost 25% of its staff and numbers fell below 1,000. Key scientists went to work abroad in a new "brain drain". Roald Sagdeev went to the United States, where he began to tell the story of the space program. Oleg Vaisberg went on two assignments to the United States, including the South West Institute, which was well known for its space instrumentation, although he then returned to Moscow.

Dr Lev Zelenyi, Director of IKI

Many others, though, went for much longer periods or did not return. It was difficult to attract new, young graduates into space institutes when no fresh research was undertaken, although some did come and reworked old data with success [3].

Not until the early years of the new century was there a real (as distinct from inflationary) increase in the real space science budget, 2005 marking the first increase in value since the 1980s. With the Federal Space Plan, 2006–2015, came an attempt to put future missions on a more organized basis with, thanks to improvements in the Russian economy, the prospect that they would actually take place. The new missions attempted to avoid over-ambition and instead use common platforms and instruments so as to save costs, launch on smaller rockets (Soyuz rather than Proton), and focus on distinct areas of science, notably astrophysics [4]. First, we look at the missions that continued to be planned during the post-Soviet period and then the promise of new missions outlined in the plan. These were the Spektr observatory and missions to return to the Moon and Mars (Luna Resurs and Phobos Sample Return).

SPEKTR OBSERVATORIES

Spektr (not to be confused with the Mir module of the same name) was originally intended as a series of observatories to follow Astron, Grant, and Kvant. Had the Spektr program gone ahead in the 1980s, Russian space science would have stolen a lead over the NASA observatories of the 1990s. The concept of Spektr was approved at a conference in October 1987 attended by astrophysicists from Britain, Denmark, Finland, the GDR, and Italy. The idea was for at least three observatories for radio, ultraviolet, and X-ray/gamma-ray observations, starting with a high-apogee observatory, using X-ray oblique optics in the place of the earlier coded apertures [5].

The collapse of funding meant that Spektr made almost no progress in ten years. In 1997, it was announced that 200 million rubles would be injected into the project in the hope of generating some momentum. European Space Agency participation was sought, but when this was not forthcoming, the Russians turned to India

Spektr design

instead. 1997 marked devaluation and financial collapse and was financially the worst possible year to try to revive the project. With publication of the federal space plan, Spektr was redefined with four different spacecraft:

- Spektr RG (Röntgen Gamma);
- Spektr UV (Ultra Violet);
- Spektr R (Radioastron);
- Spektr M (Millimetron).

These observatories would be smaller and cheaper than those originally planned in the 1980s, using a new, smaller standard design called *Navigator*, designed by the Lavochkin design bureau and flying on either the Soyuz 2 Fregat or Zenit 3F (F for Fregat) for launcher. *Navigator* was an 850-kg octohedral prism with a maneuvering engine able to fire for 1,800 sec, a power supply of 1.5 kW, and a high-accuracy pointing system. Each is described.

The 2,100-kg Spektr Röntgen Gamma (RG) was conceived by Kvant astrophysicist Academician Rashid Sunyaev. This mission was repeatedly redefined and downscoped from its origin in the 1990s, the most recent iteration being announced as a Russian–German agreement in autumn 2009. It will fly from Baikonour on a Zenit Fregat in 2013 to the Lagrange 2 point some 1.5 million km from Earth. The mission was reduced to two X-ray telescopes: the German extended Röntgen Survey with Imaging Telescope Array, called eROSITA, from the Department of Extraterrestrial Physics of the Max Planck Institute, and the Russian ART-XC telescope operating in the 3–30-keV band. The mission will start with eROSITA making a whole-sky survey of up to 100,000 galaxy clusters, which will take four years, followed by pointed observations, while ART-XC will make a census of black holes. The aim is to find what is suspected to be a hidden population of hundreds of thousands of supermassive black holes and to continue the work of Granat with X-ray bursts, weak X-rays, supernova outbursts, black holes, and neutron stars, and trace hot gas in galaxies, find exploding stars and detect X-rays swirling around massive black holes. It is expected that Spektr RG will find numerous hitherto obscured black holes and low-surface-brightness objects. Some hope that as many as 3.2 million active galactic nuclei and 86,000 clusters will be found and provide fresh knowledge of binary systems, anomalous pulsars, and supernova remnants. This would be important for the studies of the evolution of the universe over time and the role of mysterious "dark energy" in this process. The predecessor of the Russian ART-XC telescope called MVN (Russian abbreviation of the "Monitor Vsego Neba" or All-Sky Monitor) would fly to the ISS prior to the launch of the main observatory. It is dedicated to an all-sky survey in X-rays, so that a diffuse X-ray background can be studied with high precision.

Spektr UV, Ultra Violet, also called the World Space Observatory, will have a 1.7-m mirror to gather information on the physics of the early universe, galaxies, hot stellar atmospheres, cool stars, the intergalactic environment, interstellar and solar system dust, gas clouds, active galactic nuclei, and distant planetary atmospheres. Weighing 2,250 kg, it will orbit from 500 to 300,000 km above the Earth and is expected to be the world's main ultraviolet telescope in orbit between the retirement of the American Hubble space telescope (2012) and the arrival of the Space Ultra Violet Observatory (2020). Appointed as project manager was Boris Shustov, head of the Institute of Astronomy at the Russian Academy of Sciences, who promised that the telescope would be one of the most important space telescopes of the decade and held out the hope that it would detect and explain large quantities of dark matter. Spektr UV's elongated orbit will give it a considerable advantage over the Hubble Space Telescope, whose low orbit was determined by the launching capacity of the space shuttle and has up to 20 times greater resolving power. Launch is expected on the Soyuz 2 Fregat.

Spektr Radioastron was conceived by Nikolai Kardashev, Leonid Matveyenko, and Gennadiy Sholomitsky, their idea being to use a high-altitude space telescope in collaboration with a ground telescope to achieve a long baseline to improve the accuracy of observations. It was hoped that the observatory would find super-massive black holes, nurseries of stars and planet formation, and clouds of interstellar plasma. The 3,295-kg Spektr Radioastron will carry a 10-m-diameter 27-blade radio telescope dish and orbit from 10,000 km to as far out as 390,000 km. An observing program of 500 objects was drawn up in the Physics Institute of the

Academy of Sciences by Academician Kardashev, covering quasars, star and planet formation regions, black holes, and active galactic nuclei. Spektr scientists hoped that it would analyze the signals from distant galaxies (Lavochkin's then director, Georgi Polischuk, voiced the opinion that it might even detect radio signals from distant civilizations). It will also carry a solar wind science payload called Plasma F to take 32 samples a second of the solar wind, ions, and energetic particles so as to determine their small-scale structure, bulk, velocity, density, and temperature.

The final project in the series is Spektr M, M for Millimetron, because this satellite will search the heavens in the 4–20-mm wavelengths, looking especially for leftover radiation from the big bang. It will carry a 12-m-diameter dish antenna and will operate in a highly elliptical orbit. The telescope will be cooled to 4 K for three years and then 50 K for the next ten and will have a pointing accuracy of 1 arc second. It will have the ability to search for terrestrial-type planets to characterize their atmospheres, as well as dark matter, interstellar cloud, megamasers, pre-galaxies, and learn how the universe was originally formed. Launch is set for 2018. In an important change in data handling, the scientific data on Spektr M will not routinely be transmitted directly to Earth. Instead, all the scientific results will be assembled in a large memory bank on the satellite itself. Scientists will connect to the internet, contact the satellite, preview the data that they are interested in, and, if they wish, download it. Data that are not downloaded in two years will be automatically cleaned off the memory to make way for new information.

LUNAR "POLYGONS"

In the post-Soviet space program, the Moon had been rarely mentioned. In 1997, IKI floated plans for a small spacecraft to be sent into lunar orbit, using a Molniya rocket to dive three 250-kg penetrators into the lunar surface at speed, burrowing seismic and heat flow instruments under the lunar surface, leaving transmitters just above the surface. With small nuclear isotopes, they would transmit for a year, operating as a three-point network to collect information on Moonquakes and heat flow. This mission acquired the title Luna Glob, or "lunar globe", presumably from the global nature of the seismometer system [6]. The idea of returning to the Moon was spurred by a series of successful Moon probes sent there in the early years of the new century by Europe (SMART), China (Chang e), Japan (Kaguya), and India (Chandrayan).

Luna Glob was continuously redefined and pushed back, with the penetrators taken off the agenda as too risky. A more immediate mission emerged, following Indian approaches for a collaborative mission that acquired the name of Luna Resurs ("lunar resource"). India was anxious to return to the Moon, moreover before the Chinese did. In 2010, Luna Resurs was given a 2013 launch date, with India providing a GSLV rocket, the lunar orbiter, and a small 15-kg electric rover of Russian design, while Russia provided the 1,200-kg landing stage.

Following this, there were plans for new sample missions: Luna Grunt ("lunar soil"). The ultimate goal was for what is called the "lunar polygon" (the Russian word "polygon" can be used for "base"), an automated lunar base of permanent telescopic and radio observatories (e.g. low-frequency array), laboratories, rovers, and sample return missions. At the time, the Americans had planned a manned base and this was a rival approach, an efficient unmanned polygon promising a high scientific return.

PHOBOS SAMPLE RETURN

As we saw in Chapter 5, during the 1970s, the Lavochkin design bureau had been charged with recovering samples from Mars, the hugely ambitious and complex project 5M. But if collecting samples from Mars was too difficult, what about Mars's small moon, Phobos? Russia had a good knowledge of Phobos from the Phobos 2 mission in 1989. Accordingly, in 1999, Lavochkin began a feasibility study of recovering rock samples from the moon Phobos and invested an initial 9 million rubles in the project. The study was reworked many times in the following years. The project was originally called "Phobos Grunt", in Russian "Phobos Ground" or "Phobos Soil". After many years, they came to the view that the words "Phobos Grunt" did not travel well into English and just as English-language speakers had got used to the term, the mission was renamed Phobos Sample Return or, for short, Phobos SR. Appointed in 2000 as mission scientist was Alexander Zakharov, a veteran of Intercosmos, Prognoz, and the earlier Phobos missions. He had come to IKI as far back as 1968 as a computer engineer but before that, he had spent a couple of years underground in a huge cave in the salt mines of the Ukraine building a neutrino laboratory.

The final mission specification was to use the Zenit 3F launcher to send the 8,100-kg spacecraft into a Martian orbit similar to that followed by Phobos 2 more than 20 years earlier, entering an observation orbit and then closing in on the little moon. Small jets would be used to press the spacecraft onto Phobos's surface in the low gravity. The lander was a low, squat, neat polygon on three legs, with a set of scientific instruments and a sampling device, which would take Phobos regolith samples and

Alexander Zakharov, Phobos Sample Return scientist

deliver them to the capsule at the return stage. The return stage was a small, box-shaped module with a solar panel on one side, the ball-shaped sample recovery cabin fitting snugly in the middle. The return stage would lift off the lander, enter Martian orbit, orientate itself, fire to Earth, transit for 280 days, and then release the small sample recovery cabin for its plunge into the atmosphere to land in the nuclear warhead testing site of Sary Sagan. The lander would continue to work on Phobos for a year after the return stage was fired back to Earth. Between 100 and 200 g of Phobos regolith were expected. The small size of the spacecraft, benefitting from recent advances in electronics, miniaturization, and materials, was a dramatic contrast to the size and complexity of Sergei Kryukov's project in the second half of the 1970s.

The intention was to improve knowledge of Phobos itself: density, mass, regolith, center of gravity, inertia, whether there was ice inside, and its precise orbit. Traces of life (paleolife) would be sought. The ultimate aim was to identify the origins of Phobos, either as a captured asteroid (the originally favored theory) or the outcome of a collision between an object and Mars. IKI pointed out that the spectral analysis of the moon carried out by Phobos 2 marked it out as significantly different from carbonaceous asteroids. As one might expect, the mission had a series of wider objectives relating to Mars itself. The Russians had by no means given up on mapping the Martian magnetic field, so two plasma spectrometers and a magnetometer were carried. The rate of erosion of the Mars atmosphere would be again measured. Attempts would be made to measure the atmospheric dynamics and climate change of the planet [7]. Instruments are listed in Table 8.1.

The project went through a further iteration in 2006, when China joined it. China had not yet sent a probe to the planets, so Phobos Sample Return provided an

Table 8.1. Phobos Sample Return instruments.

GChC Gas Chromatograph (chemical composition of Phobos soil volatiles)
TDA Thermo Difference Analyzer
MS Mass spectrometer
MSp Mossbauer spectrometer (mineralogical composition of regolith)
FOGS Gamma spectrometer (composition of regolith)
HEND Neutron spectrometer (search for water)
LASMA Laser spectrometer (regolith chemical composition)
MANAGA Secondary ions mass analyzer (regolith chemical composition)
LWPR Long-wave planetary radar (inner structure of Phobos)
SEISMO Seismometer (gravimetry of Phobos)
TV, panoramic and stereo TV system (TSNN, Panorama, Stereo)
AOST Fourier spectrometer (Martian atmosphere)
Meteor (micrometeorites)
FMPS Plasma set (ion, spectrometers, magnetometer for Martian plasma environment)
LIBRATION Stars and solar tracker (Phobos proper and forced motion)
USO Ultrastable oscillator (celestial mechanics experiment)

opportunity for China to get a spacecraft to Mars much earlier than otherwise would be the case. It was agreed that Phobos Sample Return would carry a 120-kg satellite attached to its side, called Yinghuo 1. When Phobos Sample Return arrived in its initial Mars elliptical orbit of 800–80,000 km, it would detach Yinghuo 1 to study the atmosphere and ionosphere of Mars. As Phobos Sample Return altered orbit to meet Phobos at 9,700 km, Russia and China would calibrate their instruments together and receive reports on the ionosphere from their two spacecraft simultaneously in quite different orbits, giving them an additional scientific bonus. Although Yinghuo 1 made the mission a little more complicated, this was outweighed by the scientific gain and the funding provided by the Chinese. At a late stage, specialists in the Hong Kong Polytechnic University in China contributed a 400-g device to grind Phobos rock for *in situ* analysis.

Yinghuo arrived in Moscow in time for its October 2009 launch. Although the Chinese satellite provided additional resources for the project, scientists became more and more nervous as they tried to integrate the two spacecraft in time for

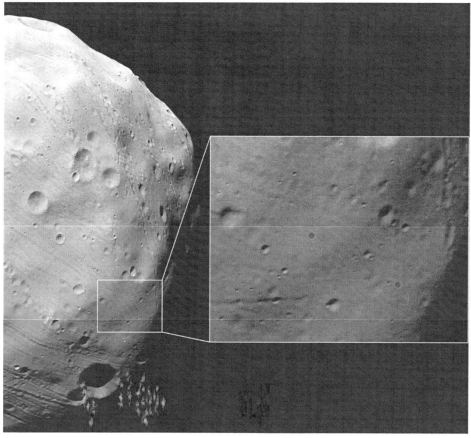

Possible landing sites for Phobos Sample Return (Credit: ESA)

launch less than two months ahead. At the last team review of the project a month before launch, it was decided to delay the project until the next launch window two years later. This was not the only such project delayed, for America's Mars Science Laboratory was similarly postponed while at an advanced stage.

One of the reasons for the delay was second thoughts about the soil sampler. To increase the reliability of the Phobos sampling, it was decided to include an additional device on the sampling complex, in case Phobos Sample Return landed on a hard surface. This device would knock off samples, which would then be seized by manipulators. In April 2010, it was learned that the Centre for Space Research in Poland was commissioned to develop a new mini-soil sampler called *Chomik* (hamster), a 2-kg recoilless jackhammer requiring only 1.5 W of electricity and modeled on the European Philae lander then en route to a comet on the Rosetta probe. *Chomik*, though, was designed to work with the manipulators, but not replace them. Lavochkin took further advantage of the delay to design an intelligent vision system whereby during the 3-hr sampling period, the manipulators would be guided to the best rocks to lift. Communications trials were made with Europe's Mars Express, then orbiting Mars, to test the deep-space relay system.

The next project was intended to see the dropping of a series of weather stations on Mars in 2016: Mars Net. The concept echoed a French–American project of the 1990s and the first illustrations appeared in 2009. Mars Net was based on the small instrument package on the VEGA balloon in the 1980s and was for a series of very small penetrator probes onto the Martian surface, the experimental package weighing only 4 kg. It was intended to carry out the project in cooperation with Finland, which had an expertise in cold weather meteorology and had contributed to the American Phoenix mission. The experimental packages would be primarily weather stations measuring, for a Martian year, temperature, pressure, wind speed, humidity, and dust from a number of locations.

Later, there was the prospect of a return to the Martian surface, with a small six-wheel rover (95 kg) and then a sample return mission (Mars Grunt or "Mars Soil"). Designs were drawn up by the Lavochkin design bureau, showing how a dome-shaped lander would enter the Martian atmosphere protected within an inflatable rubber braking cone and fire rockets for the final stage. Once a robotic arm had selected and retrieved samples, a small rocket in the top of the dome would blast Earthward.

Ultimately, cosmonauts would go to Mars, almost certainly as part of an international project along the lines of the International Space Station. In the meantime, ground work could be done on the psychological aspects of the missions. From the 1960s, habitats were built in the Institute for Biomedical Problems in Moscow, starting with year-long missions in 1968. These were not biospheres designed to develop closed-cycle systems (Chapter 6), but served the different purpose of testing human interactions in an isolated environment simulating a long space journey. The most ambitious was the simulation of a 520-day journey to Mars called Mars 500, which began in June 2010, with seven volunteers from Russia, Europe, and China.

SPACE SCIENCE IN THE FEDERAL SPACE PLAN, 2006–2015

The federal space plan, published July 2006, was an attempt to put Russian space planning on a more systematic basis in what was now a more orderly political and economic environment. A striking feature was the reinstatement of space science as a cornerstone of the program. The plan included not only the missions mentioned earlier (Spektr observatories, Luna Glob, and Phobos Sample Return), but also a return to Venus.

Here, proposals were revived for a long-duration Venus lander, Venera D (D for *dolgozhivuschaya*, or long-living), in 2016, on the powerful new Soyuz 2.1.b launcher, able to send 1,800 kg to Venus. The original Venera D concept was a mission first planned in the early 1980s to survive on the boiling-hot surface for up to a month with the objective of detecting seismic activity. A camera would carry live television during the descent. The new Venera D mission went through numerous evolutions in the late 2000s, the most recent iteration being a lander, orbiter, balloon, and *vetrolet* (or kite) operating at 45 km. Early discussions of scientific objectives suggested that the aim of the orbiter would be to study atmospheric composition, structure, dynamics, cloud system, and lightning; the balloons would study the clouds *in situ* during a nighttime journey, while the lander would drill the surface, listening for seismic activity and thunderstorms, and watch for lightning for much longer than any of the earlier landers. Sixteen prospective landing sites were provisionally selected, chosen from a mixture of terrain, aiming at those considered flattest and most survivable. It was calculated that even though the mysterious *tesseræ* looked uneven, it should still be possible to find an even spot there [8].

By the early 2010s, the following missions were under discussion, some funded by the federal space plan and others from different sources. This list is necessarily partly speculative, but gives an idea of the range of missions considered:

- *Chibis* (a Russian word for a lapwing), the name of both a unified platform developed for 50-kg microsatellites and the microsatellites themselves; currently, there are two different *Chibis* microsatellites planned to study lightning phenomena from orbit and monitor greenhouse gases, respectively;
- *Intergelizond*, a solar probe based on the Phobos Sample Return designed to fly by Venus and Mercury and then approach the Sun's poles (the current suggestion is to within 42 million km at 38° to its equator); it will study the Sun with a combination of optical, ultraviolet, and X-ray instruments to measure the solar atmosphere and structures, flares, coronal mass ejections, energetic particles, plasma, solar dust, and the solar wind; there will be a focus on the polar regions of the Sun;
- *Gamma 400*, a space observatory to investigate space gamma-ray emission 0.1–3,000 GeV, search for and study the gamma-ray bursts and registration of electrons, positrons, and nuclei with energies more then 0.1 GeV; some of these tasks may hold the clue to the nature of dark matter;
- *OSIRIS*, a star-mapping mission, part of a project called *Astrometriya*; *OSIRIS* is an abbreviation of the *Opticheski Zvezdnyi Interferometr,*

Razmeschaemyi na Iskusstvennom Sputnike (Optical Star Interferometer based
on the Artificial Satellite); the name of the spacecraft is Celesta;
- *Terion F2*, a 300-km-altitude geophysics mission, both planned for 2018.

Just as the *Navigator* platform was developed for deep-space missions, so a small
platform was also developed by Lavochkin for some of the new science missions, the
Karat platform, with a weight of 96 kg including 60 kg of scientific payload. Five
candidate missions were put forward:

- *Zond PP*, a remote sensing mission to assess climate and environmental
  change, humidity, salinity, and energy exchange;
- *Monika*, to study solar cosmic rays;
- *Relek*, to study electron precipitation;
- *Strannik*, a solar weather observer to fly in Earth polar orbit to study the
  processes at the magnetosphere boundary;
- *Resonans*, four small satellites built by Lavochkin to study the Earth's
  magnetic field as a model for extra-solar Earthlike planets with a magnetic
  field, due for launch in 2014; they follow in the footsteps of the *Troika* idea
  and the Interball series and will operate in pairs with intersecting elliptical
  orbits of 500–28,000 km, 63.4°, also called magneto-synchronous orbits;
  general objectives will be to study the evolution of the magnetic field, the ring
  current, storms, and plasma dynamics, while a particular objective will be to
  study "magnetospheric cyclotron resonance masers", believed to be the
  dynamic forces that shape the radiation belts [9].

The interest in small satellites reflected a desire to get around the problems
associated with large, modern scientific projects. In the 1960s, spacecraft could be
designed, built, and flown within periods as short as three months and instruments
could even be added on the launch pad. By the 1990s, although computers enabled
much more sophisticated instrumentation and data flow, missions took over two
years to get under way, required a steady line of funding, and were much more
vulnerable to political and economic changes. Smaller, less expensive satellites were
less vulnerable to these pressures.

Several missions are in consideration for the next federal space plan, such as a
Europa lander and a mission to the asteroid Apophis. The Europa mission is
currently intended to fly along with a joint NASA ESA project, formerly *Laplace*,
for two orbiters to study Europa and Ganymede (with a possible Japanese spacecraft
to study the Jovian magnetosphere). Russian participation will focus on a Europa
lander to survive on the surface of the planet as long as possible. Preliminary work
has begun on designing and building instruments, such as a new type of spectroscope
called ATR (Attenuated Total Reflection) to analyze soil (and possible microorgan-
isms therein). Various options for drilling into the thin crust of its icy ocean were
considered, such as a pulsing fiber laser.

HISTORICAL OVERVIEW

Having looked briefly towards the future, now may be the appropriate point to look back, review, and analyze the meaning of the story so far. The preceding chapters have focused on the space science missions and, indeed, their substantial achievements. Now is the time to analyze the trajectory of Soviet and Russian space science and some of the key points in its story. Chapters 1–7 narrated the progress of Soviet and Russian space science. In Table 8.2, it is possible to summarize the main lines of development and programs.

As may be seen, the most consistent line of development was the exploration of the Earth's environment, its radiation, and magnetic fields. Next came human space science, space biology, and deep space exploration. By contrast, space physics and astrophysics were later developers. Chapter 1 noted the analyses of the Central Intelligence Agency of the 1950s, which pointed to Soviet expertise in geophysical sciences and a high level of interest in biology, with a relative underdevelopment of astrophysics. This analysis was astute in predicting the relative priorities of the program over the subsequent years, with the great observatories program representing a determined effort to catch up for lost ground from the 1970s onwards.

It might seem strange that the many scientists within the space program actually considered that science was a low priority in the Soviet space program. This was certainly a dominant discourse of analysis – one shaped by Sagdeev [10]. Before entering the discussion, we should remember that this has been a refrain of space scientists in other countries, too, even in the well funded American program. Having said that, Sagdeev was, as director of the Institute for Space Research, IKI, from 1974 to 1990, in a good position to judge. His general view was that science was a low priority in a program dominated by political rivalry with the United States, defense, and applications, in that order, fueled by vain leaders, whose outcomes were determined by the machinations of the many camps of the overweight Soviet government and party bureaucracy. Scientists, he said, were the "poor relatives of the rich space czars" and competed badly with the admirals and the generals. While some had pet projects – for example, Ustinov was interested in space telescopes, as was Chelomei at one stage – most made clear their contempt for "your damned space science". Except in the very early years, science never accounted for more than 16% of the space program – over 50% was gobbled up by the manned program, a lament shared by space scientists the world over.

IKI space scientists, like many of their colleagues in other countries, favored robotic exploration, which they felt gave a higher return on their investment. Sagdeev admitted that IKI was "not particularly fond of the manned flight program", but politically could not afford not to participate in it, accordingly joining in discussions with program managers about scientific instruments or payloads "for the cosmonauts to play with". Although he attended many meetings in Mishin's design bureau to plan the scientific program of Salyut, he felt that "it left no room for serious science". His successor, Valentin Glushko, he felt was interested only in *grands projets* or *coups* against the United States that would show the Soviet Union in general and himself in particular in favorable light.

Table 8.2. Main lines of development and highlights of Soviet and Russian space science.

	Earth's environment	Earth–Sun	Human spaceflight and space biology	Space physics and astrophysics	Deep space
1930s	Balloon missions USSR and OSOAVIAKHIM				
1940s	*Akademik* series				
1950s	Sounding rocket programs First Earth satellites, exploration of radiation belts		First dog flights (sounding rockets and orbital)		First Moon probes Discovery of solar wind Mapping of lunar farside and orbital)
1960s	Mapping of Earth's radiation belts (Elektron) Cosmos program		First manned flights Ability of humans to withstand 0 G Earth orbit as platform for Earth observations	Proton series	Determination of conditions on Venus Nature of lunar surface, environment
1970s	Intercosmos program *Ionozond, Ellipse, Magik, Oval* missions	Prognoz series	First orbital stations: gardens *(Oasis)*, materials processing Start Bion program	*Orion* observatories *Efir, Energiya* series	Nature of lunar rock Completion of maps of Moon Descent through atmosphere of Mars, characterization Martian environment Atmosphere, environment of Venus
1980s	APEX, *Aktivny*	*Intershock* mission	Permanent orbital stations and prolonged endurance	Relikt Astron Kvant	Analysis of Venus soil, mapping Venus, interception of comet Halley, Phobos
1990s	Interball	Koronas	Space gardens (Svet)		Granat, Gamma
2000s					

His criticism of the manned space program is well founded if we look at the various attempts that were made to put scientists into space. Three doctors flew – Boris Yegorov (Voskhod), Oleg Atkov (Salyut 7), and Valeri Poliakov (Mir) (and, much later, Boris Morukov on the shuttle) – but no scientists. In spring 1965, on the initiative of Mstislav Keldysh, the first group of scientist cosmonauts was recruited. Twenty-four candidates were considered and five approved: Rudolf Gulyayev, Ordinard Kolomitsev, Mars Fatkullin (IZMIRAN), Valentin Yershov (Institute of Applied Mathematics), and Georgi Katys (Institute for Automatics and Telematics). Plasma scientist Oleg Vaisberg applied and made good progress until he was put into a gyrating rotating chair, which swung the unhappy candidate through three dimensions to test for space sickness, normally quite effectively: "The chair got me," he admitted. The rest began a year's training in 1967 and, from July 1968, awaited assignment to missions. They were joined in 1967 by a group of military scientists: Vladimir Aleksyev, Mikhail Burdayev, and Nikolai Provatakin. Later, a number of other scientists qualified for or enlisted in cosmonaut training: Zyyadin Abuzyarov (oceanographer), Gurgen Ivanyan (geologist), and Irina Latysheva (electronics). In 1994, the restored Russian Academy of Sciences approved the recruitment of a scientist–cosmonaut group, but little progress was made. The idea of scientist–cosmonauts turned out to be a sad one: the concept never got the institutional backing necessary and none ever flew [11]. So, is it fair to conclude that the role of science was poorly regarded in the Soviet space program? To answer this question, we need to look at the key decision makers, decisions, and institutes in the space science program.

THE KELDYSH ASCENDANCY

Decision-making within the Soviet space program was a complex process, the outcome of discussion, argument, and debate between the party, government, the political leadership, the military, the design bureaus, and the scientific institutes. The original scientific institute was, we noticed in Chapter 1, Peter the Great's Academy of Sciences and the many institutes connected to it. The Academy of Sciences proved itself to be remarkably resilient to political changes. In the 1920s, the new Soviet government had attempted to replace it with a rival "Communist Academy" (of which Tsiolkovsky accepted membership) – but this venture got nowhere and was quietly abandoned. Indeed, not until 1951 was the president of the Academy even a member of the Communist Party (Alexander Nesmeyanov).

At one level, the Academy of Sciences was a powerful body, for it had overall responsibility for the country's research institutes. Whereas in the United States, most research was conducted in universities, in the Soviet Union (and much of continental Europe), research was conducted in institutes while the universities were primarily teaching bodies. At the time of Sputnik, there were about 4,000 institutes in the Soviet Union, growing to over 5,300 by the early 1970s. Many were small – the average institute had 270 staff – but a number were dominant and had large staffs [12]. On the other hand, the Academy had no direct leverage over the design bureaus

Table 8.3. Post-war presidents of the Soviet Academy of Sciences.

1945–1951	Sergei Vavilov
1951–1961	Alexander Nesmeyanov
1961–1975	Mstislav Keldysh
1975–1986	Anatoli Alexandrov
1986	Vladimir Kotelnikov
1986–1990	Yuri Marchuk
1990–1991	Yuri Osipov

– indeed, a leading commentator described it as relatively weak and peripheral organ of the state [13]. Despite that, it played a critical role in the science of the space program. Way back in the 1950s, the chief designer needed the Academy's blessing for his sounding rocket and later Earth satellite projects to succeed. Although political support within the government and party was more important and ultimately decisive, Korolev could not afford to have the Academy oppose him. The Academy comprised a vast range of disciplines, often far distant from spaceflight, but, fortunately for him, there was a critical mass of supporters and sympathizers, even if there were a few who were dismissive of the idea of spaceflight. Korolev appears to have devoted a considerable amount of time and energy to building up support for space projects in the Academy, in the course of time becoming an academician himself and then bringing with him colleagues to consolidate his position. Both Korolev and Glushko became corresponding members in the first post-Stalin election of new members (October 1953) and Alexander Nesmeyanov spoke publicly in favor of the satellite project in Vienna the following month. Korolev won support at a practical level by inviting academicians to fit their experiments on the *Akademik* missions (the title was no accident either), so it worked to the advantage of both. Having a sympathetic president in the Academy was essential for the prosperity of space science and Table 8.3 names the presidents during this period.

Korolev's luckiest break was the up-and-coming star of the Academy, Mstislav Keldysh. Not only was he sympathetic to spaceflight, but he had the political skills to steer spaceflight projects past his colleagues for endorsement. His Institute of Applied Mathematics was organizationally part of the Academy and, because of his work there, he reached the praesidium in 1953, becoming vice-president in 1960. Over the years, more scientists sympathetic to the spaceflight project were elected and the praesidum elected in February 1957 included, besides Keldysh, supporters such as Anatoli Blagonravov, Peter Kapitsa, and Igor Kurchatov. In February 1958, Sergei Korolev, Valentin Glushko, and Georgi Petrov were elected academician, while new corresponding members included leading designers Vladimir Barmin, Mikhail Ryazhansky, Nikolai Pilyugin, and Vasili Mishin. Korolev was himself elected to the praesidium in 1960 and finally the space program's main supporter, Mstislav Keldysh, was appointed president by the government on 19th May 1961.

Not one to treat the position as a sinecure, he used his position to reinforce Soviet studies of basic science, while at the same time pushing back the frontiers of new sciences, such as quantum electronics, holography, genetics, and molecular biology. Keldysh had a unique combination of hands-on science (he personally oversaw the calculation of the trajectories of interplanetary spaceships), management, administration, politics, and leadership, and was frequently asked by the government to head up expert commissions to adjudge the wisdom of individual space projects. Even as other personalities in the space program were banished to anonymity, Keldysh was permitted to continue as the public face of Soviet science, speaking to journalists in Moscow and touring many international conferences abroad, visiting Britain and the United States. The graying Keldysh was an important interlocutor between the design bureaus on the one hand and the party and government on the other. His title of "chief theoretician" reflected his huge scientific contribution.

THE EMERGENCE OF IKI

Having such an interlocutor was important, for the key decisions of the early space program were the outcome of often lengthy proposals, negotiations, and bargaining between the Academy, government and party, design bureaus, and institutes. When Western analysts first investigated the papers and records of the space program in the 1990s, they were struck by how ad hoc these decisions were, contrary to Western impressions that a great long-term master plan was in place. The most recent study of the origins of the Soviet space program, Asif Siddiqi's *The Red Rockets' Glare*, takes this a step further, for he found that far from being top-down and state-directed, the program was rooted in networks of enthusiasts who were able, from time to time, to attract, capture, and sustain the interest of an inconsistent and sometimes uninterested government [14]. In the early period, we can identify the main decisions of early space science (Table 8.4), with Keldysh as a moderator of many of them.

The replacement of Nikita Khrushchev in October 1964 brought in the new leadership of Leonid Brezhnev. Although this period is not now well regarded – many histories now call it "the period of stagnation and decline" – it is important to recognize that important reforms were made that had a crucial bearing on the development of the space program. The charge sheet against Khrushchev included a lack of collegiality in his leadership, impulsiveness, and adventurism, so one of the objectives of his successors was to put the decision-making arrangements of party

Table 8.4. Key dates and key decisions in early Soviet space science.

30 Jan 1956	Object D
15 Feb 1957	PS 1 and PS 2
9 May 1960	Elektron series
8 Aug 1960	DS satellites, Cosmos 2 rocket
Oct 1962	Second round of DS satellites
3 Aug 1964	*Ionosfernaya* and other programs

and government on a more orderly and predictable basis, in this case the relationship between government and the space program in general and space science in particular. The reforms also included an opening to the socialist countries and the West, so 1965 became a pivotal year for Soviet space science.

All the space-related design bureaus and other institutes were brought within a new Ministry of General Machine Building (1965), with a minister in charge (Sergei Afanasayev), while a formal institute was established for space science. Such a body had been sought by both Keldysh and Korolev as far back as 1959. Keldysh wrote a formal proposal in July 1963, so it was under consideration even during the Khrushchev period and arguably might have happened even had he stayed longer. The idea of a special space research institute that would coordinate automated space studies and manage its own design bureau for scientific instruments was embodied in the Institute for Space Research (IKI abbreviation from Russian *Institut kosmicheskih issledovanyi*) and was approved by party and government on 15th May 1965, with a formal date of investiture of 14th July 1965 [15]. The person responsible for establishing the institute was Gennadiy Skuridin. Formally, it remained part of the Academy of Sciences and its first director was academician and mechanics expert Georgi Petrov from the aviation laboratory, TsIAGI. IKI was built in south-western Moscow, where an old village had formerly stood – appropriately enough – on the road to Kaluga.

In no time, the surrounding forest was cleared and the area became a vast building site, eventually completed in 1973. IKI itself had a number of important divisions, §1 being Keldysh's own Institute of Applied Mathematics, §2 for chemistry, §3 for space astronomy (Shklovsky), with a space physics department founded in 1969 under Albert Galeev. Early IKI was a mixture of old hands (e.g. Shklovsky) and sometimes cheeky young scientists, lectures and subsequent debates being robust experiences [16]. It united "space" divisions from other institutes, which sometimes went to IKI in their whole. Over the next two decades, staff numbers grew to 1,500 people and research ships operated in its name (e.g. *Academician Mstislav Keldysh, Academician Boris Petrov*). From 1967, IKI had its own design bureau or OKB in Frunze, now Bishkek, Kirghizia, which mainly made specialized instruments but in some cases entire spacecraft (e.g. Gamma); its own terminal in the Yevpatoria tracking center in the Crimea and from 1978 had its own special design bureau for space device engineering devices and equipment (SKB) in the town of Tarusa, near Kaluga. Construction of this greenfield site of workshops and apartments amidst birch woodland outside Kaluga was finished in 1986 and the center had its own receiving dish. The core of the institute's work became its ten departments and three laboratories, supplemented by engineering, testing (e.g. vacuum, electromagnetic), control, and data-processing services.

Managing the many diverse elements of IKI took their toll on the first director, Georgi Petrov. Although well regarded, some of those unhappy about the new configuration managed to get their reviews reflected in the first, five-year evaluation, so Georgi Petrov decided to call it a day [17]. Mstislav Keldysh replaced him with atom scientist Roald Sagdeev, from *Akademgorodok*, the science city in Novosibirsk in western Siberia, where the crafty Khrushchev had given scientists a high level of

IKI during its early years

scientific and political freedom, far enough away not to cause him problems. Sagdeev, though, brought his critical mind with him, quickly questioning some of the underlying assumptions of the space program, starting the "war of the worlds" (Chapters 4 and 5) and opening international cooperation.

It took some time for the new institute to build its role, but, by the early 1970s, IKI was able to define the research priorities of the Soviet space program, not least the interplanetary program, commissioning the design bureaus and overseeing the missions. IKI became a real center where missions were proposed, prioritized, planned, and developed, and had a specific role in the selection of instrumentation. Important decisions in later space science continued to be made through government and party decision, but IKI was able to put thinking on a more strategic basis. Direct point-to-point competition with the Americans was replaced by a strategy of concentrating on particular areas of expertise (e.g. Venus). Missions were now selected by a science council in IKI. Missions go through two phases, phase A being proof-of-concept (80% completed this successfully), phase B being a more rigorous selection of priorities before approval as a funded project. During the 1960s, the science council would meet to discuss future projects every two weeks, initially chaired by Sergei Vernov, with Mstislav Keldysh attending from time to time. Table 8.5 details the main governmental decisions from the early IKI period onwards.

IKI also fulfilled an important but unadvertised role in the space program, that of its public face. The Soviet space program had no publicly identifiable agency, like NASA in the United States, and the system of design bureaus was a closely kept secret, not only to outsiders, but within the USSR itself. Most design bureaus operated under what were called "post box addresses" – officially, they did not have street addresses – while some cities, like those that housed OKB-586 (Dnepropetrovsk) and OKB-10 (Krasnoyarsk), were closed to foreigners, ZATO (*Zakrytie Administrativno-Territorialnye Obrazovania*). Western visitors to the Academy of

Table 8.5. Later Soviet space science: key decisions from the early IKI period.

15 Apr 1965	Cooperation in space with socialist countries
15 May 1965	Establishment of IKI, with investiture in July 1965
22 Jun 1965	Approval of 18 DS-U satellites
9 Feb 1970	Decision on orbital stations (Salyut)
Jul 1970	Request by Mishin for proposals for instruments for Salyut
Feb 1971	Mishin Academy discussion on science on long-term orbital stations
26 Jun 1972	AUOS design
17 Feb 1976	Gamma

Sciences main building could see for themselves that not very much spaceflight actually happened there. From the period of Sputnik, graying academicians of the Academy of Sciences (e.g. Leonid Sedov) had been presented to the Western media as the guiding forces of the space program, but before long, it became apparent that they had little managerial responsibility [18]. Now, IKI was now able to fulfill this visible, public role, this time with some conviction.

Not only that, but the space program became more open. Leonid Brezhnev recognized the costliness of the Soviet Union attempting to build up its own expertise in every sphere of scientific and industrial endeavor: better to share distinct areas of expertise with both the socialist countries and, where possible or desirable, Western countries. Cooperation would be more economical than using foreign currency to buy in and copy Western technology in those areas in which there were deficits. This would inevitably mean more open cooperation, visiting, and exchange – a process for which IKI was now ideally placed. It is no coincidence that the construction of IKI took place at the same time as the formation of Intercosmos (Chapter 7) and cooperation with selected Western nations (e.g. France). There were limits, though. Iosif Shklovsky, who had only been abroad once before, to follow the 1947 solar eclipse off the coast of Brazil, was allowed to travel to the United States, but his travel permit was revoked in 1973 when he defended dissident Andrei Sakharov.

Rules about publication abroad were also relaxed. Traditionally, scientists could only publish in conference papers for events that they had received permission to attend, such permissions being given cautiously. In 1968, new and more liberal rules were introduced, but by 1971, only 2% of Russian scientific papers were published outside the country. It took some time for the rule change to take effect. The liberalization process was painfully slow and many very senior scientists were refused permission to travel, reasons never being given. Vladimir Kurt recalled many years later that scientists in IKI were still expected to fulfill their full range of duties to the state and the party: political training sessions, voluntary work on Saturdays (*subbotniks*), beet-harvesting brigades, and standing out on the streets at the appointed place to cheer visiting friendly African heads of state [19].

As for Mstislav Keldysh, he suffered ill-health from an operation in 1971, relinquished his post as president in November 1975, developing in his spare time an interest in impressionist paintings. He died suddenly in his car while still in its driveway on 24th June 1978 and was deservedly buried in the Kremlin Wall. That

Roald Sagdeev addressing an international meeting of scientists at the time of the VEGA mission

Soviet space science achieved as much as it did was largely due to his interest, skills, guiding hand, commitment, and combination of practical knowledge and ability to deal with the political establishment [20].

The great reorganization of 1965, coupled with the ascendancy of Keldysh, should have ensured that space science retained a high priority within the space program. Korolev, Yangel, and Keldysh were very conscious of the need to ensure that a substantial proportion of the space program be devoted to space science, but once they were gone, it was harder for it to find a place. It may well be the case that from the late 1970s, science lost some ground in favor of both the manned program and the extensive military program. In despair in the post-Keldysh period for the lack of interest in or support for science, Sagdeev contemplated resignation on a number of occasions. Despite that, IKI had some good moments, too, perhaps its finest being in March 1986, when hundreds of foreign scientists assembled in IKI to watch the interception of comet Halley by one of the most international missions of the time: VEGA. Sagdeev stayed as director until 1988, when Albert Galeev took over, with leadership passing in the new century to Lev Zelenyi.

KEY INSTITUTES AND PERSONALITIES IN THE SPACE SCIENCE PROGRAM

The architecture of Soviet space science set down in the great Brezhnevite reorganization in 1965 is largely that which operates to this day. Many of the different institutes and bodies involved in space research were brought together under the IKI umbrella in the Academy of Sciences, but others continued independently, such as:

- the Lebedev physics institute, its formal title being the PN Lebedev Physics Institute of the Academy of Sciences, with departments for astronomy and optics, the largest of all the institutes, with over 900 scientists at one stage;
- IZMIRAN, or the NV Pushkov Institute of Terrestrial Magnetism, Ionosphere and Radio Wave Propagation of the Academy of Sciences, which took the lead in many of the missions concerned with the Earth's environment, such as Interball;
- the Vernadsky Institute, or the VI Vernadsky Institute for Geochemistry and Analytical Chemistry of the Academy of Sciences; here, a Planetary Geochemistry Laboratory was established in 1961, headed by one of Russia's most famous scientists, Yuri Surkov, who shaped the instrumentation program of the Venera, Mars, and Luna probes and wrote extensively about the results; this was the laboratory that received and analyzed the Moon samples;
- the PK Sternberg Astronomical Institute of Moscow State University, where the lunar and planetary physics department took responsibility for mapping the Moon and the planets; the Institute was based on Moscow's observatory, which had followed its earlier cousin in St Petersburg in Pulkovo; it was named after the Bolshevik astronomer and later education minister Pavel Sternberg, who died of typhus in 1920;
- Institute for Astronomy of Russian Academy of Sciences, formerly Astrosoviet; the first network of satellite tracking stations were established under the umbrella of this organization (see Chapter 1).

Some of these were old institutes, to which space science departments were added. One of the oldest was IZMIRAN, originally founded in 1939 in Leningrad as the Scientific Research Institute for Geomagnetism (NIIZM). It was evacuated to the Urals during the war, returning in 1946, but relocated to Troitsk (also called Troitskoe) in the old, classic-style yellow-painted but damaged meteorological observatory. Renamed IZMIRAN in 1959, it kept branches in St Petersburg and the magnetic ionospheric observatory in Kaliningrad. Its earliest equipment was the non-magnetic schooner *Zarya*, built in Finland but fitted with Soviet equipment [21]. Directors were N.V. Pushkov (1940–1969), Vladimir Migulin (1969–1988), and then Viktor Oraevsky, and by the end of the Soviet period, it had 1,200 staff. The Lebedev Institute, also called FIAN (Russian for Physics Institute of the Academy of Sciences), was based on the old Physical Laboratory of the Academy of Sciences in St Petersburg, later removed to Moscow, reorganized by A.F. Ioffe in 1921, turned into the Physical Institute in 1934 by the optical physicist Sergei Vavilov, eventually named after P.N. Lebedev. It became possibly the most prestigious institute of the Soviet Union, attracting such scientists as L.I. Mandelstam, I.E. Tamm, Andrei Sakharov, and Vitaly Ginzburg. A center of both basic and applied research, its space work focused on the Earth's atmosphere (e.g. ozone), the solar atmosphere, dark matter, cosmic ray physics, and gamma-ray astronomy.

The Sternberg Institute was a long-standing body that was brought into the space program. Here, encouraged by the success of the Automatic Interplanetary Station in 1959, Sergei Korolev proposed the establishment of a Department of Lunar and Planetary Physics within the Sternberg Institute, with a specific brief to prepare maps in advance of a lunar landing by cosmonauts. Appointed its first director in 1964 was the leading Moon mapper Yuri Lipsky, with a team comprising Vladislav Shevchenko, Zhanna Radionova, and V.I. Chikmachev. Two years later, though, when Korolev died, it fell on harder times. His successor, Vasili Mishin, did not appear interested in the publication and dissemination of lunar maps and withdrew funding for the project. Rescue came from an unexpected source, for rocket engine designer Valentin Glushko stepped in and, allocating resources from his Gas Dynamics Laboratory, provided sufficient money for the Zond 3 mapping to be brought to a successful conclusion.

This episode was just one example of the interplay of personalities and institutes that was a feature of the Soviet space program. As noted earlier, one of the great surprises to Western analysts of the Soviet space program was the rivalry between design institutes and the complex and fluctuating relationships between designers, institutes, the military, and political leadership. Scientists have never been seen as part of this equation, but this book indicates that they were, too. It seems that teams of scientists competed in general for the attention of design institutes and in particular for individual missions. Some groups and institutes established a relatively continuous line of missions (e.g. Yuri Galperin), while others managed to obtain some individual missions (e.g. Konstantinov). Some teams appeared to have competed around individual institutes (e.g. Vernov around OKB-1, Naum Grigorov around OKB-52).

There is nothing unusual about this. In the other space-faring countries, such as the United States and Europe, it is quite normal for teams of scientists to vie for the attention of space agencies and promote particular missions and for rivalry to be quite adversarial. In the West, the person who comes to lead the mission is normally called the "principal investigator" and has an overall responsibility for its scientific outcomes. This term does not appear to have been used at the time in the Soviet Union, but in later years, the comparable term *Glavny experimentator* (literally "chief experimenter") was applied.

In the early years of the Soviet space program, some of the principal experimenters are now evident. Their names were not well known outside the Soviet Union, although they would have been familiar to the scientific community. The key personalities, we now know, were, Sergei Vernov, Konstantin Gringauz, and Yuri Galperin, and, in the case of deep-space exploration, Vasili Moroz, Yuri Surkov, and Mikhail Marov. In addition, others have a strong association with individual fields, such as Shmaia Dolginov (magnetism), Tatiana Nazarova (meteorites), and Leonid Ksanformaliti (lightning), just to name a few. During the Soviet period, little attention was given to these scientists in the popular press that publicized space science and the political atmosphere of the time discouraged what might be perceived as "the cult of personality". Their roles, though, were evident in the papers published in the international scientific papers. Vasili Moroz was

Table 8.6. MSU principal investigators, selected missions.

Pavel Shargin	Cosmos 4, 7, 9, 15, 41
P.V. Vakulov	Cosmos 17
Yevgeni Gorchakov	Cosmos 53, 127, 219
Yuri I. Logachev	Cosmos 159
Pavel Shargin and Yevgeni Gorchakov	Cosmos 137, 143
E.A. Pryakhin	Cosmos 208, 228
Elmar Sosnovets	Cosmos 256, 480
S.N. Kuznetsov	Cosmos 378, 426, 721
G.I. Pugacheva	Cosmos 428, 490
O.R. Grigoryan	Cosmos 484
Yevgeni Gorchakov and Elmar Sosnovets	Cosmos 900
Naum Grigorov and V.Y. Shestoperov	Cosmos 1543, 1713
Naum Grigorov	Intercosmos 6
S.P. Ryumin	Cosmos 1686
A.F. Titenkov	Cosmos 1870
G.Y. Kolesov and A.N. Podorolsky	Cosmos 2344
P.V. Vakulov, S.N. Kuznetsov, and V.A. Kuznetsova	Intercosmos 3, 5
V.G. Stolpovsky	Intercosmos 13
Naum Grigorov, A.A. Gusev, and A.F. Titenkov	Intercosmos 17
Yuri V. Minseev	Intercosmos 19
M.V. Telitsov	Intercosmos *Bulgaria 1300*

recognized in 2007, three years after his death, when a large Martian crater basin was named after him, fittingly just to the left of the Mars 6 landing site. Hopefully, they will all receive the credit they deserve.

Principal investigators and their teams can be inferred from the names attached to the scientific papers subsequently arising from these missions. Only in recent years have principal investigators been specifically identified and Table 8.6 lists some of these associated with the DV Skobeltsyn Institute for Nuclear Physics at Moscow State University. As may be seen, it opens the door to names and personalities who may not be as prominent as those listed above, but whose contribution was nonetheless important.

SOVIET AND RUSSIAN SPACE SCIENCE IN A GLOBAL CONTEXT

As noted in Chapter 1, the results from the early Sputniks were shared widely within the International Geophysical Year, with many exchanges of views and information between American and Soviet scientists. Unfortunately, the Cold War was to get the better of these exchanges. American scientists maintaining contacts with Soviet colleagues some time after the IGY found themselves quizzed by their intelligence services as to whom they knew, whom they met, and their assessment of Soviet technical capacities (presumably, the KGB kept watch, too). The Cold War climate of the 1960s was inimical to contact between scientists of the two nations and the Russians soon noted that the Americans had stopped reading their research publications. They learned that young American scientists found that citing Soviet publications as part of grant applications went against them.

To their exasperation, Soviet scientists began to read about American scientists announcing as geophysical "discoveries" in the 1970s and 1980s things that they had themselves discovered and publicized 20 years earlier, but about which the Americans were clearly in ignorance. We noted earlier how the polar jet and the inner radiation belt were "discovered" by the Americans – but after earlier Soviet discoveries. When the Americans grew plants on the Mir space station in the mid 1990s, they hailed it as a breakthrough, unaware that the Soviet Union had carried out such experiments as far back as the 1970s, the outcomes of which had been circulated at the time in a book on botany in orbital stations [22]. As late as 2009, it was announced that water was "discovered" on the Moon – even though Salyut 5 had indicated a substantial volume there in the 1970s. But nobody seemed to know, or remember.

Although it had been quite common, Russian scientists stopped visiting the United States and after the Moon landing, American scientists came to the conclusion that the USSR was so backwards, it had nothing more to offer in any case. With few exceptions in the West, nobody bothered to assemble Soviet research results. One important exception was NASA, which commissioned extensive translations of selected material, called the TTF series, but we have little sense of the degree to which it was utilized by the American scientific community. But nobody formally brought cooperation and sharing to an end: it withered under the weight of the Cold War and changing perceptions of the national roles of both protagonists. Exchanges between the Russian world and the Western world became limited to certain defined international fora, such as the annual International Astronautical Federation (IAF) conferences and COSPAR meetings of scientists.

The low point in cooperation was probably 1965–1972. In autumn 1963, John F. Kennedy had moved quite close to turning the man-on-the-Moon project into a collaborative venture with the Soviet Union, Nikita Khrushchev gradually coming around to the idea. That came to an end first with Kennedy's assassination and his replacement by Lyndon Johnson, who had no such interest or disposition. In October 1964, Brezhnev began a long-term military build-up that had little scope for cooperation with its main adversary [23]. Cooperation only became possible again when the Soviet Union recognized, following the Apollo landings, that it could not compete head to head with the United States. This opened the door to the Apollo–Soyuz mission and other possible future joint endeavors, but even this rapprochement turned out to be short-lived. The Afghanistan invasion in 1979 ushered in a long stand-off and President Reagan lapsed the cooperation agreement in 1982. Under-the-radar cooperation was permitted in limited fields (e.g. biology) or overlooked on a case-by-case basis.

As a result, Soviet space science and Western science developed, to use an astronomical metaphor, in two parallel universes, the two intersecting one another only at limited points. It was the antithesis of the heady promise of the International Geophysical Year, which had emphasized sharing, cross-fertilization of ideas, and the comparing of results between nations. Even in an area of cooperation, the parallel universe existed. American scientists were surprised in the late 1980s at Soviet progress in biospheres, but they should not have been, for Iosif Gitelson regularly reported on progress, including the first half-year ground mission by three men. Cordiality between scientists across international boundaries gave way to the challenging of Soviet results, as the prolonged dispute over planetary magnetic fields, Earthquake detection, and the spat over Venusian lightning illustrated. Students of Western scientific journals alleged that long after the Cold War had ended, a bias against publishing articles on Russian science persisted. Presumably, the journals would explain this on the basis that there was no useful science to publish in the first place [24].

Many of the scientific results of the Soviet missions were little known in the West, although, as we know through the references, they were not only presented to international meetings, but were often in English as well. Popular versions were publicized through booklets made available through embassies and specialized bookshops. Despite this, Soviet media were not able to match the strength of American domination of the printed and visual world and there was no equivalent of, for example, the *National Geographical Magazine*, with its high production values. In the visual area, the Soviet Union did itself no favors, for reproductions of images were often poor. In 2009, almost 40 years after they were taken, the Vernadsky Institute released the original Lunokhod images and their high quality and density were immediately apparent.

This was a problem of which scientists were themselves aware. As far back as 1960, during the annual meeting of the Academy of Sciences, Academician I.I. Artobolevsky complained that although books had been written by outstanding specialists, they were monographs for the "well-trained reader" and "did not have the character of popular publications". Some scientists were able to popularize science for "the broad mass of readers" and he cited Vladimir Vernadsky, but publishers "must consider that there are books of various scientific levels designed for different types of readers":

> "Recently, when serving as a jury member for international awards for scientific popular books, I read approximately 40 popular books published abroad. Among them I discovered the most wonderful forms of popularization. The most difficult questions of astronomy, biology, theory of relativity, cybernetics, were explained in extremely colourful and presentable form. Perhaps we should consider the translation of some of the popular books which have been published abroad." [25]

Nauka Press published the outcomes of missions in Russian textbooks from the 1960s onwards, many being beautifully presented albeit in a style old-fashioned to a Western reader. Sadly, Nauka Press was not well known in the West and few copies found their way Westward. The main journal of Soviet space science was *Kosmicheskie Issledovaniya*. Here, language should not have been a barrier, for it was also published in English as *Cosmic Research*. There is little evidence, though, of Western libraries taking out subscriptions and in Britain, for example, only one set is known to be extant, held in the Science Museum library in Swindon and infrequently visited.

CONCLUDING REMARKS

Hence, the importance of setting the record straight. Even if the priority given to space science disappointed Sagdeev and his colleagues, there was a solid record of achievement and discovery that dated back to the balloon flights of the 1930s. The golden years were a romantic pioneering period, from the first dog flights to parachute descents to Venus, from the travels of Lunokhod across the Moon to the interception of comet Halley. These prestigious missions were accompanied by a host of experiments, from the mundane but invaluable exploration of the effects of zero gravity and the Earth's own environment to the unusual and bizarre, such as searching for anti-matter. It is time that story be told.

REFERENCES

[1] These events are described in more detail by this author in *Russia in Space – the failed frontier?* Praxis, Chichester, 2002, and *The Resurgent Russian Space Program.* Praxis, Chichester, 2008.

[2] Zelenyi, Lev: *Is the Golden Age of Russian Space Science Still Ahead?* Novosti, Moscow, 2006; Zak, Anatoli: Russia tries to chart its space future. Posting on *www.russianspaceweb.com*, July 2008.

[3] Graham, Loren; Dezhina, Irina: *Science in the New Russia – crisis, aid, reform.* Indiana University Press, Bloomington, 2008; Galeev, Albert; Tamkovich, G.M., eds: *35th Anniversary of the Institute of Space Research of the Russian Academy of Sciences.* Author, Moscow, 1999.

[4] Afanasiev, Igor: Russia's new space science projects, in David M. Harland; Brian Harvey, eds: *Space Exploration, 2008.* Praxis, Chichester, with Springer, 2007.

[5] Polishchuk, Georgi: Perspective of Russian space activities for scientific research, in Olga Zakutnyaya; Odintsova, D.: *Fifty Years of Space Research.* Institute for Space Research, Moscow, 2009; Zaitsev, Yuri: Little green men will have to wait for a long time. *Space Daily*, 21 February 2008.

[6] Polishuk, Georgi, *et al.*: Russian robotic lunar exploration program. Paper presented to International Astronautical Congress, Glasgow, October 2008; Galimov, E.M.: Luna Glob project in the context of the past and present lunar exploration by Russia. *Journal of Earth Sciences*, No. 114, 6, December 2005.

[7] Zelenyi, Lev, *et al.*: Phobos Sample Return mission, in Olga Zakutnyaya; Odintsova, D.: *Fifty Years of Space Research.* Institute for Space Research, Moscow, 2009; Zak, Anatoli: Phobos Grunt development in 2008-9. *www.russianspaceweb.com*, June 2010.

[8]   Perminov, V.; Morozov, N.: Mission of the long-lived Venusian station. *Novosti Kosmonautiki*, No. 8, 223, 2001; Zak, Anatoli: Russia plots return to Venus. BBC, 7 October 2009; Zabalueva, E.V.; Ivanov, M.A.; Basilevsky, Alexander T.: *Characteristics of Small Scale Roughness in the Proposed Landing Sites in Venus*. Vernadsky Institute, Moscow, 2009; Ivanov, M.I.: *Comparison of RMS Slopes of Terrestrial Example and Tessera Terrain*. Vernadsky Institute, Moscow, 2009; Khartov, V.V., *et al.*: Russian program of Venus exploration by means of automated spacecraft, heritage and perspectives – the Venera D project. Paper presented to International Astronautical Federation Congress, Prague, September 2010.

[9]   Polishuk, Georgi, *et al.*: Russian robotic lunar exploration program. Paper presented to International Astronautical Congress, Glasgow, October 2008.

[10]   Sagdeev, Roald: *The Making of a Soviet Scientist*. John Wiley & Sons, New York, Chichester, Brisbane, Toronto, and Singapore, 1994.

[11]   Marinin, Igor; Lissov, Igor: Russian scientist cosmonauts – raw deal for science in space. *Spaceflight*, Vol. 38, No. 11, November 1996.

[12]   Graham, Loren; Dezhina, Irina: *Science in the New Russia – crisis, aid, reform*. Indiana University Press, Bloomington, 2008.

[13]   Barry, Willam: The missile design bureaux and Soviet manned space policy, 1953–1970. Ph.D. thesis, University of Oxford, 1996.

[14]   Siddiqi, Asif: *The Red Rockets' Glare – spaceflight and the popular imagination in Russia, 1857–1957*. Cambridge University Press, Cambridge, 2010; see also Medvedev, Zhores: *Soviet Science*. McMillan, London, 1979, p. 106.

[15]   Galeev, Albert; Tamkovich, G.M., eds: *35th Anniversary of the Institute of Space Research of the Russian Academy of Sciences*. Author, Moscow, 1999.

[16]   Zelenyi, Lev: Fiztech, IKI – further on, in Zakutnyaya, Olga, ed., *Space, the First Step*. IKI, Moscow, 2007.

[17]   Marov, Mikhail: Discovered space, in Olga Zakutnyaya, ed., *Space, the First Step*. IKI, Moscow, 2007.

[18]   Burchett, Wilfred; Purdy, Anthony: *Cosmonaut Yuri Gagarin – first man in space*. Panther, London, 1961.

[19]   Kurt, Vladimir: The first steps in our space astronomy, in Zakutnyaya, Olga, ed., *Space, the First Step*. IKI, Moscow, 2007.

[20]   Eneev, Timur: *MV Keldysh – chief theoretician*, in Zakutnyaya, Olga, ed., *Space, the First Step*. IKI, Moscow, 2007.

[21]   Dolginov, Shmaia: The first magnetometer in space, in Haerendel, G., *et al.*, eds: *40 Years of COSPAR*. COSPAR, Paris, 1998.

[22]   Nechitailo, Galina; Mashinsky, A.L.: *Space Biology – studies on orbital stations*. Mir Publishers, Moscow, 1993.

[23]   Sagdeev, Roald; Eisenhower, Susan: United States – Soviet space cooperation during the Cold War, in Zakutnyaya, Olga; Odintsova, D.: *Fifty Years of Space Research*. Institute for Space Research, Moscow, 2009.

[24]   Zak, Anatoli: Phobos Grunt development in 2008–9. *www.russianspaceweb.com*, June 2010.

[25]   Academy of Sciences of the USSR: *Annual Meeting, 24–6 February 1960,*

Minutes. US Department of Commerce, Office of Technical Services, Washington, DC, 29 August 1960.

RUSSIAN-LANGUAGE REFERENCES

1. Курт В.Г. Точка бифуркации отечественной программы внеатмосферной астрономии – Троицкий вариант, №15, 28.10.2008.
2. Космический астрометрический эксперимент ОЗИРИС. Под ред. Л.В. Рыхловой и К.В. Куимова. Фрязино: «Век 2», 2005, 350 с.
3. Келдыш М.В. Избранные труды. Ракетная техника и космонавтика. М.: «Наука», 1988.

Annexe: Summary of Soviet and Russian space science missions

The following is a summary listing of Soviet and Russian space science missions discussed in the text under a set of useful headings. In the case of lunar, Mars, and Venus probes, missions that were unsuccessful or that returned little or no data are not included.

The balloon program

30 Sep 1933	*USSR 1*
30 Jan 1934	*OSOAVIAKHIM*
26 Jun 1935	*USSR 1*
12 Oct 1938	*Komsomol*

Sputnik program

4 Oct 1957	Sputnik (PS)
3 Nov 1957	Sputnik 2
15 May 1958	Sputnik 3

1MS series

6 Apr 1962	Cosmos 2

2MS series

26 Apr 1962	Cosmos 3
28 May 1962	Cosmos 5

Elektron series

30 Jan 1964	Elektron 1, 2
10 Jul 1964	Elektron 3, 4

DS series

16 Mar 1962	Cosmos 1	DS 2
18 Aug 1962	Cosmos 11	DS A1
22 May 1963	Cosmos 17	DS A1
18 Mar 1964	Cosmos 26	DS MG

6 Jun 1964	Cosmos 31	DS MT
24 Oct 1964	Cosmos 49	DS MG
10 Dec 1964	Cosmos 51	DS MT
20 Jan 1965	Cosmos 53	DS A1
2 Jul 1965	Cosmos 70	DS A1

DS U series

19 Oct 1965	Cosmos 93	DS U2 V
4 Nov 1965	Cosmos 95	DS U V
26 Nov 1965	Cosmos 97	DS U2 M
12 Feb 1966	Cosmos 108	DS U1 G
24 May 1966	Cosmos 119	DS U2 I
12 Dec 1966	Cosmos 135	DS U2 MP
21 Dec 1966	Cosmos 137	DS U2 D
14 Feb 1967	Cosmos 142	DS U2 I
3 Mar 1967	Cosmos 145	DS U2 M
21 Mar 1967	Cosmos 149	DS MO *Opticheski*
5 Jun 1967	Cosmos 163	DS U2 MP
6 Jun 1967	Cosmos 166	DS UZ S
19 Dec 1967	Cosmos 196	DS U1 G
26 Dec 1967	Cosmos 197	DS U2 V
20 Feb 1968	Cosmos 202	DS U2 V
19 Apr 1968	Cosmos 215	DS U1 A
20 Apr 1968	Cosmos 219	DS U2 D
12 Jun 1968	Cosmos 225	DS U1 Ya
5 Jul 1968	Cosmos 230	DS UZ S
14 Dec 1968	Cosmos 259	DS U2 I
20 Dec 1968	Cosmos 261	DS U2 GK
26 Dec 1968	Cosmos 262	DS US GF
16 Jan 1970	Cosmos 320	DS MO *Opticheski*
20 Jan 1970	Cosmos 321	DS U2 MG
24 Apr 1970	Cosmos 335	DS U1 R
13 Jun 1970	Cosmos 348	DS U2 GK
10 Aug 1970	Cosmos 356	DS U2 MG
17 Nov 1970	Cosmos 378	DS U2 IP
4 Jun 1971	Cosmos 426	DS U2 K
2 Dec 1971	Cosmos 461	DS U2 MT

AUOS series

30 Mar 1977	Cosmos 900	AUOS Z R O *Oval*
18 Dec 1986	Cosmos 1809	AUOS Z I E *Ionozond*
2 Mar 1994	Koronas I	AUOS SM KI
31 Jul 2001	Koronas *Fyzika*	AUOS SM KF

Intercosmos program DS series

14 Oct 1969	Intercosmos 1	DS UZ IK 1
25 Dec 1969	Intercosmos 2	DS U1 IK 1
7 Aug 1970	Intercosmos 3	DS U2 IK 1
14 Oct 1970	Intercosmos 4	DSUZ IK 2

2 Dec 1971	Intercosmos 5	DS U2 IK 2
30 Jun 1972	Intercosmos 7	DS UZ IK 3
1 Dec 1972	Intercosmos 8	DS U1 IK 2
19 Apr 1973	Intercosmos 9	DS U2 IK 8 *Kopernik*
30 Oct 1973	Intercosmos 10	DS U2 IK 3
17 May 1974	Intercosmos 11	DS UZ IK 4
31 Oct 1974	Intercosmos 12	DS U2 IK 4
12 Mar 1975	Intercosmos 13	DS U2 IK 5
11 Dec 1975	Intercosmos 14	DS U2 IK 6
27 Jul 1976	Intercosmos 16	DS UZ IK 5

Intercosmos program AUOS series

19 Jun 1976	Intercosmos 15	AUOS Z T IK
24 Sep 1977	Intercosmos 17	AUOS Z R E IK *Ellipse*
24 Oct 1978	Intercosmos 18	AUOS Z M IK *Magik*
		Magion 1
27 Feb 1979	Intercosmos 19	AUOS Z I IK *Ionozond*
1 Dec 1979	Intercosmos 20	AUOS Z R P IK *Priroda*
6 Feb 1981	Intercosmos 21	AUOS Z R IK *Priroda*
24 Sep 1989	Intercosmos 24	AUOS Z AV IK *Aktivny* (with Magion 2)
18 Dec 1991	Intercosmos 25	AUOS AP IK APEX (with Magion 3)

Omega **series**

13 Apr 1964	Cosmos 14
13 Dec 1963	Cosmos 23

Meteor design

7 Aug 1981	Intercosmos 22 *Bulgaria 1300*
30 Jan 2009	Koronas *Foton*

Aureole series

27 Dec 1971	Aureole 1	DS U2 GKA
26 Dec 1973	Aureole 2	DS U2 GKA
21 Sep 1981	Aureole 3	AUOS Z MA IK

Prognoz series

14 Apr 1972	Prognoz 1
29 Jun 1972	Prognoz 2
15 Feb 1973	Prognoz 3
22 Dec 1975	Prognoz 4
22 Nov 1976	Prognoz 5
22 Sep 1977	Prognoz 6
30 Oct 1978	Prognoz 7
25 Dec 1980	Prognoz 8
1 Jul 1983	Prognoz 9 *Relict*
16 Apr 1985	Prognoz 10 *Intershock*

Interball/Prognoz M **series**

22 Aug 1995	Interball 1 (with Magion 4)

29 Aug 1996   Interball 2 (with Magion 5)

Ionosfernaya
2 Dec 1970   Cosmos 381

Energiya, Efir series
Energiya 13KS
7 Apr 1972   Intercosmos 6
2 Jul 1978   Cosmos 1026

Efir 36KS
10 Mar 1984   Cosmos 1543
27 Dec 1985   Cosmos 1713

Proton 1 series
16 Jul 1965   Proton 1
2 Nov 1965   Proton 2
6 Jul 1966   Proton 3

Proton 2 series
16 Nov 1968   Proton 4

Observatories
23 Mar 1983   Astron
31 Mar 1987   Kvant
30 Nov 1989   Granat
11 Jul 1990   Gamma

Bion
31 Oct 1973   Cosmos 605/Bion 1
23 Oct 1974   Cosmos 690/Bion 2
21 Nov 1975   Cosmos 782/Bion 3
3 Aug 1977   Cosmos 936/Bion 4
25 Sep 1979   Cosmos 1129/Bion 5
14 Dec 1983   Cosmos 1514/Bion 6
11 Jul 1985   Cosmos 1667/Bion 7
26 Sep 1987   Cosmos 1887/Bion 8
15 Sep 1989   Cosmos 2044/Bion 9
29 Dec 1992   Cosmos 2229/Bion 10
24 Dec 1996   Bion 11

Foton
6 Apr 1985   Cosmos 1645
21 May 1986   Cosmos 1744
24 Apr 1987   Cosmos 1841
14 Apr 1988   Foton 1
24 Apr 1989   Foton 2
11 Apr 1990   Foton 3
4 Oct 1991   Foton 4

8 Oct 1992	Foton 8
14 Jun 1994	Foton 9
15 Feb 1995	Foton 10
9 Oct 1997	Foton 11
7 Sep 1999	Foton 12
31 May 2005	Foton M-2
14 Sep 2007	Foton M-3

Pion

9 Jun 1989	Pion 1, 2
7 Aug 1989	Pion 3, 4
2 Sep 1992	Pion 5, 6

Moon

2 Jan 1959	First Cosmic Ship
12 Sep 1959	Second Cosmic Ship
4 Oct 1959	AIS
18 Jul 1965	Zond 3
31 Jan 1966	Luna 9
31 Mar 1966	Luna 10
24 Aug 1966	Luna 11
22 Oct 1966	Luna 12
21 Dec 1966	Luna 13
7 Apr 1968	Luna 14
15 Sep 1968	Zond 5
14 Nov 1968	Zond 6
8 Aug 1969	Zond 7
12 Sep 1970	Luna 16
20 Oct 1970	Zond 8
10 Nov 1970	Luna 17/Lunokhod
28 Sep 1971	Luna 19
14 Feb 1972	Luna 20
8 Jan 1973	Luna 21/Lunokhod 2
2 Jun 1974	Luna 22
9 Aug 1976	Luna 24

Venus

12 Feb 1961	AIS/Venera 1
1 Apr 1964	Zond 1
12 Nov 1965	Venera 2
16 Nov 1965	Venera 3
12 Jun 1967	Venera 7
5 Jan 1969	Venera 5
10 Jan 1969	Venera 6
17Aug 1970	Venera 7
27 Mar 1972	Venera 8
8 Jun 1975	Venera 9
14 Jun 1975	Venera 10
9 Sep 1978	Venera 11

14 Sep 1978	Venera 12
30 Oct 1981	Venera 13
4 Nov 1981	Venera 14
2 Jun 1983	Venera 15
7 Jun 1983	Venera 16
15 Dec 1984	VEGA 1
21 Dec 1984	VEGA 2

Mars

1 Nov 1962	Mars 1
30 Nov 1964	Zond 2
19 May 1971	Mars 2
28 May 1971	Mars 3
21 Jul 1973	Mars 4
25 Jul 1973	Mars 5
5 Aug 1973	Mars 6
9 Aug 1973	Mars 7
7 Jul 1988	Phobos 1
12 Jul 1988	Phobos 2

Korabl Sputnik biological missions

19 Aug 1960	Korabl Sputnik 2	Belka, Strelka
1 Dec 1960	Korabl Sputnik 3	Pchelka, Mushka (destroyed)
9 Mar 1961	Korabl Sputnik 4	Chernushka
25 Mar 1961	Korabl Sputnik 5	Zvedochka

The early manned and unmanned missions with biological objectives

12 Apr 1961	Vostok	Yuri Gagarin
6 Aug 1961	Vostok 2	Gherman Titov
11 Aug 1962	Vostok 3	Andrian Nikolayev
12 Aug 1962	Vostok 4	Pavel Popovich
14 Jun 1963	Vostok 5	Valeri Bykovsky
18 Jun 1966	Vostok 6	Valentina Tereshkova
12 Oct 1964	Voskhod	Vladimir Komarov, Konstantin Feoktistov, Boris Yegorov
18 Mar 1965	Voskhod 2	Pavel Belyayev, Alexei Leonov
22 Feb 1966	Cosmos 110	Dogs Uterok and Ugolyok

Early Soyuz missions with scientific objectives

26 Oct 1968	Soyuz 3	Georgi Beregovoi
14 Jan 1969	Soyuz 4	Vladimir Shatalov
15 Jan 1969	Soyuz 5	Boris Volynov, Alexei Yeliseyev, Yevgeni Khrunov
11 Oct 1969	Soyuz 6	Georgi Shonin, Valeri Kubsov
12 Oct 1969	Soyuz 7	Anatoli Filipchenko, Vladislav Volkov, Viktor Gorbtako
13 Oct 1969	Soyuz 8	Vladimir Shatalov, Alexei Yeliseyev
1 Jun 1970	Soyuz 9	Andrian Nikolayev, Vitally Sevastianov

Orbital stations

19 Apr 1971	Salyut
25 Jun 1974	Salyut 3
26 Dec 1974	Salyut 4
22 Jun 1975	Salyut 5
29 Sep 1977	Salyut 6
19 Apr 1982	Salyut 7
20 Feb 1986	Mir
20 Nov 1998	ISS (Zarya)

Expedition crews of orbiting space stations
Listing crews that occupied orbital stations for extended periods. Unsuccessful and visiting missions not included.

Salyut

6 Jun 1971	Soyuz 11	Geogi Dobrovolski, Vladislav Volkov, Viktor Patsayev	23 days

Salyut 3

3 Jul 1974	Soyuz 14	Pavel Popovich, Yuri Artyukin	15 days

Salyut 4

10 Jan 1975	Soyuz 17	Alexei Gubarev, Georgi Gechko	29 days
26 May 1975	Soyuz 18B	Pyotr Klimuk, Vitally Sevastianov	63 days

Salyut 5

6 Jul 1976	Soyuz 21	Boris Volynov, Vitally Zholobov	49 days
7 Feb 1977	Soyuz 24	Viktor Gorbatko, Yuri Glazhkov	15 days

Salyut 6

10 Dec 1977	Soyuz 26	Georgi Geechko, Yuri Romanenko	96 days
15 Jun 1978	Soyuz 29	Vladimir Kovalyonok, Alexander Ivanchenkov	139 days
25 Feb 1979	Soyuz 32	Vladimir Lyakhov, Valeri Ryumin	175 days
9 Apr 1980	Soyuz 35	Leonid Popov, Valeri Ryumin	185 days
27 Nov 1980	Soyuz T-3	Leonid Kizim, Oleg Makarov, Gennadiy Strekhalov	13 days
12 Mar 1981	Soyuz T-4	Vladimir Kovalyonok, Viktor Savinyikh	75 days

Salyut 7

13 May 1981	Soyuz T-5	Anatoli Berezovoi, Valentin Lebedev	211 days
27 Jun 1983	Soyuz T-9	Vladimir Lyakhov, Alexander Alexandrov	150 days
8 Feb 1984	Soyuz T-10	Leonid Kizim, Vladimir Solovyov, Oleg Atkov	239 days
6 Jun 1985	Soyuz T-13	Vladimir Dzhanibekov, Viktor Savinyikh	108 days

| 17 Sep 1985 | Soyuz T-14 | Vladimir Vasyutin, Alexander Volkov, Georgi Grechko | 56 days |
| 6 May 1986 | Soyuz T-15 | Leonid Kizim, Vladimir Solovyov | 50 days |

Kizim and Solovyov flew from Mir and returned there.

Mir

13 Mar 1986	Soyuz T-15	Leonid Kizim, Vladimir Solovyov	125 days
6 Feb 1987	Soyuz TM-2	Yuri Romanenko, Alexander Laveikin	174 days
21 Dec 1987	Soyuz TM-4	Vladimir Titov, Musa Manarov	366 days
26 Nov 1988	Soyuz TM-7	Alexander Vokov, Sergei Krikalev	151 days
6 Sep 1989	Soyuz TM-8	Alexander Viktorenko, Alexander Serebrov	166 days
11 Feb 1990	Soyuz TM-9	Anatoli Soloviev, Alexander Balandin	179 days
1 Aug 1990	Soyuz TM-10	Gennady Manakov, Gennady Strekhalov	130 days
2 Dec 1990	Soyuz TM-11	Viktor Afanasayev, Musa Manarov	175 days
18 May 1991	Soyuz TM-12	Anatoli Artsebarski, Sergei Krikalev	144 days
2 Oct 1991	Soyuz TM-13	Alexander Volkov	175 days
17 Mar 1992	Soyuz TM-14	Alexander Viktorenko, Alexander Kaleri	145 days
27 Jul 1992	Soyuz TM-15	Anatoli Soloviev, Sergei Avdeev	188 days
24 Jan 1993	Soyuz TM-16	Gennady Manakov, Alexander Polishuk	179 days
1 Jul 1993	Soyuz TM-17	Vasili Tsibliev, Alexander Serebrov	196 days
8 Jan 1994	Soyuz TM-18	Viktor Afanasayev, Yuri Usachov, Valeri Poliakov	182 days
1 Jul 1994	Soyuz TM-19	Yuri Malencnenko, Talgat Musabayev	125 days
4 Oct 1994	Soyuz TM-20	Alexander Viktorenko, Elena Kondakova	169 days
14 Mar 1995	Soyuz TM-21	Vladimir Dezhurov, Gennady Strekhalov	115 days
3 Sep 1995	Soyuz TM-22	Yuri Gidzenko, Sergei Avdeev, Thomas Reiter	179 days
21 Feb 1996	Soyuz TM-23	Yuri Onufrienko, Yuri Usachov	192 days
17 Aug 1996	Soyuz TM-24	Valeri Korzun, Alexander Kaleri	196 days
10 Feb 1997	Soyuz TM-25	Vasili Tsibliev, Alexander Lazutkin	184 days
5 Aug 1997	Soyuz TM-26	Anatoli Soloviev, Pavel Vinogradov	197 days
29 Jan 1998	Soyuz TM-27	Talgat Musabayev, Nikolai Budarin	207 days
13 Aug 1998	Soyuz TM-28	Gennadiy Padalka, Sergei Avdeev	198 days
20 Feb 1999	Soyuz TM-29	Viktor Afanasayev, Jean-Pierre Hagnere	188 days
4 Apr 2000	Soyuz TM-30	Sergei Zalotin, Alexander Kaleri	72 days

Bibliography

This bibliography details some of the main sources used. By its nature, the subject matter has been reported through a combination of books, scientific articles, gray literature, internet, and even radio sources. Individual sources have already been cited in the end-of-chapter references.

For geographical and linguistic convenience, sources tend to break down into Russian sources – both in the Russian and English languages – and Western sources, principally in English but some also in French. The primary sources for the outcomes of Soviet and Russian space science missions are the papers published by the principal investigators and mission scientists. These may be found in the main journal of Soviet space science, *Kosmicheski Issledovaniya*, in papers presented at international spaceflight gatherings, and in the subsequent papers issued by COSPAR and the International Astronautical Federation conferences. Other papers were published in more specialized journals addressing specific scientific disciplines, such as *Geomagnetism & Aeronomy*. In Russia itself, the principal book publisher of the scientific outcomes of space missions was Nauka Press. Science missions were covered in magazines such as *Zemlya i Vsellenie* and *Novosti Kosmonautiki*. Popular English-language summaries of the early achievements of Soviet space science were published extensively by the Novosti Press Agency (APN) in its *Soviet Booklets* series and the magazine *Science in the USSR*. Western publications of scientists of their discoveries are rare, exceptions being Surkov, published by this publisher, Praxis, in 1997, and Konstantin Gringauz and Mikhail Marov (Cambridge and Yale University Presses, respectively). From a relatively early stage, NASA made translations of Russian technical articles, called the TTF series. Some early-period TTFs may be found in libraries, while more recent ones have been digitized by NASA in Portable Document Format (PDF).

Although a substantial body of work has been undertaken by Western analysts to describe and analyze the Soviet and Russian space programs, little has been done to focus specifically on the scientific aspects. In books, those authors who have written most about the scientific instrumentation and outcomes are Gatland and Dubelaar and they and others are listed below. In articles, those who have concentrated on the scientific aspects most prominently are Salmon and Powell and a selection of the most focused such articles are given below. In addition, a number of other analyses of Soviet and Russian space programs have, in passing, shed considerable light on scientific programs, notably those by Phil Clark.

JOURNALS

Advances in Space Research and *Space Research* (*COSPAR*)
Air & Cosmos
Aviation Week & Space Technology
Flight International
Kosmicheski Issledovatl (English-language version, *Cosmic Research*)
NASA *TTF* series
Novosti Kosmonautiki
Novosti *Soviet Booklets* series
Science in the USSR
Soviet Weekly
Spaceflight
Zemlya i Vsellenie (*Earth and the Universe*)

BOOKS

Burchett, Wilfred; Purdy, Anthony: *Cosmonaut Yuri Gagarin – first man in space.* Panther, London, 1961.
Burgess, Colin; Dubbs, Chris: *Animals in Space – from research rockets to the space shuttle.* Praxis/Springer, Chichester, 2007.
Chertok, Boris: *Rockets and People. Vols I–III*, series editor Asif Siddiqi. NASA, Washington, DC, 2006–2009.
Cross, Jack L.: *A Guide to the Russian Academy of Sciences*, 2nd edn. Cross Associates, Austin, Texas, 1997.
Dubbelaar, Bart: *The Salyut Project.* Progress Publishers, Moscow, 1986.
Gatland, Kenneth: *Manned Spaceflight.* Bladford, London, 1971.
— *Robot Explorers.* Blandford, London, 1972.
Gitelson, Iosif I.; Lisovsky, G.M.; MacElroy, R.D.: *Manmade Closed Ecological Systems.* Taylor & Francis, London, 2002.
Harland, David M.: *The Story of Space Station Mir.* Springer/Praxis, Chichester, 2005.
Krieger, F.J., ed.: *Behind the Sputniks – a survey of Soviet space science.* Rand Corporation, Washington, DC, 1960.
Lemaire, J.F.; Gringauz, Konstantin: *The Earth's Plasmasphere.* Cambridge University Press, Cambridge, 1998.
Marov, Mikhail Y.; Grinspoon, David H.: *The Planet Venus.* Yale University Press, New Haven and London, 1998.
Medvedev, Zhores: *Soviet Science.* McMillan, London, 1979.

Sagdeev, Roald: *The Making of a Soviet Scientist*. John Wiley & Sons, New York, Chichester, Brisbane, Toronto, and Singapore, 1994.

Shklovsky, Iosif: *Five Billion Vodka Bottles to the Moon – tales of a Soviet scientist*. Norton, New York and London, 1991.

Siddiqi, Asif: *The Rockets' Red Glare – spaceflight and the popular imagination in Russia, 1857–1957*. Cambridge University Press, Cambridge, 2010.

Sidorenko, A.V., ed.: *Poverkhnost Marsa*. Nauka Press, Moscow, 1980.

Soviet Writings on Earth Satellites and Space Travel. McGibbon & Kee, 1959.

Sternfeld, Ari: *Soviet Space Science*, 2nd edn. Basic Books, 1953.

Stoiko, Michael: *Soviet Rocketry – the first decade of achievement*. David & Charles, Newton Abbot, 1970.

Surkov, Yuri: *Exploration of Terrestrial Planets from Spacecraft – instrumentation, investigation, interpretation*, 2nd edn. Praxis Publishing with John Wiley & Sons, 1997.

ARTICLES, CHAPTERS, AND THESES

Hendrickx, Bart: Elektron – the Soviet response to Explorer. *Quest*, Vol. 8, No. 1, 2000.

Kidger, Neville: The Intercosmos program, 1967–80, an overview. *Spaceflight*, Vol. 23, No. 6, June 1981.

Kulikov, Stanislav: Top priority space projects. *Aerospace Journal*, November–December 1996.

Marinin, Igor; Lissov, Igor: Russian scientist cosmonauts – raw deal for science in space. *Spaceflight*, Vol. 38, No. 11, November 1996.

Pillet, Nicolas: Le program Intercosmos. *www.kosmonavtika.com*, accessed 9 April 2007.

— Le program Vertikal. *www.kosmonavtika.com*, accessed 9 April 2007.

Powell, Joel W.: Nauka modules. *Journal of the British Interplanetary Society*, Vol. 41, No. 3, March 1988.

— Research from Soviet satellites – the hidden side of Soviet space research. *Spaceflight*, Vol. 25, No. 1, January 1983.

— Soviet space science. *Journal of the British Interplanetary Society*, Vol. 36, No. 10, October 1983.

Sagdeev, Roald, ed.: The principal phases of space research in the USSR in Yuri Gagarin – to mark the 25th anniversary of the first manned space flight. *Social Sciences Today*, Moscow, 1986.

Salmon, Andy: Mir – workshop and laboratory, in Hall, Rex, ed.: *The History of Mir, 1986–2000*. British Interplanetary Society, London, 2000.

— Research in orbit, in Hall, Rex, ed.: *The International Space Station – from imagination to reality*. British Interplanetary Society, London, 2002.

— Science on board the Mir space station, 1986–1994. *Journal of the British Interplanetary Society*, Vol. 50, No. 8, August 1997.

— Science on ISS. Paper presented to the British Interplanetary Society, 7 June 2003.

Siddiqi, Asif: A secret uncovered – the Soviet decision to land cosmonauts on the Moon. *Spaceflight*, Vol. 46, No. 5, May 2004.

Vereschetin, V.; Rimsha, M.: Intercosmos – twenty years on. *Science in the USSR*, 1987, No. 6, November–December 1987.

Walsh, Tom C.: Communicating science in the Sputnik era. Unpublished Master's degree, Dublin City University, 2002.

Zaitsev, Yuri: Russian space science fuels up with new ideas for Earth sciences and more. *Novosti*, 7 November 2006.

Zelenyi, Lev: Is the golden age of Russian space science still ahead? *Novosti*, September 2006.

RADIO

Radio Moscow, English-language service

WEBSITES

Mitchell, Don: *www.mentallandscape.com*
Wade, Mark: *www.astronautix.com*
Zak, Anatoli: *www.russianspaceweb.com*

Index